P9-CAU-620

CALCULUS

A Short Course with Applications

SECOND EDITION

CALCULUS

A Short Course with Applications

SECOND EDITION

Gerald Freilich
Queens College of the City University of New York

Frederick P. Greenleaf
New York University

HARCOURT BRACE JOVANOVICH, PUBLISHERS
San Diego New York Chicago Atlanta Washington, D.C.
London Sydney Toronto

To the memory
of my parents.

—G. F.

To my parents.

—F. G.

ISBN: 0-15-505746-4

Library of Congress Catalog Card Number: 84-81513

Printed in the United States of America

PREFACE

This textbook offers an intuitive approach to calculus with a strong and creative emphasis on applications. It is written for the student whose mathematical background may include only a working knowledge of high-school algebra. Our goal is to give students a working knowledge of calculus as well as an awareness of its important applications in today's world.

The Second Edition of *Calculus: A Short Course with Applications* incorporates improvements based on five years of classroom experience at New York University and Queens College (CUNY) as well as improvements suggested by reviewers, colleagues, and readers. Although the NYU course evolved to meet the needs of students in the management sciences, we are mindful of the needs of students in the biological and social sciences and have included a wide range of applications in these areas. The revisions for this edition have been extensive, and we are hopeful they will prove beneficial to students and instructors alike. In this Second Edition, we have

- Doubled the number of exercises
- Created many new figures and examples
- Highlighted major concepts with vivid color type
- Emphasized with color the graphic statements of more than 260 figures
- Discussed concepts in an applied context
- Compiled a checklist of key topics at the end of each chapter
- Assembled a full set of review exercises to complete each chapter

Answers to the even-numbered exercises appear at the end of the text, with complete solutions to those exercises available in a Solutions Manual.

The basic philosophy of the book remains the same: well-motivated discussion presented in a clear, lively way. Here are some of the features we emphasize as we present our treatment of calculus—in a compact 390 pages.

•**Applications** *Theoretical concepts make sense when developed in an understandable applied context.* Realistic applications are woven into the development of all major topics, and are illustrated by ample sets of examples and exercises. The stage is set early (in Section 1.3) with a self-contained account of terminology from the management sciences that enhances the review material in Chapter 1. Many important topics are introduced with simplified case studies that lead into routine worked examples and exercises. These case studies are labeled "Illustration" in the text.

•**Examples and exercises** *Examples should be structured to develop mathematical skills and an intuitive feel for theoretical concepts.* Every new topic begins with routine problems, worked out in complete detail. These are followed by applied problems with a more verbal orientation. We take pains to give applied examples a realistic feeling, so students become accustomed to seeing the mathematical problem emerge from a description in words. Common sense dictates that this be done gradually, paving the way for full-scale word problems in optimization. In every chapter, exercises are arranged in the same order as the topics are discussed, to make assignments easier.

•**Relevance** *The course should prepare students to deal with calculus in their chosen disciplines.* Several features are designed to meet the real needs of the management sciences. The "marginal concept"—the use of differentials as an approximation principle—is clearly presented as part of the discussion of derivatives, is used again to promote the discussion of optimization problems, and is the basis of many exercises. The theory of several variables is given careful attention. Level curve diagrams are used extensively to promote discussions of partial derivatives, optimization, boundary extrema, and constrained optimization (Lagrange multipliers). For students in the biological and social sciences, growth and decay models are introduced early (in Chapter 1) and then reappear in discussions of both exponential functions and differential equations. There is also a brief, self-contained introduction to probability density functions in the chapter on integration.

•**Level of rigor** *The concept of limit should be developed in a meaningful context, rather than as a mathematical abstraction.* Derivatives are discussed immediately after the review material in Chapter 1, and the necessary notions of limit are developed intuitively. Average rates of change are introduced first, in various applied situations; the derivative is then presented as the limit value of such averages—as an instantaneous rate of change—accompanied by numerical evidence and applied interpretations. Once limits have become familiar, their formal properties are summarized in a final optional section. Integrals are developed in terms of antiderivatives. Their more complicated interpretation as limits of Riemann sums is placed in a single optional section.

•**Review of prerequisites** *The basic prerequisites are reviewed in Chapter 1 and are keyed to the applications that follow.* Chapter 1 may be read by students on their own. Examples are used throughout to show how this material will be involved in future applications. Section 1.3 introduces most of the terminology used in applications and should be read before students go on to the study of derivatives. An appendix reviews the elementary topics of real numbers, inequalities, and exponent laws.

•**Calculators** *Students should be encouraged to use calculators in and out of class.* Many examples and exercises involve realistic numbers, for which the use of a calculator is natural. We also exploit the availability of calculators in dealing with exponentials and logarithms. A full theory is given in Chapter 4, but there is currently no obstacle to introducing these useful functions early and informally, so students can get used to calculations in which they appear.

•**Course duration** *This textbook has been designed for a one-semester or a two-quarter course.* However, we include many optional topics—more than could be covered in one semester. Certain sections in Chapters 1–3 form the core of the book; the remaining sections may be added according to the needs of the instructor. The later chapters on exponentials and logarithms (Chapter 4), integrals (Chapters 5–6), and several variables (Chapter 7) are completely independent; these topics may be covered in any order or even omitted. In particular, the theory of several variables may be taken up early if the audience is heavily oriented toward the management sciences. The important sections in these chapters are listed below.

The core sections in Chapters 1–3 are:

Chapter 1: Sections 1–3, 5. (Students may read this chapter on their own.)
Chapter 2: Sections 1–6.
Chapter 3: Sections 1–8.

The important sections in the remaining chapters are:

Chapter 4: Sections 1, 2, 4.
Chapter 5: Sections 1–3, 5.
Chapter 6: This chapter is optional.
Chapter 7: Sections 1–5, 7.

We would like to express our appreciation to the reviewers whose comments contributed so much to the text. These include the reviewers for the Second Edition: Richard A. Brualdi (University of Wisconsin, Madison), Daniel S. Drucker (Wayne State University), Richard Semmler (Northern Virginia Community College), and Bruce H. Edwards (University of Florida, Gainesville); as well as the reviewers for the original edition: Frank Warner (University of Pennsylvania), Paul Knopp (University of Houston), Larry Goldstein (University of Maryland), and Richard Joss (California State University, Long Beach).

We would also like to acknowledge our debt to Richard Wallis, our acquisitions editor at Harcourt Brace Jovanovich, for his invaluable advice and guidance during the preparation of the Second Edition, and to the staff at Harcourt Brace Jovanovich for their fine work. We are particularly grateful for the efforts of Marji James as our production editor.

Gerald Freilich
Frederick P. Greenleaf

CONTENTS

CHAPTER 1

Preliminary Concepts

1.1 Functions and Their Graphs

Economics, business administration, biology, chemistry, and physics are frequently concerned with the dependence of one quantity upon others. For example, a manufacturer might want to know how profit varies with production level. Or in a typical biological problem, we might study the dependence of the size of a bacterial population on time, temperature, or the concentration of certain nutrients.

Mathematicians have abstracted from these problems the concept of function. The quantities to be measured and compared are called **variables**. To keep things simple, in this chapter we shall compare just two variables. Functions involving several variables make their appearance in Chapter 7. A **function** is a rule or formula that gives the value of one variable if the value of the other variable is specified. For example, temperatures measured on the Celsius (centigrade) and Fahrenheit scales are connected by a simple formula

$$F = \frac{9}{5}C + 32, \qquad [1]$$

where F is degrees Fahrenheit, and C is degrees Celsius.

If a newspaper reports that the temperature in Paris was 20°C, this means that the Fahrenheit temperature was $(\frac{9}{5} \times 20) + 32 = 68$°F—a pleasant day, rather than one on which we should worry about our antifreeze. Here the variables are F and C; the function is the rule [1] giving the value of F when C is specified.

In a simple manufacturing problem, the variables are the profit P (which might be measured in dollars per month) and the production level x (the number of units produced per month). A factory manager can set production at various levels. Once the production level has been set, the profit is determined. If the manager changes the number of units produced per month, then of course the monthly profits change, too. The variables P and x play rather different roles here. The production level x may be set as the manager wishes, within reasonable limits. This is (unfortunately) not true of the profit variable. Asymmetry in the roles of the variables is typical of most functions. Usually, one variable may be varied freely, while the other is determined by the function. The "free" variable is called the **independent variable** and the other the **dependent variable**. In [1] the independent variable is C, and F depends upon it. In a manufacturing problem the production level x is the independent variable, and the profit P is the dependent variable.

Example 1 Apples are priced at 49¢ per pound. Describe the relationship between the cost and the quantity bought.

Solution The natural independent variable x is the number of pounds bought. The dependent variable then, is the price paid, which we denote by y (measured in cents). The function relating y to x is concisely expressed by the formula

$$y = 49x \qquad \text{for } x \geq 0.$$

Notice that this relationship has meaning in this situation only for $x \geq 0$. ■

There is a standard notation for functions. If we let x stand for the independent variable and y for the dependent variable, and if f is some function that relates x to y, the symbol

$$f(x) \quad \text{(which we read "} f \text{ of } x\text{")}$$

stands for the value of the dependent variable y when the independent variable is equal to x. This notation saves a lot of verbiage. Thus if $y = 4$ when $x = -3$, we can just write $4 = f(-3)$. If we are told that $1 = f(2)$, then y has the value 1 when $x = 2$.

Many functions are given by an algebraic formula. For example:

$$f(x) = x^2 - x + 1$$

$$f(x) = \frac{9}{5}x + 32$$

$$f(x) = \frac{1}{x}$$

$$f(x) = \frac{1}{1 + x^2}$$

$$f(x) = 1 \qquad \text{(constant function; } y = 1 \text{ for all } x\text{)}$$

Example 2 If $f(x) = x^2 - x + 1$, what is the value of this function when $x = 2$? When $x = -1$? When $x = 0$? When $x = \frac{1}{2}$?

Solution We find the value by inserting the designated x value into the formula. Thus,

$$f(2) = (2)^2 - 2 + 1 = 4 - 2 + 1 = 3$$

$$f(-1) = (-1)^2 - (-1) + 1 = 1 + 1 + 1 = 3$$

$$f(0) = (0)^2 - (0) + 1 = 1$$

$$f\left(\frac{1}{2}\right) = \left(\frac{1}{2}\right)^2 - \left(\frac{1}{2}\right) + 1 = \frac{1}{4} - \frac{1}{2} + 1 = \frac{3}{4} \quad \blacksquare$$

Example 3 Pure antifreeze (ethylene glycol) freezes at $-11.5°C$ and boils at $198°C$. Use the conversion formula [1]

$$F(C) = \frac{9}{5}C + 32$$

to determine the corresponding freezing and boiling points on the Fahrenheit scale.

Solution Setting $C = -11.5$ in this function, we get the freezing point

$$F(-11.5) = \frac{9}{5}(-11.5) + 32 = 11.3°F$$

Similarly, the boiling point is

$$F(198) = \frac{9}{5}(198) + 32 = 388.4°F \quad \blacksquare$$

Although mathematicians often let x stand for the independent variable and y for the dependent variable, they occasionally use more suggestive symbols, such as C and F in Example 3. After all, the names we give the variables do not really matter. What counts is the way one variable depends on the other.

Sometimes the independent variable can take on only restricted values. In Example 1, the cost of apples makes little sense for negative x, so attention is restricted to $x \geq 0$. In manufacturing problems, negative levels of production would not have much meaning. The set of "allowable" or "feasible" values for the independent variable is called the **feasible set** or **domain of definition** of the function.† In Example 1 the domain consists of all numbers $x \geq 0$.

Example 4 A study conducted by the XYZ Manufacturing Company shows that their monthly profit P depends on the number x of units produced per month according to the formula

$$P(x) = -3800 + 15x - 0.005x^2$$

† Mathematicians favor "domain of definition" or simply "domain," but "feasible set" is commonly used and sounds more natural in economics. We will use both terms interchangeably so the reader will feel at home with either.

Operating full time, the plant can produce at most 1000 units per month. What is the feasible set of production levels? Find P when $x = 300$. When $x = 0$. When $x = 1000$.

Solution Feasible production levels must satisfy the conditions

(i) $x \geq 0$ (negative production levels make no sense)
(ii) $x \leq 1000$ (full-time plant capacity is 1000 units per month)

so the feasible set is the interval $0 \leq x \leq 1000$. When $x = 300$ the profit is

$$\begin{aligned} P = P(300) &= -3800 + 15(300) - 0.005(300)^2 \\ &= -3800 + 4500 - 450 \\ &= \$250 \end{aligned}$$

Similarly,

$$P(0) = -3800 + 15(0) - 0.005(0)^2 = \$-3800$$

(negative profit means they are losing money for the month), and

$$\begin{aligned} P(1000) &= -3800 + 15(1000) - 0.005(1000)^2 \\ &= -3800 + 15{,}000 - 5000 \\ &= \$6200 \end{aligned}$$

■

If $f(x)$ is given by a formula without any mention of a domain of definition, the domain is taken to be all x where the formula makes sense. Thus, if $f(x) = x^2 - x + 1$, the domain is all x. But if

$$f(x) = \frac{1}{x}$$

this makes sense only if $x \neq 0$. So the domain is all $x \neq 0$. Similarly, if

$$f(x) = \sqrt{x}$$

this makes sense only if $x \geq 0$.

Not all functions are given by algebraic formulas. Here is one example.

Example 5 U.S. postage for a first-class letter depends on its weight w. The cost c is 20¢ for the first ounce (or less) and 17¢ for each additional ounce or fraction thereof. Describe the function $c(w)$. Find the cost of a 3.5-ounce letter.

Solution The weight w (in ounces) is the independent variable. It must be positive, so the feasible set is all $w > 0$. For small letters, say under 4 ounces, the function may be described by a table of values

$$c = \begin{cases} 20 & \text{if } 0 < w \leq 1 \\ 37 & \text{if } 1 < w \leq 2 \\ 54 & \text{if } 2 < w \leq 3 \\ 71 & \text{if } 3 < w \leq 4 \end{cases}$$

From this we see that $c(3.5) = 71$¢. ■

It may be an overstatement to say that "a picture is worth a thousand words," yet a picture of a function is very useful. This picture is called the graph of the function. In a plane, let us draw horizontal and vertical coordinate axes so that these perpendicular axes meet at the point corresponding to zero on each line (see Figure 1.1). The point where they meet is called the **origin**. In this **coordinate plane** we may label each point by a pair of numbers (a, b) called the **coordinates** of the point. These coordinates tell us how to find the point: The first number a tells us to move a units horizontally, starting from the origin. The second number b tells us then to move b units parallel to the vertical axis. If either a or b is negative, we must move parallel to the appropriate axis—but in the *negative* direction. For example, the origin O has coordinates $(0, 0)$. To locate the point Q with coordinates $(1, -2)$ as shown in Figure 1.1, we move 1 unit to the right along the horizontal axis, and then -2 units (2 units *down*) parallel to the vertical axis.

The **graph** of a function f is the set of points (x, y) in the plane such that $y = f(x)$. The general idea is shown in Figure 1.2, where we indicate the graph points corresponding to $x = -1, 0, 1, 2$. There is one point on the graph above

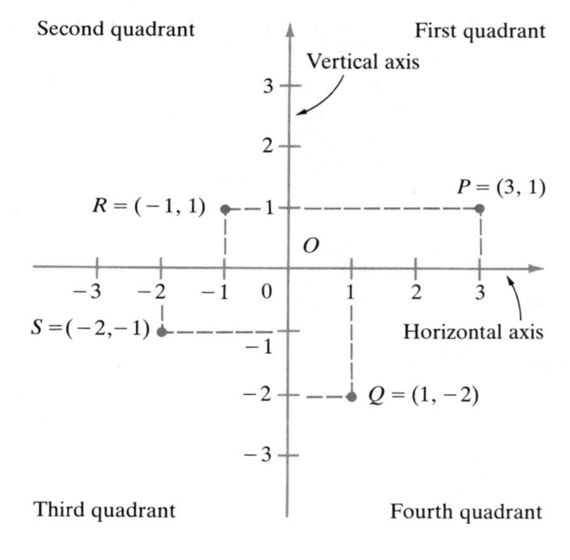

Figure 1.1
The point P in the coordinate plane has coordinates $(3, 1)$ because we reach it by moving $+3$ units parallel to the horizontal axis and $+1$ units parallel to the vertical axis. The origin O has coordinates $(0, 0)$. Coordinates of other points Q, R, S are indicated. The coordinate axes divide the plane into four pieces called *quadrants*, which are labeled as shown.

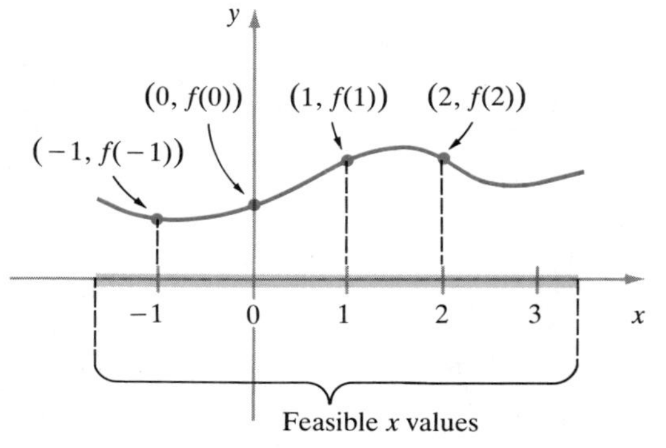

Figure 1.2
The graph of $f(x)$ consists of all points of the form $(x, y) = (x, f(x))$, where x is in the feasible set (indicated by the shaded interval on the x axis). Graph points for $x = -1, 0, 1, 2$ are shown as solid dots.

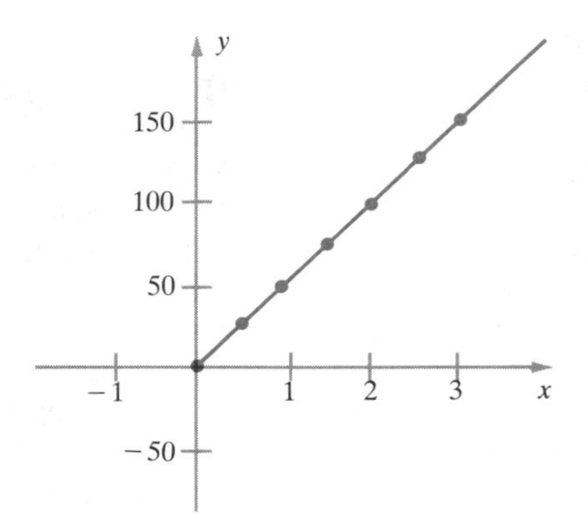

x	$y = 49x$
0	0.0
$\frac{1}{2}$	24.5
1	49.0
$\frac{3}{2}$	73.5
2	98.0
$\frac{5}{2}$	122.5
3	147.0

Figure 1.3
The function $y = 49x$ is defined for $x \geq 0$. For selected values of x the values of $y = 49x$ are tabulated at the right. The line indicates the location of *all* points on the graph. For convenience we use different scales of length on the two coordinate axes. The unit length on the vertical axis is much smaller than the unit length on the horizontal axis. This strategy allows us to sketch the graph in a reasonable amount of space.

(or below) each x in the domain of definition; there are *no* graph points corresponding to other x values. As an example, let us draw the graph of the function $f(x) = 49x$ (for $x \geq 0$) discussed in Example 1. We do this by plotting enough points on the graph to suggest its shape; then we draw a smooth curve through these points. In Figure 1.3 we have listed several allowable values of x and have computed the corresponding values of $y = 49x$. Each entry in the table gives the coordinates $(x, y) = (x, 49x)$ of a point on the graph. For example, $y = 98$ if $x = 2$, so $(2, 98)$ is on the graph. These graph points are indicated by solid dots in the figure. The complete graph is part of a straight line beginning at the origin. Because the function is not defined for $x < 0$, there are no graph points lying above or below negative values of x on the horizontal axis.

In sketching graphs we always associate the independent variable with the horizontal axis and the dependent variable with the vertical axis, as in Figure 1.3.

Example 6 Sketch the graph of the function $f(x) = x^2$.

Solution First plot the graph points for a few values of x. The solid dots in Figure 1.4 correspond to the tabulated values of x and $y = x^2$. A few additional remarks will help us make an accurate sketch.

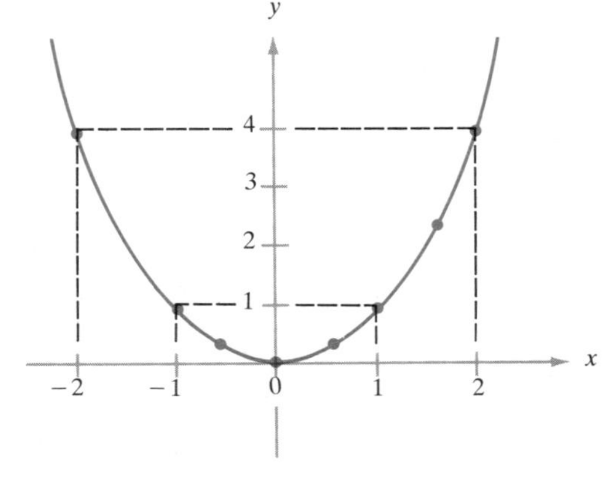

x	$y = x^2$
−2	4.0
−1	1.0
−0.5	0.25
0	0.0
0.5	0.25
1	1.0
1.5	2.25
2	4.0

Figure 1.4
Graph of $y = x^2$. Points on the graph above $+x$ and $-x$ have the same height above the x axis, namely $y = x^2$. This is shown for $x = \pm 1$ (which correspond to $y = 1$) and $x = \pm 2$ (which correspond to $y = 4$). Thus the graph is symmetric about the vertical axis.

The square $y = x^2$ of any number x is positive, whether x is positive or negative. Thus the points on the graph all lie in the upper half of the coordinate plane. And $y = f(x)$ assumes the same value at $+x$ and $-x$, so that the points on the graph above $+x$ and $-x$ have the same height above the x axis, namely $y = x^2 = (-x)^2$. This means that the graph is symmetric with respect to the y axis, as shown in Figure 1.4. Finally, as x moves right along the x axis, taking on large positive values (or to the left, taking on large negative values), the dependent variable $y = x^2$ assumes large positive values. With these general observations in mind we can sketch the curve shown in Figure 1.4. ■

Example 7 Draw the graph of the postage function in Example 5, relating c = cost to w = weight.

Solution Here the variables are labeled w and c instead of x and y. If w lies in the range $0 < w \leq 1$, the related value of c is 20. Thus, as w moves from 0 to 1 along the horizontal axis, the corresponding points (w, c) on the graph all have the same height $c = 20$ above the horizontal axis; they lie on a horizontal line segment 20 units above the horizontal axis. If $1 < w \leq 2$, then $c = 37$; so the corresponding points on the graph lie on a horizontal line segment 37 units above the w axis. A similar analysis shows that the graph consists of horizontal segments, as shown in Figure 1.5. ■

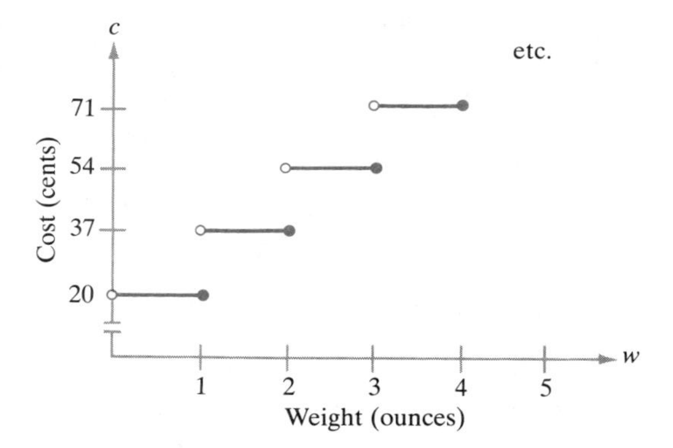

Figure 1.5
The graph of the postage function $c = c(w)$ in Example 7 is a "step function." An open dot indicates that the point is *not* included in the graph; solid dots *are* on the graph. Note the sudden breaks in the graph at $w = 1$, $w = 2$, and $w = 3$. Compare this with the smoothness of the graph in Figure 1.4. There is no algebraic formula for the postage function, though there is a perfectly well-defined logical rule for calculating $c(w)$.

General observations about the behavior of $f(x)$ as x moves far to the left or right can be very useful in making an accurate sketch from a small number of plotted points. In the next example, we shall use a simple fact about reciprocals:

$$\frac{1}{\text{(small positive)}} = \text{(large positive)}$$
$$\frac{1}{\text{(large positive)}} = \text{(small positive)} \qquad [2]$$

For example, 200,000 is quite a larger number, but $1/200{,}000 = 0.000005$ is very small. For negative values, the outcome, determined by the rule of signs $(+)/(-) = (-)$, is similar.

Example 8 Sketch the graph of the function $y = 1/x$.

Solution The function is not defined at $x = 0$; there is no point on the graph above or below $x = 0$. Generally, a graph has some sort of singular behavior wherever an algebraic formula for the function has denominator zero. It is a good idea to plot several points near such values of x (here $x = 0$). The general principle [2] can also be helpful. In Figure 1.6 we have tabulated some values of

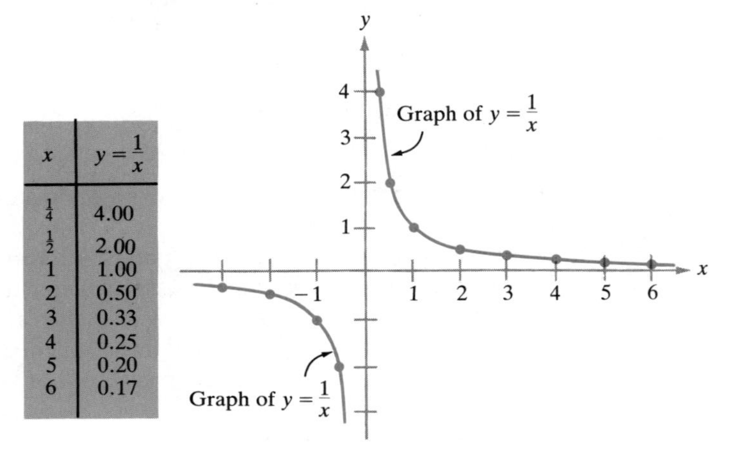

x	$y = \frac{1}{x}$
$\frac{1}{4}$	4.00
$\frac{1}{2}$	2.00
1	1.00
2	0.50
3	0.33
4	0.25
5	0.20
6	0.17

Figure 1.6
The graph of $y = 1/x$, defined for $x \neq 0$, consists of two separate curves. Values of $y = 1/x$ are tabulated for selected positive x; the value at $-x$ is (-1) times the value at $+x$, so there is no need to make a separate table for negative x.

$y = 1/x$, taking several values of x near the troublesome point $x = 0$. We can save ourselves the trouble of calculating $f(x)$ for negative values of x by noticing that $f(-x) = (-1) \cdot f(x)$; the value at $-x$ is the negative of the value at x. Let us see what happens for small positive or small negative values of x.

$$x \text{ small positive; } y = \frac{1}{x} \text{ large positive} \quad \text{(graph point high above } x \text{ axis)}$$

$$x \text{ small negative; } y = \frac{1}{x} \text{ large negative} \quad \text{(graph point far below } x \text{ axis)}$$

This behavior is shown in the final sketch (Figure 1.6). What happens as x moves far to the right? The corresponding graph point approaches the x axis, but is slightly above it because $y = 1/(\text{large positive}) = (\text{small positive})$. ■

Exercises 1.1

1. If $f(x) = 3x - 2$, find the values $f(0)$, $f(-1)$, $f(1)$, and $f(8)$. For what value of x is $f(x) = 4$?

2. If $f(x) = x^2 - 2$, what is the value of $y = f(x)$ when
(i) $x = -1$ (ii) $x = 2$ (iii) $x = 0$ (iv) $x = \sqrt{2}$

Find the values $f(-2)$, $f(0)$, $f(\frac{1}{2})$, $f(1)$, and $f(5)$ for the following functions.

3. $f(x) = 4 - 13x$

4. $f(x) = 5 - x + 2x^2$

5. $f(x) = 1 + x - x^2$

6. $f(x) = x^3 - 3x + 2$

7. $f(x) = x^4 + 1$

8. $f(x) = 4000 + 200x - 2.5x^3$

9. $f(x) = \sqrt{x + 2}$

10. $f(x) = \sqrt{100 - 4x^2}$

Find the values $y = f(x)$ for the following functions at $x = -2, x = -1, x = 1, x = 10$.

11. $\dfrac{1}{x}$

12. $x - \dfrac{1}{x}$

13. $\dfrac{4}{x - 3}$

14. $\dfrac{1}{1 + x^2}$

15. $(2 - x)^2$

16. $\dfrac{2x}{1 + x^2}$

17. $(x + 3)^{-2}$

18. $(x + 2)^{1/2}$

19. $x(x - 1)$

20. $(x + 2)^{3/2}$

21. Temperatures measured in degrees Fahrenheit are converted to temperatures on the Celsius scale by the formula

$$C = \frac{5}{9}(F - 32)$$

obtained by solving [1] for C as a function of F. Convert the following commonly encountered Fahrenheit temperatures to their Celsius values.

(i) -5

(ii) 0

(iii) 32 (freezing)

(iv) 70 (room temperature)

(v) 90

(vi) 98.6 (body temperature)

(vii) 212 (boiling)

22. Rent-a-Heap of Los Angeles offers to rent its cars for $90 per week plus 15¢ per mile. If we use the car for a week, the cost C depends on the mileage x.

(i) Write the formula for cost in terms of mileage.
(ii) Find the costs if we cover 550 and 800 miles.

23. The weight w (in pounds) of men with a certain body type varies with their height h (in feet) according to the formula $w = 0.9h^3$. (Can you think of a reason for the third power?) Two men with this build have heights 5′8″ and 6′0″. What are their weights? What is the ratio of their weights? The ratio of their heights?

24. A rectangular box has a square base, with each side of the base s inches long. If the height of the box is three times this base length, give a formula for the volume V of the box as a function of s (see Figure 1.7).

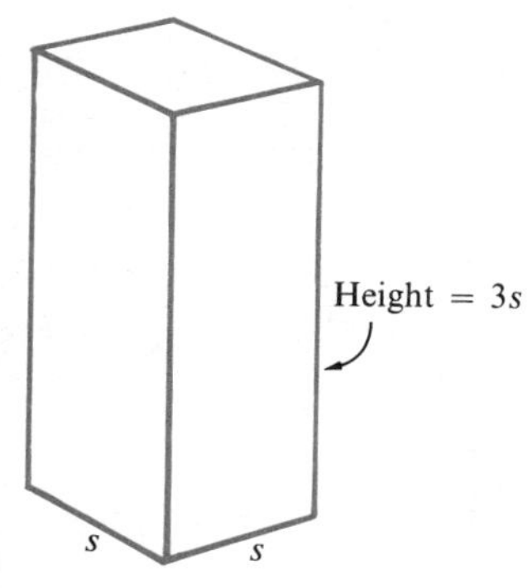

Figure 1.7
The box in Exercises 24–26. The base is square, each edge of length s.

25. Suppose the box in Exercise 24 is reinforced with fiberglass tape along each edge. How many edges are there? Give a formula for the length l of tape required as a function of s.

26. If the tape in Exercise 25 costs $2.80 per 100 feet, find the cost of reinforcing 8000 such boxes as a function of s. (*Note*: Be careful about converting inches to feet!)

27. A cable television company estimates that the number of subscribers in Modesto will grow according to the formula

$$N(x) = 16{,}000\frac{x}{x + 5}$$

where x is the number of years since the start of operations in January 1981. If this is true, how many subscribers will there be after 2 years? After 5 years? By January 1984?

28. Surly Steve, Inc., is an electronics store in Manhattan. Steve estimates that his monthly profit P on the sale of telephone answering sets depends on the number x of sets sold according to the formula

$$P(x) = -600 + 40x - 0.1x^2$$

Calculate his monthly profit if he sells

(i) $x = 0$ (ii) $x = 15$ (iii) $x = 20$ (iv) $x = 100$

sets per month.

29. A large manufacturer of television sets has found that his yearly profit P depends on the number of sets x produced per year according to the formula

$$P(x) = -5{,}000{,}000 + 61x - 0.000098x^2 \quad \text{dollars per year.}$$

Calculate the profit if

(i) $x = 0$ (ii) $x = 500{,}000$ (iii) $x = 125{,}000$ (iv) $x = 311{,}000$

30. Use graph paper to plot the following points in the coordinate plane.

(i) $(0, 0)$
(ii) $(1, 0)$
(iii) $(0, 1)$
(iv) $(-3, 0)$
(v) $(-1, -2)$
(vi) $(3, -2)$
(vii) $(-2, 3)$
(viii) $(\frac{3}{2}, \frac{2}{3})$
(ix) $(2, 1 + \sqrt{2}) = (2, 2.414)$

31. A ball is dropped from the roof of a tall building. Let s (in feet) be the distance fallen t seconds after release. The following measurements were made:

t	0	0.25	0.5	1.0	1.5	1.75	2.0	2.25	2.5
s	0	1	4	16	36	49	64	81	100

Plot these points in the coordinate plane, and sketch a reasonable approximation to the graph of $s = s(t)$. From the graph, estimate the value of s when $t = 1.25$ seconds. For which value of t is $s = 70$ feet?

In Figure 1.8 we show the graph of a certain function $f(x)$. Answer the following questions by inspecting the graph.

32. What is $f(1)$?

33. What is $f(2)$?

34. What is $f(-1)$?

35. What is $f(0)$?

36. Is $f(3)$ positive or negative?

37. For which value(s) of x is $f(x) = 2$?

38. For which value(s) of x is $f(x) = 0$?

39. For which x is $f(x) \geq 0$?

40. What is the feasible set for $f(x)$?

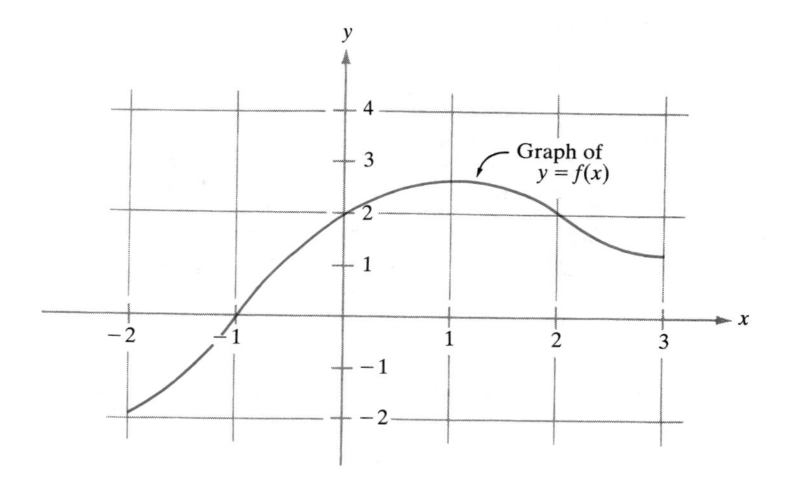

Figure 1.8
The graph of the function $f(x)$ in Exercises 32–40.

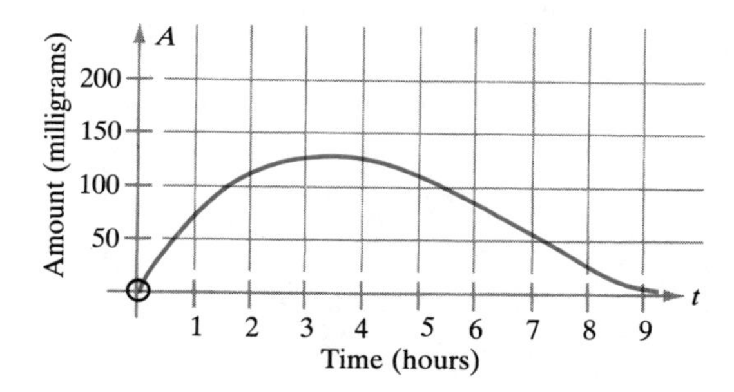

Figure 1.9
Amount of aspirin in the bloodstream as a function of time t (see Exercises 41–44).

A time-release capsule slowly releases aspirin into the bloodstream until it is exhausted. In a typical patient, the amount A of aspirin in the blood t hours after taking the capsule is described by the graph shown in Figure 1.9. Answer the following questions by inspecting this graph.

41. How much aspirin is present after 1 hour?

42. How much aspirin is present after 7 hours?

43. After how many hours is the maximum amount of aspirin present?

44. What is the largest value of A?

Using graph paper, sketch the graphs of the following functions. Plot a few points and fill in a smooth curve.

45. $f(x) = 1 - 3x$

46. $f(x) = x - 2$

47. $f(x) = x$

48. $f(x) = 3$ (a constant function)

49. $f(x) = -2x + 3$

50. $f(x) = x^2 - 1$

51. $f(x) = 2 - x - x^2$

52. $f(x) = x^3$

53. Calculate the values of the functions

(i) $y = 2^x$ (ii) $y = 5 \cdot (2^x)$

for $x = -3, -2, -1.5, -1.0, -0.5, 0.0, +0.5, +1.0, +1.5, 2.0$, and 3.0. Plot these values on graph paper, and make a sketch of the graphs. Choose a vertical scale so that the graphs fit conveniently on the paper. (*Hint*: Write exponents as fractions where necessary. You may also want to recall the exponent laws reviewed in the Appendix.)

54. Sketch the graph of the *square-root function* $y = \sqrt{x}$. Sketch at least the part of the graph for which $0 \leq x \leq 10$. Plot several points such that $0 \leq x \leq 1$, which is the hardest part of the graph to plot correctly. (*Note:* $y = \sqrt{x}$ is defined only for $x \geq 0$, so the graph lies over the positive part of the x axis.)

55. Sketch the graph of $f(x) = 1/x^2$. This function is not defined at $x = 0$. As in Example 8, special care should be taken for values of x near zero. What happens as x gets small? Large positive? Large negative?

56. Sketch the graph of $f(x) = 1/(1 + x)$, taking $0 \leq x \leq 5$ as the domain of definition.

57. Sketch the graph of $f(x) = 1/(1 - x)$ defined for all $x \neq 1$. Pay special attention to the shape near the exceptional value $x = 1$. What happens as x becomes large positive/negative? What happens as x approaches 1 from the right/left?

58. Describe the natural domain of definition for each function.

(i) $\dfrac{1}{x - 1}$ (iv) $\dfrac{1}{\sqrt{x}}$

(ii) $\dfrac{1}{x + 3}$ (v) $\dfrac{1}{\sqrt{4 - x}}$

(iii) $\dfrac{x}{x - 1}$ (vi) $\dfrac{1}{x(x + 1)}$

In Figure 1.10, we show the graph of the U.S. price index for the years 1910–1974 plotted from government statistics. Here, t = year number, and the index $I = I(t)$ tells you how

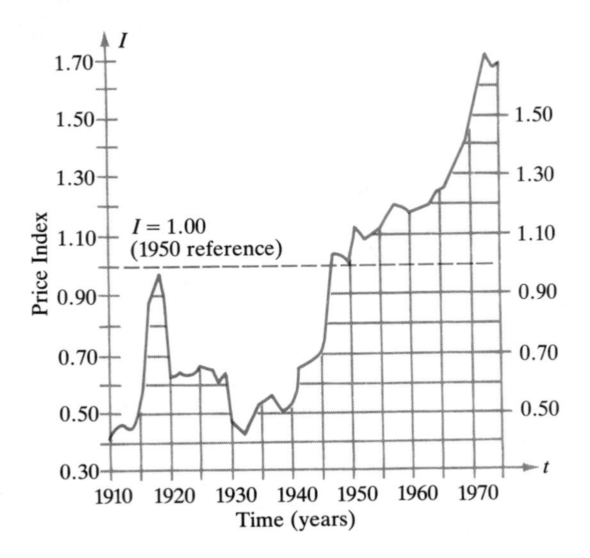

Figure 1.10
Graph of U.S. wholesale price index (taking $I = 1.00$ in base year 1950). (Source: Bureau of Labor Statistics).

many dollars in the year t would buy the same amount of goods \$1.00 would buy in the base year of 1950. (Conversely, the reciprocal $1/I(t)$ tells you how much a dollar in year t is worth in 1950 dollars.) A 1950 dollar is worth $I(t)$ dollars in the year t; $I = 1$ when $t = 1950$, which corresponds to the point (1950, 1) on the graph. Answer the following questions by inspecting this graph.

59. What was the value of the price index in 1965? In 1930?

60. In which year(s) was the price index equal to 1.50? What does $I = 1.50$ tell us about the value of the dollar in such years?

61. In which years was the value of the dollar $\frac{3}{4}$ that of a 1950 dollar?

62. In which years was the value of the dollar twice that of a 1950 dollar?

63. If a salary was \$20,000 in 1975, what was its equivalent in the year 1950? In the year 1938?

1.2 The Straight Line

The simplest curve in the plane is a straight line. In this section we will review basic facts about the slope of a line and the equations that describe the line. If a line is not vertical, its direction with respect to the coordinate axes in the plane is described by a number called the **slope** of the line. Slope is measured by marking two points P_1 and P_2 on the line and computing the ratio

$$\text{slope}(L) = \frac{\text{change in } y \text{ coordinate}}{\text{change in } x \text{ coordinate}}$$

as we move from P_1 to P_2. For the reader's convenience, we have outlined an orderly procedure for computing the slope. It is standard to use the symbol Δ to mean "change in" some quantity. Thus Δy means "the change in y," and Δx means "the change in x." The meaning of Δy and Δx in computing the slope is shown in Figure 1.11.

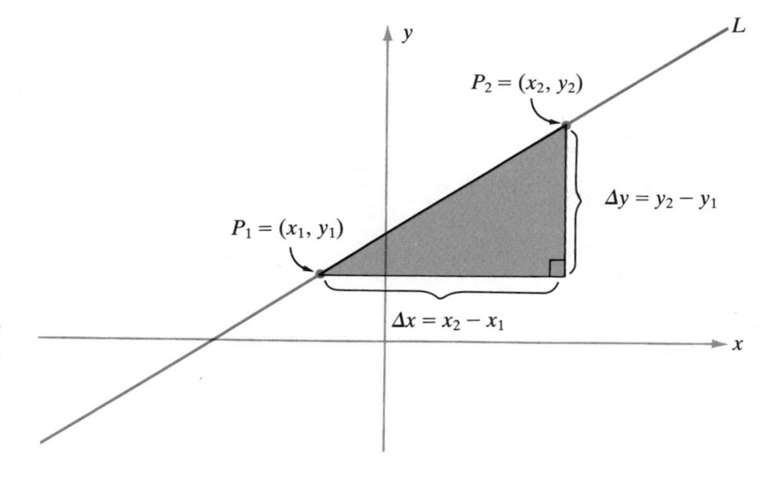

Figure 1.11
Given points $P_1 = (x_1, y_1)$ and $P_2 = (x_2, y_2)$ on a line L, the slope is $\Delta y/\Delta x = (y_2 - y_1)/(x_2 - x_1)$. If we take a different pair of points on L, the resulting triangle would be similar to the one shown and would yield the same value for the slope. Thus, the slope does not depend on the particular pair of points chosen.

CALCULATING THE SLOPE OF A NONVERTICAL LINE L

Step 1 Choose two points on L and write down their coordinates.

$$P_1 = (x_1, y_1) \qquad P_2 = (x_2, y_2)$$

Step 2 Compute the change in x as we move from P_1 to P_2.

$$\Delta x = x_2 - x_1$$

Step 3 Compute the corresponding change in y.

$$\Delta y = y_2 - y_1$$

Step 4 Compute the ratio of these changes.

$$\text{slope}(L) = \frac{\Delta y}{\Delta x} = \frac{y_2 - y_1}{x_2 - x_1}$$

The slope of L does not depend on which pair of points we choose on L (see Figure 1.11), nor does it matter which point we label as P_1 and which as P_2 (see comments following Example 9 and in the caption to Figure 1.11). The slope is not defined for a vertical line: $x_1 = x_2$ and $\Delta x = 0$ for any two points on a vertical line, so that $\Delta y/\Delta x$ is undefined.

Example 9 Find the slope of line L if the points $P_1 = (1, -1)$ and $P_2 = (3, 0)$ lie on L. Find the slope of the line M passing through $(1, -3)$ and $(-2, 3)$.

Solution (See Figure 1.12.) To compute slope(L) we may use the points $(x_1, y_1) = (1, -1)$ and $(x_2, y_2) = (3, 0)$. Then $x_1 = 1$, $y_1 = -1$ and $x_2 = 3$,

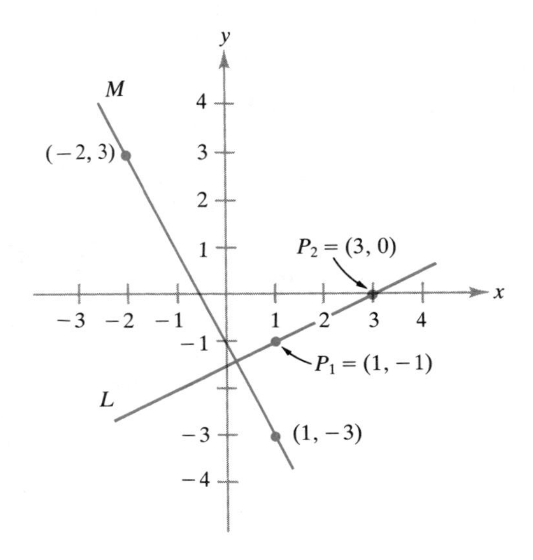

Figure 1.12
Lines L and M in Example 9. The line L contains $P_1 = (1, -1)$ and $P_2 = (3, 0)$. Its slope is $\frac{1}{2}$. Observe that L rises as we move to the right, which is true for any line with positive slope. Line M has negative slope -2 and falls as we move to the right.

$y_2 = 0$, and we obtain

$$\Delta y = y_2 - y_1 = 0 - (-1) = 0 + 1 = 1$$
$$\Delta x = x_2 - x_1 = 3 - 1 = 2$$
$$\text{slope}(L) = \frac{\Delta y}{\Delta x} = \frac{1}{2}$$

Similarly, the points $(x_1, y_1) = (1, -3)$ and $(x_2, y_2) = (-2, 3)$ lie on M so that

$$\text{slope}(M) = \frac{3 - (-3)}{-2 - (1)} = \frac{6}{-3} = -2$$ ■

What if we reverse the roles of initial and final points in Example 9, labeling $P_1 = (x_1, y_1) = (3, 0)$ and $P_2 = (x_2, y_2) = (1, -1)$? In that case we obtain

$$\Delta y = y_2 - y_1 = -1 - 0 = -1$$
$$\Delta x = x_2 - x_1 = 1 - 3 = -2$$
$$\frac{\Delta y}{\Delta x} = \frac{-1}{-2} = \frac{1}{2}$$

The two minus signs attached to Δy and Δx cancel, giving exactly the same value for slope(L) as before. Thus the labeling of points P_1 and P_2 is immaterial.

Many geometric statements may be translated into statements about the numerical value m of the slope. The slope of L is positive ($m > 0$) if the line rises as we move to the right in the plane (see Figure 1.13). Similarly, m is negative if the line falls as we move to the right. The slope is zero for horizontal lines. In general, the larger the slope, the more steeply the line rises. Two nonvertical lines L_1 and

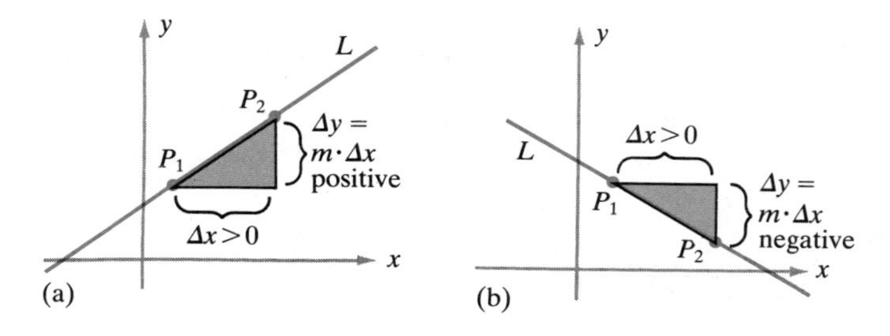

Figure 1.13
A rising line L has positive slope, and a falling line has negative slope. In (a), (b), and (c), Δx is positive. In (a), Δy is positive, and so is the slope m. In (b), Δy is negative, and so is m. In (c), Δy and m are zero; because $\Delta y = m \cdot \Delta x = 0$, all points on L have the same y coordinate, and L must be horizontal.

L_2 with slopes m_1 and m_2 are

(i) parallel if and only if $m_1 = m_2$;
(ii) perpendicular if and only if $m_2 = -1/m_1$ or, equivalently, if $m_1 \cdot m_2 = -1$. [3]

Statement (i) is obvious; we omit proof of the more subtle fact (ii). Using [3ii], we see that the lines L and M in Example 9 are perpendicular. It is not even necessary to sketch the lines to see this. Just compute and compare their slopes.

If we want to represent a straight line as the graph of some function, how do we find the appropriate function? First notice that the slope of L is not enough to determine its position in the plane (any two parallel lines have the same slope); but if we also know one point on the line, then its position is determined. Suppose L has slope(L) $= m$ and that L passes through some point $P = (x_1, y_1)$. If $Q = (x, y)$ is any other point on L, then by definition of slope

$$m = \frac{\Delta y}{\Delta x} = \frac{y - y_1}{x - x_1} \quad [4]$$

This equation may be rewritten; multiplying both sides by $(x - x_1)$ we get

$$(y - y_1) = m \cdot (x - x_1) \quad [5]$$

A point $Q = (x, y)$ in the plane lies on L if and only if its coordinates x, y satisfy Equation [5]. This identifies L as the solution set of [5], which is called the **point-slope** form of the equation for L.

Example 10 Find the equation of the line L that passes through $P = (-2, -1)$ with slope(L) $= 3$. Sketch the line.

Solution Because $P = (-2, -1)$ lies on L and the slope is $m = 3$, every other point $Q = (x, y)$ on L must satisfy the equation

$$3 = m = \frac{\Delta y}{\Delta x} = \frac{y - (-1)}{x - (-2)} = \frac{y + 1}{x + 2}$$

This is equivalent to (has the same solution set as) the equation

$$y + 1 = 3 \cdot (x + 2) = 3x + 6$$

or

$$y = 3x + 5 \quad [6]$$

To determine the position of L in a sketch, we need the coordinates of one more point on the line other than $P = (-2, -1)$. We can get other points on L by setting fixed values of x in [6]. If we let $x = -1$, we find that $y = +2$, so $Q = (-1, 2)$ lies on L. We can now sketch L by drawing the straight line through P and Q, as in Figure 1.14. ■

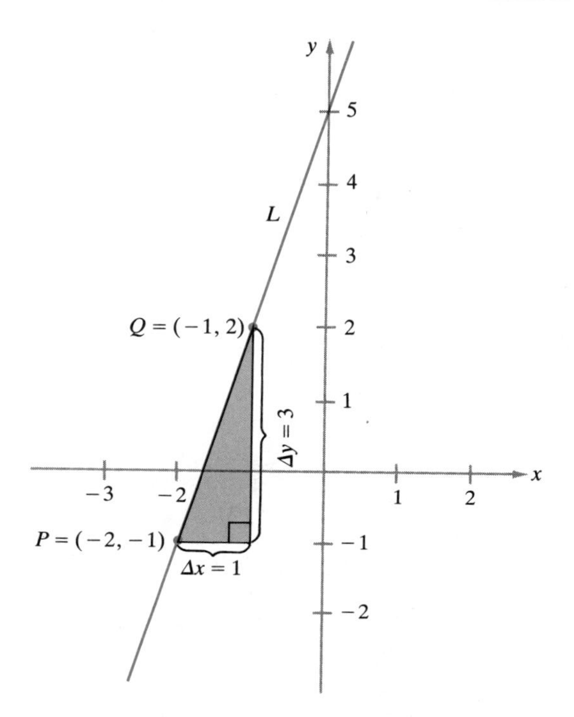

Figure 1.14
The line L has slope $m = +3$ and passes through $P = (-2, -1)$, indicated by a solid dot. Its equation is $y = 3x + 5$; that is, L is the solution set of this equation. The point $Q = (-1, 2)$ also lies on L, so we may draw L as the line through P and Q.

In Example 10 we could have used other x values such as $x = 0$ or $x = 2$ to find the second point Q on the line.

Here are some word problems that use a straight line to represent data.

Example 11 A bicycle store has varied its Saturday rental price p with the following effect on demand d = (number of rentals).

Price p	**Demand d**
\$4.00	86
\$5.50	50

As a first guess, we would reasonably expect that demand varies linearly with price, at least for prices between these extremes. (That is, the graph of the demand function $d = d(p)$ should be a straight line.) Determine the equation of this line and sketch it. What is the expected demand when the price is set at \$5.00?

Solution Label the coordinate axes in the plane as follows

p represented by the horizontal axis
d represented by the vertical axis.

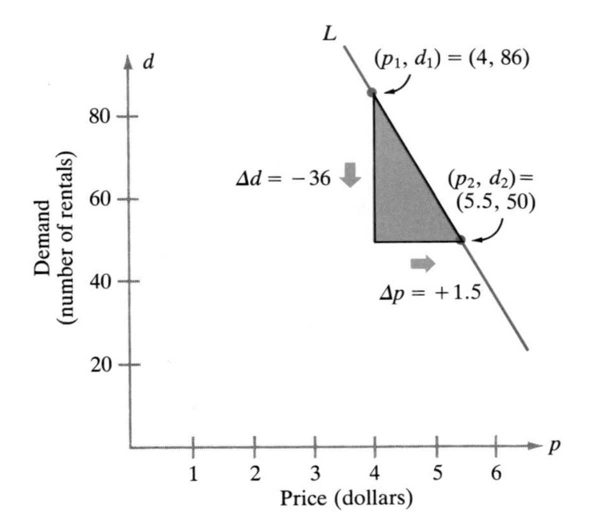

Figure 1.15
Straight line graph used to estimate the demand function in Example 11. Solid dots are the given data points. Equation of the line is $d = -24p + 182$. It slopes downward, demand decreasing with increasing rental price.

The appropriate line L passes through the points $(p_1, d_1) = (4, 86)$, $(p_2, d_2) = (5.5, 50)$. Its slope is therefore

$$m = \frac{\Delta d}{\Delta p} = \frac{d_2 - d_1}{p_2 - p_1} = \frac{50 - 86}{5.5 - 4.0} = = -\frac{36}{1.5} = -24$$

Because L passes through (4, 86), a point-slope equation [5] for L is $(d - 86) = m(p - 4) = -24(p - 4)$, or

$$d = -24(p - 4) + 86 = -24p + 182 \qquad [7]$$

We may sketch L by plotting the two given points and drawing a line through them, as in Figure 1.15. Setting $p = 5.00$ in Formula [7] for the demand, we find that $d = -24(5) + 182 = 62$ rentals. ■

Example 12 A physician starting practice has a library of medical books worth \$4800. For tax purposes, its value is assumed to *depreciate linearly* over a five-year interval: The graph of V = (value after t years) is a straight line with

$$V = 4800 \quad \text{when } t = 0 \qquad V = 0 \quad \text{when } t = 5$$

Find the equation of this line and the value after 3 years.

Solution In the coordinate plane let

t be represented by the horizontal axis (independent variable)

V be represented by the vertical axis (dependent variable)

The line we seek passes through $(t_1, V_1) = (0, 4800)$ and $(t_2, V_2) = (5, 0)$, as in

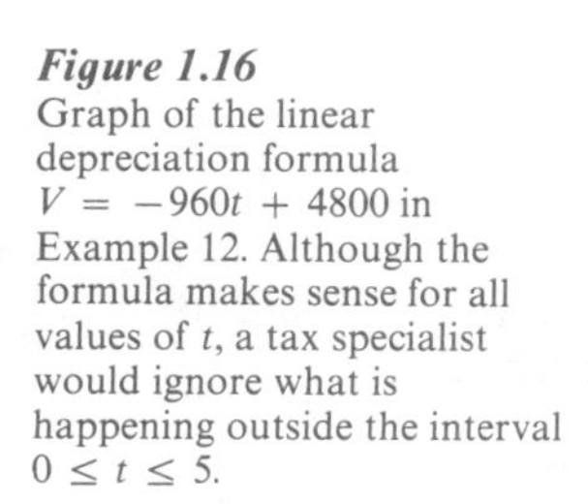

Figure 1.16
Graph of the linear depreciation formula $V = -960t + 4800$ in Example 12. Although the formula makes sense for all values of t, a tax specialist would ignore what is happening outside the interval $0 \leq t \leq 5$.

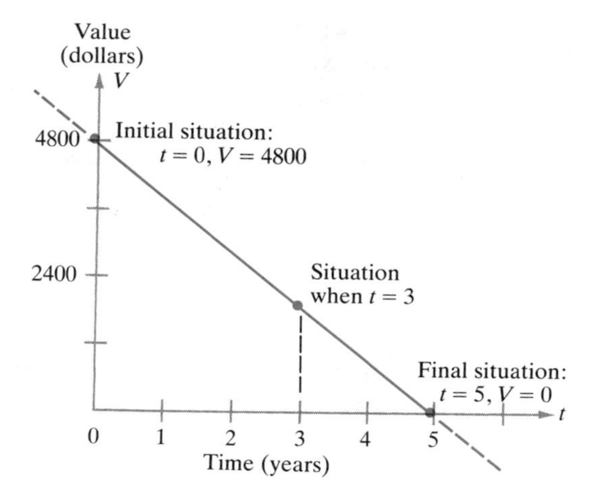

Figure 1.16, so its slope is

$$m = \frac{\Delta V}{\Delta t} = \frac{V_2 - V_1}{t_2 - t_1} = \frac{0 - 4800}{5 - 0} = -\frac{4800}{5} = -960$$

Because the line passes through (5, 0), we may write its equation as

$$(V - 0) = -960(t - 5) \quad \text{or} \quad V = -960t + 4800.$$

Clearly, $V = -960(3) + 4800 = \$1920$ after $t = 3$ years. ■

So far we have found the equation of a line from geometric information. We could also go the other way. First we note (without proof) which equations describe a line in the plane.

> **Equation of a Line** If at least one of the constants A and B is nonzero in the equation
>
> $$Ax + By + C = 0 \qquad (A, B, C \text{ constants}) \qquad [8]$$
>
> then the solution set of this equation is a straight line. Furthermore, every possible straight line is the solution set of an equation of this kind.

For example, the equation $y = (\frac{2}{3})x + (\frac{5}{3})$ is equivalent to $2x - 3y + 5 = 0$, which has the desired form.

What if we are given an equation such as [8] and are asked to sketch the line? Obviously, it is enough to locate any two points on the line. This is easy to do: Give one variable a definite value, and use the equation to find the corresponding value of the other variable.

Example 13 Sketch the line L determined by the equation $3x + 4y - 5 = 0$. Where does L cross the x axis? The y axis? What is the slope of L?

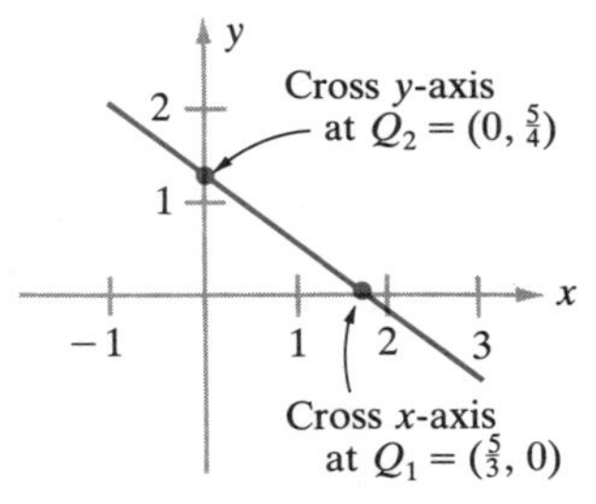

Figure 1.17
The line determined by equation $3x + 4y - 5 = 0$. Coordinates of the crossover points Q_1, Q_2 are determined in Example 13.

Solution The point Q_1 where L crosses the x axis must have y coordinate $y = 0$. Thus we may compute x: $3x + 4(0) - 5 = 0$, or $3x = 5$. Thus $x = \frac{5}{3}$ and the crossover point is $Q_1 = (\frac{5}{3}, 0)$ as shown in Figure 1.17. Similarly, where L crosses the y axis we must have $x = 0$, so that $y = \frac{5}{4}$; the crossover occurs at $Q_2 = (0, \frac{5}{4})$. Now that we know the coordinates of two points on L, we may compute the slope

$$\text{slope}(L) = \frac{\Delta y}{\Delta x} = \frac{\frac{5}{4} - 0}{0 - \frac{5}{3}} = -\frac{3}{4}$$

We could find other points on L by fixing the value of x (or of y) and using the equation for L to solve for the value of the other coordinate. ■

Exercises 1.2

1. Compute the changes Δx and Δy in x and y as we move from $P_1 = (-2, 3)$ to $P_2 = (0, -1)$. Find the slope of the line joining these two points.

Find the slopes of the lines joining the following pairs of points:

2. $(1, -3)$ and $(-1, 3)$

3. $(-1, -1)$ and $(-4, 8)$

4. $(4, 2)$ and $(6, 2)$

5. $(2, 2)$ and $(5, 3)$

6. $(-1, -4)$ and $(1, 2)$

7. $(1, 0)$ and $(7, 2)$

8. Find the equation of the line L that passes through the points $P = (-1, 2)$ and $Q = (3, -1)$. What is the slope of L?

9. Find the equation of the line that passes through $P = (2, -1)$ and has slope $m = \frac{1}{3}$.

In Exercises 10–15, determine equations for the following lines.

10. The line through the points $(1, 2)$ and $(-1, 0)$.

11. The line through the points $(-4, 5)$ and $(1, -2)$.

12. The line through $(1, 4)$ that has slope $m = -\frac{1}{2}$.

13. The line through $(3.10, 2.37)$ with slope $m = -\frac{1}{2}$.

14. The line with slope $m = -\frac{1}{3}$ that crosses the y axis at $y = +5$.

15. The line with slope $m = \frac{2}{7}$ that crosses the x axis at $x = 0$.

16. The costs C in a manufacturing operation vary linearly with production level q. Suppose that $C = \$15{,}000$ when $q = 100$, and that $C = \$16{,}100$ when $q = 120$. From these data find the formula for C as a function of q. (*Hint*: The data specify two points on the graph of this function (a straight line).)

17. A tract of land with producing oil wells is, at present, valued at \$1,300,000. In 12 years the wells will be depleted and then the land alone will be worth \$200,000. Find a formula for the value of the tract, assuming linear depreciation over the 12 years.

18. A new house is bought for \$55,000. The owner estimates that the value will increase to \$70,000 in 10 years. Find a linear formula that gives the value of the house over the next 10 years.

19. Based on past experience a manufacturer knows that the price p at which he will sell q items is a linear function $p = p(q)$. He has found that

(i) To sell $q = 1000$ units he can ask \$90 per unit.
(ii) To sell $q = 1600$ units he must reduce his price to \$80 per unit.

Find the equation of his "demand function" $p = p(q)$. What price should he charge if he wants to sell 1450 units? If the price is set at \$87 per unit, how many units will he sell?

20. Through market surveys, a bicycle store has estimated the potential demand q for June at various selling prices p (in dollars), as indicated in the table. Plot these points on

p	180	185	190	195	200
q	860	770	680	590	500

graph paper, taking p as the horizontal axis. Do they lie along a straight line? If so, find the equation $q = q(p)$ of this line. For what range of values of p would you have reasonable confidence in the predictions of this "demand equation"?

21. Find the slope of the line whose equation is $2x + 3y = 6$. Where does it cross the x axis? Where does it cross the y axis?

22. Find the coordinates of the point where the line $x + 2y = 200$ crosses the x axis; likewise for the y axis. What is its slope?

Sketch the lines determined by the following equations. Find their slopes.

23. $4x + 3y = 0$

24. $x + 2y - 3 = 0$

25. $x - y - 4 = 0$

26. $y = 4x - \sqrt{5}$

27. $y - 2 = 0$

28. $-3x + 2y = 7$

29. $0.67x - 0.80y = 0.59$

30. $350x + 420y - 500 = 0$

31. As prices rise, cattlemen are willing to put more beef on the market; when prices fall, they tend to cut back. Suppose this relationship between supply and price is given by the formula

$$p = \frac{1}{30}q + \frac{1}{2} \qquad (\text{for } 50 \leq q \leq 150)$$

where p = price (dollars per pound) at which producers are willing to supply q million pounds per week. Plot the graph of $p = p(q)$.

32. If L is the line determined by the equation

$$Ax + By + C = 0 \qquad (A, B, \text{ and } C \text{ fixed constants; } B \text{ nonzero})$$

show that

$$\text{slope}(L) = -\frac{A}{B}$$

Hint: Locate two points on L by taking $x = 0$ and $x = 1$ and solve for the corresponding values of y.

33. On a single piece of graph paper, sketch the lines whose equations are

(i) $3x - 4y = -1$
(ii) $3x - 4y = 0$
(iii) $3x - 4y = +1$
(iv) $3x - 4y = +2$

by plotting the points where the lines cross the coordinate axes. What are their slopes? How are the lines related to each other geometrically? What can you say about the family of all lines L_c (c is a real number) given by the equation

$$3x - 4y = c?$$

How does the position of L_c (the solution set of this equation) vary as the constant c increases?

1.3 Some Functions from Economics

Suppose a firm produces a single product, such as refrigerators (the nature of the product does not really matter). If you are managing this firm, you face many decisions. One of the major decisions is to set the rate of production. How can you do this? One way is to guess—if you guess correctly often enough you will be praised for your business acumen. If not, you may have plenty of time to ponder what went wrong. In that case you might realize that instead of, or perhaps in addition to, guessing you should have attempted to analyze the problem.

This suggests that you study the relationship between the rate of production and the **profit** P, which is just the difference between what comes in and what goes out. We call the last two variables the **revenue** R and the **cost** (or **operating cost**) C; thus

$$P = R - C \qquad [9]$$

Each of the variables P, R, and C depends on the level of production, which is the independent variable.

In economic theory we fix a "short-run period of time": a week, a month, or some other convenient period. The **level of production**, indicated by x, is the number of units produced during this period. The plant can be run at different levels of production. At one extreme it can be left idle ($x = 0$ units); at the other extreme it can be run at full capacity. Operating costs are indicated by C, as in [9]. Certain costs are independent of production level and must be paid even if there is *no* production at all. These are the **fixed costs** indicated by the symbol FC. They include such items as rent, interest paid to investors, amortization of the firm's equipment, and your salary during the short-run period. Other costs result from actual production, such as labor costs, costs of raw materials, and so on. These **variable costs** are indicated by VC; $C = FC + VC$.

Within certain limits, the variable x is under direct control of an economic strategist. Once the level of production x is set, the operating costs C are determined by the price of raw materials and labor. Thus x is a natural independent variable, and the cost C may be expressed as a function of x, giving us the **cost function** $C(x)$. In actual operations, determining the cost function would require detailed analysis of operating procedures, costs of labor and raw materials, and so on. Sometimes the variable costs VC are the same for each unit produced, say A dollars per unit. Then the variable costs are proportional to x and are given by a very simple formula

$$VC = A \cdot x \quad \text{dollars} \qquad (A = \text{cost per unit produced})$$

and the total cost function is given by

$$C(x) = FC + VC = FC + A \cdot x \qquad \text{for } x \geq 0$$

This is what happens in the next few examples.

Example 14 A retail gasoline station sells a single grade of fuel. The weekly costs of labor (salaries of three regular employees), insurance, property taxes, upkeep, and interest on invested capital amount to \$1080. A large supplier provides gasoline to the station at \$1.10 per gallon, regardless of sales volume. Describe the weekly costs of this operation. Find the cost if the station dispenses 6000 gallons in one week.

Solution The natural production variable is x = (number of gallons sold). Fixed costs amount to \$1080 for the week. Because the station buys its fuel at \$1.10 per gallon, the variable cost is $VC = 1.10x$ for any level of sales $x \geq 0$. Therefore the total cost is

$$C(x) = FC + VC = 1080 + 1.10x \qquad \text{for } x \geq 0 \tag{10}$$

If $x = 6000$ gallons per week, the total cost is

$$\begin{aligned} C(6000) &= 1080 + 1.10(6000) \\ &= \$7680 \end{aligned}$$

■

The graph of the cost function $C(x)$ in [10] is shown in Figure 1.18; the point marked on the graph corresponds to sales of $x = 6000$ gallons per week.

Revenue can be analyzed in various ways. In simple situations the entire output is sold at a fixed price, say p dollars per unit. Then the revenue is

$$\begin{aligned} R &= (\text{selling price per unit}) \cdot (\text{number of units produced}) \\ &= p \cdot x \end{aligned} \tag{11}$$

where x is the production level, and the profit is

$$P(x) = R - C = p \cdot x - C(x) \tag{12}$$

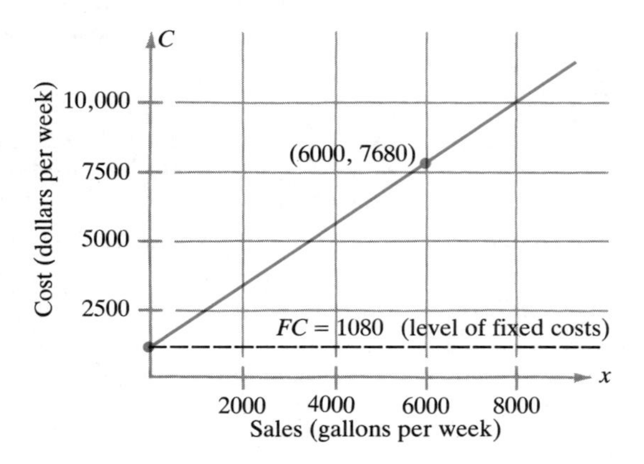

Figure 1.18
Total weekly costs for the gasoline station in Example 14. If $x = 0$, then $C(0) = FC + 0 = \$1080$ is just the fixed cost. The cost when $x = 6000$ is indicated by the solid dot. The graph of $C(x) = 1080 + 1.10x$ is a straight line. The dashed horizontal line indicates the level of fixed costs: $FC = \$1080$ per week.

Example 15 Suppose the gasoline station in Example 14 sells all its fuel at \$1.30 per gallon. Describe the weekly revenue and profit in terms of the number x of gallons sold.

Solution We just analyzed the costs and found that

$$C(x) = 1080 + 1.10x$$

Each gallon sells for \$1.30, so the revenue is

$$R(x) = 1.30x$$

and the profit is

$$\begin{aligned} P(x) &= R(x) - C(x) = 1.30x - (1080 + 1.10x) \\ &= 0.20x - 1080 \end{aligned}$$

when x gallons are sold. ■

Example 16 The Veniero Bakery has decided to produce fruit tarts for the month of July. Fixed costs for the month are \$1100. Ingredients and labor amount to \$4.50 per pound produced. All the tarts are eagerly consumed at a selling price of \$6.75 per pound. Find the cost, revenue, and profit functions. How many pounds must be sold to break even (cost = revenue)?

Solution The natural production variable is x = (pounds produced). The cost at production level x is then

$$C(x) = FC + VC = 1100 + 4.5x \quad \text{dollars}$$

and the revenue is

$$R(x) = 6.75x \quad \text{dollars}$$

Thus the profit for the month is

$$\begin{aligned} P(x) &= R(x) - C(x) = 6.75x - (1100 + 4.5x) \\ &= 2.25x - 1100 \end{aligned}$$

Breakeven occurs when $R(x) = C(x)$, which is the same as saying that $P(x) = 0$:

$$0 = 2.25x - 1100$$

$$x = \frac{1100}{2.25} = 488.9 \quad \text{pounds}$$

Once sales exceed $x = 489$ pounds for the month, the bakery is making money. ■

In complex, large-scale operations the variable costs are not proportional to x. As x increases from $x = 0$, the operation at first tends to become more efficient. Manpower is employed more effectively. And by investing capital in research, more efficient products may be devised and labor-saving devices introduced. Such effects, referred to as **economies of size**, reduce the cost per unit. After x increases beyond a certain level, we encounter **diseconomies of size**, which cause the cost per additional unit to rise. These effects might occur if some scarce raw materials were used or if substantial overtime wages were paid. Figure 1.19 shows the graph of a typical cost function for a large range of x. In the next situation, variable costs are *not* proportional to the production level: Sometimes buying in large quantities is cheaper.

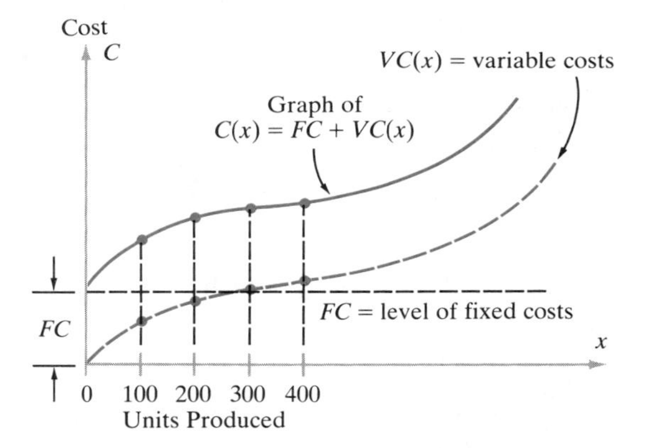

Figure 1.19 A realistic cost function. The dashed horizontal line represents the fixed costs: $C(x) = FC$ when $x = 0$. The solid curve represents the total cost of producing x units; it is the graph of $C(x) = FC + VC$. As x begins to increase from zero, the costs $C(x)$ rise less and less steeply (the graph begins to flatten out) because economies of size predominate. For very large x, costs once again begin to rise steeply, as diseconomies of size become overwhelming.

Example 17 Suppose the gasoline station in Example 14 has weekly fixed costs of \$1080, as before. But now the wholesale supplier offers to provide fuel at a discount according to the formula

$$(\text{wholesale cost per gallon}) = 1.10 - 0.00001x$$

if we order x gallons for the week.† Assuming that we sell fuel for \$1.30 per gallon, find the cost, revenue, and profit functions. What is the cost if we sell 6000 gallons for the week? What is the profit?

† At low volume we pay about \$1.10 per gallon. If we order 10,000 gallons we are charged $1.10 - 0.00001x = 1.10 - 0.10 = \$1.00$ per gallon, making a total charge of $1.00(10{,}000) =$ \$10,000 for the week.

Solution If we sell x gallons for the week, our *cost per gallon* is $1.10 - 0.00001x$ dollars. Multiplying this by the number of gallons x, we obtain the variable cost

$$\begin{aligned} VC &= (\text{cost per gallon}) \cdot (\text{number of gallons}) \\ &= (1.10 - 0.00001x) \cdot x \\ &= 1.10x - 0.00001x^2 \end{aligned} \quad [13]$$

The fixed costs are \$1080, so the total cost per week at sales level x is

$$C(x) = FC + VC = 1080 + 1.10x - 0.00001x^2 \quad [14]$$

This cost function is graphed in Figure 1.20 (solid curve). If $x = 6000$, the cost in the discount situation is obtained by setting $x = 6000$ in Formula [14]

$$\begin{aligned} C(6000) &= 1080 + 1.10(6000) - 0.00001(6000)^2 \\ &= \$7320 \end{aligned}$$

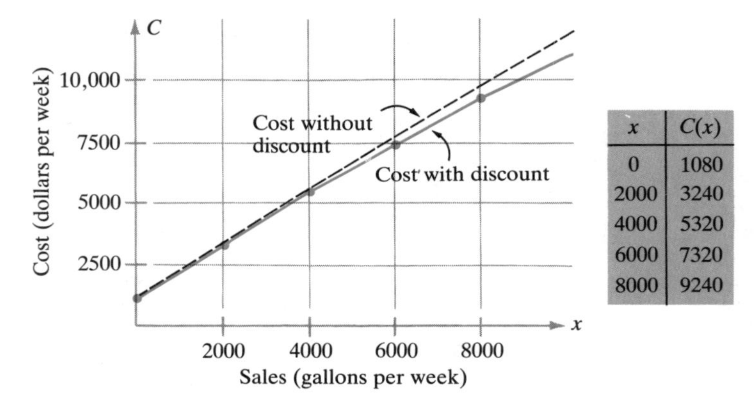

x	$C(x)$
0	1080
2000	3240
4000	5320
6000	7320
8000	9240

Figure 1.20
The cost function $C(x) = 1080 + 1.10x - 0.00001x^2$ in Example 17 includes a discount. Its graph is the solid curve; dots correspond to values shown in the table. The dashed line is the cost function without the discount (recall Example 14).

Compare this with the cost of \$7680 in the no-discount situation discussed in Example 14.

For any x, the revenue is $R(x) = 1.30x$, so the profit is

$$\begin{aligned} P(x) &= R(x) - C(x) = 1.30x - (1080 + 1.10x - 0.00001x^2) \\ &= -1080 + 0.20x + 0.00001x^2 \end{aligned} \quad [15]$$

Setting $x = 6000$, we find that our weekly profit is

$$\begin{aligned} P(6000) &= -1080 + 0.2(6000) + 0.00001(6000)^2 \\ &= \$480. \end{aligned}$$

At this level of sales the station is just beginning to break even. ■

Just as the variable costs need not be proportional to production level in large-scale operations, so too for revenue. To deal with this we must introduce another important tool in economic analysis, the **demand function**. To discuss it, we must consider the **price** p at which we sell our product. We could set p (dollars per unit) anywhere between 0 and $+\infty$, but an arbitrary choice will not guarantee optimum profit, or even any profit at all. To make a sensible choice we must test the market conditions to see how demand for the product varies with asking price.

If a manufacturer produces a certain number of units per year and if the price per unit is set too low, the demand will be very great, leaving many unsatisfied customers willing to buy at this (low) price. He should raise the price in this case. On the other hand, if the price is too high, there will be too few customers and he should obviously lower his price to avoid accumulating unsold stock. Once the production level x is set, there is a selling price $p(x)$ at which exactly x units will be sold, leaving no unsatisfied demand. We call this function $p(x)$ the **demand function**† because the price $p = p(x)$ directly reflects the market demand for the product. A typical demand function is shown in Figure 1.21.

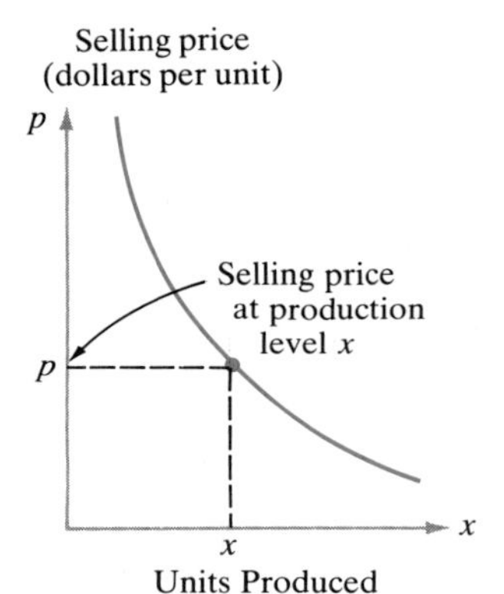

Figure 1.21
A typical demand function. The price is very large if the supply is small. (Point p on the y axis is the selling price at production level x.) Intuitively, the asking price $p(x)$ at which we will sell x units must decrease steadily as production increases. After all, is there any case in which increased supply leads to increased prices, all other factors being held constant? At high values of x the price must be kept very low to remain competitive and to sell this many units in a glutted market.

We could determine $p(x)$ by making market surveys or by extrapolating existing sales figures. This price is independent of the particular manufacturing methods used to produce the item. It reflects only the performance of the finished item on the sales market. In contrast, the cost function $C(x)$ gives the manufacturer's cost of producing x items, without regard to whether or not he can sell this many items, or sell them at an acceptable profit. Thus $p(x)$ and $C(x)$ measure independent but equally important relationships between a manufacturing business and the market in which it operates. With both in hand we can calculate many things of interest.

For one thing, they determine the revenue at production level x, and hence also the profit. We get the total revenue $R(x)$ by multiplying the number of units x by the price $p(x)$ at which we can sell this many units. Thus

$$R(x) = x \cdot p(x) \qquad [16]$$

and the profit is

$$P(x) = R(x) - C(x) = x \cdot p(x) - C(x) \qquad \text{for } x \geq 0 \qquad [17]$$

once the cost function $C(x)$ and function $p(x)$ have been determined.

In simple situations we sell all of our output at a single price, regardless of production level. Then $p(x)$ is constant, say $p(x) = A$ dollars per unit, and the revenue $R(x) = x \cdot p(x) = A \cdot x$ is proportional to the number of units produced. This was the case in Examples 16 and 17; it is not realistic in other situations where the demand behaves as shown in Figure 1.21. Here are some examples of prices varying with quantity to be sold.

Example 18 The Aerobic Bicycle Rental store determined that the number of rentals per day in June varies with the rental price. The resulting demand

† An alternative definition of the demand function occurs in some economics texts, where the asking price p is taken as the independent variable. Then $x = x(p) =$ (quantity sold at price p) reflects the market demand for the product and is regarded as the demand function. Actually, $x(p)$ is the "inverse" of our demand function $p(x)$. It is obtained by taking our function $p = p(x)$ and solving for x as a function of p (see Exercise 25 for an example). Given either one of these functions, the other can be determined. In the final analysis, either one can be taken as the demand function.

function is

$$p(x) = 20 - 0.05x \quad \text{dollars} \qquad (x = \text{number of rentals}) \qquad [18]$$

The store has fixed costs of \$325 per day, and variable costs amount to \$3.50 per bicycle rented (to cover maintenance and periodic replacement as cycles become hopelessly worn out).

(i) Find the cost, revenue, and profit functions for this operation.
(ii) If the store wants to rent out 185 bikes, what rental fee should they charge?
(iii) On June 25 they rented 185 bikes. What was their revenue? Their profit?

Solution Variable costs are $VC = 3.50x$, so the cost function is

$$C(x) = FC + VC = 325 + 3.50x \quad \text{dollars}$$

Revenue is given by

$$R(x) = x \cdot p(x) = x(20 - 0.05x) = 20x - 0.05x^2$$

so the profit function is

$$\begin{aligned} P(x) &= R(x) - C(x) = (20x - 0.05x^2) - (325 + 3.50x) \\ &= -325 + 16.5x - 0.05x^2 \end{aligned}$$

if x bikes are to be rented.

The price at which they will be able to rent out 185 bikes is obtained by setting $x = 185$ in the demand function; it is

$$p(185) = 20 - 0.05(185) = 10.75 \quad \text{dollars}$$

The revenue and profit at this sales level are

$$\begin{aligned} R(185) &= 20(185) - 0.05(185)^2 = \$1988.75 \\ P(185) &= -325 + 16.5(185) - 0.05(185)^2 = \$1016.25 \end{aligned}$$ ■

Example 19 A manufacturer of television sets can set his production anywhere between 0 and 500,000 units per year. Detailed analysis of plant operations leads to the following estimate of costs.†

$$C(x) = 5{,}000{,}000 + 89x + 0.000012x^2 \qquad \text{for} \quad 0 \le x \le 500{,}000$$

Market analyses yield an estimate for the demand function at various levels of production

$$p(x) = 150 - 0.000086x \qquad \text{for} \quad 0 \le x \le 500{,}000$$

Using this information,

(i) determine the price at which they can sell 500,000 sets per year.

† Fixed costs (cost when $x = 0$) equal \$5,000,000 yearly; the other terms, involving x, give the variable costs. At low production levels the cost per set is about \$89, so that $VC = 89.00x$ (approximately). The extra term involving x^2 reflects diseconomies of size; because x^2 is small when x is small, these become substantial only for large x.

(ii) if the manager decides to price his sets at \$130 each, determine how many he will be able to sell at this price.
(iii) calculate the costs, revenue, and profit if $x = 500{,}000$ sets per year.

Then sketch the graph of the profit function $P(x)$, and use the graph to estimate the production level x yielding the highest profit. What is the amount in dollars of this maximized profit? What price should be charged at this optimal production level?

Solution To sell 500,000 sets, the price should be

$$p(500{,}000) = 150 - 0.000086(500{,}000) = 150 - 43$$
$$= \$107 \text{ per set}$$

If the price is arbitrarily set at $p = \$130$, we find the number x of sets that can be sold at this price by placing $p = 130$ on the left side of the demand function $p = p(x)$ and solving for x:

$$130 = 150 - 0.000086x$$
$$x = \frac{20}{0.000086} = 232{,}558 \quad \text{units per year}$$

Yearly revenue at production level x is $R(x) = x \cdot p(x)$

$$R(x) = x(150 - 0.000086x) = 150x - 0.000086x^2$$

The profit function is given by

$$P(x) = R - C = -5{,}000{,}000 + 61x - 0.000098x^2$$

for $0 \leq x \leq 500{,}000$. In particular, if $x = 500{,}000$ we obtain

$$C(500{,}000) = 5{,}000{,}000 + 89(500{,}000) + 0.000012(500{,}000)^2$$
$$= \$52{,}500{,}000 \quad \text{per year}$$
$$R(500{,}000) = 150(500{,}000) - 0.000086(500{,}000)^2$$
$$= \$53{,}500{,}000 \quad \text{per year}$$
$$P(500{,}000) = R - C = \$1{,}000{,}000 \quad \text{per year}$$

Notice that we need not substitute $x = 500{,}000$ into the profit function once R and C have been determined, because $P = R - C$.

The graph of $P(x)$ is plotted in Figure 1.22. A visual estimate indicates that maximum profit is achieved at $x = 311{,}000$. The corresponding selling price is obtained from the demand function

$$p = p(311{,}000) = 150 - 0.000086(311{,}000)$$
$$= \$123.25$$

The maximum profit is $P(311{,}000) = \$4.49$ million per year. Notice that $P(x)$ is negative for some production levels (manufacturer operates at a loss), and profits actually decline for high values of x because of diseconomies of size. ■

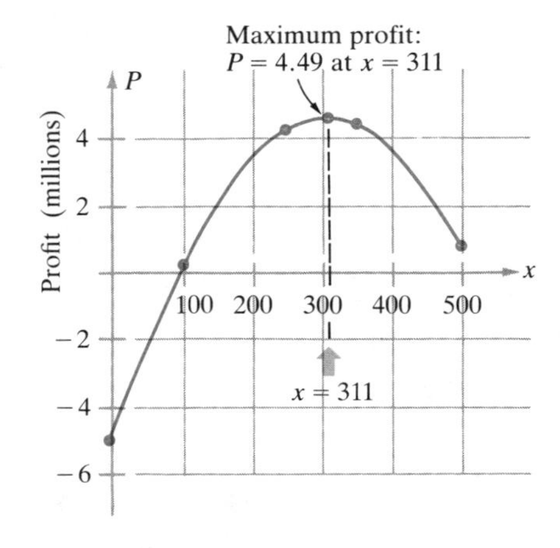

Figure 1.22
Graph of the profit function $P(x)$ in Example 19. By visually estimating the coordinates of the highest point on the graph, maximum profit occurs when $x = 311{,}000$ sets per year. Then $P = \$4.49$ million per year. At the maximum feasible production level $x = 500{,}000$, the profit is lower; $P = \$1.00$ million per year because of diseconomfes of size. Maximum production does not necessarily yield maximum profit.

x (thousand)	$P(x)$ (million)
0	−5.00
100	0.12
250	4.13
311	4.49
350	4.35
500	1.00

A Final Note: In many situations the production variable x can only take on integer values because the firm produces a whole number of units. To simplify the mathematics we adopt the fiction that x takes on all real values $x \geq 0$. We "smooth" the data in graphs by plotting the points that correspond to integer values of x, then drawing a smooth curve through these points.

Exercises 1.3

A mail-order supplier operating out of a rented office sells vitamin packs. To sell x packs per week involves the following costs:

(i) Office rent: $250 per week
(ii) Mailing costs: $0.85 per package
(iii) Interest on money invested in inventory: $85 per week
(iv) Each pack costs the supplier $4.85
(v) Phone bill: $102 per week.

Answer Exercises 1–4 using these facts.

1. Which are fixed costs and which are variable costs?

2. Find a formula for the weekly costs in terms of x.

3. If each pack sells for $10.98 (postage included), find the cost, revenue, and profit functions.

4. Find the costs, revenue, and profit if 147 packs are sold in one week.

5. Rent-a-Heap of Colorado will rent you a car for $14 per day plus 13¢ per mile. Find
(i) the cost of renting a car for a day trip of 350 miles
(ii) the cost of renting a car for the day and driving x miles

6. A builder of prefabricated houses has fixed costs $FC = \$750{,}000$ per year and variable costs $VC = 29{,}000q$ proportional to the number q of units constructed per year. Find the cost function $C(q)$. Find the cost of producing 50 houses in a year.

7. Repeat Exercise 6, except that now the variable costs $VC = 29{,}000q - 100q^2$ include a quadratic term that reflects an economy of size. Find the cost function. Find the cost of producing 50 houses in a year.

8. The fixed costs of an encyclopedia publisher are \$10,000 per week, and each encyclopedia costs \$60 to produce. Find the cost function $C(x)$, where x is the number of sets produced per week. Find $C(0)$, $C(10)$, $C(500)$, $C(1000)$. For what value of x is $C(x) = \$19{,}000$ per week?

9. A manufacturing operation has a linear cost function $C(x) = A + B \cdot x$, where x is the production level. Past records show that

$$C = \$15{,}000 \quad \text{when} \quad x = 100$$
$$C = \$16{,}100 \quad \text{when} \quad x = 120$$

From these data find the values of the constants A and B in the cost formula. What is the fixed cost FC for this operation?

10. Rent-a-Heap offers the following deal on a rented car: \$14 per day plus 200 free miles; beyond 200 miles we pay 17¢ per mile. Find the formula for the cost $C(x)$ of renting the car for the day and driving x miles:
(i) if $0 \leq x \leq 200$ (ii) if $x > 200$
What are our costs if we drive 380 miles that day?

11. Graph the cost function $C(x)$ determined in Exercise 10. (*Note:* $C(x)$ is given by different formulas for $0 \leq x \leq 200$ and $x > 200$. What happens to the graph at $x = 200$?)

12. The cost function for the gasoline station in Example 14 has been shown in Figure 1.18. Make a copy on graph paper. Then plot the graph of the revenue function $R(x) = 1.30x$ on the same sketch. Use the resulting sketch to estimate the sales level x where the cost and revenue curves cross. (This is the breakeven sales level, where $R = C$.)

13. A group of studio apartments that rent for \$400 per month yields a profit (over and above expenses) of \$65 per month from each occupied apartment. But each unoccupied apartment causes a net loss of \$35 per month. There are 60 apartments in the complex.
(i) Express the monthly profit P as a function of x, the number of occupied apartments. What is the feasible range of x in this problem?
(ii) Express the monthly revenue R as a function of x.
(iii) Find the breakeven value of x at which P becomes positive.
(iv) Find the cost function $C(x)$ using the relation $P = R - C$.
Sketch the graph of the profit function, showing the limited feasible set of values for x.

A small-scale Mercury smelter has fixed costs of \$5000 per week and variable costs (mostly fuel, labor, and cost of raw ore) $VC = 27x$, proportional to the number x of pounds produced per week. They sell their refined Mercury for \$37 per pound. Answer the questions in Exercises 14–18 using these facts.

14. Find the cost, revenue, and profit functions for this smelter.

15. At what production level does the operation break even?

16. Write out a formula for the *average cost per pound* produced

$$A(x) = \frac{C(x)}{x} = \frac{\text{(total cost)}}{\text{(total number of pounds produced)}}$$

17. Evaluate the total cost C and the average cost per pound A for $x = 10, 50, 100, 200, 500$, and 1000 pounds per week.

18. Sketch the graph of $A(x)$. What happens to the average cost $A(x)$ as x gets very small? As x gets very large?

19. Here is a more realistic wholesale discount arrangement than the one discussed in Example 17. Suppose a retail gasoline station operates under the following conditions:
(i) *Low volume* ($0 \le x \le 8000$ gallons per week). Three employees needed; fixed costs (including wages) are \$1080 per week. Price paid to supplier is \$1.10 per gallon.
(ii) *High volume* ($x > 8000$ gallons per week). An additional employee must be hired to handle the work: weekly salary \$240. But the supplier offers a discount: all purchases beyond 8000 gallons per week are made at \$1.00 per gallon.
Give (separate) formulas for the cost function during low- and high-volume operations. Find the weekly costs when $x = 6000, 8000$, and $12{,}000$ gallons per week. Sketch the graph of $C(x)$, paying particular attention to what happens when $x = 8000$. (*Hint:* $C(x)$ will be given by different formulas for $0 \le x \le 8000$ and $x > 8000$.)

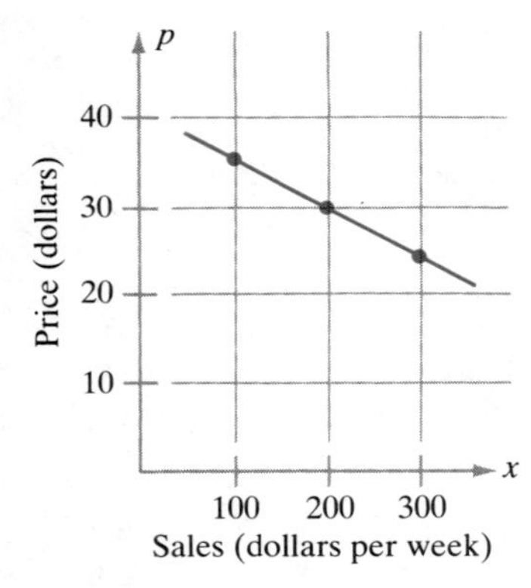

Figure 1.23
The demand for jeans (Exercises 20–23).

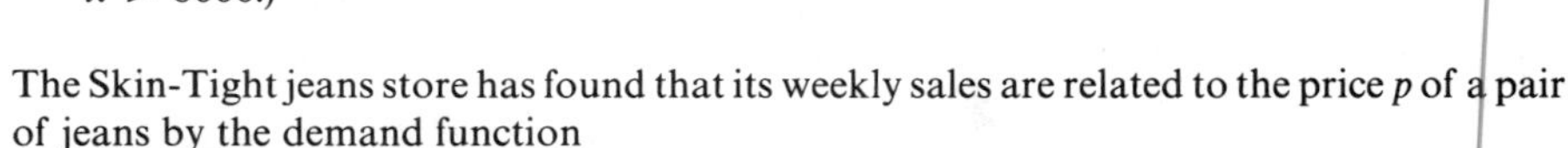

The Skin-Tight jeans store has found that its weekly sales are related to the price p of a pair of jeans by the demand function

$$p(x) = -0.05x + 40 \quad \text{dollars}$$

where x is the number of jeans they wish to sell in a week. Answer Exercises 20–23 using these facts. (*Note:* For reference, the graph of this demand function is shown in Figure 1.23.)

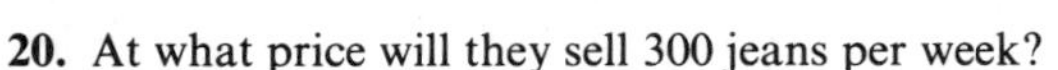

20. At what price will they sell 300 jeans per week?

21. At what price will they sell 150 per week?

22. If they want to sell a recent shipment of 280 jeans during the next week, what price should they charge?

23. If they charge \$30 for a pair of jeans, how many will they sell?

24. A cut-rate gasoline station has found that its demand function is

$$p(x) = 0.8667 + \frac{1066}{x} \quad \text{dollars}$$

where x is the number of gallons sold per day, and p is the price per gallon.
(i) At what price will they sell 2000 gallons per day?
(ii) If they want to sell 8000 gallons per day, what price should they charge?
(iii) If they charge \$1.25 per gallon, how many gallons will they sell?
(*Note:* For reference, the graph of this demand function is shown in Figure 1.24.)

25. Take the demand function

$$p(x) = 0.8667 + \frac{1066}{x} \quad \text{dollars}$$

of Exercise 24 and solve for x in terms of p. (*Note:* This gives the alternative version of the demand function, in which the price p is regarded as the independent variable, and $x = x(p)$.)

26. A manufacturer has fixed costs $FC = \$1500$ per week and variable costs $VC = 35q$ proportional to the level of production q. The demand function is given by $p(q) = 48 - 0.01q$. Determine
(i) the cost function
(ii) the revenue function
(iii) the profit function
as functions of q. Calculate the revenue and profit for $q = 100, 300, 800$, and 1000.

27. A manufacturer has fixed costs $FC = \$1000$ per month and variable costs of \$100 for each unit produced. If $x =$ (number of units produced per month), find a formula for the monthly costs $C(x)$. Suppose that demand for his product is described by

$$p(x) = 300 - 2x \qquad \text{for} \quad 0 \leq x \leq 40$$

Find the revenue function $R(x)$ and profit function $P(x)$. What are the costs, revenue, and profit when $x = 30$?

28. In Exercise 27, sketch the graph of the profit function, and estimate the maximum profit over the feasible set of production levels $0 \leq x \leq 40$. What level of production yields maximum profit?

29. The cost function for a retail gasoline station selling x gallons of fuel per week is $C(x) = 1080 + 1.10x$, and the demand function is given by

$$p(x) = 1.50 - 0.00002x \quad \text{dollars}$$

Find the revenue function $R(x)$ and the profit function $P(x)$. What is the profit when the station sells 10,000 gallons per week? At what price did the station have to offer fuel to sell 10,000 gallons per week?

30. Sketch the graph of the profit function determined in Exercise 29. What level of sales x yields the maximum profit, assuming the feasible sales levels are $0 \leq x \leq 15{,}000$ gallons per week? At what price should the station sell fuel to achieve the optimal sales level just determined?

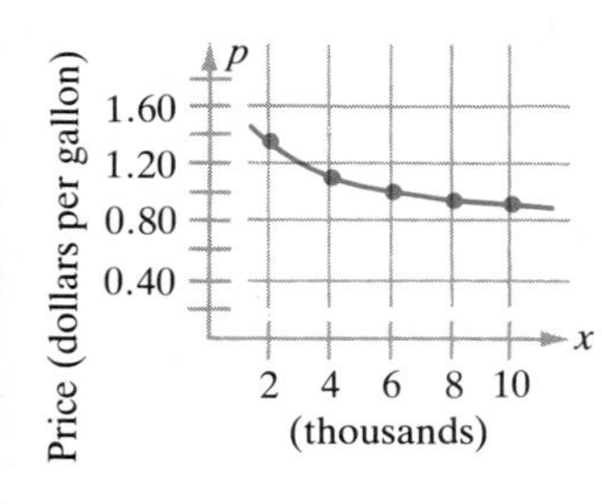

Figure 1.24
Demand function for the cut-rate gasoline station (Exercise 24).

The television manufacturer of Example 19 has updated his cost–demand analysis. In 1984 he expects the cost and demand functions to be given by

$$\left.\begin{aligned} C(x) &= 5{,}000{,}000 + 109x + 0.000012x^2 \\ p(x) &= 170 - 0.00011x \end{aligned}\right\} \quad \text{for } 0 \leq x \leq 500{,}000$$

Use this information to answer Exercises 31–35.

31. Find the profit function $P(x)$.

32. Sketch the graph of the profit function, plotting at least the points corresponding to production levels $x = 0, 100, 200, 250, 300$, and 500 thousand.

33. Using the graph, estimate the production level yielding maximum profit.

34. In Exercise 33 we found the optimum production level. What price will ensure that this many units are sold?

35. Using the graph, estimate the maximum feasible profit (the value of P at the optimum production level).

1.4 Exponential Functions and Population Growth

In one more application of the concept of function we shall describe some models of biological population growth. Later, we will see that the same mathematical methods can be used to describe many other phenomena: growth of money in investment schemes, radioactive decay, degradation of pesticide residues in the environment, and so on. Exponential functions appear in all such problems. The following examples are based on exponential functions of the form $y = 2^{kx}$, where k is a fixed constant. Computing their values amounts to calculating various powers of 2. For example, if $k = 1/30$, then the function $y = 2^{x/30}$ has values

$$
\begin{array}{ll}
x = 0 & y = 2^{0/30} = 2^0 = 1.000 \\
x = 15 & y = 2^{15/30} = 2^{1/2} = 1.414 \\
x = 30 & y = 2^{30/30} = 2^1 = 2.000 \\
x = 120 & y = 2^{120/30} = 2^4 = 16.000
\end{array}
$$

and so on. We will have something to say about more general exponential functions at the end of this section; we will study them in detail in Chapter 4.

Population growth problems have two common features:

(i) There is an initial population level, which is the population when we begin our observations.
(ii) There is a "doubling time" T: Starting at any time, the population will have doubled T years later. [19]

For example, suppose Country X has a population of 25 million that is increasing so fast that it will double every 30 years. Knowing the initial population and the doubling time $T = 30$ years, we can compute some values of P:

$$
\begin{array}{lll}
\text{When } t = 0, & P(0) = 25{,}000{,}000 & \text{(initial population)} \\
\text{When } t = 30, & P(30) = 50{,}000{,}000 & \\
\text{When } t = 60, & P(60) = 100{,}000{,}000 & \\
\text{When } t = 90, & P(90) = 200{,}000{,}000 &
\end{array}
\qquad [20]
$$

and so on. Each value is double the previous one, because 30 years have elapsed. These values are listed and plotted on a graph in Figure 1.25.

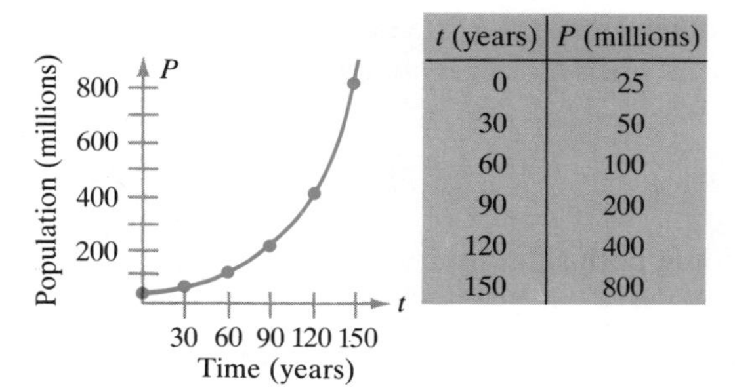

t (years)	P (millions)
0	25
30	50
60	100
90	200
120	400
150	800

Figure 1.25
Graph of the function $P(t) = 25{,}000{,}000 \cdot 2^{t/30}$, which fits the data in [20]. This function has the general form $A \cdot 2^{kt}$ where A and k are constants; here the exponent kt varies, and the base 2 remains fixed. Exponential functions appear in all kinds of growth problems. They will be discussed thoroughly in Chapter 4.

What kind of function $P = P(t)$ will describe the population after t years, for any t? The following formula, involving the exponential function $2^{t/30}$ generates the correct P values for the times listed in [20].

$$P(t) = 25{,}000{,}000 \cdot 2^{t/30} \quad \text{for} \quad t \geq 0 \qquad [21]$$

This is easily checked out: For example, when $t = 60$, we obtain $2^{60/30} = 2^2 = 4$, and the formula gives $P(60) = 25{,}000{,}000 \cdot (4) = 100{,}000{,}000$ as desired. In fact, as long as the growth rate remains constant, Formula [21] gives the size of the population for *all* values of $t \geq 0$. Thus, after 15 years this model predicts a population of

$$\begin{aligned} P(15) &= 25{,}000{,}000 \cdot 2^{15/30} = 25{,}000{,}000 \cdot 2^{1/2} \\ &= 35{,}355{,}000 \end{aligned}$$

and similarly for other values of t.

Using methods of calculus we can demonstrate an important fact about all growth problems: If the growth rate remains constant, the population can always be expressed as an exponential function of t

$$P(t) = A \cdot 2^{kt} \quad \text{for } t \geq 0 \qquad [22]$$

where A and k are constants to be determined from the specific facts about the population. How do we find A and k? When $t = 0$, we have $2^0 = 1$, so

$$P(0) = A \cdot (2^{k \cdot 0}) = A \cdot (2^0) = A \cdot 1 = A$$

In other words

$$A = \text{the population size at time } t = 0 \text{ (the initial population)} \qquad [23]$$

The other constant k is determined from the doubling time T by noting what happens when $t = T$. By definition of doubling time, $P(T) = 2 \cdot P(0)$, so

$$2 = \frac{P(T)}{P(0)} = \frac{A \cdot 2^{kT}}{A \cdot 2^{k \cdot 0}} = \frac{A \cdot 2^{kT}}{A \cdot 1} = 2^{kT}$$

Dividing both sides by 2 we obtain

$$1 = \frac{2^{kT}}{2} = 2^{(kT-1)}$$

But $x = 0$ is the only value of x for which $2^x = 1$; thus we conclude that $kT - 1 = 0$, or $kT = 1$, which means that

$$k = \frac{1}{T} \qquad \text{where } T \text{ is the doubling time} \qquad [24]$$

For Country X, $T = 30$ and $A = 25{,}000{,}000$, so the growth law [21] yields the formula whose graph is shown in Figure 1.25.

Example 20 If the population of a country doubles every 15 years, and the population in 1970 was 17,000,000, find the law describing the population t years after 1970. What is the predicted population in the year 2000? In the year 2030?

Solution The population growth is modeled by a law of the form $P = A \cdot 2^{kt}$. Here, $A = 17{,}000{,}000$ (the initial population in 1970, when $t = 0$) and the doubling time is $T = 15$ years. Because $k = 1/T = 1/15$ we obtain

$$P(t) = 17{,}000{,}000 \cdot (2^{t/15}) \quad \text{for } t \geq 0 \qquad [25]$$

In the year 2000, $t = 30$ and $2^{t/15} = 2^{30/15} = 2^2 = 4$, so

$$P(30) = 17{,}000{,}000\,(2^{30/15}) = 68{,}000{,}000$$

In the year 2030, $t = 60$ and

$$P(60) = 17{,}000{,}000\,(2^{60/15}) = 272{,}000{,}000$$ ■

It is unlikely that this rate of population growth could actually be maintained for long. Nevertheless, as long as the assumptions about constant growth rate prove correct, Formula [25] accurately describes the population as a function of time. Other examples of exponential population growth occur in experiments with bacteria or one-celled animals.

Example 21 A certain type of green algae doubles every 6 hours with an adequate supply of sunlight and nutrients. Suppose we start with an initial colony of 5000 cells. Find a function of the form $N = A \cdot 2^{kt}$ describing the size of the population after t hours. What is the size of the colony 3 days later?

Solution Since the initial population is $N(0) = 5000$, we have $A = 5000$. The doubling time is $T = 6$ hours, so $k = 1/T = 1/6$ and the correct formula is

$$N(t) = 5000 \cdot 2^{t/6} \quad (t \text{ in hours}; t \geq 0)$$

After 3 days, $t = 72$ hours, there are

$$N(72) = 5000 \cdot (2^{72/6}) = 5000\,(2^{12})$$
$$= 5000(4096) = 20.48 \text{ million cells.}$$

■

D. Lack provides another interesting example of exponential growth.† In 1937, two male and six female pheasants were released on an isolated island where they had not been found previously, and where migration to or from the island was geographically impossible. Under conditions that must be presumed conducive to the pheasants' tastes, their population grew from 8 in the spring of 1937 to 1325 in the spring of 1942. The actual pattern of growth is shown in Figure 1.26 and is rather well fitted by an exponential function of the general form [22].

We conclude with some comments on more general exponential functions of the form $y = a^x$, where $a > 0$. They too are useful in describing growth and decay phenomena (see Chapter 4). The following examples indicate the shapes of their graphs.

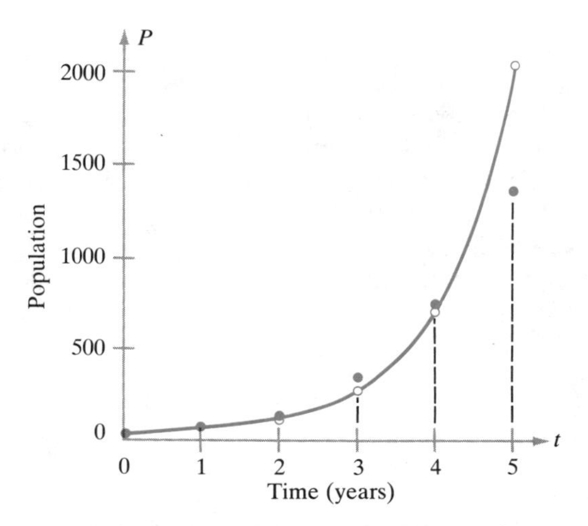

Year	1937	1938	1939	1940	1941	1942
t (years)	0	1	2	3	4	5
Population	8	30	81	282	705	1325
$8(2^{1.6t})$	8	24	73	222	676	2048

Figure 1.26
Tabulated values for the pheasant population in each year are plotted as solid dots. The population is approximately given by the exponential law $P = 8(2^{1.6t})$. The graph of $P = 8(2^{1.6t})$ is the solid curve; tabulated values of $8(2^{1.6t})$ are plotted as open dots. There are several reasons why the actual population begins to lag behind the predicted value $8(2^{1.6t})$. For example, the number of pheasants may be getting large enough to stress the limited environmental resources of the island. Once this happens, the assumptions leading to exponential growth are no longer valid.

† D. Lack, *The Natural Regulation of Animal Numbers* (New York: Oxford University Press, 1954).

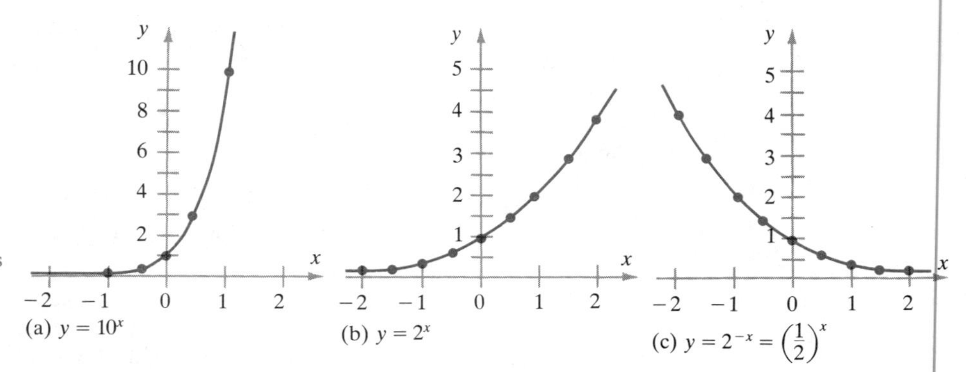

Figure 1.27
Graphs of $y = a^x$ for various choices of a. Note that the vertical scale in Figure 1.27a differs from that in Figures 1.27b and 1.27c.

The base a in an exponential function $y = a^x$ is always a fixed positive number $a > 0$. The function is formed by letting the exponent x vary. (In contrast, the powers $x, x^2, x^3, \ldots$ that appear in polynomials are formed by letting the base vary, keeping the exponent fixed.)

Example 22 Sketch the graphs of the exponential functions

(i) $f(x) = 10^x$ (base $a = 10$)
(ii) $f(x) = 2^x$ (base $a = 2$)
(iii) $f(x) = (\frac{1}{2})^x$ (base $a = \frac{1}{2}$)

Solution The numbers a^x are defined for all x and a^x is always positive, so the graph stays above the horizontal axis. For simple values of x, such as $x = 0, \pm\frac{1}{2}, \pm 1, \ldots$ it is easy to compute the value of a^x and plot the corresponding graph points. The values are tabulated in Table 1.1, and the points are plotted in Figure 1.27. Notice that the shape of the graph depends strongly on whether $a > 1$ or $a < 1$. When $a > 1$ the values of a^x increase very rapidly as x moves to the right, and become very small (approaching zero) as x moves to the left. When $a < 1$ the behavior is just the opposite; compare the graphs of $f(x) = 2^x$ and $f(x) = (\frac{1}{2})^x$. ■

	***x* values**								
	$x = -2.0$	$x = -1.5$	$x = -1.0$	$x = -0.5$	$x = 0.0$	$x = 0.5$	$x = 1.0$	$x = 1.5$	$x = 2.0$
$10^x =$	0.0100	0.0316	0.100	0.3162	1.000	3.1622	10.000	31.6227	100.000
$2^x =$	0.2500	0.3535	0.500	0.7071	1.000	1.4142	2.000	2.8284	4.000
$(\frac{1}{2})^x =$	4.0000	2.8284	2.000	1.4142	1.000	0.7071	0.500	0.3535	0.250

Table 1.1
Values of a^x for the bases $a = 10, 2, \frac{1}{2}$, and selected values of x. Evaluation requires taking fractional powers of the base a; for example, $10^{1.5} = 10^{3/2} = (1000)^{1/2} = 31.6227$ and $10^{-1.5} = 1/10^{1.5} = 1/31.6227 = 0.0316$.

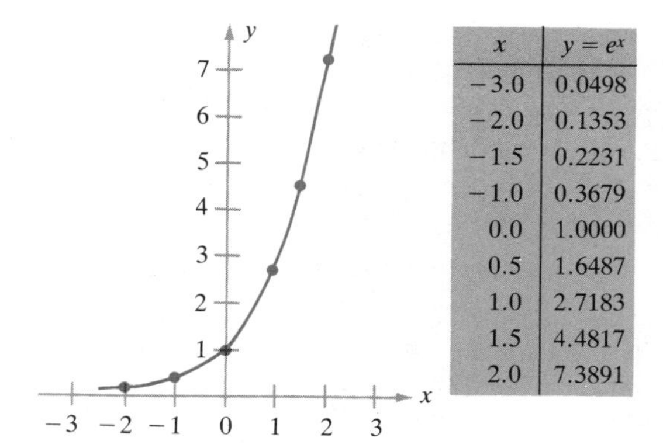

x	$y = e^x$
−3.0	0.0498
−2.0	0.1353
−1.5	0.2231
−1.0	0.3679
0.0	1.0000
0.5	1.6487
1.0	2.7183
1.5	4.4817
2.0	7.3891

Figure 1.28
Graph of the function $y = \exp(x) = e^x$. The tabulated values are plotted as solid dots. Because e^x is always positive, the graph lies above the x axis. It rises very rapidly as x moves to the right, and approaches the x axis as x moves to the left.

Using the laws of exponents (Appendix 1) we could deduce the following general properties for any exponential function $f(x) = a^x$.

(i) The values are positive, $a^x > 0$ for all x, so the graph stays above the horizontal axis.

(ii) When $x = 0$, $a^0 = 1$, so the graph passes through the point $P = (0, 1)$ no matter what the base a.

(iii) For all x

$$a^{-x} = \frac{1}{a^x}$$

so the value at $-x$ is the reciprocal of the value at x.

The exponential with base $a = 1$ is of little interest, because we obtain the degenerate constant function $f(x) = 1^x = 1$ for all x.

There is a special number $e = 2.71828\ldots$ whose corresponding exponential function $f(x) = e^x$ has such remarkable properties that all modern discussions of exponential functions to any base a are carried out in terms of this particular exponential function. With a pocket calculator we can easily compute values of this function. For example

$$e^{0.5} = \sqrt{2.71828} = 1.6487 \qquad e^{-2} = \frac{1}{e^2} = \frac{1}{(2.71828)^2} = 0.1353$$

The graph of $f(x) = e^x$ has the same general features as any exponential function a^x with $a > 1$. For future reference, we show the graph and a few computed values of e^x in Figure 1.28. The number $e = 2.71828\ldots$ plays a special role in mathematics, much like that of $\pi = 3.141596\ldots$, with which you may be more familiar. The number e arises naturally in continuous growth problems, as we shall explain in Chapter 4.

Exercises 1.4

1. Calculate the values of the functions

(i) $y = 2^x$ (ii) $y = 80 \cdot 2^x$

for $x = -3.0, -2.5, -2.0, \ldots, 2.0, 2.5, 3.0$. Plot these on graph paper, and sketch their graphs. (*Hint*: Use an appropriate vertical scale for (ii).)

2. The population of California in 1980 was about 22 million. If it increases at such a rate that it will double in 50 years, find the function $P(t)$ describing the population P in terms of the elapsed time t (years after 1980). What will the population be in 1995? What will happen in 2020?

3. In Exercise 2, sketch the graph of the population function $P = P(t)$. From the graph, estimate when the population will hit 60 million.

4. The population of Sand County, Nebraska, is growing very slowly. It is estimated that it will take 100 years to double from its 1950 value of 45,000. Find the population P in terms of years t elapsed since 1950. Estimate the population in 1975. In 1990. (*Hint*: You will need a pocket calculator to find the appropriate value of 2^x in your 1990 estimate.)

5. Country Y is troubled by a high growth rate. Between 1970 and 1984 the population went from 10 million to 20 million. If this continues, what are the prospects for the years 1991, 1998, 2000, and 2020?

6. Certain types of laboratory mice will double their population in 5 months. If we start with 500 mice, how many will there be after 7.5 months? 10 months? One year? Two years?

7. A culture of *E. coli* bacteria with a good supply of nutrients doubles in size every three hours. If we start with $N = 16{,}000$ bacteria, determine the formula for the population size $N(t)$ after t hours. How many bacteria are there after one day (24 hours)? After 2 days? After 10 days?

8. Yeast cells reproduce very rapidly. If the population size doubles every hour, and we start with 1000 cells, how many will we have after 8 hours? After 1 day? After 2 days?

9. Sketch the graph of $y = 2^{-x}$, showing at least the part corresponding to $-3 \le x \le 3$. (*Note:* This function has the form $A \cdot 2^{kt}$, where $A = 1$ and $k = -1$. It turns up in decay problems, such as those discussed next.)

Radioactive isotopes decay at a steady rate. Each has a characteristic **half-life** T, a period of time over which half the initial amount decays into other elements. In this situation the amount Q of isotope remaining after elapsed time t is given by the exponential formula

$$Q(t) = A \cdot 2^{kt}$$

Now, the constants A and k are given by:

$$A = Q(0) \quad \text{is the initial amount of isotope, present when } t = 0$$

$$k = -\frac{1}{T} \quad \text{where } T \text{ is the half-life of the isotope}$$

Use these facts to answer the following questions about the decay of radioactive isotopes.

10. Radium has a half-life of 1600 years. Find $Q(t)$ if we start with 5 grams of radium. How much is left after 800 years? After 2000 years?

11. The artificial element plutonium 239 has a half-life of 24,400 years. Find $Q(t)$ for a 1-kilogram sample. How much will be left after 1000 years? After 100,000 years? After 1 million years?

12. Potassium 40 is a natural radioactive isotope present in many rocks and is sometimes used to date certain rock structures. It has a long half-life of 1.28 billion years (1.28×10^9 years). If we started with 10 pounds of potassium 40 five billion years ago (roughly the age of the Earth), how much would be left now? What *fraction* of the original amount would be left now?

In an experiment, a tissue culture has been subjected to ionizing radiation. It was found that the number N of undamaged cells depends on the exposure time t (in hours) according to the formula

$$N(t) = 10{,}000 \cdot 2^{-2t} \qquad \text{for } t \geq 0$$

Use this formula to answer the following questions.

13. How many cells were present initially when $t = 0$?

14. How many cells survive a 1-hour exposure? A 15-minute exposure?

15. Find the elapsed time T after which only half the original cells survive (i.e., $N(T) = 5000$).

Here is a more realistic model of population growth, which uses the exponential function $y = 2^x$ in a more subtle way. When the growth of a population is limited by external factors, it starts to grow exponentially but eventually it levels off at some equilibrium level. In one model of such "limited growth," the number of bacteria in a culture medium deficient in nutrients is

$$N = \frac{1000 \cdot 2^t}{9 + 2^t} \qquad (t \text{ in hours})$$

Use this formula to answer the following questions:

16. How many bacteria are present initially?

17. How many bacteria are present after 4 hours? After 7.5 hours?

18. Construct a table of values for $t = 0, 1, 2, \ldots, 6, 7, 8$ hours. Then sketch the graph of the function $N = N(t)$.

19. From the graph in Exercise 18, estimate the values of t for which $N = 500$ and $N = 750$. Do you think the population will ever reach the value $N = 1000$?

1.5 Intersections of Graphs; the Quadratic Formula

How do we decide when a business is breaking even? If we know the costs $C(x)$ and revenue $R(x)$, this happens when

$$\text{costs} = \text{revenue}$$

so we want to find the production level x where $C(x) = R(x)$. This occurs precisely when the graphs of these functions cross, or intersect, plotting them on the same sketch as in Figure 1.29.

There are many other situations in which we must find the **point of intersection** for the graphs of two functions drawn in the same coordinate plane.

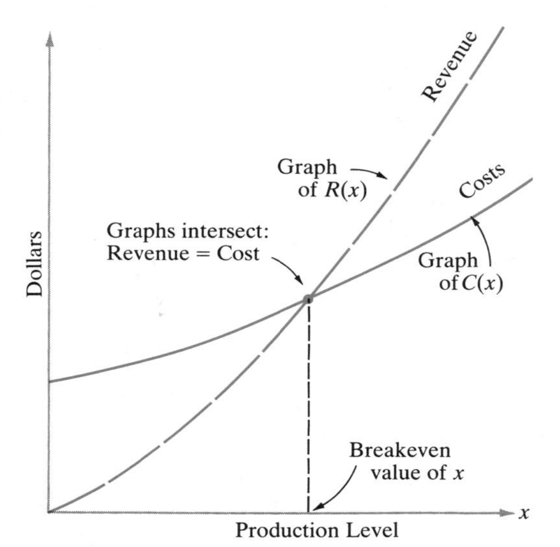

Figure 1.29
The intersection of cost and revenue curves for a business operation determines when the operation breaks even. The breakeven production level x corresponds to the point where the curves meet; there, the curves have the same height above the horizontal axis.

In this section we describe techniques for solving this problem, along with some applications.

The basic idea is to convert the geometric problem into one where we solve an algebraic equation. If $f(x)$ and $g(x)$ are two functions, their graphs meet above or below a point x on the horizontal axis if the graphs have the same height. This is the same as saying that $f(x) = g(x)$. Thus

Graphs intersect if $f(x) = g(x)$

Finding the intersection points amounts to finding the x value(s) for which $f(x) = g(x)$; that is, we want to solve the equation

$$f(x) - g(x) = 0 \qquad [26]$$

The solutions of such an equation are called its **roots** or **zeros**. If x is any root of Equation [26], we may determine the value of y either from the function $y = f(x)$ or from $y = g(x)$ to obtain the coordinates of the point (x, y) where the graphs intersect. Ultimately, our success with the algebraic approach depends on our ability to find the roots of Equation [26]. In the next two examples we shall deal with linear functions; then it is easy to find the roots.

Example 23 A retail gasoline station has weekly costs and revenue

$$\left.\begin{aligned} C(x) &= 1080 + 1.10x \\ R(x) &= 1.30x \end{aligned}\right\} \quad \text{for } x \geq 0$$

At what level of sales x (gallons per week) does the station break even?

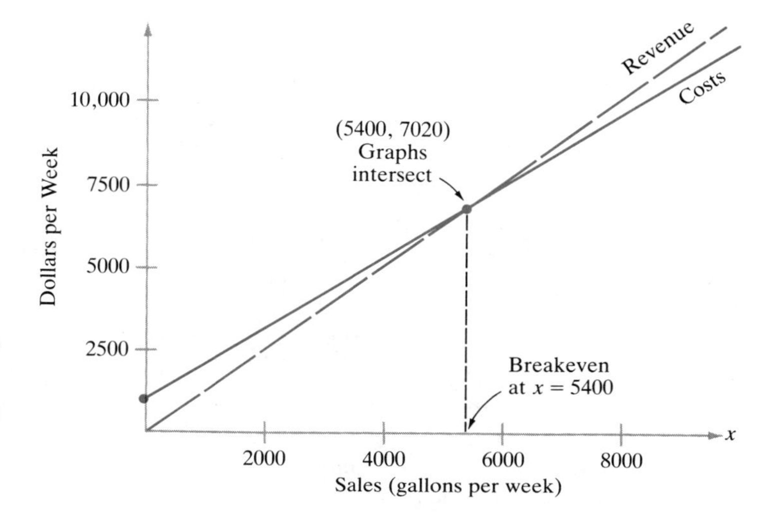

Figure 1.30
The cost and revenue functions in Example 23, plotted on the same coordinate plane. The curves intersect above the sales level $x = 5400$. At this level of sales the operation breaks even: cost = revenue.

Solution The graphs are straight lines, as shown in Figure 1.30; we must find where they meet. This happens when $R(x) = C(x)$, or

$$1.30x = 1080 + 1.10x$$
$$0.20x = 1080$$
$$x = \frac{1080}{0.2} = 5400 \quad \text{gallons per week}$$

For this value of x the graphs have the same height and intersect. ■

Here is another important application: The equilibrium price of a commodity in a competitive market is determined by the well-known condition *supply equals demand.* More precisely, this equilibrium point is the intersection of the supply and demand curves for the commodity.

Example 24 As prices rise, cattlemen are willing to put more beef on the market; when prices fall, they tend to cut back. Suppose this *supply relationship* for beef is given by the formula

$$p = \frac{1}{60}q + \frac{1}{2} \quad \text{for} \quad 0 \le q \le 120 \qquad [27]$$

For each price p (dollars per pound), q is the quantity producers are willing to supply at this price, in millions of pounds per week. But consumer action produces opposite effects. Consumers buy less beef as the price increases and more as the price decreases. Suppose that this *demand relationship* is described by

$$p = 3 - \frac{1}{40}q \quad \text{for} \quad 0 \le q \le 120 \qquad [28]$$

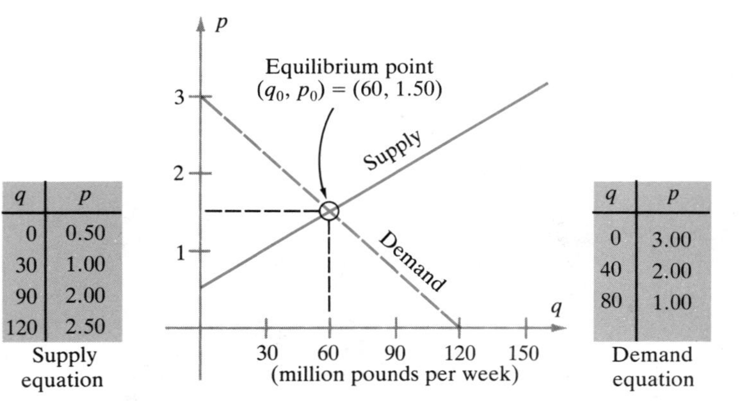

q	p
0	0.50
30	1.00
90	2.00
120	2.50

Supply equation

q	p
0	3.00
40	2.00
80	1.00

Demand equation

Figure 1.31
Both the supply function $p = (\frac{1}{60})q + (\frac{1}{2})$ and the demand function $p = 3 - (\frac{1}{40})q$ are linear, so their graphs are straight lines. If the price is $p = \$1.00$, then suppliers are willing to produce only 30 million pounds per week, but consumers demand 80. This imbalance forces the price to rise. If the price were $p = \$2.00$, consumer demand would be $q = 40$, and producers would be willing to supply $q = 90$, so the price would fall. At the equilibrium point $(q_0, p_0) = (60, 1.50)$ where the graphs meet, the price $p_0 = \$1.50$ per pound corresponds to equal supply and demand, $q_0 = 60$, and the price is stable.

The public will buy q million pounds of beef per week if the price is p dollars per pound. Find the point of intersection of the supply and demand curves. That is, find the price and quantity of beef at which the amount produced matches the amount consumed.

Solution Both the supply function [27] and demand function [28] are linear functions of q. Hence their graphs are straight lines, as shown in Figure 1.31. On the graphs, $(q_0, p_0) = (60, 1.50)$ is their point of intersection. Economists refer to this as the **equilibrium point** and to $p_0 = \$1.50$ per pound as the equilibrium price. At this price, the quantity demanded by consumers—$q_0 = 60$ million pounds per week—equals the quantity producers are willing to supply.

To find the equilibrium point by algebraic means we observe that at the equilibrium level $q = q_0$ the supply and demand curves intersect and have the same p values. Thus, at the equilibrium price we have

$$\text{(supply price)} = \text{(demand price)}$$

$$\frac{1}{60}q + \frac{1}{2} = 3 - \frac{1}{40}q$$

or

$$\left(\frac{1}{60} + \frac{1}{40}\right)q = 2.50$$

$$\frac{1}{24}q = 2.50$$

$$q = 60 \text{ million pounds}$$

Once we know the equilibrium value $q_0 = 60$, we find the equilibrium price p_0 by substituting q_0 into either the supply or demand equation to obtain

$$p_0 = \frac{1}{60}q_0 + \frac{1}{2} = \frac{1}{60}(60) + \frac{1}{2} = \$1.50 \text{ per pound}$$

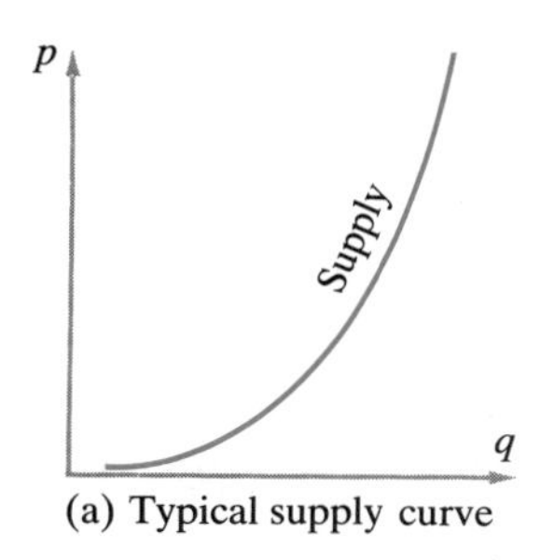

(a) Typical supply curve

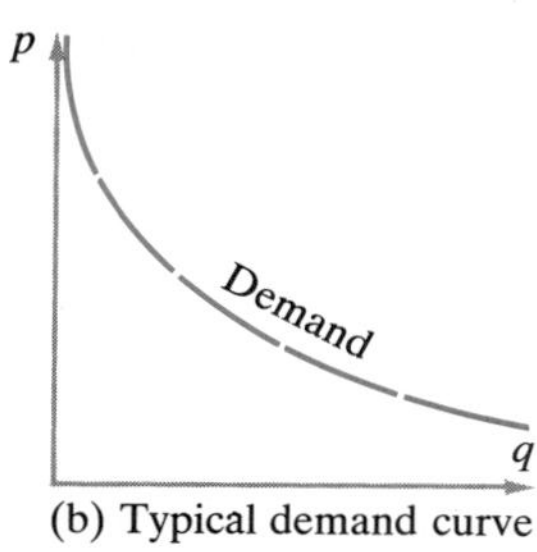

(b) Typical demand curve

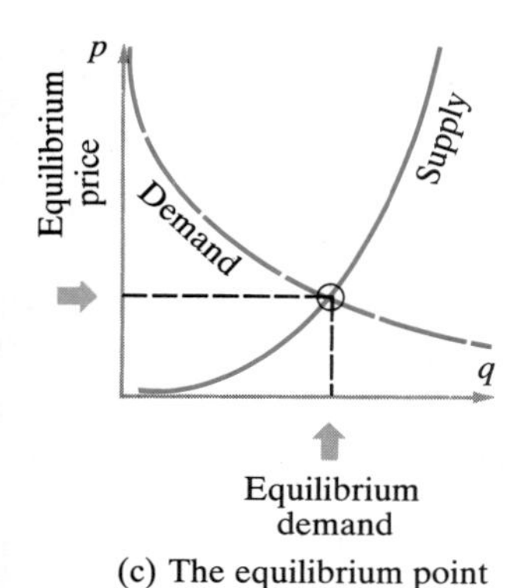

(c) The equilibrium point

Figure 1.32
Typical supply and demand curves.

In general, supply and demand curves will not be linear, as they were in the last example. Nevertheless, they will have the general shape shown in Figures 1.32a and 1.32b for the simple reason that supply increases as the price increases, but demand steadily decreases. We find the equilibrium price by determining where the curves intersect, as in Figure 1.32c.

Finding the intersection point of two curves can be difficult. But we can always find it if the curves are linear or quadratic. In this case, finding the intersection amounts to solving a linear or a **quadratic equation** of the form

$$Ax^2 + Bx + C = 0 \qquad \text{(where } A, B, \text{ and } C \text{ are constants; } A \neq 0) \qquad [29]$$

We can find the roots of a quadratic equation by applying the **quadratic formula**. This formula from basic algebra states that the roots are given by

$$x = \frac{-B + \sqrt{B^2 - 4AC}}{2A} \qquad x = \frac{-B - \sqrt{B^2 - 4AC}}{2A} \qquad [30]$$

These two formulas differ only in the choice of a (+) or (−) sign in front of the radical, so they are often written together in a single expression

$$x = \frac{-B \pm \sqrt{B^2 - 4AC}}{2A}$$

The quadratic formula tells us explicitly how to find the roots of the quadratic equation from the given data, namely the coefficients A, B, and C. If the term $B^2 - 4AC$ under the radical is negative, the square root makes no sense (at least in the real number system). When this happens, the appropriate conclusion is that the equation $Ax^2 + Bx + C = 0$ *has no solutions* (in the real number system).

Here are some examples based on the quadratic formula.

Example 25 Find all solutions of the equation $x^2 - 2x - 3 = 0$.

Solution Here $A = 1$, $B = -2$, and $C = -3$, so the roots are

$$x = \frac{-(-2) \pm \sqrt{(-2)^2 - 4(1)(-3)}}{2(1)}$$

$$= \frac{2 \pm \sqrt{4 + 12}}{2}$$

$$= \frac{2 \pm \sqrt{16}}{2} = \frac{2 \pm 4}{2}$$

Taking the (+) sign we obtain one root: $x = \frac{6}{2} = 3$; taking the (−) sign we obtain a second root: $x = -\frac{2}{2} = -1$. ■

Example 26 Find the points of intersection of the graphs of $y = 3 - x - x^2$ and $y = x + 1$.

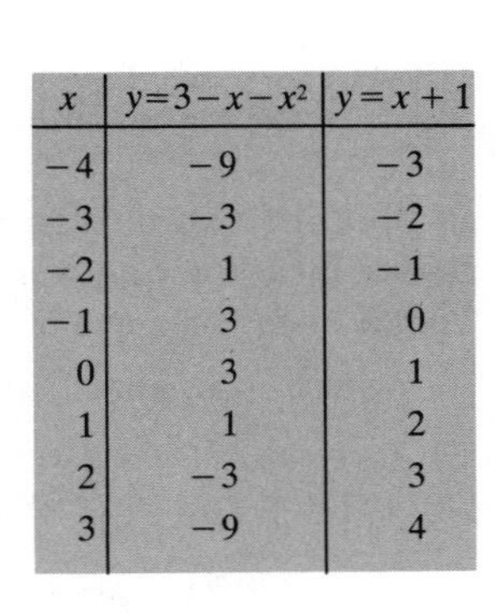

x	$y=3-x-x^2$	$y=x+1$
−4	−9	−3
−3	−3	−2
−2	1	−1
−1	3	0
0	3	1
1	1	2
2	−3	3
3	−9	4

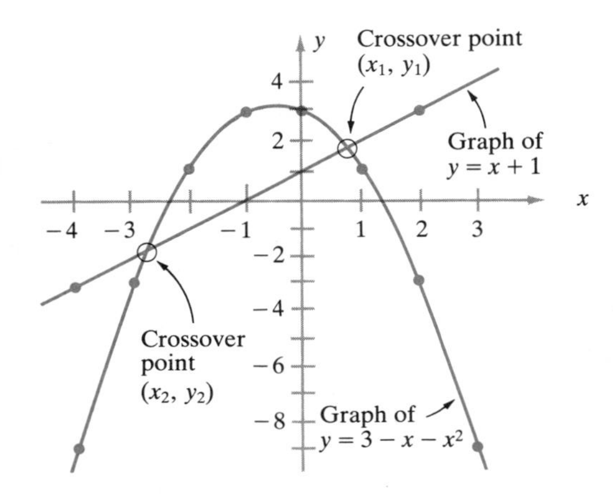

Figure 1.33
A table of values has been computed for each of the functions $y = 3 - x - x^2$ and $y = x + 1$. The corresponding points on the graphs are indicated by solid dots. The points of intersection of the graphs (represented by circles) may be estimated by inspecting the graphs: $(x_1, y_1) \approx (0.7, 1.7)$ and $(x_2, y_2) \approx (-2.7, -1.7)$. Using the algebraic approach we may calculate the exact intersection points: $(x_1, y_1) = (-1 + \sqrt{3}, +\sqrt{3})$ and $(x_2, y_2) = (-1 - \sqrt{3}, -\sqrt{3})$.

Solution The graphs are shown in Figure 1.33. The graphs intersect above x values where the two functions are equal:

$$3 - x - x^2 = x + 1$$

Shifting all terms to the right hand side we obtain a quadratic equation

$$0 = x^2 + 2x - 2$$

with $A = 1$, $B = 2$, and $C = -2$. We find its solutions using the quadratic formula; there are two solutions:

$$x_1 = \frac{-B + \sqrt{B^2 - 4AC}}{2A} = \frac{-2 + \sqrt{2^2 - 4(1)(-2)}}{2}$$
$$= \frac{-2 + \sqrt{12}}{2} = -1 + \sqrt{3} = 0.732\ldots$$

and

$$x_2 = \frac{-B - \sqrt{B^2 - 4AC}}{2A} = \frac{-2 - \sqrt{2^2 - 4(1)(-2)}}{2}$$
$$= \frac{-2 - \sqrt{12}}{2} = -1 - \sqrt{3} = -2.732\ldots$$

For each of the roots x_1 and x_2 the corresponding y values where the original curves intersect are obtained by substituting these x values into either of the functions. It is easiest to insert them into $y = x + 1$, so for $x_1 = -1 + \sqrt{3}$ we get the corresponding y value

$$y_1 = x_1 + 1 = (-1 + \sqrt{3}) + 1 = \sqrt{3} = 1.732\ldots$$

Similarly for $x_2 = -1 - \sqrt{3}$ we get

$$y_2 = x_2 + 1 = (-1 - \sqrt{3}) + 1 = -\sqrt{3} = -1.732\ldots$$

Therefore the points of intersection shown in Figure 1.33 have coordinates

$$(x_1, y_1) = (-1 + \sqrt{3}, \sqrt{3})$$
$$(x_2, y_2) = (-1 - \sqrt{3}, -\sqrt{3})$$ ■

Example 27 The supply and demand curves shown in Figure 1.34 are given by the functions

$$\left.\begin{array}{ll}\text{(supply)} & p(q) = q^2 + 20q \\ \text{(demand)} & p(q) = 1200 - 10q - q^2\end{array}\right\} \quad \text{for} \quad 0 \le q \le 30$$

where q = (thousands of units). Find the equilibrium price for this commodity. How many units are sold/purchased at this price?

Solution We must find the equilibrium point where the curves meet. At the equilibrium price p, supply equals demand and we have

$$q^2 + 20q = 1200 - 10q - q^2$$
$$2q^2 + 30q - 1200 = 0$$

We solve the latter equation using the quadratic formula. Here $A = 2$, $B = 30$, $C = -1200$ and its roots are

$$q = \frac{-30 \pm \sqrt{(30)^2 - 4(2)(-1200)}}{2(2)}$$

$$q = \frac{-30 \pm \sqrt{900 + 9600}}{4}$$

$$= \frac{-30 \pm \sqrt{10{,}500}}{4} = \frac{-30 \pm 102.47}{4}$$

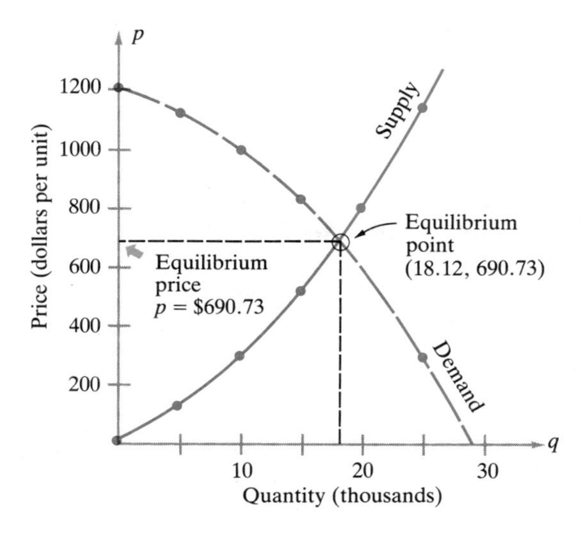

Figure 1.34
The supply and demand curves in Example 27 are given by quadratic functions of q, and so are not straight lines. The coordinates (18.12, 690.73) of the equilibrium point are found using the quadratic formula.

or

$$q_1 = \frac{-30 + 102.47}{4} = 18.12 \qquad q_2 = \frac{-30 - 102.47}{4} = -33.12$$

Only the feasible root $q_1 = 18.12$ is relevant to our problem. The curves cross above $q = 18.12$. Inserting this q value into either the supply or the demand equation gives the same value for p; from the (simpler) supply equation we obtain

$$p(18.12) = (18.12)^2 + 20(18.12) = 690.73 \text{ dollars per unit}$$ ■

Having seen what can be done for quadratic equations, we must point out that if a polynomial equation $h(x) = 0$ has degree three or higher there is no easy algebraic formula for its roots. This would be the case for a third degree equation such as

$$x^3 - x + 1 = 0$$

Simple polynomials can sometimes be handled by factoring the polynomial $h(x)$, writing it as the product of several polynomials of lower degree. For example, we may write

$$x^3 - 3x^2 = x^2(x - 3) = x \cdot x \cdot (x - 3)$$
$$x^3 - 2x^2 - x + 2 = (x + 1)(x^2 - 3x + 2)$$
$$= (x + 1)(x - 2)(x - 1)$$

To find the roots we use the simple fact: a product can equal zero if and only if one of its factors is zero. Thus, the only way we can have

$$0 = x^3 - 2x^2 - x + 2 = (x + 1)(x - 2)(x - 1)$$

is to have

$$x = -1 \qquad \text{(first factor zero)}$$
$$x = 2 \qquad \text{(second factor zero)}$$

or

$$x = 1 \qquad \text{(third factor zero)}$$

We conclude that the roots of $h(x) = x^3 - 2x^2 - x + 2 = 0$ are $x = 1$, 2, and -1.

Every quadratic polynomial $Ax^2 + Bx + C$ can be factored if it has real roots (see Exercises 41–48). A general polynomial can always be factored if we can guess one of its roots by trial and error. Once the polynomial has been factored, we can replace the original problem by one involving a polynomial of lower degree. We try to continue this process until we arrive at a quadratic equation, which we can always solve using the quadratic formula.

Example 28 Find all roots of the cubic (third degree) equation $h(x) = x^3 - 8x - 3 = 0$, given that $x = 3$ is a root.

Solution The root $x = 3$ might be found by trial and error, inserting a few simple values of x into $h(x)$ to see if we obtain zero; for $x = 3$, $h(3) = (3)^3 - 8(3) - 3 = 0$

Because $x = 3$ is a root, it follows that $(x - 3)$ may be factored out of $h(x)$; by long division we obtain

$$\begin{array}{r|l} & x^2 + 3x + 1 \\ \hline x - 3 & x^3 \qquad - 8x - 3 \\ & x^3 - 3x^2 \\ \hline & \quad 3x^2 - 8x \\ & \quad 3x^2 - 9x \\ \hline & \qquad x - 3 \\ & \qquad x - 3 \\ \hline & \qquad\quad 0 \end{array}$$

Thus $h(x) = x^3 - 8x - 3 = (x - 3)(x^2 + 3x + 1)$. This product is zero if and only if one or the other of the factors $(x - 3)$, $(x^2 + 3x + 1)$ is zero. If $x - 3 = 0$ we obtain the root $x = 3$. If $x^2 + 3x + 1 = 0$, we obtain two more roots by applying the quadratic formula. The roots of the equation $x^2 + 3x + 1 = 0$ (with $A = 1$, $B = 3$, and $C = 1$) are

$$x_2 = \frac{-3 - \sqrt{5}}{2} = -2.618\ldots \qquad x_3 = \frac{-3 + \sqrt{5}}{2} = -0.382\ldots$$

We conclude that the roots of the equation $h(x) = 0$ consist of the three numbers

$$x_1 = 3, \quad x_2 = \frac{-3 - \sqrt{5}}{2}, \quad x_3 = \frac{-3 + \sqrt{5}}{2}$$

■

To summarize: if $x = r$ is a root of a polynomial equation $h(x) = 0$, then $(x - r)$ will always divide exactly into $h(x)$, and we may factor $h(x)$ in the form

$$h(x) = (x - r) \cdot k(x)$$

where $k(x)$ has lower degree than $h(x)$. In the above example $r = 3$, $k(x) = x^2 + 3x + 1$, and we factored $h(x) = x^3 - 8x - 3 = (x - 3)(x^2 + 3x + 1)$.

Example 29 Find the roots of the cubic polynomial $p(x) = x^3 - x^2 - x + 1$.

Solution By trying various choices of x we find one root: $p(x) = 0$ if $x = 1$. Thus $p(x) = (x - 1) \cdot (x^2 - 1)$ because long division yields

$$\begin{array}{r|l} & x^2 \qquad\quad - 1 \\ \hline x - 1 & x^3 - x^2 - x + 1 \\ & x^3 - x^2 \\ \hline & \qquad 0 \; - x + 1 \\ & \qquad\quad - x + 1 \\ \hline & \qquad\qquad 0 \end{array}$$

To check, multiply out $(x - 1)(x^2 - 1)$. But we may factor $x^2 - 1$ as follows

$$x^2 - 1 = (x - 1)\cdot(x - (-1)) = (x - 1)\cdot(x + 1)$$

and obtain a complete factorization of $p(x)$

$$x^3 - x^2 - x + 1 = (x - 1)(x^2 - 1) = (x - 1)(x - 1)(x + 1)$$

As usual, the product can be zero only if one of the factors is zero. Thus the roots of $p(x)$ are $r = +1$ and $r = -1$. Notice that two of the factors correspond to the same root $r = 1$; mathematicians would regard $r = 1$ as a "double root" of the polynomial because it appears twice. ■

Exercises 1.5

1. Draw the graphs of the functions $y = 3x - 2$ and $y = 6 - 2x$ on a single piece of graph paper. Estimate the coordinates of the point of intersection by examining the sketch. Then, using algebra, calculate both coordinates of the point of intersection.

For each of the following pairs of lines, find both coordinates of the point of intersection.

2. $y = 4x + 10$ and $y = -3x - 5$
3. $y = 75 - 45x$ and $y = -100 + 75x$
4. $y = 58.3 - 27.4x$ and $y = 40x - 61.2$
5. $x - 3y + 1 = 0$ and $y = -x + 5$
6. $-x + 3y = 600$ and $2x + 2y = 500$
7. Suppose a retail gasoline station has weekly cost and revenue functions

$$C(x) = 1500 + 1.25x \qquad R(x) = 1.38x$$

Find the breakeven sales level x.

8. Suppose the daily demand for unleaded gasoline in Los Angeles during September 1982 was given by (q in millions of gallons per day)

$$p(q) = \frac{73 - q}{50} \text{ dollars per gallon}$$

and the supply curve was given by

$$p(q) = \frac{q + 100}{80} \text{ dollars per gallon}$$

where p is the price in dollars. Find the equilibrium price. What was the level of sales at this price?

9. Consider supply and demand curves for beef slightly different from those in Example 24,

$$\text{(supply)} \qquad p(q) = \frac{q + 40}{75}$$

$$\text{(demand)} \qquad p(q) = \frac{130 - q}{35}$$

where q is measured in millions of pounds per week. (These revised figures might represent the situation a year later.) Find the equilibrium price p and the equilibrium level of sales q.

10. If the supply and demand curves for seedless grapes sold in New York City during July are

$$\text{(supply)} \quad p(q) = \frac{q + 1.10}{2} \qquad p = \text{dollars per pound}$$

$$\text{(demand)} \quad p(q) = \frac{4.42 - q}{2.8} \qquad q = \text{millions of pounds per week}$$

what was the equilibrium price? How many pounds per week were sold at this price?

Use the quadratic formula to find all roots of the following equations. (*Note:* If $B^2 - 4AC$ is negative, the term $\pm\sqrt{B^2 - 4AC}$ in the quadratic formula does not make sense as a real number. Remember that, in this event, the proper conclusion is that there are *no* (real) roots of the equation at hand.)

11. $x^2 - 3x + 1 = 0$

12. $x^2 + x + 1 = 0$

13. $x^2 + 4x - 8 = 0$

14. $400x^2 - 1 = 0$

15. $4x^2 - 2x - 7 = 0$

16. $8x^2 + 37 = 0$

17. $500{,}000 - 200x + 0.01x^2 = 0$

18. $500{,}000 - 20x + 0.01x^2 = 0$

19. Sketch the graphs of the functions $y = 2x + 6$ and $y = x^2 - 9$ on a single piece of graph paper. Estimate the coordinates of each intersection point by examining your sketch. Then find the coordinates exactly using the quadratic formula.

20. Sketch the curves $y = 1 - x^2$ and $y = x^2 - 4x + 4$ on a single piece of graph paper. Do the curves intersect at all? How is this revealed when you try to find the intersection point(s) using the quadratic formula?

Find both coordinates of each intersection point for the following pairs of curves. Use the quadratic formula.

21. $y = x^2 - 2x + 1$ and $y = x$

22. $y = x^2 + x + 1$ and $y = x + 1$

23. $y = x^2 - x - 2$ and $y = 4 + 3x - x^2$

24. $y = 2x + 7$ and $y = x^2 - 9$

25. $y = x^2 + 40x$ and $y = 500 - 30x - x^2$

26. $y = 1 - x^2$ and $y = x^2 - 4x + 4$

27. Suppose the supply and demand curves for beef shown in Figure 1.35 are given by

$$\text{(supply)} \quad p(q) = \frac{(q + 30)}{60}$$

$$\text{(demand)} \quad p(q) = \frac{245}{3q}$$

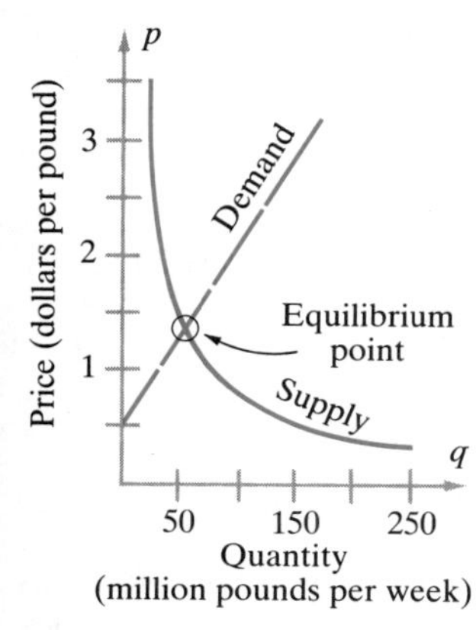

Figure 1.35
A more realistic set of supply and demand curves for beef (see Exercise 27). Here, q = (millions of pounds per week) and p = (dollars per pound).

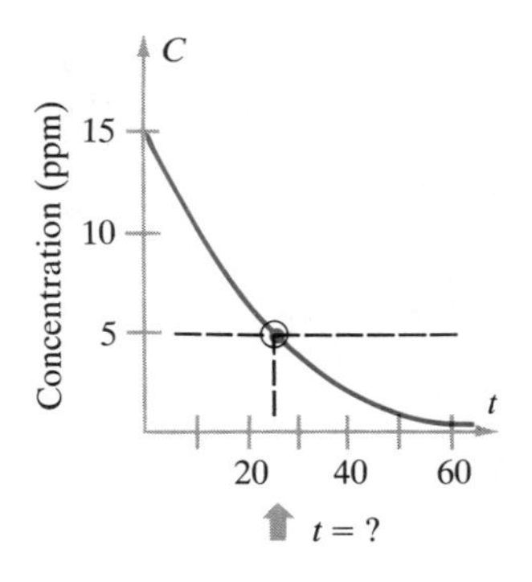

Figure 1.36
Concentration of pesticide as a function of t (in years) in Exercise 28; $C(t) = 15 - \frac{1}{2}t + \frac{7.5}{1800}t^2$ for $0 \leq t \leq 60$.

where p = (dollars) and q = (millions of pounds per week). Find the level of demand q and the price p when supply equals demand.

28. A polluted lake has been found to contain DDT residues of 15 parts per million (ppm). Natural processes slowly degrade the pesticide. Suppose the concentration C decreases with time according to the formula

$$C(t) = 15 - \frac{1}{2}t + \frac{7.5}{1800}t^2 \text{ parts per million} \qquad (t \text{ in years; } 0 \leq t \leq 60)$$

whose graph is shown in Figure 1.36. After how many years does C fall below an acceptable level of 5 ppm?

29. In Example 17 of Section 1.3 we discussed a gasoline station whose weekly cost and revenue functions were

$$C(x) = 1080 + 1.10x - 0.00001x^2$$
$$R(x) = 1.30x$$

Find the breakeven sales level x. (*Note:* The graph of $C(x)$ has been sketched in Figure 1.20. It is not a straight line.)

30. Suppose that the monthly costs of a manufacturing plant are given by

$$C(x) = 50{,}000 + 100x - 0.1x^2$$

where x = (number of units produced per month). Suppose all units are sold at a price of \$110, regardless of production level. Calculate the revenue $R(x)$ and the profit $P(x)$. At what production level do they break even?

Find all roots of the following polynomial equations. Use the quadratic formula wherever possible. Try to guess a root, and factor the polynomials of degree higher than two.

31. $x^2 + 1 = 0$ **32.** $x^2 - 3x + 1 = 0$ **33.** $x^2 - 2x + 1 = 0$

34. $x^3 - 4x = 0$ **35.** $x^3 + x - 2 = 0$ **36.** $x^3 - x^2 - x + 1 = 0$

37. $x^3 - 8 = 0$ **38.** $x^4 - 8x = 0$ **39.** $x^4 - 1 = 0$

40. $x^3 + 1 = 0$

Here is a useful fact: If we know that the equation $Ax^2 + Bx + C = 0$ has real roots r_1 and r_2, then we can immediately write down a factorization of the polynomial $p(x) = Ax^2 + Bx + C$. In fact

$$p(x) = A(x - r_1)(x - r_2) \qquad [31]$$

If $p(x)$ has no real roots, the polynomial cannot be factored into smaller pieces; as examples of this we have $x^2 + 1$ and $x^2 - x + 1$, which have no real roots. Use this idea to give complete factorizations for the following polynomials.

41. $x^2 + 4x - 21$ **42.** $x^2 - 2x - 2$ **43.** $x^2 + x - 12$

44. $x^2 - 2$ **45.** $x^2 - 2x + 1$ **46.** $2x^2 + 5x - 3$

47. $x^2 - 4x + 1$ **48.** $x^2 - x + 1$

Checklist of Key Topics

Function
Feasible set and domain of definition
Coordinate plane
Graph of a function
Slope of a line
Finding the equation of a straight line
Cost function
Revenue function
Profit function
Demand function
Population growth law
Exponential functions 2^{kx}
Exponential function e^x
Intersections of graphs
Quadratic formula
Supply and demand equilibrium

Chapter 1 Review Exercises

In Exercises 1–4, sketch the graph of the function. Find the exact coordinates of the point where the graph intersects the vertical axis.

1. $f(x) = 4$ (constant function)

2. $f(x) = -3$ (constant function)

3. $f(x) = -3x - 4$

4. $f(x) = \frac{1}{2}x + 2$

In Exercises 5–11, sketch the graph of the function.

5. $f(x) = 3 - 2x$

6. $f(x) = 3x + 5$

7. $f(x) = 1 + 2x - x^2$

8. $f(x) = x^2 + 3x + 1$

9. $f(x) = 1 + \dfrac{1}{x^2}$ (domain: all $x > 0$)

10. $f(x) = x + \dfrac{2}{x}$ (domain: all $x > 0$)

11. $f(x) = 2x - \dfrac{1}{x}$ (domain: all $x > 0$)

12. The point $(3, a)$ lies on the graph of the function $f(x) = x + \dfrac{1}{x}$. Find a.

13. The point $(-1, b)$ lies on the graph of the function $f(x) = \dfrac{1}{1 + 3x^2}$. Find b.

14. The point $(c, 2)$ lies on the graph of the function $f(x) = x + \dfrac{1}{x}$. Find c.

15. The point $(d, 1)$ lies on the graph of the function $f(x) = \dfrac{1}{1 + 3x^2}$. Find d.

In Exercises 16–19, determine the equation of the straight line.

16. The line through the points $(-3, -2)$ and $(-1, 0)$

17. The line through the points $(3, 1)$ and $(1, 3)$

18. The line with slope $m = \frac{1}{2}$ that crosses the x axis at $x = -4$.

19. The line with slope $m = -2$ that crosses the y axis at $y = 0$.

In Exercises 20–25, find all roots of the equation.

20. $x^2 = 3 - 2x$

21. $x^2 - 4x = 5$

22. $x^3 + x^2 - x = 0$

23. $2x^3 + 2x^2 - x = 0$

24. $x^3 - 3x^2 + 4 = 0$

25. $x^3 - x^2 - 4x + 4 = 0$

26. A housing contractor has fixed costs $FC = \$115{,}000$ per year. Assume that the cost function $C(x)$ is linear, where x is the yearly construction level, and that $C(50) = \$765{,}000$. Find the cost function. What are the costs when $x = 40$?

27. A gasoline station has a linear demand function $p(x)$, where x is the number of gallons sold per day. If we know that $p(2000) = \$1.40$ and $p(5000) = \$1.25$, find the equation of the demand function. Then find $p(6000)$. Find the revenue when $x = 6000$.

28. The weekly cost function for a supplier who assembles a certain type of pocket radio is $C(x) = 3500 + 16x$ dollars per week, where x is the number produced. The demand function for this item is $p(x) = 40 - 0.02x$. Find the weekly profit function $P(x)$.

29. The monthly cost of manufacturing and producing x calculators per month is $C(x) = 7500 + 14x$ and the demand function is $p(x) = 65 - 0.03x$. Find the monthly profit function.

30. The population of a certain city is 350,000 in 1985 and is increasing at such a rate that the population will double in 20 years. Find the function $P(t)$ describing the population in terms of years t elapsed since 1985. Draw the graph of $P(t)$ over the domain $0 \leq t \leq 40$.

31. The amount Q of a radioactive isotope remaining after t years is given by an exponential formula $Q(t) = A \cdot 2^{-kt}$ where A, k are constants. If $Q = 6$ milligrams initially (when $t = 0$), and if $Q = 3$ milligrams after 6000 years, find the values of A and k. Then make a rough sketch of the graph of $Q(t)$ over the domain $0 \leq t \leq 12{,}000$.

In Exercises 32–33, find the points where the graphs intersect. Give both coordinates of each point of intersection.

32. $f(x) = 3x^2 - 4 \qquad g(x) = 5x - 3$

33. $f(x) = x^2 - 5 \qquad g(x) = 2x - 3$

34. The daily supply and demand curves for blueberry muffins produced at the Waverly bakery are

$$p(x) = \frac{-55}{2000}x + 120 \qquad \text{cents (demand curve)}$$

$$p(x) = \frac{35}{2000}x + 30 \qquad \text{cents (supply curve)}$$

Find the equilibrium production level and the equilibrium price. Make a sketch showing both curves and the equilibrium point where they intersect.

CHAPTER 2

Derivatives and the Concept of Limit

2.1 Average Rates of Change

If you drive 165 miles between 1:00 and 4:00 P.M., your average velocity for the trip is

$$\frac{\text{distance traveled}}{\text{time elapsed}} = \frac{165 \text{ miles}}{3 \text{ hours}} = 55 \text{ miles per hour}$$

If a portfolio of stocks originally worth \$4800 increases in value to \$8700 in 2.5 years, the average rate of growth during this period is

$$\frac{\text{change in value}}{\text{time elapsed}} = \frac{8700 - 4800}{2.5} = 1560 \text{ dollars per year}$$

Or consider a manufacturer's weekly costs C, which depend on the production level x. If costs rise from \$48,500 to \$57,800 when production is increased from 500 to 650 units, the average rate of change of costs is

$$\frac{\text{change in } C}{\text{change in } x} = \frac{57{,}800 - 48{,}500}{650 - 500} = \frac{9300}{150} = 62 \text{ dollars per unit}$$

On the average, costs increase by \$62 for each additional unit produced when production is raised from 500 to 650 units.

Any time one variable y depends on another variable x we may examine what happens to y in response to changes in x. If x increases from an initial value x_1 to a new value x_2, the net change in x is $x_2 - x_1$. If the corresponding y values are y_1 and y_2, the net change in y is $y_2 - y_1$. Just as in the preceding situations, we define the **average rate of change** of y to be

$$\text{average rate of change} = \frac{\text{change in } y}{\text{change in } x} = \frac{y_2 - y_1}{x_2 - x_1}$$

This average provides a useful summary of what happens to y as a result of the change in x.

There is a common notation for describing changes in a variable. We shall use it throughout this book. If x goes from x_1 to x_2, this change is called an **increment** in x and is indicated by the symbol

$$\Delta x = (\text{final value}) - (\text{initial value}) = x_2 - x_1$$

The symbol Δx is read as "delta x". The corresponding increment in the dependent variable y is

$$\Delta y = y_2 - y_1$$

and the formula for the average rate of change can be rewritten

$$\text{average rate of change} = \frac{\Delta y}{\Delta x} \quad [1]$$

using this concise notation. Notice that we *always* take (final value) minus (initial value) in discussing an increment in some variable and in forming averages!

Example 1 If $y = x^3$, what is the increment Δy as x increases from $x_1 = 1$ to $x_2 = 3$? Find the average rate of change $\Delta y/\Delta x$ corresponding to this change in x.

Solution The increment in x is

$$\Delta x = x_2 - x_1 = 3 - 1 = 2$$

The corresponding values of y at x_1 and x_2 are

$$y_1 = (x_1)^3 = (1)^3 = 1$$
$$y_2 = (x_2)^3 = (3)^3 = 27$$

Hence, the increment for y is

$$\Delta y = y_2 - y_1 = 27 - 1 = 26$$

and the average rate of change is

$$\frac{\Delta y}{\Delta x} = \frac{26}{2} = 13$$

■

Example 2 If $y = f(x) = 2 + x + x^2$, find the average rate of change in y as x increases from 0 to 1.

Solution Clearly, $\Delta x = x_2 - x_1 = 1 - 0 = 1$. Meanwhile

$$y_1 = f(x_1) = 2 + (0) + (0)^2 = 2$$
$$y_2 = f(x_2) = 2 + (1) + (1)^2 = 4$$

so the corresponding increment in y is

$$\Delta y = y_2 - y_1 = 4 - 2 = 2$$

The average rate of change is therefore

$$\frac{\Delta y}{\Delta x} = \frac{2}{1} = 2$$

■

In specific applications, the variables and the averages $\Delta y/\Delta x$ take on specific and important meanings, as they will in the following examples. Notice that whatever names we give the variables, the average rate of change is always the increment in the dependent variable divided by the increment in the independent variable.

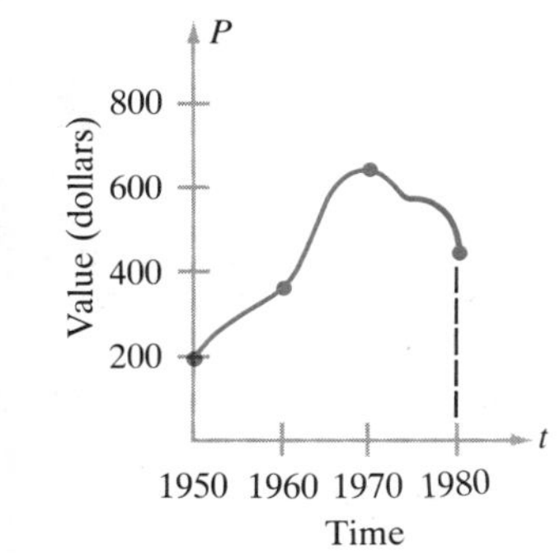

t	P
1950	200
1960	380
1970	650
1980	450

Figure 2.1
Value of a share in Bankruptcy Mutual Fund for the period 1950–1980 (see discussion in Example 3).

Example 3 The graph in Figure 2.1 shows the price per share of Bankruptcy Mutual Fund (BMF) for the period 1950–1980. Here, t is the year, and $P = P(t)$ is the price per share at time t. What was the average rate of change in P as t changed in the period from 1950 to 1980? What was the average rate of change between 1970 and 1980?

Solution From 1950 to 1980 the change in t is $\Delta t = 30$ years, and P increased from 200 to 450; so $\Delta P = 250$. The average growth rate for this period was

$$\frac{\Delta P}{\Delta t} = \frac{250}{30} = 8.33 \quad \text{dollars per year}$$

Between 1970 and 1980, $\Delta t = 10$, and $\Delta P = 450 - 650 = -200$ (a loss!), so the average growth rate for this period was

$$\frac{\Delta P}{\Delta t} = \frac{-200}{10} = -20 \quad \text{dollars per year}$$

The minus sign indicates a *decrease* or loss in value. ■

Remark: Advertisers of BMF could legitimately claim that the average yearly growth in value of a share in the fund was $8.33 during the 30-year period from 1950 to 1980. Equally true, but probably unpublicized by BMF, is the fact that the value of a share declined on an average $20 per year during the 10-year period from 1970 to 1980. This example illustrates a practical use of average rates of change and emphasizes the strong dependence of the average on the particular period being considered.

Example 4 Suppose the weekly profit P of a manufacturer is related to the production level x by the function

$$P(x) = 50x - 0.04x^2 - 10{,}000$$

Find the increment ΔP if the production level is increased from 500 to 600 units. Find the average rate of change $\Delta P/\Delta x$. What does it mean?

Solution At the initial production level $x_1 = 500$ the profit is

$$\begin{aligned} P_1 = P(x_1) = P(500) &= 50(500) - 0.04(500)^2 - 10{,}000 \\ &= 5000 \quad \text{dollars} \end{aligned}$$

At the final production level $x_2 = 600$ the profit is

$$\begin{aligned} P_2 = P(x_2) = P(600) &= 50(600) - 0.04(600)^2 - 10{,}000 \\ &= 5600 \quad \text{dollars} \end{aligned}$$

Hence the increment in P is

$$\Delta P = P_2 - P_1 = 5600 - 5000 = 600 \quad \text{dollars}$$

if production is raised by $\Delta x = 100$ units from 500 to 600 units per week. In general, $\Delta P > 0$ means that the production level x_2 is more profitable than the initial production level x_1; $\Delta P < 0$ means that production level x_2 is less profitable. The average rate of change

$$\frac{\Delta P}{\Delta x} = \frac{600}{100} = 6 \quad \text{dollars per unit}$$

tells us that, on the average, profit increases by \$6 for each additional unit produced when production is raised from 500 to 600 units. ■

Example 5 A demographic study shows that the number y of residents in Mono County is described by the function $y = f(x)$

$$f(x) = 100{,}000 \cdot 2^{x/10}$$

where $x =$ (years elapsed since 1950). Find the average rate of change in y (the average population growth rate) over the period from 1965 to 1975.

Solution The years 1965 and 1975 correspond to x values $x_1 = 15$ and $x_2 = 25$, so the increment in x is

$$\Delta x = x_2 - x_1 = 25 - 15 = 10 \quad \text{years}$$

The corresponding y values and increment Δy are

$$\begin{aligned} y_1 &= f(x_1) = 100{,}000(2^{15/10}) = 100{,}000 \cdot 2^{3/2} = 100{,}000(2\sqrt{2}) = 282{,}800 \\ y_2 &= f(x_2) = 100{,}000(2^{25/10}) = 100{,}000 \cdot 2^{5/2} = 100{,}000(4\sqrt{2}) = 565{,}600 \\ \Delta y &= y_2 - y_1 = 282{,}800 \end{aligned}$$

so the average growth rate over this particular 10-year period is

$$\frac{\Delta y}{\Delta x} = \frac{282{,}800}{10} = 28{,}280 \quad \text{residents per year} \qquad [2]$$

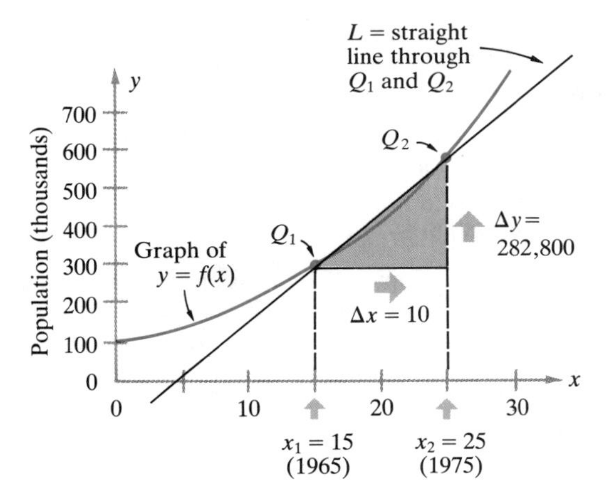

Figure 2.2
The graph of the population function $f(x) = 100{,}000\ 2^{t/10}$ in Example 5. The points Q_1 and Q_2 correspond to the situation in 1965 (when $x = 15$) and 1975 (when $x = 25$). The average rate of change of $y = f(x)$ over this period is the slope of the straight line L that passes through Q_1 and Q_2.

On the average, the population grew by this amount each year; forming the average [2] amounts to spreading the total growth equally over each of the intervening years. ■

There is a graphical interpretation of the average rate of change, which we will illustrate using the last example. The graph of the function is shown in Figure 2.2. The initial and final values of x are marked on the horizontal axis. Over them sit the points $Q_1 = (x_1, y_1)$ and $Q_2 = (x_2, y_2)$ corresponding to the initial and final situations in 1965 and 1975. The fact that $\Delta x = 10$ means that x moves 10 units to the right on the horizontal axis. Because y values represent the height of the graph above the x axis, the increment $\Delta y = 282{,}800$ is the amount the graph rises as a result of this change in x. If we draw a straight line L between Q_1 and Q_2, the average rate of change $\Delta y/\Delta x$ is just the *slope of this straight line.*

This geometric interpretation is true for any function $y = f(x)$. The average rate of change $\Delta y/\Delta x$ is always equal to the slope of the straight line joining the corresponding initial and final points on the graph

$$\text{average rate} = \frac{\Delta y}{\Delta x} = \frac{\text{amount graph rises (or falls)}}{\text{horizontal distance covered}} \qquad [3]$$

just as in Figure 2.2.

Exercises 2.1

1. Three friends drive straight through from Chicago to New York in 15 hours, covering a distance of 810 miles. Find the average speed for the trip in miles per hour (mph).
2. In Exercise 1, a total of 36 gallons of gasoline were consumed during the trip. Find the average rate of fuel consumption in miles per gallon—that is, find the average rate of change of distance travelled with respect to amount of fuel consumed.

3. An Arctic cold front passed through Minneapolis, dropping the temperature from 36°F at 8:00 P.M. to 5°F at 10:30 P.M. Find the average rate of change in temperature over this period.

4. Referring to Example 3 and Figure 2.1, find the average rate of change in price of a BMF share during the following periods.
   (i) from 1950 to 1960
   (ii) from 1960 to 1970
   (iii) from 1950 to 1970
   (iv) from 1960 to 1980

5. The number of copies N sold weekly by a stock market newsletter varies with the amount x spent on advertising (Figure 2.3). From the data presented in this graph find the average rate of change in N as the weekly advertising budget x is raised from

   (i) \$0 to \$500 (ii) \$500 to \$1000 (iii) \$1000 to \$1500

   Why do the average rates decrease even though the increment $\Delta x = \$500$ is the same?

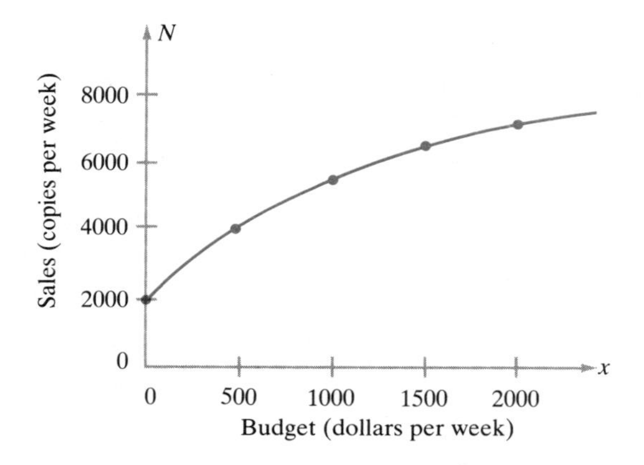

x	N
0	2000
500	4000
1000	5500
1500	6500
2000	7000

Figure 2.3
Weekly sales N versus the amount x spent per week on advertising (see Exercise 5).

6. Calculate the change Δy in the function $y = x^2 - x + 1$ as x changes from
   (i) $x_1 = 0$ to $x_2 = 1$
   (ii) $x_1 = 1$ to $x_2 = 2$
   (iii) $x_1 = 1$ to $x_2 = -1$
   (iv) $x_1 = 2$ to $x_2 = 2 + \Delta x$

7. In Exercise 6 calculate the average rates of change $\Delta y/\Delta x$.

8. Let $y = 1/x$. Taking initial point $x_1 = 2$ and final point $x_2 = 2 + \Delta x$, find the increment Δy if
   (i) $\Delta x = 1$
   (ii) $\Delta x = 0.5$
   (iii) $\Delta x = -1$
   (iv) $\Delta x = -0.5$

9. Let $s = 4 - 3t$. Find the average rate of change of s with respect to t between $t_1 = 1$ and $t_2 = 3$. What is the average rate of change for *arbitrary* t_1 and t_2?

In Exercises 10–17 find the average rate of change in y as x changes from x_1 to x_2.

10. $y = x^2; x_1 = 1, x_2 = 3$

11. $y = x^2; x_1 = 0, x_2 = 2$

12. $y = x^2; x_1 = 2, x_2 = 0$

13. $y = 3x^2; x_1 = 2, x_2 = 7$

14. $y = 2 - 3x^2; x_1 = 0, x_2 = 3$

15. $y = x^2 - 4x; x_1 = 1, x_2 = 2$

16. $y = \dfrac{8}{x}; x_1 = 2, x_2 = 8$

17. $y = 3^x; x_1 = 0, x_2 = 4$

18. The daily revenue R of a cafeteria is related to the number x of customers by the formula $R = x(7 - 0.01x)$. Find the average rate of change $\Delta R/\Delta x$ as x increases from 300 to 400 and from 200 to 300.

19. If the cost per week of manufacturing an inexpensive transistorized audio amplifier is given by

$$C(q) = 50{,}000 + 100q - 0.1q^2$$

where q is the number of units produced per week, find the cost of producing (i) 500 units per week and (ii) 600 units per week. What is the average rate of change $\Delta C/\Delta q$ of C with respect to q if production is raised from $q_1 = 500$ to $q_2 = 600$ units per week?

20. A stone is dropped into a well. The distance s that it falls in t seconds is given by the formula $s = 16t^2$ feet. Calculate the average velocity $\Delta s/\Delta t$ between $t_1 = 1$ and $t_2 = 3$ seconds.

21. A ball thrown straight up into the air with an initial speed of 64 feet per second will have height $h(t) = 64t - 16t^2$ after t seconds. Its average velocity over any time interval is just $\Delta h/\Delta t$. Calculate the average velocity over the following time intervals.

(i) $t_1 = 0$ to $t_2 = 2$
(ii) $t_1 = 0$ to $t_2 = 3$
(iii) $t_1 = 2$ to $t_2 = 3$
(iv) $t_1 = 0$ to $t_2 = 4$

(*Note*: Positive velocities correspond to a net increase in height h, negative velocities to a net decrease.)

22. A psychologist running memory tests finds that the number of words a certain person can memorize in t minutes is given (approximately) by the formula $N = 10\sqrt{t}$. Find the average rate of change of N as t changes from

(i) 0 to 1 minute
(ii) 0 to 10 minutes
(iii) 10 to 20 minutes
(iv) 20 to 30 minutes

Why do you think the averages (ii)–(iv) are decreasing, even though the time increment $\Delta t = 10$ is the same in each case?

23. Referring to Example 5, find the average rate of population growth in Mono County over the periods: (i) 1970–1975, (ii) 1950–1970.

24. Consider $y = x^2 + x$ and the initial value $x_1 = 0$. Find x_2 if the average rate of change in y between x_1 and x_2 is $\Delta y/\Delta x = 4$.

2.2 Instantaneous Rate of Change

Average rates of change are important, but in many situations they are not quite appropriate. For example, in deciding whether to invest in a mutual fund, its average growth over the past 30 years is not as relevant as its growth right now (see Example 3, Section 2.1). Here is another example.

If we drive 150 miles in 3 hours, our average speed for the trip is 50 mph. But our speedometer measures instantaneous speed at any moment, which could have been 70 mph some of the time and 40 mph at other moments. If we try to argue our way out of a speeding ticket, the fact that our average speed for the past 3 hours was 50 mph, even if provable, does not interest the police officer. He is concerned with our speed at the moment he spotted our car. These examples indicate the need for a concept of *rate of change at a single moment*; the present time in the case of the mutual fund and the time of apprehension in the case of the speeding violation. In this section we shall describe this concept—the **instantaneous rate of change** of one variable with respect to another.

Suppose the relationship between two variables is given by some function $y = f(x)$. If x_1 is a fixed "base value" of the independent variable x, consider what happens as x changes by an amount Δx from x_1 to $x_2 = x_1 + \Delta x$. *For most functions the average rates of change $\Delta y/\Delta x$ will be about the same as long as Δx is small.* In fact, as the increment Δx is made smaller and smaller, the averages tend toward a "limit value" interpreted as the instantaneous rate of change of y when $x = x_1$. Put another way, the average rates hover about the instantaneous rate and get closer to it as Δx gets close to zero. The idea will be illustrated in the following examples.

Example 6 Let $f(x) = x^2$, and consider the base value $x_1 = 1$. Find an expression for the average rate $\Delta y/\Delta x$ as x changes from $x_1 = 1$ to $x_2 = 1 + \Delta x$. Make a table of values of $\Delta y/\Delta x$ for the specific increments $\Delta x = 0.1, 0.05, 0.01, 0.005$, and 0.001. What happens to $\Delta y/\Delta x$ as the increment Δx is made smaller and smaller?

Solution The change in x is $x_2 - x_1 = \Delta x$. The corresponding y values and increment Δy are

$$y_1 = f(x_1) = (1)^2 = 1$$
$$y_2 = f(x_2) = (1 + \Delta x)^2 = 1 + 2(\Delta x) + (\Delta x)^2$$

and

$$\Delta y = y_2 - y_1 = 2(\Delta x) + (\Delta x)^2$$

The formula for the average rate of change

$$\frac{\Delta y}{\Delta x} = \frac{2(\Delta x) + (\Delta x)^2}{\Delta x}$$

simplifies if we cancel Δx in the numerator and denominator:

$$\frac{\Delta y}{\Delta x} = 2 + \Delta x \qquad [4]$$

for any nonzero increment. As we can see from the table of values for $\Delta y/\Delta x$ listed in Table 2.1, the averages are all close to 2 if Δx is small. In fact, Formula [4] shows directly that the averages $\Delta y/\Delta x$ approach the value 2 as the increment Δx approaches zero; the constant term 2 stays fixed while the second term in [4]

Δx	Δy	$\dfrac{\Delta y}{\Delta x} = 2 + \Delta x$
0.1	0.21	2.100
0.05	0.1025	2.050
0.01	0.0201	2.010
0.005	0.010025	2.005
0.001	0.002001	2.001
⋮	⋮	↓
		2.0000...

Table 2.1 The increments and average rates of change $\Delta y/\Delta x$ in Example 6. As we take smaller and smaller nonzero increments Δx, the averages $\Delta y/\Delta x$ approach the limiting value 2.

becomes smaller and smaller. Thus we are led to assign the value 2 as the limit value of the averages $\Delta y/\Delta x$. This limit value tells us how fast y is changing when $x = 1$. ■

If we form averages $\Delta y/\Delta x$ by taking small increments Δx away from some base value of x, this limit value—if it exists—is indicated by the symbol

$$\lim_{\Delta x \to 0} \frac{\Delta y}{\Delta x}$$

In the preceding example, the averages $\Delta y/\Delta x = 2 + \Delta x$ tend toward 2 as Δx gets smaller and smaller; we express this concisely by writing

$$\lim_{\Delta x \to 0} \frac{\Delta y}{\Delta x} = 2 \quad \text{or} \quad \lim_{\Delta x \to 0} (2 + \Delta x) = 2$$

Here is another example in which there is a limit value for the average rates.

Example 7 Find the instantaneous rate of change $\lim_{\Delta x \to 0} \dfrac{\Delta y}{\Delta x}$ for the function $y = 3x^2 - x + 1$ when $x = 2$.

Solution As x changes from the base value $x_1 = 2$ to a nearby value $x_2 = 2 + \Delta x$, the increment in x is Δx. The initial and final y values are

$$\begin{aligned} y_1 &= f(x_1) = 3(2)^2 - 2 + 1 = 11 \\ y_2 &= f(x_2) = 3(2 + \Delta x)^2 - (2 + \Delta x) + 1 \\ &= 3(4 + 4(\Delta x) + (\Delta x)^2) - (2 + \Delta x) + 1 \\ &= 11 + 11(\Delta x) + 3(\Delta x)^2 \end{aligned}$$

so that the increment in y is

$$\Delta y = y_2 - y_1 = 11(\Delta x) + 3(\Delta x)^2$$

Hence for any nonzero increment Δx, the average rate of change is

$$\frac{\Delta y}{\Delta x} = \frac{11(\Delta x) + 3(\Delta x)^2}{\Delta x} = 11 + 3(\Delta x)$$

Some values of $\Delta y/\Delta x$ are shown in Table 2.2. This numerical evidence suggests that the limiting value is 11. As in the previous example, you can see that this is so without resorting to calculations by directly examining the expression $\Delta y/\Delta x = 11 + 3(\Delta x)$. As Δx gets smaller and smaller, so does the term $3(\Delta x)$; meanwhile, the constant term 11 remains fixed, so the averages approach the limit value 11 as Δx approaches zero:

$$\lim_{\Delta x \to 0} \frac{\Delta y}{\Delta x} = \lim_{\Delta x \to 0} \left(11 + 3(\Delta x)\right) = 11$$

■

Taking these examples as a guide, we are ready to define the instantaneous rate of change of any function $y = f(x)$ to be the limit value of the average rates $\Delta y/\Delta x$.

INSTANTANEOUS RATE OF CHANGE (THE "DELTA PROCESS") Given a function $y = f(x)$ and some base point x, consider the values of f at nearby points $x + \Delta x$ for small nonzero increments Δx. Then compute

Step 1 The value of f at the base point, namely $y_1 = f(x)$
Step 2 The value of f at the nearby point, namely $y_2 = f(x + \Delta x)$
Step 3 The corresponding change in y

$$\Delta y = y_2 - y_1 = f(x + \Delta x) - f(x)$$

Step 4 The average rate of change going from x to $x + \Delta x$

$$\frac{\Delta y}{\Delta x} = \frac{y_2 - y_1}{\Delta x} = \frac{f(x + \Delta x) - f(x)}{\Delta x}$$

Step 5 The limit value

$$\lim_{\Delta x \to 0} \frac{\Delta y}{\Delta x}$$

as the increment Δx approaches zero.

We call this limit value the **instantaneous rate of change** of y at the base point x.†

This limit value can often be computed for an arbitrary base point x with about the same amount of effort as for particular base points such as $x = 2$ or $x = -3$. This is why we state the procedure for an arbitrary base point. In

† For most functions we consider, the limit defining the instantaneous rate of change will exist. If it does not exist, the instantaneous rate of change is not defined at that base point.

Δx	Δy	$\frac{\Delta y}{\Delta x} = 11 + 3(\Delta x)$
0.1	1.13000	11.3000
0.05	0.55075	11.0150
0.01	0.11003	11.0030
0.005	0.05501	11.0015
0.001	0.01100	11.0003
⋮	⋮	↓
		11.0000...

Table 2.2. For $f(x) = 3x^2 - x + 1$, we consider small increments $\Delta x \neq 0$ away from the base point $x_1 = 2$. Values of Δx, Δy, and the average rate of change $\Delta y/\Delta x$ are shown. The averages $\Delta y/\Delta x$ approach the limit value 11 as Δx gets smaller and smaller.

carrying out the delta process, notice that x and Δx stand for quite different things: x is the base point, Δx is the increment away from this base point, and our objective is to see what happens to the averages $\Delta y/\Delta x$ as Δx is made smaller and smaller.

The next example, for arbitrary base point x, should be compared with Example 6 where we considered the same function $y = x^2$ at the particular base point $x = 1$.

Example 8 Calculate the instantaneous rate of change of the function $y = x^2$ at an arbitrary base point x.

Solution At base point x and nearby point $x + \Delta x$, the y values are

$$y_1 = x^2$$
$$y_2 = (x + \Delta x)^2 = x^2 + 2x(\Delta x) + (\Delta x)^2$$

and the increment in y is

$$\begin{aligned}\Delta y = y_2 - y_1 &= \cancel{x^2} + 2x(\Delta x) + (\Delta x)^2 - \cancel{x^2} \\ &= 2x(\Delta x) + (\Delta x)^2\end{aligned}$$

Therefore the average rate of change is

$$\frac{\Delta y}{\Delta x} = \frac{2x(\Delta x) + (\Delta x)^2}{\Delta x} = 2x + (\Delta x)$$

for any nonzero increment Δx. As Δx gets smaller and smaller, the base point x does not vary, so the first term $2x$ stays fixed. The second term becomes very small as Δx approaches zero, so the instantaneous rate of change is given by the limit value

$$\lim_{\Delta x \to 0} \frac{\Delta y}{\Delta x} = \lim_{\Delta x \to 0} (2x + \Delta x) = 2x \qquad [5]$$

for any base point x. If $x = 1$ the rate of change has the particular value $2(1) = 2$, as in Example 6. At another base point, say $x = 3$, we get a different rate of change, namely $2(3) = 6$. ■

The rate of change of a function $y = f(x)$ depends on the base point being considered. At some base points f is changing rapidly while at others it changes slowly, just as in the last example.

The next examples provide additional practice in carrying out the delta process.

Example 9 Calculate the instantaneous rate of change of $y = x^3$ at an arbitrary base point.

Solution Taking base point x and nearby point $x + \Delta x$, we compute the corresponding y values[†]

$$\begin{aligned} y_1 &= x^3 \\ y_2 &= (x + \Delta x)^3 = x^3 + 3x^2(\Delta x) + 3x(\Delta x)^2 + (\Delta x)^3 \end{aligned}$$

Hence

$$\begin{aligned} \Delta y = y_2 - y_1 &= \cancel{x^3} + 3x^2(\Delta x) + 3x(\Delta x)^2 + (\Delta x)^3 - \cancel{x^3} \\ &= 3x^2(\Delta x) + 3x(\Delta x)^2 + (\Delta x)^3 \end{aligned}$$

and the average rate of change is

$$\frac{\Delta y}{\Delta x} = \frac{3x^2(\Delta x) + 3x(\Delta x)^2 + (\Delta x)^3}{\Delta x} = 3x^2 + 3x(\Delta x) + (\Delta x)^2$$

for all small increments Δx. Now let Δx get smaller and smaller. The base point x does not vary, so the first term $3x^2$ stays fixed. The last two terms involve Δx and obviously become smaller and smaller as Δx approaches zero, so that

$$\text{instantaneous rate of change} = \lim_{\Delta x \to 0} \frac{\Delta y}{\Delta x} = 3x^2$$

at any base point x. ■

[†] If a and b are two numbers, powers of their sum such as $(a + b)^2, (a + b)^3, \ldots$ are computed by multiplying and then combining similar terms:

$$\begin{aligned} (a + b)^2 &= a^2 + 2ab + b^2 \\ (a + b)^3 &= (a + b)(a + b)^2 = (a + b)(a^2 + 2ab + b^2) = a^3 + 3a^2b + 3ab^2 + b^3 \end{aligned}$$

and so on. If we let $a = x$ and $b = \Delta x$, we obtain the formula for y_2, which we have already given. The formula telling us how to multiply out an arbitrary power

$$(a + b)^n = a^n + na^{n-1}b + \frac{n(n-1)}{2}a^{n-2}b^2 + \cdots + b^n$$

is well known in algebra as the "binomial formula."

Example 10 Suppose a manufacturer's weekly costs are related to the level of production x by the formula $C = 5000 + 150x - 0.01x^2$. How rapidly are his costs changing (instantaneous rate of change) with x? How fast are they changing at the particular production level $x = 2000$?

Solution The problem is to compute the instantaneous rate of change of the cost function $C(x) = 5000 + 150x - 0.01x^2$ at an arbitrary base point x. Considering the base point x and nearby production levels $x + \Delta x$, we first calculate the corresponding values of C:

$$\begin{aligned} C_1 &= C(x) = 5000 + 150x - 0.01x^2 \\ C_2 &= C(x + \Delta x) = 5000 + 150(x + \Delta x) - 0.01(x + \Delta x)^2 \\ &= 5000 + 150(x + \Delta x) - 0.01(x^2 + 2x(\Delta x) + (\Delta x)^2) \\ &= 5000 + 150x - 0.01x^2 + 150(\Delta x) - 0.01(2x)(\Delta x) - 0.01(\Delta x)^2 \end{aligned}$$

The resulting increment in C is therefore

$$\Delta C = C_2 - C_1 = 150(\Delta x) - 0.02x(\Delta x) - 0.01(\Delta x)^2$$

and the average rate of change is

$$\frac{\Delta C}{\Delta x} = 150 - 0.02x - 0.01(\Delta x)$$

Because x is fixed, the average approaches the limit value $150 - 0.02x$ as Δx gets smaller and smaller. Thus, the instantaneous rate of change at any production level x is

$$\text{instantaneous rate of change} = 150 - 0.02x$$

At the particular production level $x = 2000$, this rate of change is $150 - 40 = 110$ dollars per unit. ■

The next example is a little more complicated; it introduces some new ideas for dealing with limits.

Example 11 If $y = 1/x$, find its instantaneous rate of change at an arbitrary base point using the delta process.

Solution The function is undefined at $x = 0$, so it is meaningless to discuss the instantaneous rate of change there. For base points $x \neq 0$, we must examine the behavior of $\Delta y/\Delta x$ for small nonzero increments Δx. We have

$$y = \frac{1}{x} \quad \text{at the base point } x$$

$$y = \frac{1}{x + \Delta x} \quad \text{at the nearby point } x + \Delta x$$

so that

$$\frac{\Delta y}{\Delta x} = \frac{\left[\dfrac{1}{x + \Delta x} - \dfrac{1}{x}\right]}{\Delta x} \quad \text{for all small increments } \Delta x$$

The limit value of $\Delta y/\Delta x$ is not at all apparent from this formula. But if we simplify the expression algebraically, some of the Δx cancel, and it is easier to see what happens as Δx gets small.

$$\begin{aligned} \frac{\Delta y}{\Delta x} &= \frac{1}{\Delta x} \cdot \left[\frac{1}{x + \Delta x} - \frac{1}{x}\right] = \frac{1}{\Delta x} \cdot \left[\frac{x - (x + \Delta x)}{x(x + \Delta x)}\right] \\ &= \frac{1}{\Delta x} \cdot \left[\frac{-(\Delta x)}{x^2 + x(\Delta x)}\right] \\ &= \frac{\Delta x}{\Delta x} \cdot \left[\frac{-1}{x^2 + x(\Delta x)}\right] \\ &= \frac{-1}{x^2 + x(\Delta x)} \end{aligned} \quad [6]$$

As Δx approaches zero, all terms involving Δx become very small, and the remaining terms stay fixed. With this in mind, notice how the denominator in [6] behaves:

$$\lim_{\Delta x \to 0} \left(x^2 + x(\Delta x)\right) = x^2$$

Because the numerator stays equal to -1 and the denominator approaches the value x^2, the quotient approaches $-1/x^2$ as Δx approaches zero. Thus, we have determined the instantaneous rate of change.

$$\lim_{\Delta x \to 0} \frac{\Delta y}{\Delta x} = \frac{-1}{x^2}$$

■

Exercises 2.2

1. For each of the following statements, decide whether it refers to an average or instantaneous rate of change.
   (i) When the car skidded off the road, it must have been doing 60 mph.
   (ii) This model will give you 30 miles to the gallon in city driving.
   (iii) Stick to this diet, and you will lose 5 pounds a month.
   (iv) After launch, the spacecraft gradually accelerated, achieving a velocity of 3000 mph at altitude 15 miles.

In Exercises 2–3, consider the function $f(x) = x^2 - 2x$ and a fixed base point $x_1 = 3$.

2. Find an expression for the average rate of change $\Delta y/\Delta x$ as x changes from $x_1 = 3$ to $x_2 = 3 + \Delta x$. Make a table of values of $\Delta y/\Delta x$ for the specific increments $\Delta x = 0.01$, 0.001, 0.0001.

3. Find the instantaneous rate of change of y at the base point $x_1 = 3$.

4. Find the instantaneous rate of change of $y = x^2 - 2x$ at an arbitrary base point x.

5. Use the delta process to calculate the instantaneous rate of change of the function $f(x) = x$ at an arbitrary base point.

6. Use the delta process to show that the instantaneous rate of change of a constant function $f(x) = c$ is zero at any base point. (*Note*: Intuitively, this conclusion is quite reasonable: The value of the function is everywhere the same, so its rate of change should be zero.)

7. Calculate the instantaneous rate of change $\lim_{\Delta x \to 0} \Delta y/\Delta x$ for the function $y = 4 - x^2$ at an arbitrary base point x. What is the value of this instantaneous rate at the particular base point $x = 2$? At $x = -3$? For which base points is the instantaneous rate equal to 4? To zero?

8. Repeat Exercise 7 for the function $y = 2x^2 - 3x + 1$.

In Exercises 9–20 use the delta process to compute the instantaneous rate of change of $y = f(x)$ at an arbitrary base point x.

9. $y = 4x - 3$

10. $y = 2 - 3x$

11. $y = 1 + x^2$

12. $y = x^2 - x - 1$

13. $y = 2x^2 - 3x$

14. $y = x + x^3$

15. $y = x^2 - 4x + 1$

16. $y = x^3 - 1$

17. $y = x^4$

18. $y = 7x^2 - \dfrac{x}{2}$

19. $y = \dfrac{1}{x^2}$

20. $y = 100 - 2x + 0.1x^2$

21. Use the delta process to show that a linear function $y = Ax + B$ (A, B being constants) has instantaneous rate of change

$$\lim_{\Delta x \to 0} \frac{\Delta y}{\Delta x} = A \quad \text{at any base point } x$$

That is, the instantaneous rate is equal to the slope of the straight line determined by the equation $y = Ax + B$.

22. Suppose a manufacturer's costs are $C(x) = 50{,}000 + 100x - 0.1x^2$ dollars per week at production level x. How rapidly are the costs changing (instantaneous rate) with x at the particular production level of 250 units per week? At an arbitrary production level x?

23. If a rumor spreads to $N(t) = 50t^2$ people in t days, how rapidly is the rumor spreading (instantaneous rate of change of N) when $t = 3$? How many people have heard the rumor after 3 days?

2.3 The Derivative of a Function

The instantaneous rate of change of a function $y = f(x)$ is so important it is given a special name. It is called the **derivative** of $f(x)$ and is denoted by the symbol $f'(x)$. The derivative measures how fast the value of $f(x)$ is changing at the base point being considered. At some base points, f is changing rapidly; at others, it changes slowly. Thus the derivative $f'(x)$ depends on the base point x, and so should be thought of as a new function "derived from" the original function $f(x)$. In Section 2.2 we have already computed the derivatives of some simple functions using the delta process:

If $f(x) = x^2$, its derivative (instantaneous rate of change) is $f'(x) = 2x$

If $f(x) = x^3$, its derivative (instantaneous rate of change) is $f'(x) = 3x^2$

> **DEFINITION** Given a function $y = f(x)$, we define the **derivative** $f'(x)$ at base point x to be the limit value of averages
>
> $$f'(x) = \lim_{\Delta x \to 0} \left(\frac{f(x + \Delta x) - f(x)}{\Delta x} \right) = \lim_{\Delta x \to 0} \frac{\Delta y}{\Delta x}$$
>
> provided this limit value exists. In that case we say f is **differentiable** at x.

Various symbols other than $f'(x)$ are used for the derivative of $y = f(x)$. In this book we shall use the most popular symbols

$$f'(x) \qquad \frac{df}{dx} \qquad \frac{dy}{dx} \qquad \frac{d}{dx}(f(x)) \qquad Df \tag{7}$$

interchangeably. We could have used just one symbol throughout, but certain formulas are easier to remember if we use the appropriate notation. Besides, when readers turn to other books, they should be experienced enough to recognize that all the symbols [7] stand for the same thing—the derivative function.

The basic method for calculating derivatives is the delta process of Section 2.2, which can be time consuming. We shall apply the delta process once and for all to obtain differentiation rules that allow us to write down at a glance the derivatives of simple functions such as polynomials. Thereafter, we will appeal to these rules rather than to the delta process wherever possible.

The first rule gives the derivative of a simple power of x

$$f(x) = x^r \qquad (r \text{ a fixed constant})$$

Two special cases should be mentioned first. If $r = 0$, then $f(x) = x^0 = 1$ (constant function everywhere equal to 1), and if $r = 1$ then $f(x) = x^1 = x$. The derivatives of these functions are very easy to find by the delta process (see Exercises 5–6, Section 2.2):

If $f(x) = 1$, then $f'(x) = 0$ (constant function everywhere zero)

If $f(x) = x$, then $f'(x) = 1$ (constant, everywhere equal to 1)

We also worked out the derivatives of a few other powers x^r:

r	Function $y = x^r$	Derivative
0	$f(x) = 1$ everywhere	$f'(x) = 0$ everywhere
1	$f(x) = x$	$f'(x) = 1$ everywhere
2	$f(x) = x^2$	$f'(x) = 2x$
3	$f(x) = x^3$	$f'(x) = 3x^2$

A regular pattern seems to be emerging. If $f(x) = x^4$, we might guess (correctly) that $f'(x) = 4x^3$. More generally, if $f(x) = x^r$ (r a positive integer), it seems reasonable to conjecture that $f'(x) = rx^{r-1}$, because this formula yields the right result for $r = 0, 1, 2, 3, 4$. This formula is actually valid for *any* value of r, whether or not it is a positive integer.

POWER RULE If $f(x) = x^r$, its derivative is

$$\frac{d}{dx}(x^r) = rx^{r-1} \qquad [8]$$

for any fixed exponent r.

We will justify this rule at the end of this section; meanwhile, here are some examples of its use.

Example 12 Use the power rule to differentiate the functions

(i) $f(x) = 1$ (constant function everywhere equal to one)
(ii) $f(x) = x$
(iii) $f(x) = x^5$
(iv) $f(x) = \dfrac{1}{x^2}$
(v) $f(x) = \dfrac{1}{\sqrt{x}}$

Solution In (i), $f(x) = x^0 = 1$. Here, $r = 0$, so the power rule yields

$$\frac{d}{dx}(1) = \frac{d}{dx}(x^0) = 0 \cdot (x^{0-1}) = 0 \cdot x^{-1} = 0 \quad \text{everywhere}$$

In (ii) $r = 1, f(x) = x^1$, and

$$\frac{d}{dx}(x) = \frac{d}{dx}(x^1) = 1 \cdot (x^{1-1}) = 1 \cdot (x^0) = 1 \cdot 1 = 1 \quad \text{everywhere}$$

In (iii), $r = 5$ so that

$$\frac{d}{dx}(x^5) = 5(x^{5-1}) = 5x^4$$

In (iv) and (v), rewrite the formula using negative exponents: $1/x^2 = x^{-2}$ (here $r = -2$) and $1/\sqrt{x} = x^{-1/2}$ (here $r = -1/2$). Then apply the power rule to obtain

$$\frac{d}{dx}\left(\frac{1}{x^2}\right) = \frac{d}{dx}(x^{-2}) = (-2)x^{(-2)-1} = (-2)x^{-3} = \frac{-2}{x^3}$$

$$\frac{d}{dx}\left(\frac{1}{\sqrt{x}}\right) = \frac{d}{dx}(x^{-1/2}) = \left(-\frac{1}{2}\right)x^{(-1/2)-1} = -\frac{1}{2}x^{-3/2}$$

$$= -\frac{1}{2}\frac{1}{x^{3/2}} = \frac{-1}{2x\sqrt{x}}$$

■

Example 13 Find the derivative of $f(x) = x^{20}$. What is the value of the derivative when $x = 1$?

Solution Applying the power rule to $f(x) = x^{20}$ we obtain

$$f'(x) = 20(x^{20-1}) = 20x^{19}$$

When $x = 1$, the derivative has the particular value

$$f'(1) = 20(1)^{19} = 20 \cdot (1) = 20$$

■

Example 14 If $y = \sqrt{x}$, how fast is y increasing (instantaneous rate of change) when $x = 4$?

Solution The rate of change is just the value of the derivative when $x = 4$. Now $y = \sqrt{x} = x^{1/2}$, so for any base point x the derivative is

$$f'(x) = \frac{d}{dx}(x^{1/2}) = \frac{1}{2}x^{(1/2)-1} = \frac{1}{2}x^{-1/2} = \frac{1}{2}\frac{1}{x^{1/2}} = \frac{1}{2\sqrt{x}}$$

At the particular base point $x = 4$ the instantaneous rate of change is

$$f'(4) = \frac{1}{2\sqrt{4}} = \frac{1}{4}$$

■

Next we show how to compute the derivative of a combination of functions in terms of the derivatives of the functions we started with. If we multiply a function $f(x)$ by a constant k we obtain a new function $k \cdot f(x)$. For example if $f(x) = 1/x$ and $k = -2$, then $k \cdot f(x) = -2/x$; or if $f(x) = x^3$ and $k = 5$, then

$k \cdot f(x) = 5x^3$. The derivative of this new function $k \cdot f(x)$ is just the constant k times the derivative df/dx of the original function. This rule will be proved at the end of this section.

DERIVATIVE OF CONSTANT MULTIPLE $k \cdot f(x)$ If k is any constant and $f(x)$ any differentiable function, then

$$\frac{d}{dx}(k \cdot f(x)) = k \cdot \frac{d}{dx}(f(x)) \qquad [9]$$

Example 15 Find the derivatives of the following functions.

(i) $y = -7x^5$

(ii) $y = 4\sqrt{x}$

(iii) $y = \dfrac{-1}{x^2}$

(iv) $y = 6$ (constant function)

Solution

$$\frac{d}{dx}(-7x^5) = (-7)\frac{d}{dx}(x^5) = -7(5x^4) = -35x^4$$

$$\frac{d}{dx}(4\sqrt{x}) = 4\frac{d}{dx}(\sqrt{x}) = 4\frac{d}{dx}(x^{1/2}) = 4\left(\frac{1}{2\sqrt{x}}\right) = \frac{2}{\sqrt{x}}$$

$$\frac{d}{dx}\left(\frac{-1}{x^2}\right) = (-1)\frac{d}{dx}\left(\frac{1}{x^2}\right) = (-1)\frac{d}{dx}(x^{-2})$$

$$= (-1)(-2)x^{(-2)-1} = 2x^{-3} = \frac{2}{x^3}$$

$$\frac{d}{dx}(6) = 6\frac{d}{dx}(1) = 6(0) = 0 \quad \text{(constant function)}$$

■

It may be worth remembering that, as in Part iv (Example 15), *the derivative of any constant function $f(x) = k$ is identically zero*, since $k = k \cdot 1$ and

$$\frac{d}{dx}(k) = k\frac{d}{dx}(1) = k \cdot 0 = 0 \quad \text{everywhere}$$

Another way to combine two functions $f(x)$ and $g(x)$ is to form their sum $f + g$, adding their values for each x:

$$(f + g)(x) = f(x) + g(x)$$

Thus if $f(x) = x^2$ and $g(x) = x$, we obtain $f + g = x^2 + x$; or if $f(x) = 5x$ and $g(x) = -1/x$, we obtain $f + g = 5x - (1/x)$. The derivative of a sum is the sum of the derivatives:

Derivative of a Sum $f(x) + g(x)$ If $f(x)$ and $g(x)$ are differentiable functions then

$$\frac{d}{dx}(f(x) + g(x)) = \frac{df}{dx} + \frac{dg}{dx} \qquad [10]$$

A proof will be given at the end of this section.

These rules can be used in combination, as in the latter parts of the next example.

Example 16 Find the derivatives of

(i) $y = x + x^2$ (ii) $y = 1 + x^3$ (iii) $y = x + \frac{1}{x}$

(iv) $y = x - \frac{2}{x^2}$ (v) $y = 3x^2 - 2$ (vi) $y = 4\sqrt{x} + 7x^5$

Solution Using the sum rule and then the power rule we obtain

$$\frac{d}{dx}(x + x^2) = \frac{d}{dx}(x) + \frac{d}{dx}(x^2) = 1 + 2x$$

$$\frac{d}{dx}(1 + x^3) = \frac{d}{dx}(1) + \frac{d}{dx}(x^3) = 0 + 3x^2 = 3x^2$$

$$\frac{d}{dx}\left(x + \frac{1}{x}\right) = \frac{d}{dx}(x) + \frac{d}{dx}\left(\frac{1}{x}\right) = 1 + \frac{d}{dx}(x^{-1})$$

$$= 1 + (-1)x^{-2} = 1 - \frac{1}{x^2}$$

Successively using all of the preceding rules we obtain

$$\frac{d}{dx}\left(x - \frac{2}{x^2}\right) = \frac{d}{dx}(x) + \frac{d}{dx}\left(\frac{-2}{x^2}\right) = \frac{d}{dx}(x) + (-2)\frac{d}{dx}(x^{-2})$$

$$= 1 + (-2)(-2)x^{-3} = 1 + \frac{4}{x^3}$$

$$\frac{d}{dx}(3x^2 - 2) = \frac{d}{dx}(3x^2) + \frac{d}{dx}(-2) = 3\frac{d}{dx}(x^2) + (-2)\frac{d}{dx}(1)$$

$$= 3(2x) + (-2)(0) = 6x$$

$$\frac{d}{dx}(4\sqrt{x} + 7x^5) = \frac{d}{dx}(4x^{1/2}) + \frac{d}{dx}(7x^5) = 4\frac{d}{dx}(x^{1/2}) + 7\frac{d}{dx}(x^5)$$

$$= 4\left(\frac{1}{2}\right)x^{-1/2} + 7(5x^4)$$

$$= 2x^{-1/2} + 35x^4 = \frac{2}{\sqrt{x}} + 35x^4$$

■

A **polynomial** is any linear combination of the powers 1, x, x^2, x^3,...

$$P(x) = k_0 + k_1x + k_2x^2 + \cdots + k_nx^n \quad (k_0, \ldots, k_n \text{ constants})$$

For example, the profit function $P = -10{,}000 + 50x - 0.04x^2$ in Example 4 is a polynomial, with $k_0 = -10{,}000$, $k_1 = 50$, and $k_2 = -0.04$; and so are such functions as

$$\begin{aligned} y &= x^3 + 3x - 7 \\ y &= -6x^4 + 20x^3 - x + 5 \\ y &= 3 - x \\ y &= -14 \quad (\text{constant function; } k_0 = -14) \end{aligned} \qquad [11]$$

By repeated use of the differentiation rules we can differentiate any polynomial.

Example 17 Calculate the derivatives of the polynomial functions [11].

Solution

$$\frac{d}{dx}(x^3 + 3x - 7) = \frac{d}{dx}(x^3) + 3\frac{d}{dx}(x) - 7\frac{d}{dx}(1) = 3x^2 + 3$$

$$\begin{aligned} \frac{d}{dx}(-6x^4 + 20x^3 - x + 5) &= -6\frac{d}{dx}(x^4) + 20\frac{d}{dx}(x^3) - \frac{d}{dx}(x) + 5\frac{d}{dx}(1) \\ &= -24x^3 + 60x^2 - 1 \end{aligned}$$

$$\frac{d}{dx}(3 - x) = 3\frac{d}{dx}(1) - \frac{d}{dx}(x) = -1 \quad (\text{constant function})$$

$$\frac{d}{dx}(-14) = -14\frac{d}{dx}(1) = 0 \quad (\text{constant function})$$

■

Example 18 Weekly sales revenue R for a stock market newsletter depends on the amount x spent for advertising according to the formula $R(x) = 25{,}000 + 150\sqrt{x}$. Find the instantaneous rate of change of revenue with respect to x when $x = \$2500$.

Solution The instantaneous rate of change is just the derivative

$$\begin{aligned} \frac{dR}{dx} &= \frac{d}{dx}(25{,}000 + 150\sqrt{x}) = 25{,}000\frac{d}{dx}(1) + 150\frac{d}{dx}(x^{1/2}) \\ &= 25{,}000(0) + 150(\tfrac{1}{2})x^{-1/2} \\ &= \frac{75}{\sqrt{x}} \end{aligned}$$

If $x = 2500$, we obtain $R'(2500) = 75/\sqrt{2500} = 75/50 = 1.50$. Later, we will see that this means revenue rises by about \$1.50 for each \$1.00 increase in x if the advertising budget is raised slightly from its present level $x = 2500$. ■

Function	Derivative
$y = e^x$	$\frac{dy}{dx} = e^x$
$y = \ln x$	$\frac{dy}{dx} = \frac{1}{x}$

Table 2.3 Derivatives of the special functions e^x, $\ln x$.

Two special functions turn up frequently in applications.

The exponential function: $y = e^x$

The natural logarithm function: $y = \ln x$

These will be reviewed extensively in Chapter 4, where they appear naturally in problems of growth and compound interest. For the time being, we shall take a pragmatic view of these useful functions, regarding them as special functions whose values are found using a calculator or by referring to the tables at the end of this book. We shall handle their derivatives by invoking the differentiation rules listed without proof in Table 2.3. We need not know anything more about these functions to understand our occasional use of them in the next few chapters.

For readers who have not encountered these functions before, we remark that e refers to the special number $e = 2.71828\ldots$; $y = e^x$ is just e raised to the x power and is similar to the exponential function $y = 2^x$, whose values are just powers of the number 2. Here, $\ln x$ is the *natural logarithm*, with base $e = 2.71828\ldots$. You may have encountered the *logarithm to base* 10, $y = \log_{10}(x)$, which is a somewhat different function. Some calculators are equipped to compute both kinds of logarithms. The symbol $\ln x$ is almost universally used for the natural logarithm; $\log_{10}(x)$ is sometimes indicated by the symbol $\log x$. If you are using a calculator, be sure you know which keys yield the natural logarithm before attempting any computations involving $\ln x$.

Here are some examples of how the differentiation rules for these functions are handled.

Example 19 Find

(i) the derivative of $f(x) = 10x^2 - 13e^x + 11 \ln x$

(ii) the derivative of $g(x) = -7e^x + \sqrt{x}$

(iii) the value $h'(2.5)$ if $h(x) = \frac{1}{x} + \ln x - e^x$

Solution For (i) and (ii)

$$f'(x) = 10\frac{d}{dx}(x^2) - 13\frac{d}{dx}(e^x) + 11\frac{d}{dx}(\ln x)$$

$$= 10(2x) - 13(e^x) + 11\left(\frac{1}{x}\right) = 20x - 13e^x + \frac{11}{x}$$

$$g'(x) = -7\frac{d}{dx}(e^x) + \frac{d}{dx}(x^{1/2}) = -7e^x + \frac{1}{2}x^{(1/2)-1}$$

$$= -7e^x + \frac{1}{2}x^{-1/2}$$

For (iii) we must first find the derivative $h'(x)$, and then set $x = 2.5$:

$$h'(x) = \frac{d}{dx}\left(\frac{1}{x}\right) + \frac{d}{dx}(\ln x) - \frac{d}{dx}(e^x)$$

$$= \frac{d}{dx}(x^{-1}) + \frac{1}{x} - e^x$$

$$= (-1)x^{(-1)-1} + \frac{1}{x} - e^x$$

$$= -x^{-2} + \frac{1}{x} - e^x$$

$$= \frac{-1}{x^2} + \frac{1}{x} - e^x$$

Setting $x = 2.5$, we obtain $e^{2.5} = 12.1825$, using a calculator or the tables. Thus

$$h'(2.5) = \frac{-1}{(2.5)^2} + \frac{1}{2.5} - e^{2.5} = -11.9425$$ ■

Proofs of the Differentiation Rules Using the delta process we will now show that

$$\frac{d}{dx}(x^r) = r \cdot x^{r-1}$$

for the particular exponents $r = 0, 1, 2, \ldots$. (The rule is valid for any exponent r, but the proof is much more technical, so we omit it.) Fixing a base point x and increment Δx, the corresponding increment in $y = x^r$ is

$$\Delta y = (x + \Delta x)^r - x^r$$

We expand $(x + \Delta x)^r$ by multiplying it out, concentrating attention on the first two terms in the expansion.

$$(x + \Delta x)^1 = x + \Delta x$$
$$(x + \Delta x)^2 = x^2 + 2x(\Delta x) + (\Delta x)^2$$
$$(x + \Delta x)^3 = x^3 + 3x^2(\Delta x) + 3x(\Delta x)^2 + (\Delta x)^3$$
$$\vdots$$
$$(x + \Delta x)^r = x^r + rx^{r-1}(\Delta x) + \frac{r(r-1)}{2}x^{r-2}(\Delta x)^2 + \cdots + (\Delta x)^r$$

We have not written out the terms indicated by (. . .). All we need to know about them is that they all have $(\Delta x)^2$ as a common factor. Thus

$$\Delta y = (x + \Delta x)^r - x^r = rx^{r-1}(\Delta x) + \frac{r(r-1)}{2}x^{r-2}(\Delta x)^2 + \cdots + (\Delta x)^r$$

and the average rate of change $\Delta y/\Delta x$ is

$$\frac{\Delta y}{\Delta x} = \frac{rx^{r-1}(\Delta x) + \cdots + (\Delta x)^r}{\Delta x}$$

$$= rx^{r-1} + \frac{r(r-1)}{2}x^{r-2}(\Delta x) + \cdots + (\Delta x)^{r-1}$$

The omitted terms (. . .) each involve Δx, and so approach zero as Δx gets small. The first term rx^{r-1} *does not* have Δx as a factor and remains fixed, so $\Delta y/\Delta x$ approaches the limit value rx^{r-1}. Thus

$$f'(x) = \lim_{\Delta x \to 0}\left(\frac{\Delta y}{\Delta x}\right) = rx^{r-1}$$

for any base point x, as required. ■

Proof of Formula [9]: If x is a fixed base point and Δx a nonzero increment, the average rate of change for $y = k \cdot f(x)$ is

$$\frac{\Delta y}{\Delta x} = \frac{k \cdot f(x + \Delta x) - k \cdot f(x)}{\Delta x} = k \cdot \frac{f(x + \Delta x) - f(x)}{\Delta x} = k \cdot \frac{\Delta f}{\Delta x}$$

By definition of the derivative $f'(x)$, the averages $\Delta f/\Delta x$ approach $f'(x)$ as Δx gets small. Clearly, $k \cdot (\Delta f/\Delta x)$ must then approach $k \cdot f'(x)$, so that

$$\frac{dy}{dx} = \lim_{\Delta x \to 0}\frac{\Delta y}{\Delta x} = \lim_{\Delta x \to 0} k \cdot \frac{\Delta f}{\Delta x} = k \cdot f'(x) = k \cdot \frac{df}{dx}$$

This proves Formula [9]. ■

Proof of Formula [10]: If $y = f(x) + g(x)$, x is a fixed base point, and Δx a nonzero increment, we obtain

$$\frac{\Delta y}{\Delta x} = \frac{[f(x + \Delta x) + g(x + \Delta x)] - [f(x) + g(x)]}{\Delta x}$$

$$= \frac{f(x + \Delta x) - f(x)}{\Delta x} + \frac{g(x + \Delta x) - g(x)}{\Delta x}$$

$$= \frac{\Delta f}{\Delta x} + \frac{\Delta g}{\Delta x}$$

By definition of the derivatives $f'(x) = df/dx$ and $g'(x) = dg/dx$, $\Delta f/\Delta x$ approaches $f'(x)$ and $\Delta g/\Delta x$ approaches $g'(x)$ as Δx gets small. Clearly, their sum must approach $f'(x) + g'(x)$

$$\frac{dy}{dx} = \lim_{\Delta x \to 0} \frac{\Delta y}{\Delta x} = \lim_{\Delta x \to 0} \left(\frac{\Delta f}{\Delta x} + \frac{\Delta g}{\Delta x} \right) = f'(x) + g'(x)$$

which is precisely Formula [10]. ■

Exercises 2.3

In Exercises 1–18, use the power rule to find the following derivatives.

1. $\frac{d}{dx}(x^4)$

2. $f'(x)$ if $f(x) = x^7$

3. $\frac{d}{dx}(x^3)$

4. $f'(x)$ if $f(x) = x^{12}$

5. $\frac{dy}{dx}$ if $y = \frac{1}{x^3}$

6. $\frac{dy}{dx}$ if $y = \sqrt[3]{x}$

7. $\frac{d}{dx}(x^{5/3})$

8. $\frac{d}{dx}\left(\frac{1}{\sqrt{x}}\right)$

9. $\frac{d}{dx}\left(\frac{1}{\sqrt[3]{x}}\right)$

10. $\frac{d}{dx}(x^{118})$

11. $\frac{dy}{dx}$ if $y = \frac{1}{x^{50}}$

12. $\frac{dy}{dx}$ if $y = x^{3.2}$

13. $\frac{dy}{dx}$ if $y = \frac{1}{x^{3/2}}$

14. $\frac{dy}{dx}$ if $y = x^3 \cdot x^{-1/4}$

15. $f'(1)$ if $f(x) = x^{10}$

16. $f'(-3)$ if $f(x) = \frac{1}{x^2}$

17. $f'(9)$ if $f(x) = x\sqrt{x}$

18. $f'\left(\frac{1}{4}\right)$ if $f(x) = (\sqrt{x})^3$

In Exercises 19–40, use the differentiation rules of this section to find the following derivatives.

19. $\frac{d}{dx}(6x^5)$

20. $f'(x)$ if $f(x) = -2\sqrt{x}$

21. $\frac{d}{dx}\left(\frac{4}{x}\right)$

22. $f'(x)$ if $f(x) = \frac{x^9}{3}$

23. $\frac{dy}{dx}$ if $y = x^9 + \frac{1}{3}$

24. $\frac{dy}{dx}$ if $y = 4x^3 - \sqrt{x}$

25. $\frac{d}{dx}(4x^3 - \sqrt{2})$

26. $\frac{d}{dx}(1 - x^4)$

27. $\frac{d}{dx}(3 + 2x - x^2)$

28. $f'(-2)$ if $f(x) = x^4 - x^2 + 2x$

29. $f'\left(\frac{1}{2}\right)$ if $f(x) = 1 - x^4$

30. $f'(15)$ if $f(x) = 100 - 2x + 0.1x^2$

31. $\frac{dy}{dx}$ if $y = \sqrt{x} + \frac{3}{\sqrt{x}}$

32. $\frac{dy}{dx}$ if $y = (1 + x)(1 + x^2)$

33. $\frac{dy}{dx}$ if $y = \frac{1 - x^3}{x}$

34. $\frac{dy}{dx}$ if $y = 1 + x^2 - e^x$

35. $\frac{d}{dx}(e^x + \ln x)$

36. $\frac{d}{dx}(1 + 5\ln x)$

37. $\frac{d}{dx}(1 - x^2 + 20e^x)$

38. $\frac{d}{dx}(3e^x - 4\ln x)$

39. $f'(x)$ if $f(x) = 3e^x + 4x^3$

40. $f'(x)$ if $f(x) = 3\ln x - 5e^x + 6 + 7x$

In Exercises 41–46, we use names other than x and y for the variables. The definition of derivative and the differentiation rules are not affected; all that matters is which is the independent variable and which the dependent. For example, if r is a fixed exponent and we call the independent variable t, the power law takes the form $d/dt(t^r) = rt^{r-1}$. With this in mind, find the following derivatives.

41. $\frac{ds}{dt}$ if $s = 16t^2$

42. $\frac{d}{dt}(100\sqrt{t} + 7t^2 - t^3 + 5)$

43. $\frac{dC}{dq}$ if $C(q) = 50{,}000 + 100q - 0.1q^2$

44. $C'(q)$ if $C(q) = 4320 + 0.3q - 0.001\,q^2$

45. $\frac{dP}{dq}$ if $P(q) = -4{,}000{,}000 + 87q - 0.000102q^2$

46. $V'(r)$ if $V(r) = \frac{4}{3}\pi r^3$ (*Note*: $\pi = 3.14159\ldots$ is a constant)

47. The area of a sphere of radius r is $A = 4\pi r^2$. Find dA/dr. How fast is the area increasing with r when the radius is 5?

48. The profit of a manufacturer is $P = -5{,}000{,}000 + 61q - 0.000098q^2$ dollars if q units are produced per month. Find dP/dq. For which production levels (if any) is $dP/dq = 0$?

49. In Example 18 the weekly revenue R from a newsletter depends on the amount x spent on advertising according to the formula $R(x) = 25{,}000 + 150\sqrt{x}$. Find the value of dR/dx when $x = \$4000$ and when $x = \$16{,}000$. For which choice of x is $dR/dx = 1$? (*Note*: $dR/dx = 1$ means, roughly, that each additional dollar in advertising yields only \$1 in revenue; a point of diminishing returns has been reached.)

50. If the number N of bacteria in a culture varies with time t according to the formula

$$N = 2000 + 300t + 18t^2 \quad (t \text{ in minutes})$$

how fast is the population growing when $t = 10$ minutes?

51. The cost of producing x transistorized pocket radios in a day is $C(x) = 5000 + 28x + 0.01x^2$. The "unit cost" is

$$A(x) = \frac{\text{total cost}}{\text{total number of units}} = \frac{C(x)}{x} \quad \text{dollars per unit}$$

Find the derivative of the unit cost function A. For which production levels is $dA/dx = 0$?

52. Suppose the monthly profit, costs, and revenue of a manufacturing concern are given by differentiable functions $P(x)$, $C(x)$, and $R(x)$. What is the relationship between P and R, C? Show that $dP/dx = 0$ at precisely those production levels x where $dR/dx = dC/dx$.

2.4 Derivatives and the Slope of a Curve

If $y = f(x)$ is differentiable, there is an interesting geometric interpretation of the derivative in terms of the graph of f. Through each point on the graph there is a unique straight line that passes through the point and is "tangent" to the curve. The slope of this "tangent line" tells you how fast the curve is rising or falling. The slope turns out to be the derivative of f.

The concepts of slope and tangent line have roots in everyday experience. If you think of skiing down a hillside, you certainly have an intuitive idea of the slope of the hill: large slope where it is steep and small slope where it is relatively flat. Your skis would indicate the direction of the tangent line to the profile of the hill, as in Figure 2.4. From the figure it is clear that the slope and tangent line vary from place to place along the hill.

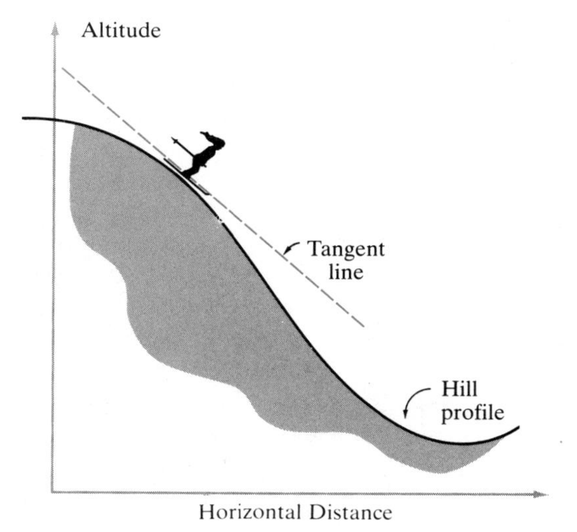

Figure 2.4
A skier at some point on a hill responds to the slope (steepness) of the hillside. Her skis indicate the direction of the tangent line.

Our task is to make a mathematically sound definition of slope and tangent line that captures this intuition. Consider a curve such as the one shown in Figure 2.5a, which is the graph of a function $y = f(x)$. We may fix a point P on the graph and consider nearby points Q; letting Q approach P we ask whether the straight line L_{PQ} joining P to Q approaches a limiting position as Q gets close to P. If it does, we regard the line L in limiting position to be the tangent line to the curve at P; if it does not, we say that there is no tangent line to the curve at P. One technical detail remains. We must make precise the notion "approaches a limiting position as Q gets close to P." Each of the lines L_{PQ} passes through P, and so is completely determined by its slope. We say that L_{PQ} approaches a limit line L if slope(L_{PQ}) approaches slope(L) as Q gets closer and closer to P. Thus the **tangent line** through P is defined to be the straight line L passing through P, whose slope is given by the limiting value

$$\text{slope}(L) = \lim_{Q \to P} \text{slope}(L_{PQ}) \qquad [12]$$

To compute this limit value, suppose the curve is the graph of $y = f(x)$ and that P is the point on the graph corresponding to some base value x_0 on the horizontal axis, as in Figure 2.5b. Then P has coordinates $P = (x_0, f(x_0))$. Nearby points Q on the graph correspond to values $x_0 + \Delta x$ on the horizontal axis, near x_0. They have coordinates $Q = (x_0 + \Delta x, f(x_0 + \Delta x))$. Next we compute slope($L_{PQ}$). Because L_{PQ} passes through P and Q, as in Figure 2.5b, we

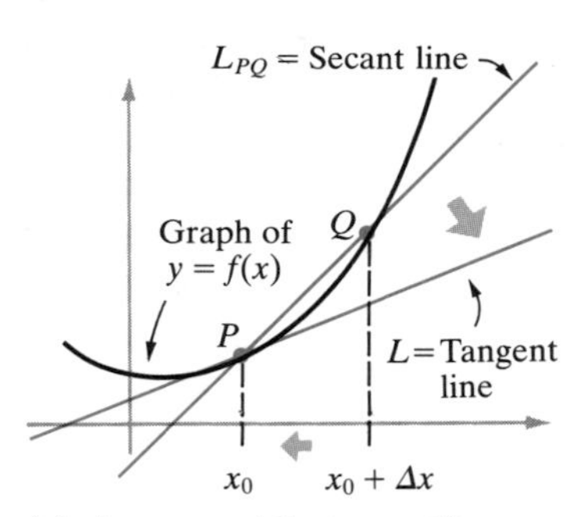

(a) A curve and the tangent line L through a point P on the curve; a typical secant line L_{PQ} is shown.

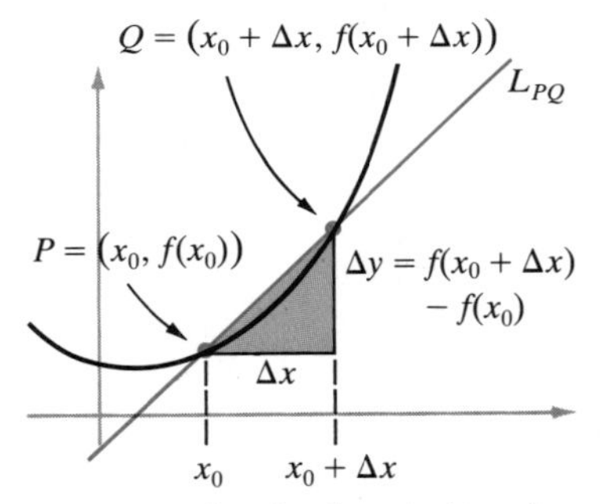

(b) Computing the slope $\Delta y/\Delta x$ of a typical secant line L_{PQ} from the coordinates of P and Q.

Figure 2.5
In Figure 2.5a, as Q moves along the graph toward P, the line L_{PQ} through P and Q approaches a limiting position, indicated by the line L. This line in limiting position is called the *tangent line* to the graph at P. The lines L_{PQ} through P and nearby points Q are called *secant lines* through P. In Figure 2.5b, we show how to calculate the slope of a typical secant line: P corresponds to some base point x_0 on the horizontal axis, and Q corresponds to some nearby point $x_0 + \Delta x$ with $\Delta x \neq 0$. The coordinates of P and Q are then $P = (x_0, f(x_0))$ and $Q = (x_0 + \Delta x, f(x_0 + \Delta x))$, and the slope of L_{PQ} is $\Delta y/\Delta x$, as shown.

find that

$$\text{slope}(L_{PQ}) = \frac{\Delta y}{\Delta x} = \frac{f(x_0 + \Delta x) - f(x_0)}{\Delta x}$$

That is, slope(L_{PQ}) is equal to the average rate of change in y as the independent variable goes from x_0 to $x_0 + \Delta x$. Since the increment Δx approaches zero as Q approaches P on the graph, we see that

$$\text{slope}(L) = \lim_{\Delta x \to 0} \frac{\Delta y}{\Delta x} = f'(x_0) \qquad [13]$$

We conclude that *the slope of the tangent line is just the derivative of f evaluated at the base value x_0*. This may be summarized as follows.

DERIVATIVE AS SLOPE OF THE TANGENT LINE Let $y = f(x)$ be a function, and let P be a point on its graph corresponding to $x = x_0$ on the horizontal axis. There is a well defined tangent line to the graph at P if $f(x)$ is differentiable at x_0. Furthermore

(i) The slope of the tangent line through P is equal to the derivative $f'(x_0)$.
(ii) The equation of the tangent line is determined by this slope and the fact that the tangent line passes through P.

The equation of the tangent line through P is

$$y = f(x_0) + f'(x_0) \cdot (x - x_0) \qquad [14]$$

Example 20 Find the slope of the tangent line to the curve $y = x^3$ at the point $P = (2, 8)$ on the graph.

Solution Here $f(x) = x^3$ and the base value on the horizontal axis is $x_0 = 2$. Using Formula [8] we obtain $f'(x) = 3x^2$ for any x. At the base value $x_0 = 2$, we have $f'(x_0) = f'(2) = 3(2)^2 = 12$. The slope of the tangent line is 12. ■

Example 21 If $y = x + x^2$, find the slope of the tangent line through $P = (1, 2)$ on the graph. Find the equation of the tangent line.

Solution The point P lies above base point $x_0 = 1$ on the x axis. As we have seen, the slope of the tangent line through P is equal to $f'(1)$. Standard differentiation rules yield $f'(x)$ for any x

$$f'(x) = \frac{d}{dx}(x + x^2) = \frac{d}{dx}(x) + \frac{d}{dx}(x^2) = 1 + 2x$$

If $x = 1$, we obtain $f'(1) = 1 + 2(1) = 3$. This is the slope of the tangent line L

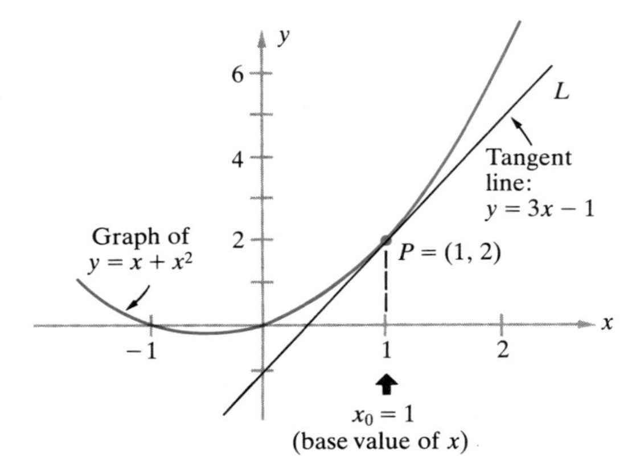

Figure 2.6
The tangent line through $P = (1, 2)$ on the graph of $y = x + x^2$ has slope $f'(1) = 3$. The equation of this line is $y = 3x - 1$.

passing through $P = (1, 2)$. Figure 2.6 shows this tangent line. Its equation, determined from Formula [14], is

$$\begin{aligned} y &= f(x_0) + f'(x_0) \cdot (x - x_0) \\ &= 2 + 3(x - 1) \end{aligned}$$

or $y = 3x - 1$. ■

Exercises 2.4

1. If $y = x^2$, find the slope of the tangent line through $P = (1, 1)$ on the graph. Find the equation of the tangent line through P. On the same piece of graph paper draw the graph of $y = x^2$ and the graph of the tangent line through P.

In Exercises 2–8 calculate the slope of the tangent line to the graph of $y = f(x)$ at the given point P. Use the differentiation rules of Section 2.3 to calculate derivatives.

2. $f(x) = 3x - 2$, $P = (2, 4)$
3. $f(x) = 1 - x^2$, $P = (-2, -3)$
4. $f(x) = 1 - x^2$, $P = (0, 1)$
5. $f(x) = -2$ (constant function), $P = (3, -2)$
6. $f(x) = x^4 - x + 1$, $P = (1, 1)$
7. $f(x) = \frac{1}{4}x^4$, $P = (-2, 4)$
8. $f(x) = 3 - x + 2x^2 - x^3$, $P = (2, 1)$

In Exercises 9–15 determine the equation of the tangent line through the given point P on the graph.

9. $f(x) = x^3$, $P = (-1, -1)$
10. $f(x) = x^3$, $P = (0, 0)$
11. $f(x) = x^3$, $P = (2, 8)$
12. $f(x) = x^2 - 2x + 3$, $P = (1, 2)$
13. $f(x) = 2x^4 + 3x^3 + x$, $P = (-2, 6)$
14. $f(x) = \frac{1}{4}x^4 - x^2$, $P = (2, 0)$

15. $f(x) = -5000 + 61x - 0.01x^2$, $P = (100, 1000)$

16. Calculate the derivative of $y = x^2 - 12x$. Then find all points P on the graph for which the tangent line is *horizontal* (slope is zero).

17. Find all points on the graph of $y = 3x - x^3$ for which the tangent line is horizontal (slope is zero).

18. Find all points on the graph of $y = x^3 - 8x + 5$ where the tangent line has slope 4. (Give both coordinates of each point.)

19. Find all points on the graph of $y = x^3 + 3x^2 - 12x + 7$ where the tangent line has slope -3. (Give both coordinates of each point.)

20. Calculate the equation of the tangent line to the curve $y = 1/x$ at the point $P = (1, 1)$ on the graph.

21. If $y = x + x^2$, find the coordinates of the points P and Q on the graph above $x = 1$ and $x = 1 + \Delta x$ in terms of Δx. Show that the secant line L_{PQ} has slope $(L_{PQ}) = 3 + \Delta x$. What does this tell you about the slope of the tangent line through P?

2.5 Some Other Interpretations of the Derivative

The notion of derivative applies to any function, whether it describes the position of a moving object, evolution of IQ scores with age, or the profit from a manufacturing operation expressed as a function of production level. This general concept allows us to unify the analysis of many seemingly unrelated phenomena. Whenever a function $y = f(x)$ describes some real-life situation, the derivative tells us how fast things are changing as the independent variable increases.

Example 22 During the first month of growth, the weight w of a certain type of laboratory mouse t days after birth is

$$w = w(t) = 10 + \frac{7}{6}t - \frac{1}{60}t^2 \quad \text{grams}$$

How fast is the weight increasing 10 days after birth? Thirty days after birth?

Solution At any time t the rate of weight increase is the derivative dw/dt, which is easily computed.

$$\begin{aligned}\frac{dw}{dt} = \frac{d}{dt}\left(10 + \frac{7}{6}t - \frac{1}{60}t^2\right) &= 10\frac{d}{dt}(1) + \frac{7}{6}\frac{d}{dt}(t) - \frac{1}{60}\frac{d}{dt}(t^2) \\ &= 10(0) + \frac{7}{6}(1) - \frac{1}{60}(2t) \\ &= \frac{7}{6} - \frac{1}{30}t \quad \text{grams per day}\end{aligned}$$

Ten days after birth the weight is $w(10) = 20$ grams and is increasing at a rate of

$$w'(10) = \frac{7}{6} - \frac{1}{30}(10) = \frac{5}{6} \quad \text{grams per day}$$

After 30 days the weight is $w(30) = 30$ grams and is increasing at a rate of

$$w'(30) = \frac{7}{6} - \frac{1}{30}(30) = \frac{1}{6} \quad \text{grams per day}$$

The weight is increasing, but not as rapidly as when the mouse was born. ■

Example 23 In testing for diabetes one has to determine how fast the body metabolizes glucose. One test involves ingesting a standard amount of glucose, waiting 1 hour, and then monitoring the concentration remaining in the blood as time passes. Suppose the amount remaining in 1 cubic centimeter of blood t hours after the glucose is administered is

$$q = 4.5t^{-1/2} \quad \text{milligrams}$$

How fast is glucose being metabolized when monitoring begins at $t = 1$ and then four hours after the glucose was taken?

Solution The rate of change in glucose level is just the derivative

$$\frac{dq}{dt} = \frac{d}{dt}(4.5t^{-1/2}) = 4.5\frac{d}{dt}(t^{-1/2})$$

$$= 4.5\left(-\frac{1}{2}\right)t^{-(1/2)-1} = -2.25t^{-3/2} \quad \text{milligrams per hour}$$

When $t = 1$ the glucose level is $q(1) = 4.5$ milligrams and the rate of change is

$$q'(1) = -2.25(1)^{-3/2} = -2.25 \quad \text{milligrams per hour}$$

The negative sign means that the glucose level is *decreasing* by 2.25 milligrams per hour; a positive sign would mean the level is increasing. Four hours after the dose was taken, $t = 4$ and the glucose level $q(4) = 2.25$ milligrams is changing at a rate of

$$q'(4) = -2.25(4)^{-3/2} = \frac{-2.25}{4^{3/2}} = \frac{-2.25}{8} = -0.2813 \quad \text{milligrams per hour}$$

There is less glucose in the blood now, and the level drops more slowly. ■

The next examples show that the instantaneous velocity of a moving object can be interpreted as a derivative.

INSTANTANEOUS VELOCITY OF A MOVING OBJECT If we drive an automobile the total distance s, measured from some fixed reference point, this total distance is some function $s = s(t)$ of the elapsed time t since the start of the trip. At any moment, our speedometer registers the instantaneous velocity v of our car. In

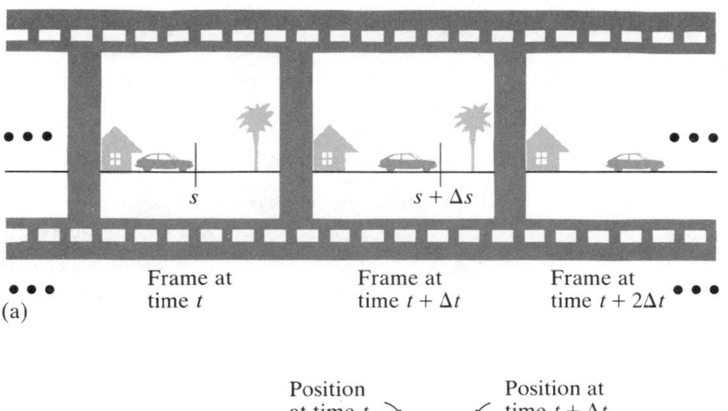

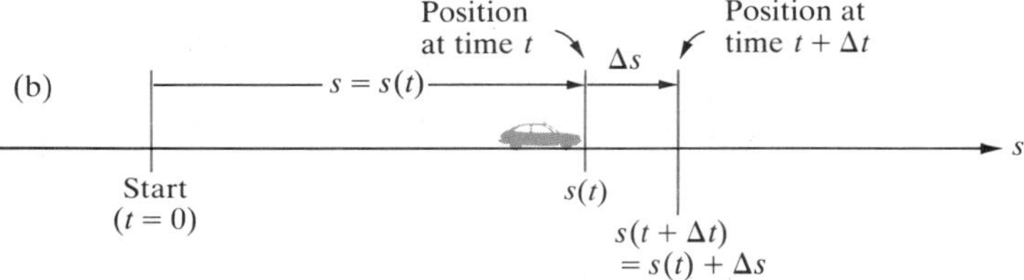

Figure 2.7
In Figure 2.7a are two successive frames of a film, the first taken at time t and the next at time $t + \Delta t$ (Δt being the time between frames). The distance the car has travelled changes from s to $s + \Delta s$ during this interval, so the average velocity over the time interval between frames is $\Delta s/\Delta t$. Figure 2.7b is a diagram abstracting the essential features of Figure 2.7a. Note that $s(t)$ is the total distance covered since the start at $t = 0$.

what follows we show that this instantaneous velocity should be interpreted as the derivative ds/dt,

$$\text{instantaneous velocity } v = s'(t) \tag{15}$$

To see why, we take our cue from motion pictures.

Suppose we film the vehicle's motion. Now compare successive frames, the first taken at time t and the next at time $t + \Delta t$ (Δt being the time interval between frames) as in Figure 2.7. The average velocity over the time interval between frames is

$$\text{average velocity} = \frac{\Delta s}{\Delta t} = \frac{\text{distance travelled between frames}}{\text{time elapsed between frames}}$$

Just as we accept motion pictures as an excellent approximation of continuous motion, similarly we are inclined to accept the average velocity between frames as an excellent approximation of the instantaneous velocity at time t. If we were to film more frames per second, making Δt smaller, we would obtain a still better approximation of the car's actual motion. (Compare the realism of modern films and those produced 70 years ago.) The corresponding average speeds, computed for smaller and smaller increments Δt starting from time t, approach a limit value. It is natural to interpret this limit value

$$\lim_{\Delta t \to 0} \frac{\Delta s}{\Delta t} \tag{16}$$

as the instantaneous velocity at time t. But this limit value is just the derivative $s'(t)$—that's how the derivative was defined! This justifies Formula [15]. The same considerations apply to *any* moving object; therefore we have the following general principle.

> If an object travels a total distance $s = s(t)$ in time t, its instantaneous velocity at any time is given by the derivative:
>
> $$\text{instantaneous velocity} = \frac{ds}{dt}$$

Example 24 A drag racer covers a distance $s(t) = 9t^2 - \frac{1}{5}t^3$ feet in t seconds during the time interval $0 \le t \le 10$ seconds. Find the instantaneous velocity 3 seconds after the start. When does the vehicle achieve a speed of 88 feet per second (60 mph)?

Solution The instantaneous velocity at any time t is given by the derivative

$$\frac{ds}{dt} = \frac{d}{dt}\left(9t^2 - \frac{1}{5}t^3\right) = 18t - \frac{3}{5}t^2 \quad \text{feet per second}$$

When $t = 3$ the velocity is

$$s'(3) = 18(3) - \frac{3}{5}(3)^2 = 48.6 \quad \text{feet per second}$$

The velocity will be 88 feet per second when

$$\frac{ds}{dt} = 18t - \frac{3}{5}t^2 \quad \text{is equal to} \quad 88$$

To find the time t when this happens, we solve the resulting equation

$$-\frac{3}{5}t^2 + 18t - 88 = 0$$

using the quadratic formula. The solutions are $t = 6.15$ and $t = 23.85$. We reject $t = 23.85$ because the formulas for $s(t)$ and $s'(t)$ are only meaningful for $0 \le t \le 10$ in our problem. Hence the velocity is 88 feet per second after 6.15 seconds. ■

Because the derivative of a function can always be interpreted graphically as the slope of the tangent line at the corresponding point on the graph, it is interesting to see what this means in a motion problem similar to the one in Example 24. The graph of $s(t)$ is shown in Figure 2.8. The tangent line corresponding to $t = 3$ has slope $s'(3) = 48.6$, which is just the velocity at this

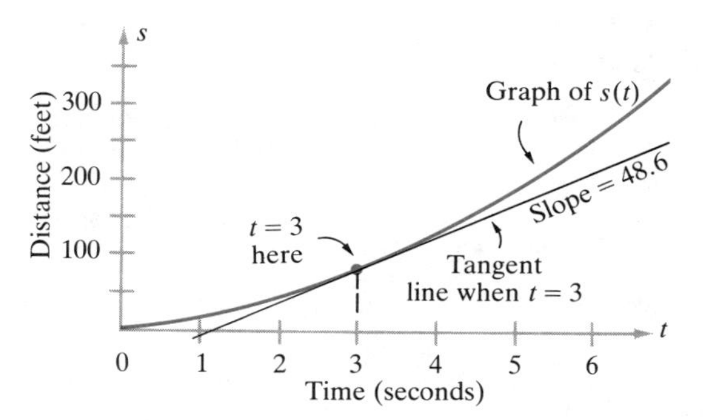

Figure 2.8
The graph of the distance function $s(t) = 9t^2 - \frac{1}{5}t^3$ in Example 24. The tangent line when $t = 3$ is shown. At any point on the graph the slope of the tangent line is equal to the velocity of the moving object.

time. When $t = 6.15$, the slope is $s'(6.15) = 88$. At any point on the graph, the slope is equal to the velocity, because $v(t) = s'(t) =$ (slope of the tangent line). The steeper the graph, the faster the car is moving.

Example 25 The motion of any freely falling object released near the Earth's surface is governed by the following law:

$$s = 16t^2 \quad \text{feet} \qquad [17]$$

where s is the distance fallen t seconds after the object has been released. Note especially that s depends only on the elapsed time and not at all on the weight of the object, a fact demonstrated by Galileo. How far will the object fall in 1 second, and how fast will it be moving? In 2 seconds?

Solution The question "how fast?" refers to the instantaneous velocity at that moment, which is

$$v = \frac{ds}{dt} = \frac{d}{dt}(16t^2) = 16\frac{d}{dt}(t^2) = 16(2t) = 32t \quad \text{feet per second}$$

at any time $t \geq 0$. After 1 second the distance covered is $s(1) = 16(1)^2 = 16$ feet and the velocity is

$$v = s'(1) = 32(1) = 32 \quad \text{feet per second}$$

After 2 seconds, the distance and velocity are

$$s(2) = 16(2)^2 = 16(4) = 64 \quad \text{feet}$$
$$s'(2) = 32(2) = 64 \quad \text{feet per second}$$

■

In certain situations the distance s can be decreasing—an automobile, for example, might be backing up—or if a ball is thrown straight up into the air and $s = s(t)$ is the distance above the ground, the ball will eventually begin to fall, and s will decrease. When this happens the velocity $v = ds/dt$ will be negative.

Example 26 If a baseball is thrown straight up with an initial velocity of 60 feet per second, its distance s above the ground after t seconds will be

$s = 60t - 16t^2$ feet. What are its height and instantaneous velocity 1 and 2 seconds after it is thrown?

Solution At any time (until the ball hits the ground) the instantaneous velocity is

$$v = \frac{ds}{dt} = \frac{d}{dt}(60t - 16t^2) = 60\frac{d}{dt}(t) - 16\frac{d}{dt}(t^2)$$
$$= 60 - 32t \quad \text{feet per second}$$

When $t = 1$ the height and velocity are

$$s(1) = 60(1) - 16(1)^2 = 44 \quad \text{feet}$$
$$v = s'(1) = 60 - 32(1) = 28 \quad \text{feet per second}$$

Because v is positive, s is increasing, and the ball is still rising. When $t = 2$

$$s = s(2) = 60(2) - 16(2)^2 = 56 \quad \text{feet}$$
$$v = s'(2) = 60 - 32(2) = -4 \quad \text{feet per second}$$

and the ball is just beginning to fall. ■

THE "MARGINAL CONCEPT" IN ECONOMICS Suppose a manufacturer has determined that his profit for producing x units is given by some function $P = P(x)$. Economists give the derivative $dP/dx = P'(x)$ a special name, referring to it as the **marginal profit** at production level x. A mathematician would call it the derivative of the profit function. Similarly, if $C(x)$ is the cost function, its derivative at a certain production level is referred to as the **marginal cost** at that production level. In other words, by force of tradition economists tend to use the phrase "marginal ... " wherever a mathematician or physicist would use the phrase "derivative of ... ." Marginal cost, marginal profit, marginal revenue, and so forth are concepts crucial to all modern discussions of economics, hence the importance of understanding the notion of derivative on which they are based. We will see in the next chapter that they also provide the key to finding optimum levels of production.

A phrase such as "marginal profit" arises in the following way. It refers to the profitability of producing additional units "on the margin" of the existing situation: in other words, of producing additional units above and beyond the existing production level. If we consider increments Δx starting from the present production level x, the derivative $P'(x)$ is the limit of averages $\Delta P/\Delta x$ as we consider smaller and smaller increments,

$$P'(x) = \lim_{\Delta x \to 0} \frac{\Delta P}{\Delta x} = \lim_{\Delta x \to 0} \left(\frac{P(x + \Delta x) - P(x)}{\Delta x} \right)$$

Thus for small increments Δx away from the base production level x we have

$$\frac{\Delta P}{\Delta x} \approx P'(x) \quad \text{or} \quad \Delta P \approx P'(x) \cdot \Delta x \qquad [18]$$

where "$\approx$" stands for "approximately equal to". An economist would put it this way:

$$\Delta P \approx (\text{marginal profit at level } x) \cdot \Delta x \qquad [19]$$

Evidently, the marginal profit $P'(x)$ does tell us (approximately) the profitability of producing additional units as long as the increment Δx is not too large. Similar remarks apply to marginal cost, marginal revenue, and so on.

Graphically, marginal profit is equal to the slope of the profit curve $P = P(x)$ for each production level x. In Figure 2.9 we show the graph of a typical profit function, indicating the tangent lines for two different production levels. Their slopes are equal to the marginal profit $P'(x)$ at these production levels. A similar graphical interpretation occurs for marginal cost, revenue, and so on.

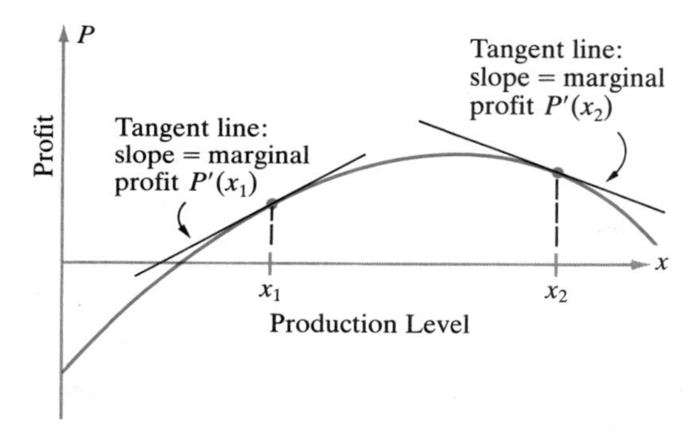

Figure 2.9 Solid curve is the graph of a typical profit function. The tangent lines to the graph corresponding to production levels x_1 and x_2 are shown. In each case, (slope of tangent line) = (marginal profit) at this production level.

■

Example 27 Suppose a manufacturer's profit depends on production level according to the formula $P = 50x - 0.04x^2 - 10{,}000$. Find the marginal profit at production level $x = 600$ and the approximate change in profit if one more unit is produced. Use this information to decide if it pays to increase the production level.

Solution When $x = 600$, the marginal profit

$$\frac{dP}{dx} = \frac{d}{dx}(50x - 0.04x^2 - 10{,}000) = 50 - 0.08x$$

has the value

$$(\text{marginal profit at level } x = 600) = P'(600) = 2.00 \quad \text{dollars per unit}$$

For small increases Δx in production, the change in profit is

$$\Delta P \approx P'(600) \cdot \Delta x = 2 \cdot (\Delta x) \qquad [20]$$

or approximately \$2 for each additional unit. In particular, the profit for producing *just one more unit* is $\Delta P \approx 2(\Delta x) = 2(1) = 2$ dollars.[†] Formula [20] shows clearly that profit increases ($\Delta P > 0$) if production is increased slightly ($\Delta x > 0$), so it does pay to increase production. ■

Example 28 A pharmaceutical manufacturer planning a warehouse to store sensitive raw materials has been informed that the annual cost C for maintaining his stock (cost of maintaining the warehouse plus recurring delivery costs) will depend on warehouse capacity x according to the formula

$$C = 5 + 4x + \frac{1600}{x} \quad \text{thousand dollars}$$

where x is in thousands of cubic feet. He was planning to let $x = 10$ thousand cubic feet. Compute the marginal cost dC/dx when $x = 10$. Use Formula [19] to see what would happen if the capacity were increased slightly by $\Delta x = 0.5$.

Solution For any capacity x the marginal cost is just the derivative

$$\frac{dC}{dx} = \frac{d}{dx}\left(5 + 4x + \frac{1600}{x}\right) = 5\frac{d}{dx}(1) + 4\frac{d}{dx}(x) + 1600\frac{d}{dx}(x^{-1})$$

$$= 4 - \frac{1600}{x^2}$$

At base level $x = 10$ the marginal cost is

$$C'(10) = 4 - \frac{1600}{(10)^2} = -12$$

thousand dollars per thousand cubic feet. For a small increment in capacity such as $\Delta x = 0.5$, the change in cost is given (approximately) by Formula [19]:

$$\Delta C \approx C'(10) \cdot \Delta x = -12(0.5) = -6 \quad \text{thousand dollars}$$

Annual inventory costs *drop* by this much if capacity is increased by $\Delta x = 0.5$ thousand cubic feet. ■

Exercises 2.5

1. The weight of a tumor in a laboratory animal is given by the formula

$$w(t) = 0.02 + 0.35t + 0.002t^2 \quad (t \text{ in days})$$

How fast is w increasing when $t = 3$? When $t = 10$?

2. The population of a bacteria culture varies with time according to the formula

$$N(t) = 2000 + 300t + 18t^2 \quad (t \text{ in minutes})$$

How fast is the population growing after 10 minutes? After 2 hours?

[†] Formula [20] is only approximate but very easy to work with. From it we find that $\Delta P \approx 2.00$ dollars if $\Delta x = 1$. The *exact* change ΔP is $\Delta P = P(601) - P(600) = 5601.96 - 5600.00 = \$1.96$, which is almost the same as the approximate value from [20] but messier to compute. More discussion of approximations such as [20] will follow in Section 2.6.

3. The braking distance B (in feet) of an automobile depends on its velocity v (in mph) according to the formula $B = v^2/10$.
(i) Find the braking distance when $v = 60$ mph.
(ii) How fast is B increasing with respect to v when $v = 60$?

4. After t years of protection, the deer population of Warren County was given by $P(t) = 3500 + 475t - 10t^2$.
(i) Find $P(10)$, the deer population after 10 years of protection.
(ii) Find $P'(10)$, the rate at which the population is increasing when $t = 10$.
(iii) When is $P'(t) = 175$ deer per year?

5. A storage vessel is gradually filled with water. Suppose the amount of water in the vessel after t minutes is $150 + 25t - 3\sqrt{t}$ cubic feet. At what rate is water entering the vessel 10 minutes after filling began? How much water is in the vessel at this time?

6. In another glucose tolerance test (see Example 23) the amount of glucose t hours after the test dose was administered was found to be

$$q(t) = 2.7 + \frac{5.3}{\sqrt{t}} \quad \text{milligrams}$$

After 4 hours, what would be the glucose level and how fast would it be changing?

7. A market research firm estimates the annual sales revenue R of a cereal manufacturer depends on the amount x spent in advertising according to the formula

$$R(x) = 17{,}000{,}000 + 4000\sqrt{x} \quad \text{dollars}$$

Find
(i) The revenue R if \$1,000,000 is spent on advertising.
(ii) The instantaneous rate of change of R with respect to x when $x = 1{,}000{,}000$.
(iii) The instantaneous rate of change of R with respect to x when $x = 9{,}000{,}000$.

8. A city's board of health observed that the influenza season lasted a month, and that the number N of sick people t days after the start of the season was given by $N = 30t^3 - t^4$.
(i) How many people were ill with influenza when $t = 10$?
(ii) How fast was the number of sick people increasing when $t = 10$?
(iii) How fast was the number of sick people changing when $t = 25$?

In Exercises 9–12 use the formula $s = 16t^2$ for a freely falling object released near the Earth's surface. Here, $s =$ distance fallen (in feet) after t seconds.

9. Calculate the instantaneous velocity of a falling object after the following elapsed times:
(i) 0.1 second
(ii) 1 second
(iii) 2 seconds
(iv) 10 seconds

10. A stone is dropped down a well 36 feet deep. Use the formula $s = 16t^2$ to determine how many seconds the stone will fall before striking bottom. What is its instantaneous velocity on impact?

11. A 150-pound object falls from the top of a 10-story building (one story = 11 feet). Use the formula $s = 16t^2$ to find out how long it will fall before impact, then calculate the velocity at impact. What would happen on a 20-story building? What if the weight of the object were doubled? Convert your answers to *miles per hour* to get a better intuitive feel for the speeds involved (one mile = 5280 feet; one hour = 3600 seconds).

12. A stone is thrown straight up from ground level. Its height h (in feet) after t seconds is given by $h(t) = 64t - 16t^2$.
   (i) Find the velocity dh/dt when $t = 1, t = 2$, and $t = 3$. Is the stone rising or falling at these times?
   (ii) After how many seconds does the stone strike the ground ($h = 0$)?
   (iii) What is its velocity when it strikes the ground?

On the surface of the Moon the motion of a freely falling body is governed by the equation

$$s = 2.66t^2 \quad (s \text{ in feet; } t \text{ in seconds}) \qquad [21]$$

which differs from the formula on Earth because the surface gravity on the Moon is weaker. Use this formula to answer Exercises 13–15.

13. What is the instantaneous velocity $s'(t) = ds/dt$ after t seconds?

14. Find a formula for the time-to-impact during a fall of s feet. (*Hint*: Solve for t in terms of s in [21].)

15. Determine the impact velocity for an object dropped from the top of a 5-story building (55 feet) on the Moon. Then convert your answer from feet per second to mph. Do you think you would survive a leap from a height of 5 stories on the Moon? (*Hint*: Use the result of Exercise 14.)

16. For the cost function $C(x) = 7500 + 32x - 0.007x^2$, calculate the marginal cost dC/dx at production level $x = 1000$. This tells us (approximately) the cost of producing one additional unit starting from the given production level $x = 1000$. Find the *exact* cost of producing the next unit, $\Delta C = C(1001) - C(1000)$ and compare with the marginal cost you have found.

17. Suppose a manufacturer's profit is given by

$$P(x) = -50{,}000 + 61x - 0.0098x^2$$

where x is the production level.
   (i) Find the marginal profit at production level $x = 500$.
   (ii) Find the exact additional profit ΔP gained by increasing production from 500 to 501 units.

18. For the profit function $P(x) = 5x - 0.004x^2 - 1000$, calculate the marginal profit dP/dx. Find the marginal profit at production level $x = 450$, and decide whether it pays to increase production slightly. Do the same for $x = 750$.

19. Suppose the demand for a certain product is described by

$$p(x) = \frac{1000}{\sqrt{x}}$$

where x is the production level and p the price at which one will sell this many items.

Find the marginal demand dp/dx and the marginal revenue dR/dx at production level $x = 100$. (*Hint*: Recall that $R = x \cdot p(x)$.)

20. The cost function of a manufacturer is $C = 100 + 100x - x^2$ and the demand function is $p = 200 - 2x$, where x is the production level; $p(x)$ is the price at which the manufacturer will sell x items.
(i) Find the marginal cost dC/dx.
(ii) Find the profit function $P(x) = x \cdot p(x) - C(x)$.
(iii) Find the marginal profit function dP/dx.

21. Suppose manufacturer's cost function is $C(x) = 1235 + 120x - x^2$ and the revenue function is $R(x) = 200x - 2x^2$. For what value of x will the marginal cost equal the marginal revenue? For this production level, what is the value of the marginal profit?

22. Referring to Example 28 and the cost function

$$C(x) = 5 + 4x + \frac{1600}{x}$$

compute the marginal cost dC/dx when $x = 30$. Use Formula [19] to estimate the change ΔC in cost if x is increased from 30.0 to 30.2 ($\Delta x = 0.2$).

2.6 The Approximation Principle

Suppose we know the value of a function f at some base point x_0 and want to compare it with values of $f(x)$ at nearby points. If the formula for f is at all complicated, this can be tedious. For example, if a manufacturer whose cost function is

$$C = f(x) = 8800 + 89x - 0.078x^2$$

knows his costs at the current production level $x = 200$ are $f(200) = \$23{,}480$, he might want to know what happens if he increases production somewhat, say to $x = 215$. Computing the precise change in costs $\Delta f = f(215) - f(200)$ is time consuming. But these calculations can be simplified if we are willing to settle for an approximate value of Δf.

If the function f is differentiable at the base point x_0, there is a simple approximation formula for the value of f at nearby points. In fact, if $x = x_0 + \Delta x$ is a nearby point determined by a small nonzero increment Δx, the averages

$$\frac{\Delta f}{\Delta x} = \frac{f(x) - f(x_0)}{x - x_0} = \frac{f(x_0 + \Delta x) - f(x_0)}{\Delta x}$$

approach the limit value $f'(x_0)$ as x gets close to x_0 and the increment Δx approaches zero; this follows from the very definition of derivative. Thus we have

$$\frac{\Delta f}{\Delta x} \approx f'(x_0) \quad \text{for all } x \text{ near } x_0 \qquad [22]$$

where "$\approx$" stands for "approximately equal to." We may rewrite this as

$$\Delta f \approx f'(x_0) \cdot \Delta x \quad \text{for all small increments} \quad \Delta x \neq 0 \qquad [23]$$

For most practical purposes we may work with the approximate values $\Delta f \approx f'(x_0) \cdot \Delta x$ instead of the exact values, as long as the nearby point is fairly close to the base point x_0. We summarize these observations as follows:

THE APPROXIMATION PRINCIPLE If $y = f(x)$ is differentiable at a base point x_0, let us consider $\Delta f = f(x) - f(x_0)$ for points $x = x_0 + \Delta x$ near x_0. This change in f is approximately

$$\Delta f \approx f'(x_0) \cdot \Delta x \qquad [24]$$

for all small increments Δx away from the base point x_0.

Obviously, if we know the value $f(x_0)$ at the base point x_0, as well as the change Δf, we can find the value $f(x_0 + \Delta x)$ at the nearby point $x_0 + \Delta x$. In fact, adding $f(x_0)$ to both sides of [24], we get a similar approximate formula for the value at $x_0 + \Delta x$:

$$\begin{aligned} f(x_0 + \Delta x) &\approx f(x_0) + f'(x_0) \cdot \Delta x \\ &= f(x_0) + f'(x_0) \cdot (x - x_0) \end{aligned} \qquad [25]$$

for all small increments Δx.

In our discussion of the marginal concept in Section 2.5 we implicitly made use of this approximation principle. There we considered a profit function $P = P(x)$ and the marginal profit at some base point $x = x_0$, and observed that

$$\Delta P \approx P'(x_0) \cdot \Delta x$$

for all small increments Δx away from the base value x_0. Here is another example.

Example 29 Suppose the weekly profit for a manufacturing operation is

$$P = -50{,}000 + 61x - 0.0098x^2 \quad \text{dollars}$$

if x units are produced. If the current production level is $x = 1000$, use the approximation principle to estimate the change in profit ΔP if production is raised by $\Delta x = 1$ unit. Compare this with the exact increment in the profit $\Delta P = P(1001) - P(1000)$.

Solution The profit function is a polynomial in x; its derivative is

$$\frac{dP}{dx} = 61 - 0.0196x$$

The base production level is $x_0 = 1000$. The nearby level is $1001 = 1000 + \Delta x$, so the increment is $\Delta x = 1$. The value of $P'(x)$ at the base level is $P'(1000) = 61 - 0.0196(1000) = 41.40$. The approximation Formula [24] tells us that

$$\Delta P \approx P'(1000) \cdot \Delta x = 41.40(\Delta x) \qquad \text{for all small increments from} \quad x_0 = 1000$$

In our case, $\Delta x = 1$ and

$$\Delta P \approx 41.40(\Delta x) = 41.40(1) = \$41.40$$

We leave the reader to calculate the exact values $P(1000) = \$1200.00$ and $P(1001) = \$1241.39$, and the exact change $\Delta P = \$41.39$. Our approximate value $\Delta P \approx 41.40$ agrees quite well, but required less computation. ■

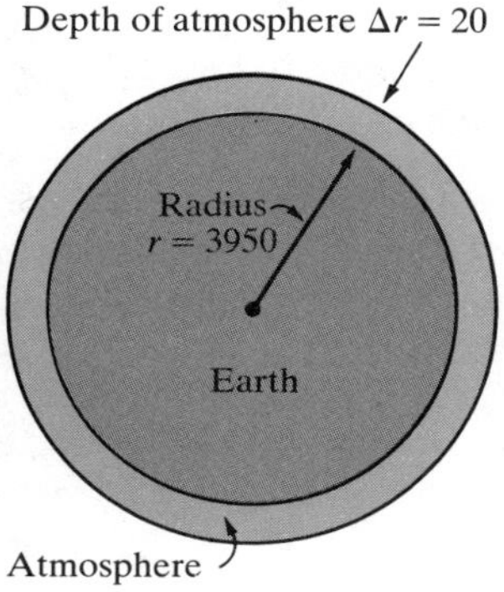

Depth of atmosphere $\Delta r = 20$

Radius $r = 3950$

Earth

Atmosphere

Figure 2.10 The Earth and its atmosphere in Example 30. The volume of a sphere of radius r is $V = \frac{4}{3}\pi r^3 = 4.189r^3$. Taking the Earth's radius $r_0 = 3950$ miles as a base value and $\Delta r = 20$, we want to estimate the volume of the atmosphere $\Delta V = V(3950 + \Delta r) - V(3950)$.

Example 30 The volume of a sphere of radius r is given by the function $V(r) = \frac{4}{3}\pi r^3$. If the radius of the Earth is 3950 miles and the effective depth of the atmosphere is 20 miles, estimate the volume of the atmosphere in cubic miles (see Figure 2.10).

Solution The atmospheric volume is the difference in volume between spheres of radius $3970 = 3950 + 20$ (Earth + atmosphere) and 3950 (Earth alone). To apply the approximation principle, take base value $r_0 = 3950$ and increment $\Delta r = 20$. The volume of the atmosphere is

$$\Delta V = V(r_0 + \Delta r) - V(r_0) = V(3970) - V(3950)$$

For any r the derivative of $V(r)$ is

$$V'(r) = \frac{4}{3}\pi \frac{d}{dr}(r^3) = \frac{4}{3}\pi(3r^2) = 4\pi r^2$$

At the base point being considered, the value of the derivative is $V'(r_0) = V'(3950) = 4\pi(3950)^2 = 1.96 \times 10^8$. By the approximation principle

$$\Delta V \approx V'(3950) \cdot \Delta r = (1.96 \times 10^8) \cdot (20) = 3.92 \times 10^9 \quad \text{cubic miles}$$

We leave calculation of the exact value

$$\Delta V = \frac{4}{3}\pi(3970)^3 - \frac{4}{3}\pi(3950)^3 = 3.941 \times 10^9 \quad \text{cubic miles}$$

to the reader. ■

Example 31 (Error analysis) A surveyor wants to find the area of a square plot of land. The measured side length s is 100 feet, so the area should be $A = s^2 = 10{,}000$ square feet. But there is a possible error of $\Delta s = \pm 0.2$ feet in his measurement. Use the formula $A = s^2$ for the area of a square to find the uncertainty ΔA in the area caused by the uncertainty Δs in the measurement of s.

Solution Consider the base value $s_0 = 100$. From the area formula $A = s^2$ we get the derivative $dA/ds = 2s$; at our base point the value of the derivative is

$$A'(100) = 2(100) = 200$$

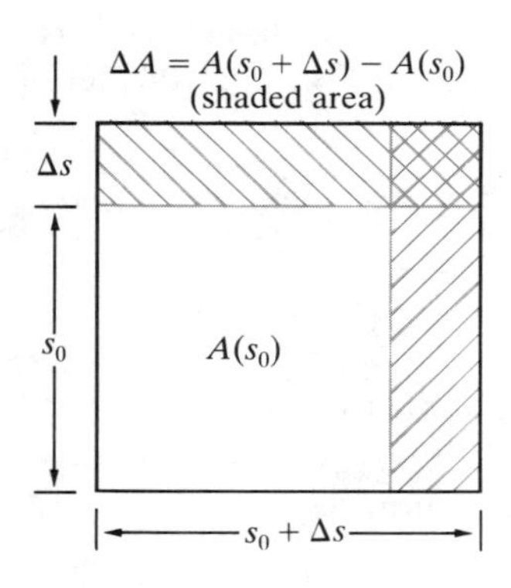

Figure 2.11
Squares with side lengths s_0 and $s_0 + \Delta s$, as in Example 31. The increment in area ΔA is the entire shaded region. According to the approximation principle, $\Delta A \approx 2s_0 \cdot (\Delta s)$.

By the approximation principle, the area changes by

$$\Delta A \approx A'(100) \cdot \Delta s = 200 \cdot \Delta s$$

as s changes from $s_0 = 100$ to a nearby value $100 + \Delta s$. Taking for Δs the maximum expected error in our measurement of s, $\Delta s = \pm 0.2$, we obtain the corresponding estimated error in A

$$\Delta A \approx 200(\pm 0.2) = \pm 40 \quad \text{square feet}$$

The situation is shown in Figure 2.11. ■

In the next example we estimate the *value* of f at a point near x_0 rather than the *change* Δf.

Example 32 The value of $f(x) = \sqrt{x}$ at the base point $x_0 = 100$ is $\sqrt{100} = 10$. Use the approximation principle to estimate the value of $\sqrt{103}$.

Solution The derivative of $f(x) = x^{1/2}$ is $f'(x) = 1/(2\sqrt{x})$. At $x_0 = 100$ its value is

$$f'(100) = \frac{1}{2\sqrt{100}} = \frac{1}{20}$$

For any nearby point $100 + \Delta x$, the change in f is

$$\Delta f \approx f'(100) \cdot \Delta x = \frac{\Delta x}{20}$$

and the value of f is

$$f(100 + \Delta x) = f(100) + \Delta f \approx 10 + \frac{\Delta x}{20}$$

Taking $\Delta x = 3$, we obtain the desired estimate

$$\sqrt{103} = f(103) \approx 10 + \frac{3}{20} = 10.1500$$

The exact value is $\sqrt{103} = 10.14889$. ■

Of course we could find $\sqrt{103}$ using a calculator. The point of this example is to illustrate the approximation principle.

Our last example shows how the approximation principle is sometimes used to estimate what is happening in the face of incomplete information about a function.

Example 33 The amount of penicillin in a patient's bloodstream 5 hours after an injection is $A = 298$ milligrams. Body metabolism is removing it at a rate of

$$\frac{dA}{dt} = -76 \quad \text{milligrams per hour}$$

at that time. Estimate the amount left 5.5 hours after the injection.

Solution We do not know the function $A(t)$ = (amount after t hours); all we know are values when $t = 5$:

$$A(5) = 298 \qquad A'(5) = -76$$

This is all we need to make a rough estimate of the value of A when $t = 5.5$ using Formula [24]. Take base point $t_0 = 5$. One-half hour later, we have

$$t = t_0 + \Delta t = 5.5$$
$$\Delta t = 0.5$$

and

$$\Delta A \approx A'(5) \cdot \Delta t = -76(0.5) = -38$$

Thus

$$A(5.5) = A(5) + \Delta A \approx 298 - 38 = 260 \quad \text{milligrams}$$ ■

Geometric Interpretation of the Approximation Principle The approximation principle says that the change Δf caused by a move from some base point x_0 to a nearby point $x_0 + \Delta x$ is well approximated by $f'(x_0) \cdot \Delta x$:

$$\Delta f \approx f'(x_0) \cdot \Delta x \quad \text{for all small increments} \quad \Delta x \qquad [26]$$

Actually, the change Δf has the precise form

$$\Delta f = f'(x_0) \cdot \Delta x + \text{Error}$$

The approximation principle works because the error becomes very small *compared to the size of the increment* Δx for small displacements from the base point. The precise statement of this "error estimate" is:

$$\text{The ratio } \frac{\text{Error}}{\Delta x} \text{ approaches zero as } \Delta x \text{ becomes small} \qquad [27]$$

Figure 2.12 gives us some idea of the relative sizes of the error and the term $f'(x_0) \cdot \Delta x$. The line L is the tangent line to the graph of f at P, which corresponds

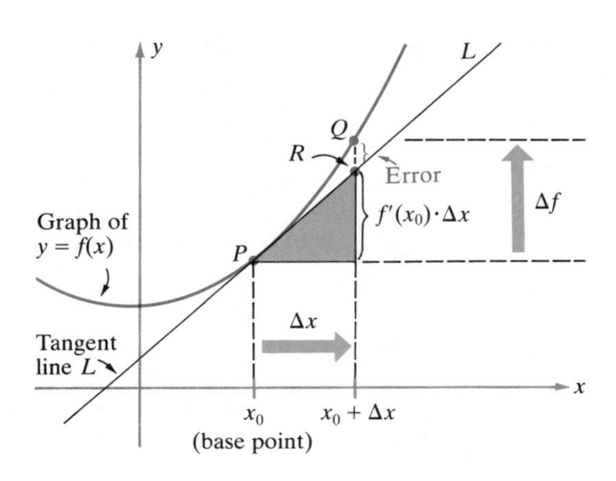

Figure 2.12
The line L, whose equation is $y = f(x_0) + f'(x_0) \cdot (x - x_0)$, is the tangent line to the graph of $f(x)$ at the point P. The tangent line closely approximates the graph near P. If we move along the horizontal axis from x_0 to nearby point $x_0 + \Delta x$, the error in approximating the true change $\Delta f = f(x_0 + \Delta x) - f(x_0)$ is the length of the segment RQ, as shown. As Δx approaches zero, this error becomes very small, even in comparison with the small increment Δx. Thus, as in the approximation principle, $\Delta f = f'(x_0) \cdot \Delta x + \text{Error} \approx f'(x_0) \cdot \Delta x$ for all small increments Δx.

to the base value x_0 on the horizontal axis. Its slope is $f'(x_0)$, so L rises $f'(x_0) \cdot \Delta x$ units as x increases by Δx; meanwhile, the actual graph of f rises by a slightly different amount $\Delta f = f(x_0 + \Delta x) - f(x_0)$. The discrepancy between these vertical displacements is the error. As long as $f'(x_0) \neq 0$, the error is negligible compared to $f'(x_0) \cdot \Delta x$ for small increments Δx away from x_0.

In this book we base several discussions on the approximation principle [26], paying little attention to the role of the error. All of these arguments can be made precise if the error estimate [27] is taken into account. This is precisely what is done in more advanced courses.

Exercises 2.6

1. The weight W of a growing puppy after 30 days is $W = 1500$ grams, and the weight is increasing by $dW/dt = 4.0$ grams per day at that time. By approximately how much will its weight increase in the next week ($\Delta t = 7$)? (*Note*: As long as an approximate answer will suffice, we need not know the formula for $W = W(t)$.)

2. In Exercise 1, the *percentage weight increase* over any time period Δt is defined as

$$\frac{\Delta W}{\text{Initial } W} \times 100 \text{ percent}$$

Find the (approximate) percentage weight increase for the time period in Exercise 1.

In Exercises 3–6, write out the approximation formula $\Delta f \approx f'(x_0) \cdot \Delta x$ for the given function f and base point x_0. Estimate the value of Δf for $\Delta x = +0.1$ and $\Delta x = -0.2$ in each case.

3. $f(x) = 1 - x^2$, $x_0 = 1$

4. $f(x) = 1 + x^2$, $x_0 = 3$

5. $f(x) = x^3 - x + 1$, $x_0 = -1$

6. $f(x) = -5000 + 72x - 0.01x^2$, $x_0 = 100$

In Exercises 7–10, write out the formula $f(x) \approx f(x_0) + f'(x_0) \cdot (x - x_0)$ for the given function f and base point x_0. Use it to estimate $f(x)$ for the given nearby point x.

7. $f(x) = x^4 + 1$, $x_0 = 1$, $x = 1.03$

8. $f(x) = x^4 + 1$, $x_0 = 1$, $x = 0.95$

9. $f(x) = 2 + 3x + 2x^2$, $x_0 = 1$, $x = 1.1$

10. $f(x) = 2 + 3x + 2x^2$, $x_0 = -1$, $x = -1.1$

11. Compute the exact value of the incremental profit $\Delta P = P(1001) - P(1000)$ for the profit function $P = -50{,}000 + 61x - 0.0098x^2$. (Compare with the approximate calculation in Example 29.)

12. At a production level of x units per week, the profit of a certain manufacturing operation is $P = -10{,}000 + 40x - 0.02x^2$ dollars. Use the approximation principle to estimate the change ΔP if production is increased from $x = 500$ to $x = 510$ units per week.

13. Repeat Exercise 12, estimating ΔP if the production level is decreased from $x = 500$ to $x = 495$ units per week.

14. Most phenomena of weather are confined to the lowest 10 miles of the Earth's atmosphere. Estimate the volume of this part of the atmosphere, in cubic miles.

15. The revenue from a weekly newsletter varies with the amount x spent during the week on advertising. The relation is given by $R(x) = 25{,}000 + 150\sqrt{x}$. The advertising budget is currently set at \$4000. Approximately how much would the revenue change if an additional \$500 were invested in advertising? If the advertising were reduced by \$500?

16. If the cost function for a manufacturing operation is

$$C(x) = 8800 + 89x - 0.078x^2 \quad \text{dollars per week}$$

and the current production level is $x = 200$ units per week, estimate the increase in costs if production is raised to 215 units per week.

17. Circular irrigation plots are common in the midwest. If a plot is to be staked out with a radius $r = 500$ yards, its area would be $A = \pi r^2$, or $\pi(500)^2 = 785{,}398$ square yards. How much would A change if there were 5% error in marking off this radius?

18. A surveyor wants to find the area of a square plot of land. His measurements show that the length of a side is 500 feet, with a possible measurement error of ± 1.2 feet. Use the formula $A = s^2$ and the approximation principle to find the uncertainty ΔA in the estimated area of 250,000 square feet.

19. Imitating Example 32, estimate $\sqrt{48}$ and $\sqrt{53}$. Use the base point $x_0 = 49$. Compare your estimates with the exact values obtained using a calculator.

20. The value of $f(x) = 1/x$ is easily calculated for $x = 1$. Use the approximation principle to estimate the value of $f(x)$ at
(i) $x = 1.10795$
(ii) $x = 0.966$
(iii) $x = 1 + h$, where h is any small positive number.

21. By approximately how much does $y = 1/x^2$ change as x goes from $x = 2$ to $x = 2.03$? Estimate the value of $y = 1/x^2$ at $x = 2.03$.

22. A large iron casting in a foundry is presently at a temperature of 680°F and is cooling at a rate of 25° per hour. What is your estimate of the temperature 2 hours later?

23. In Exercise 22, suppose you tried to use the approximation principle to estimate the temperature 40 hours later. Would you have much faith in the estimate? (What *is* the resulting estimate?)

24. Blood flows slowly along the walls of a blood vessel and more rapidly along the center. According to Poiseuille's law governing blood flow, the velocity v along the center is determined by the radius r of the vessel:

$$v = k \cdot r^2 \quad \text{centimeters per second}$$

where k is a constant related to blood pressure and viscosity. Suppose $k = 2 \times 10^4$ and r is measured under a microscope to be 0.01 centimeters, with a possible error of ± 0.001 centimeters. Find the estimated value of v and the corresponding uncertainty Δv caused by the uncertainty in measuring r.

2.7 Limits and Continuity

The concept of limit turns up in problems other than the calculation of derivatives. Here we give a general discussion. If $f(x)$ is an arbitrary function, and x_0 is some fixed base point, we are interested in the "limiting behavior" of the values of $f(x)$ as the variable x gets closer and closer to x_0.

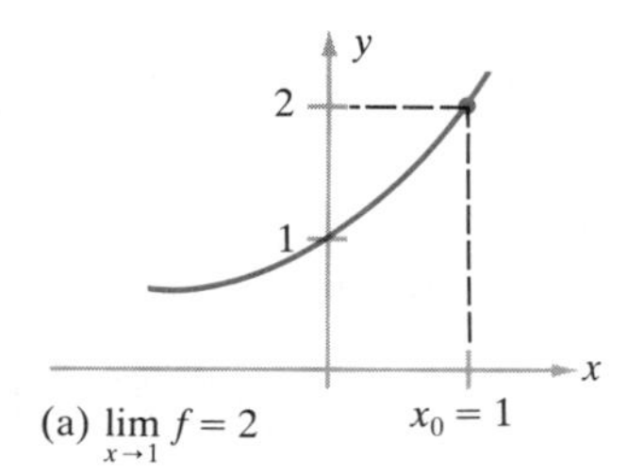

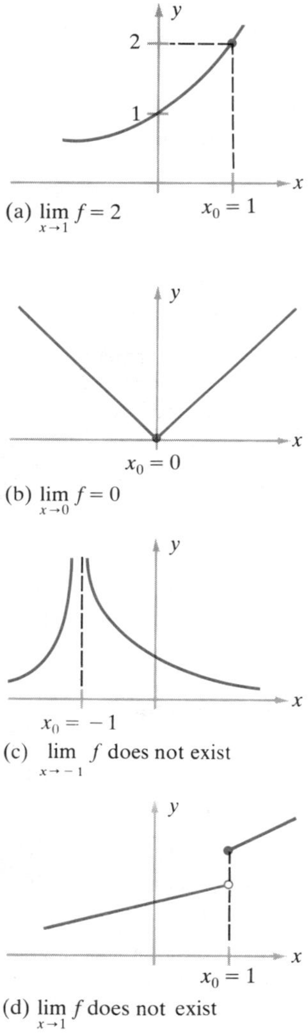

(a) $\lim_{x \to 1} f = 2$

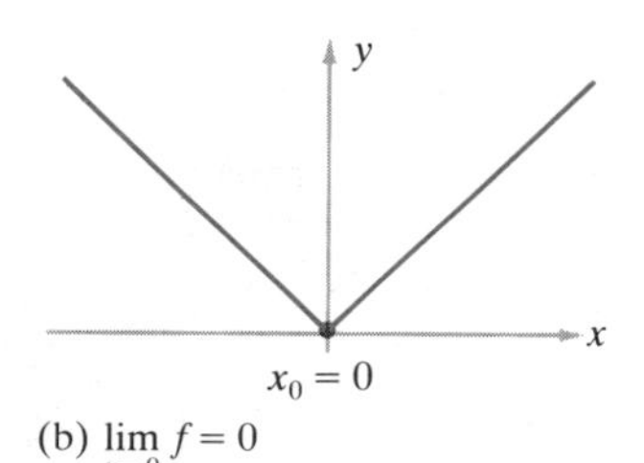

(b) $\lim_{x \to 0} f = 0$

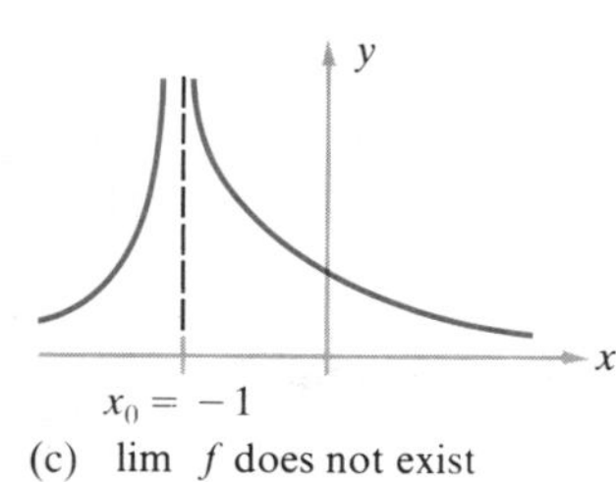

(c) $\lim_{x \to -1} f$ does not exist

$x_0 = 1$

(d) $\lim_{x \to 1} f$ does not exist

Figure 2.13
In Figures 2.13a and 2.13b the functions have limits at the indicated base points; in Figures 2.13c and 2.13d they do not.

> **THE LIMIT OF A FUNCTION** We say that a function f has a **limit** L at a base point x_0, indicated by the notation
>
> $$\lim_{x \to x_0} f(x) = L \qquad [28]$$
>
> if the values $y = f(x)$ get very close to L as x approaches x_0, keeping $x \neq x_0$.

The notion of limit of a function is fundamental; all of calculus is based on it. We have already used this notion in defining derivatives, but it has many other uses as well. The graphs in Figure 2.13 give us some idea of what it means for a function to have a limit as x approaches a base point x_0. Figures 2.13c and 2.13d have no limiting value for f at x_0. Figure 2.13d is especially interesting: For a limit

$$\lim_{x \to x_0} f(x) = L$$

to exist, the values of $f(x)$ must approach L whether x gets close to x_0 from the right *or* the left.

We have already worked with the notion of limit in discussing derivatives. Based on our experience, some limits are easily determined.

Example 34 Find $\lim_{x \to 2} (x^2 - 2x + 3)$.

Solution Here the base point is $x_0 = 2$. As x gets close to 2,

$$-2x \quad \text{gets close to } -2(2) = -4$$
$$x^2 \quad \text{gets close to } (2)^2 = 4$$

so that

$$x^2 - 2x + 3 \text{ approaches the limiting value } 4 - 4 + 3 = 3$$

Thus

$$\lim_{x \to 2} (x^2 - 2x + 3) = 3$$

■

Our experience with the delta process suggests that if $f(x)$ is complicated, we should simplify algebraically before attempting to find the limit, as we shall in the next example.

Example 35 Consider the function $f(x) = \dfrac{x^3}{x^2 + x}$. Show that

$$\lim_{x \to 0} \frac{x^3}{x^2 + x} = 0$$

Solution The function is defined near zero, so it makes sense to talk about the behavior of the values $f(x)$ as x approaches zero, keeping $x \neq 0$. Notice that the numerator and denominator both approach zero as x gets small, so it is not so clear how their quotient behaves for x near zero. But if we simplify the formula algebraically, the limiting behavior is revealed: For any $x \neq 0$, we may cancel an x in both numerator and denominator to obtain

$$f(x) = \frac{x^3}{x^2 + x} = \frac{x^2}{x + 1}$$

As x approaches zero, the expression $x + 1$ approaches 1, the expression x^2 approaches zero, and their quotient approaches the value zero. Thus

$$\lim_{x \to 0} \frac{x^3}{x^2 + x} = \lim_{x \to 0} \frac{x^2}{x + 1} = 0$$ ■

In this example—and in the definition, too—the variable x is excluded from taking the base value x_0 as we examine the limiting behavior of the values $f(x)$. We don't even assume that f is defined at x_0! The definition [28] is a statement about what $f(x)$ is doing for x *near* x_0, rather than *at* x_0 itself. There are natural reasons for building this into the definition. The function in Example 35 was not defined at $x_0 = 0$, yet it had a well-defined limit there. This sort of thing also turned up in our very first use of the limit concept, the definition of a derivative. Recall how the instantaneous rate of change of a function was defined as the limiting value of the average rates of change $\Delta y/\Delta x$, as we consider smaller and smaller increments Δx away from the base value:

$$\text{Instantaneous rate of change} = \lim_{\Delta x \to 0} \frac{\Delta y}{\Delta x}$$

Here,

Δx plays the role of the variable x in [28]
$\Delta y/\Delta x$ plays the same role as $f(x)$ in [28]
0 plays the role of the base point x_0 in [28]

The average rate of change $\Delta y/\Delta x$ cannot be defined for $\Delta x = 0$. If $\Delta x = 0$, then Δy also is zero and we obtain the meaningless ratio $\Delta y/\Delta x = 0/0$ for the average rate of change. To determine the appropriate value for the instantaneous rate of change, we had to concentrate on the values of Δx close to, but not equal to, zero. Thus, Δx can approach zero but cannot take on the value zero. Our definition of limit [28] has been framed with precisely this possibility in mind.

In the following, we list without proof the basic rules for dealing with limits. They tell us how to find the limit of a combination of functions whose limits are known.

BASIC PROPERTIES OF LIMITS Suppose that

$$\lim_{x \to x_0} f(x) \quad \text{and} \quad \lim_{x \to x_0} g(x)$$

both exist. Then

(i) If k is a constant, $\lim_{x \to x_0} k \cdot f(x) = k \cdot \left(\lim_{x \to x_0} f(x) \right)$

(ii) $\lim_{x \to x_0} (f(x) + g(x)) = \lim_{x \to x_0} f(x) + \lim_{x \to x_0} g(x)$

(iii) $\lim_{x \to x_0} f(x) \cdot g(x) = \left(\lim_{x \to x_0} f(x) \right) \cdot \left(\lim_{x \to x_0} g(x) \right)$

(iv) If $\lim_{x \to x_0} g(x) \neq 0$, then

$$\lim_{x \to x_0} \frac{f(x)}{g(x)} = \frac{\lim_{x \to x_0} f(x)}{\lim_{x \to x_0} g(x)}$$

(v) If $\lim_{x \to x_0} f(x) > 0$, then for any r

$$\lim_{x \to x_0} (f(x)^r) = \left(\lim_{x \to x_0} f(x) \right)^r$$

Example 36 Evaluate

(i) $\lim_{x \to 2} x^3 + 1$

(ii) $\lim_{x \to -1} 2x^2 - 3x$

(iii) $\lim_{x \to -1} \dfrac{x^2 + 4}{x^2 + 2}$

(iv) $\lim_{x \to 4} \dfrac{1}{\sqrt{x^2 - 12}}$

Solution In (i) the base point is 2, and we see that $\lim_{x \to 2} x = 2$. Thus,

$$\begin{aligned} \lim_{x \to 2} x^3 + 1 &= \lim_{x \to 2} x^3 + \lim_{x \to 2} 1 \quad &\text{(by Rule (ii))} \\ &= \left(\lim_{x \to 2} x \right)^3 + 1 \quad &\text{(by Rule (v); } r = 3\text{)} \\ &= (2)^3 + 1 = 9 \end{aligned}$$

In (ii), $\lim_{x \to -1} x = -1$, so

$$\begin{aligned}\lim_{x \to -1} 2x^2 - 3x &= 2\left(\lim_{x \to -1} x^2\right) - 3\left(\lim_{x \to -1} x\right) \\ &= 2\left(\lim_{x \to -1} x\right)^2 - 3(-1) \\ &= 2(-1)^2 - 3(-1) = 5\end{aligned}$$

In (iii),

$$\lim_{x \to -1} x^2 + 4 = (-1)^2 + 4 = 5$$

$$\lim_{x \to -1} x^2 + 2 = (-1)^2 + 2 = 3 \neq 0$$

and

$$\lim_{x \to -1} \frac{x^2 + 4}{x^2 + 2} = \frac{\lim_{x \to -1} x^2 + 4}{\lim_{x \to -1} x^2 + 2} = \frac{5}{3}$$

Finally, in (iv)

$$\lim_{x \to 4} x^2 - 12 = (4)^2 - 12 = 4$$

so if we take $r = -1/2$ in Rule (v), we get

$$\lim_{x \to 4} \frac{1}{\sqrt{x^2 - 12}} = \left(\lim_{x \to 4} x^2 - 12\right)^{-1/2} = 4^{-1/2} = \frac{1}{2} \qquad \blacksquare$$

In each of the preceding examples the function is actually defined at the base point where the limit is being evaluated; and in fact the limit is *equal to* the value of f at x_0

$$\lim_{x \to x_0} f(x) = f(x_0)$$

This does not always happen. The limit is determined by the behavior of $f(x)$ near, but not at, x_0. It can exist even if f is undefined at x_0 (as in Example 35). Functions with the special property

$$\lim_{x \to x_0} f(x) = f(x_0) \quad \text{for every } x_0 \text{ where } f \text{ is defined} \tag{29}$$

are given a special name: **continuous functions**. If [29] holds at a specific point x_0 we say that f is **continuous at x_0**. Continuity does not play much of a role in differentiation, but we shall touch on it in connection with integration (Chapter 5).

Intuitively, a continuous function is one whose graph may be traced without ever lifting the pencil from the paper. The simplest way a function can fail to be

Figure 2.14
Some functions with jump discontinuities. Open dots indicate points *not* on the graph. At each jump the function is defined at $x = x_0$, but because of the jump the limit $\lim_{x \to x_0} f(x)$ does not exist.

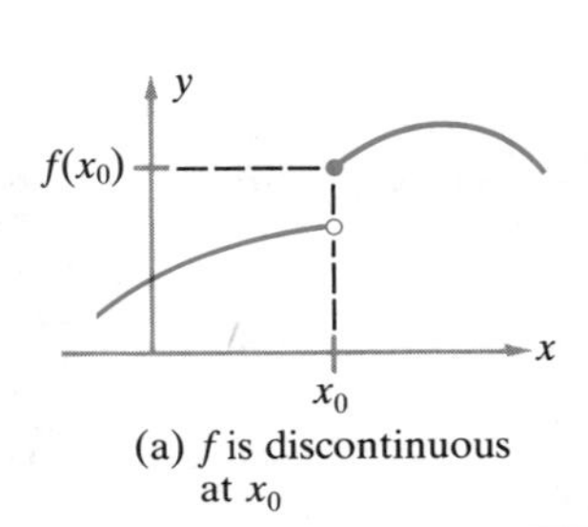

(a) f is discontinuous at x_0

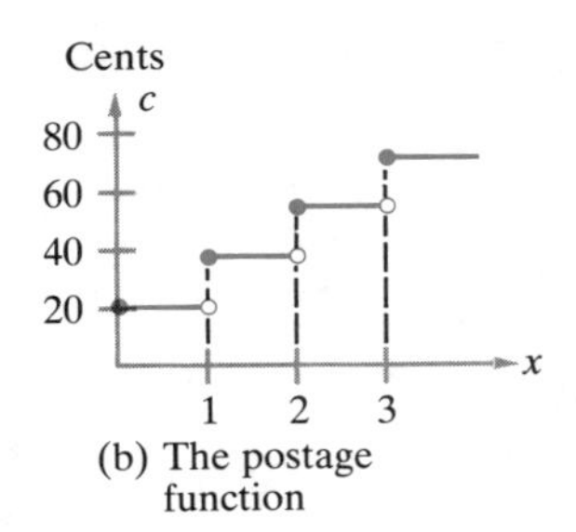

(b) The postage function

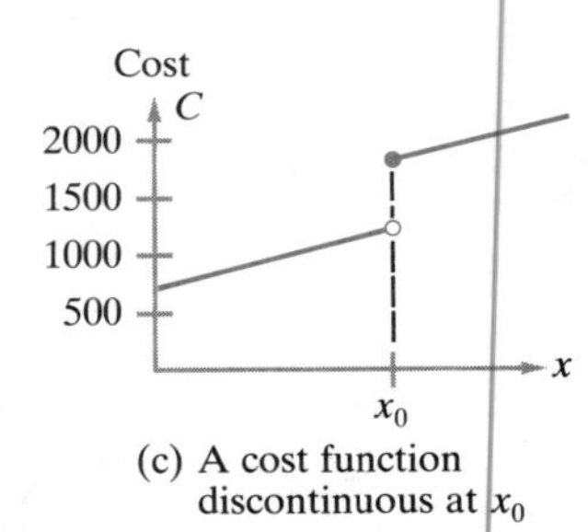

(c) A cost function discontinuous at x_0

Figure 2.15
Graphs of some functions defined near a base point x_0. The value $f(x_0)$ at x_0 is indicated by the solid dot above x_0. Consider each figure with the definition [29] in mind to see why continuity is valid or not. For example, in Figures 2.15b and 2.15f, the limit $\lim_{x \to x_0} f(x)$ does not exist. In Figure 2.15e, the limit exists, but does not coincide with the value of f at x_0: $f(x_0) = 1$, and the value of the limit is $\frac{1}{2}$.

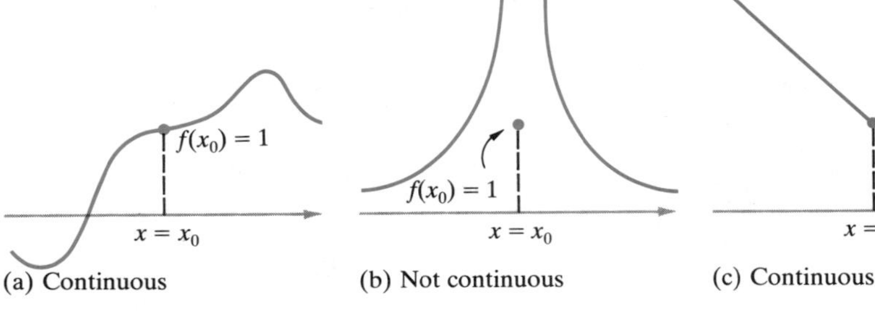

(a) Continuous (b) Not continuous (c) Continuous

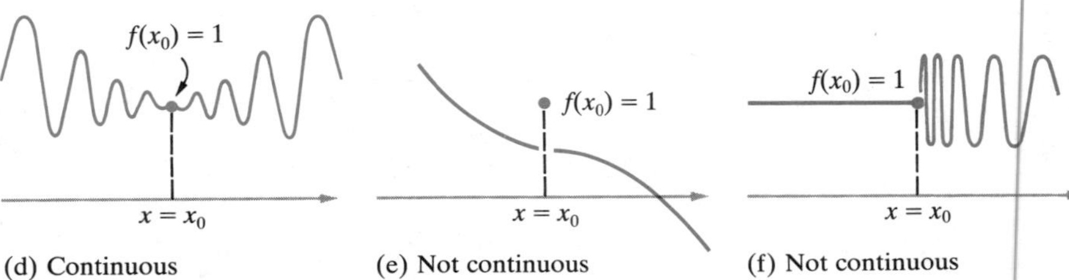

(d) Continuous (e) Not continuous (f) Not continuous

continuous is for its value to take an abrupt "jump" at some point x_0 in its domain of definition (see Figure 2.14a). Such a point is called a **jump discontinuity**. The function is discontinuous there because $\lim_{x \to x_0} f(x)$ fails to exist. The postage function (Example 5, Section 1.1) reproduced in Figure 2.14b has jump discontinuities at $x = 1$, $x = 2$, and so on; it is continuous for all other values of x. Realistic cost functions often have jump discontinuities, as shown in Figure 2.14c: Once production exceeds a certain level, new employees must be hired or a new shift started, causing a jump in weekly payroll costs.

Continuity also fails if $\lim_{x \to x_0} f(x)$ exists but fails to agree with the value of f at x_0 (see Figure 2.15e). Or, continuity might fail because the values $f(x)$ "blow up," as shown in Figure 2.15b. Figure 2.15 shows several other possibilities.

From the rules governing limits, we see that any polynomial is continuous. In fact, most functions given by algebraic formulas are continuous wherever they are defined (denominators nonzero). We shall illustrate this in the next example.

Example 37 For which points is the function $f(x) = \dfrac{x^2 + 4}{x^2 + 2x}$ continuous?

Solution If a function is to be continuous at x_0 it must be defined there, so the right side of [29] makes sense. This function is defined except when $x = 0$ or $x = -2$ (denominator zero). For all other base points x_0, the function is defined; and by the rules governing limits we obtain

$$\lim_{x \to x_0} x^2 + 4 = x_0^2 + 4$$

$$\lim_{x \to x_0} x^2 + 2x = x_0^2 + 2x_0$$

Since the limit of the denominator is nonzero, we may apply Rule (iv) to get

$$\begin{aligned} \lim_{x \to x_0} f(x) &= \lim_{x \to x_0} \frac{x^2 + 4}{x^2 + 2x} \\ &= \frac{\lim_{x \to x_0} x^2 + 4}{\lim_{x \to x_0} x^2 + 2x} \\ &= \frac{x_0^2 + 4}{x_0^2 + 2x_0} \\ &= f(x_0) \end{aligned}$$

so the function is continuous for all x except $x = 0$ and $x = 2$. ■

We note the connection between continuity and differentiability:

If f is differentiable at x_0 it must also be continuous at x_0. [30]

We shall soon be able to differentiate many functions with ease. This guarantees that all such functions are continuous. To see why [30] is true, we appeal to the approximation principle to write

$$f(x) = f(x_0) + f'(x_0)(x - x_0) + \text{error term}$$

for all x near x_0. We know that the error term becomes very small as x approaches x_0. But the main term $f'(x_0)(x - x_0)$ also becomes small because $x - x_0$ approaches zero. In the limit, only the constant term $f(x_0)$ persists, so that

$$\lim_{x \to x_0} f(x) = f(x_0) + 0 + 0 = f(x_0)$$

as required for continuity of f at x_0.

Exercises 2.7

In Exercises 1–16 evaluate the specified limits. In the first two problems, 1 and 0 stand for the constant functions $f(x) = 1$ and $f(x) = 0$.

1. $\lim_{x \to 0} 1$

2. $\lim_{x \to 1} 0$

3. $\lim_{x \to 0} x$

4. $\lim_{x \to 1} x$

5. $\lim_{x \to -2} x$

6. $\lim_{x \to -2} x^2 + 1$

7. $\lim_{x \to -2} x^2 + 3x$

8. $\lim_{x \to \sqrt{2}} x^3 - x^2 - 2x + 6$

9. $\lim_{x \to 2} \frac{x^2 + 1}{x^2 - 1}$

10. $\lim_{x \to 0} \frac{x}{x + 1}$

11. $\lim_{x \to 1} \frac{x^2 + 2x}{x}$

12. $\lim_{x \to 0} \frac{4x + 1}{x - 2}$

13. $\lim_{x \to 0} \sqrt{x^2 + 1}$

14. $\lim_{x \to 0} x\sqrt{x^2 + 1}$

15. $\lim_{x \to 2} \sqrt{\frac{x + 1}{x - 1}}$

16. $\lim_{x \to 1} \frac{1}{\sqrt{x^2 + 2x - 1}}$

In Exercises 17–24 evaluate the limits. In some cases it will help to simplify algebraically before applying the limit rules.

17. $\lim_{x \to -3} (x^2 - 7)(x^2 + x + 1)$

18. $\lim_{x \to 7} (x - 7)(x^9 + x^5 - x)$

19. $\lim_{t \to 0} \frac{3t^2 - 2t + 4}{t^2 + t - 1}$

20. $\lim_{h \to 0} \frac{h^2 - h}{h}$

21. $\lim_{t \to -1} \frac{t^2 - t - 2}{t + 1}$

22. $\lim_{x \to 0} \frac{x^3}{x^2}$

23. $\lim_{\Delta t \to 0} \frac{(\Delta t)^2 + 4(\Delta t)^3}{13 + (1 - \Delta t)^2}$

24. $\lim_{x \to 2} \frac{x^2 - 4x + 4}{x^2 - 4}$

25. Examine the graphs shown in Figure 2.16. Which functions have a limit at the base point x_0 indicated?

26. Which functions shown in Figure 2.16 are continuous at the base point x_0?

For each of the following functions, determine where the function is defined and where it is continuous.

27. $\frac{1}{x}$

28. $\frac{1}{1 - x^2}$

29. $\frac{x - 1}{x^2 + 1}$

30. $\frac{x - 1}{x + 1}$

31. $\sqrt{2x^2 - 9}$

32. $\sqrt{4x^2 + 2}$

33. The function defined by

$$f(x) = \begin{cases} 1 & \text{if} \quad x \geq 1 \\ 0 & \text{if} \quad x < 1 \end{cases}$$

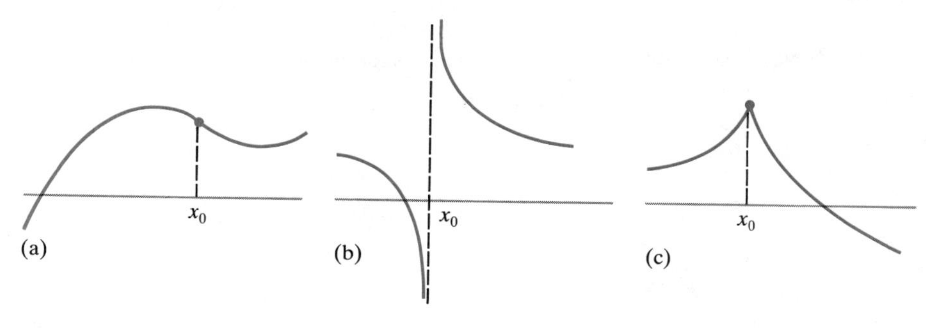

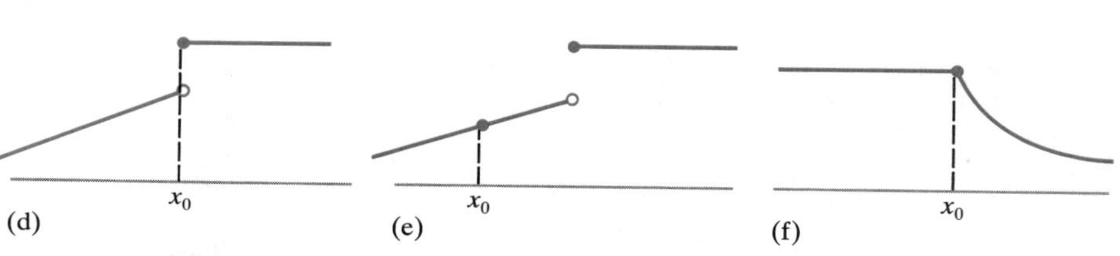

Figure 2.16
The functions in Exercises 25–26. Solid dots are on the graph, open dots are not. The base point to be considered is labeled x_0.

Graph each of these functions and determine which are continuous at the base point $x_0 = 2$. In each case, what is the value $f(x_0)$?

34. $f(x) = \begin{cases} 3 - x & \text{if} \quad x < 2 \\ x - 2 & \text{if} \quad x \geq 2 \end{cases}$

35. $f(x) = \begin{cases} 3 - x & \text{if} \quad x < 2 \\ 2 & \text{if} \quad x \geq 2 \end{cases}$

Checklist of Key Topics

Average rate of change
Instantaneous rate of change
The delta process for finding derivatives
Derivative
Differentiable function
Slope of a curve
Tangent line to a curve
Derivative as slope of the tangent line
Instantaneous velocity
The marginal concept in economics
The approximation principle:
$\Delta f \approx f'(x_0) \cdot \Delta x$
Limit of a function $\lim_{x \to x_0} f(x)$
Continuity of a function

Differentiation rules:

$$\frac{d}{dx}(x^r) = rx^{r-1} \quad \textbf{(Power rule)}$$

$$\frac{d}{dx}(e^x) = e^x$$

$$\frac{d}{dx}(\ln x) = \frac{1}{x}$$

$$\frac{d}{dx}(k \cdot f(x)) = k \cdot \frac{df}{dx}$$

$$\frac{d}{dx}(f(x) + g(x)) = \frac{df}{dx} + \frac{dg}{dx}$$

Chapter 2 Review Exercises

In Exercises 1–6, find:
(i) the average rate of change of $y = f(x)$ as x increases from x_1 to x_2.
(ii) the instantaneous rate of change at base point x_1, using the delta process.

1. $f(x) = 3x - 2, \quad x_1 = 0, x_2 = 1$

2. $f(x) = 5 - 4x, \quad x_1 = 1, x_2 = 3$

3. $f(x) = 2x^2 - 7x + 1, \quad x_1 = 0, x_2 = 2$

4. $f(x) = 13 - 12x - x^2, \quad x_1 = 1, x_2 = 3$

5. $f(x) = x - \dfrac{4}{x}, \quad x_1 = 1, x_2 = 4$

6. $f(x) = x^3 + \dfrac{1}{x^2}, \quad x_1 = 1, x_2 = 2$

7. A baseball is popped straight up. If its height after t seconds is $h(t) = 132t - 16t^2$ feet, find:
(i) the average velocity between times $t = 1$ and $t = 3$ seconds
(ii) the average velocity between times $t = 1$ and $t = 1 + \Delta t$ seconds ($\Delta t \neq 0$)
(iii) the instantaneous velocity when $t = 1$
(iv) the instantaneous velocity when $t = 5$

8. The monthly costs for a manufacturer are $C(x) = 13{,}500 + 72x - 0.01x^2$ dollars if x units are produced per month. Find:
(i) the marginal cost function dC/dx
(ii) the average rate of change in C as x increases from $x_1 = 1000$ units to $x_2 = 1100$ units

9. Find the derivative of $y = 3x^2 + 2x$ using the delta process.

10. Find the derivative of $y = 4 - x + x^3$ using the delta process.

11. A rock is dropped from a height of 144 feet. The distance it falls in t seconds is given by the formula $s = 16t^2$. Find the time it takes to hit the ground. Then find the instantaneous velocity of the rock when it hits the ground (= impact velocity).

In Exercises 12–19, find the derivatives:

12. $f'(x)$ if $f(x) = 6x^4 - \dfrac{4}{3}x^3 + \dfrac{5}{4}x^2 - 2$

13. $f'(t)$ if $f(t) = 7t^8 - \dfrac{1}{3}t^4 + \dfrac{1}{2}t^3 - t + 12$

14. $f'(1)$ if $f(x) = \ln x - x^5 + 4x^3$

15. $g'(1)$ if $g(x) = 3e^x + x^2 - 11x$

16. $g'(4)$ if $g(x) = \dfrac{x^2 + 2}{\sqrt{x}}$

17. $f'(x)$ if $f(x) = (47 - x)(2 - x + x^2)$

18. $\dfrac{d}{dx}\left(\dfrac{1}{\sqrt{x}} + 8x^{5/4} - \sqrt{3}\right)$

19. $\dfrac{d}{dx}\left(\dfrac{1}{x^3} + x^{3/2} - 10{,}000\right)$

In Exercises 20–25, find the equation of the tangent line through the point P on the graph.

20. $f(x) = \dfrac{1}{3}x^3 + x + 9, \quad P = (-3, -3)$

21. $f(x) = x^4 - \dfrac{1}{2}x + 1, \quad P = (0, 1)$

22. $f(x) = e^x, \quad P = (0, 1)$

23. $f(x) = 2 \ln x, \quad P = (1, 0)$

24. $f(x) = 3 - 2x^3$ The first coordinate of P is $x = 1$.

25. $f(x) = 2\sqrt{x}$ The first coordinate of P is $x = 4$.

26. Find *both* coordinates of all points P on the graph of $y = -3 + 6x - x^2$ at which the tangent line is horizontal.

27. Find the production levels x where the marginal cost is zero, if the cost function is given by $C(x) = 17{,}200 + 36x - 0.012x^2$.

28. The gypsy moth population of Orange County is projected to be $P(t) = 1200\, e^t$ in t years. How fast will the population be increasing 4 years from now according to this model? How large will the population be at that time?

29. The cost function of a manufacturer is $C(x) = 1200 + 89x - \frac{1}{4}x^2$ and the demand function is $p = 180 - \frac{1}{2}x$, where x is the production level.
(i) Find the marginal profit function.
(ii) Use the approximation principle to estimate the change in profit if production is increased from $x = 100$ to $x = 105$ units.

In Exercises 30–33, use the approximation principle to estimate the change $\Delta f = f(x_0 + \Delta x) - f(x_0)$.

30. $f(x) = x^4 - 3x^2 + 4, \quad x_0 = 2, \Delta x = 0.003$

31. $f(x) = x^3 - 4x^2 + 3x, \quad x_0 = 3, \Delta x = -0.01$

32. $f(x) = \dfrac{1}{x}, \quad x_0 = 3, \Delta x = -0.2$

33. $f(x) = \sqrt{x}, \quad x_0 = 64, \Delta x = 1$

34. Estimate the value of $f(1.03)$ if $f(1) = 7$ and $f'(1) = -2$.

35. Estimate $g(7.9)$ if $g(8) = -12$ and $g'(8) = 4$.

In Exercises 36–39, evaluate the limit.

36. $\lim_{x \to 3} \sqrt{x^2 + 16}$

37. $\lim_{x \to 0} \dfrac{x^2 + 2}{x + 2}$

38. $\lim_{x \to 2} \dfrac{x^2 + 3x - 4}{x^2 - 1}$

39. $\lim_{x \to 2} \dfrac{e^x + \ln x - 3}{x - 1}$

CHAPTER 3

Applications of Differentiation

3.1 The Product and Quotient Rules for Derivatives

In the next two sections we discuss rules for differentiating functions more general than those encountered in Chapter 2. These rules will suffice for all our applications.

If f and g are functions, their product $f \cdot g$ is the function

$$f(x) \cdot g(x)$$

obtained by multiplying their values for each x. Thus if $f(x) = x$ and $g(x) = x^2 - x + 1$, the product is $x(x^2 - x + 1) = x^3 - x^2 + x$. Or, if $f(x) = 1 + x^2$ and $g(x) = e^x$, then $(f \cdot g)(x) = (1 + x^2)e^x = e^x + x^2e^x$. The differentiation formula for products is more complicated than that for sums.

PRODUCT RULE If $f(x)$ and $g(x)$ are differentiable, their product $(f \cdot g)(x)$ has derivative

$$\frac{d}{dx}(f(x) \cdot g(x)) = \frac{df}{dx} \cdot g(x) + f(x) \cdot \frac{dg}{dx} \qquad [1]$$

The proof is rather technical, and is omitted. Our main concern here is the correct use of this rule.

Warning: *The derivative of a product $f(x) \cdot g(x)$ is not the product of the derivatives.* Read Formula [1] carefully before proceeding, and consider the following example. If $f(x) = x^2$ and $g(x) = x^3$, then $f(x) \cdot g(x) = x^5$; and from the power rule we can find the derivative

$$\frac{d}{dx}(f(x) \cdot g(x)) = \frac{d}{dx}(x^5) = 5x^4 \tag{2}$$

Compare this with the product of derivatives:

$$\frac{df}{dx} = 2x \qquad \frac{dg}{dx} = 3x^2 \qquad \text{and the product is } \frac{df}{dx} \cdot \frac{dg}{dx} = 6x^3$$

They don't agree! The product formula [1], used correctly, does give the right answer:

$$\begin{aligned}\frac{d}{dx}(x^2 \cdot x^3) &= \frac{d}{dx}(x^2) \cdot x^3 + x^2 \cdot \frac{d}{dx}(x^3) \\ &= (2x)(x^3) + (x^2)(3x^2) \\ &= 2x^4 + 3x^4 = 5x^4\end{aligned}$$

■

Example 1 Use the product formula to find dy/dx if $y = x^4(\sqrt{x} + 1)$. Evaluate dy/dx at $x = 1$.

Solution The factors $f(x) = x^4$ and $g(x) = \sqrt{x} + 1$ are differentiated thusly:

$$\frac{df}{dx} = \frac{d}{dx}(x^4) = 4x^3$$

$$\begin{aligned}\frac{dg}{dx} &= \frac{d}{dx}(\sqrt{x} + 1) = \frac{d}{dx}(x^{1/2}) + \frac{d}{dx}(1) \\ &= \frac{1}{2}x^{-1/2} + 0 = \frac{1}{2\sqrt{x}}\end{aligned}$$

From the product formula [1], we obtain

$$\begin{aligned}\frac{dy}{dx} &= \frac{df}{dx} \cdot g(x) + f(x) \cdot \frac{dg}{dx} \\ &= \frac{d}{dx}(x^4)(\sqrt{x} + 1) + x^4 \cdot \frac{d}{dx}(\sqrt{x} + 1) \\ &= 4x^3(\sqrt{x} + 1) + \frac{x^4}{2\sqrt{x}}\end{aligned}$$

We could simplify this answer by writing everything down as fractional powers to get $dy/dx = (\frac{9}{2})x^{7/2} + 4x^3$. At $x = 1$ we find that $dy/dx = \frac{17}{2} = 8.5$. ■

Example 2 Find dy/dx if $y = x^3e^x$.

Solution In [1] let us take $f(x) = x^3$ and $g(x) = e^x$. Using the power law and the differentiation formula for e^x (Section 2.3), we obtain

$$\begin{aligned}\frac{dy}{dx} &= \frac{df}{dx}\cdot g(x) + f(x)\cdot\frac{dg}{dx}\\ &= \frac{d}{dx}(x^3)\cdot e^x + x^3\frac{d}{dx}(e^x)\\ &= 3x^2\cdot e^x + x^3\cdot e^x\\ &= (3x^2 + x^3)e^x\end{aligned}$$

■

The **quotient** of two functions $f(x)$ and $g(x)$

$$\frac{f}{g}(x) = \frac{f(x)}{g(x)}$$

is defined by taking the quotient of their values for each x. This process yields such functions as $1/x$, where $f(x) = 1$ and $g(x) = x$, and $y = x/(x^2 + 1)$, where $f(x) = x$ and $g(x) = x^2 + 1$. The quotient function $f(x)/g(x)$ is defined wherever $g(x) \neq 0$.

As with products, there is a differentiation formula for quotients, which we state without proof in order to concentrate on its use in examples.

QUOTIENT RULE If $f(x)$ and $g(x)$ are differentiable, their quotient $f(x)/g(x)$ has derivative

$$\frac{d}{dx}\left(\frac{f(x)}{g(x)}\right) = \frac{\dfrac{df}{dx}\cdot g(x) - f(x)\cdot\dfrac{dg}{dx}}{(g(x))^2} \qquad [3]$$

for all x where $g(x) \neq 0$.

Warning: *The derivative of* $\dfrac{f(x)}{g(x)}$ *is not the quotient of derivatives.*

Example 3 Find $\dfrac{d}{dx}\left(\dfrac{1}{x^2}\right)$ using the quotient rule.

Solution Here $f(x) = 1$ and $g(x) = x^2$, so that $df/dx = 0$, $dg/dx = 2x$. Thus

$$\begin{aligned}\frac{d}{dx}\left(\frac{1}{x^2}\right) &= \frac{\dfrac{df}{dx}\cdot g(x) - f(x)\cdot\dfrac{dg}{dx}}{(g(x))^2}\\ &= \frac{(0)(x^2) - (1)(2x)}{(x^2)^2}\\ &= \frac{-2x}{x^4} = -\frac{2}{x^3}\end{aligned}$$

for all $x \neq 0$. (The function and its derivative are not defined at $x = 0$.) Notice that our answer agrees with that obtained using the rule for differentiating a power x^r, taking $r = -2$. ■

The next examples could not be done using the rules given in Chapter 2.

Example 4 Find the derivative of $y = \dfrac{1}{3x + 5}$.

Solution This function is the quotient of $f(x) = 1$ and $g(x) = 3x + 5$. We have $df/dx = 0$ and $dg/dx = 3$, so by the quotient formula

$$\frac{dy}{dx} = \frac{\dfrac{df}{dx} \cdot g(x) - f(x) \cdot \dfrac{dg}{dx}}{(g(x))^2}$$

$$= \frac{(0)(3x + 5) - (1)(3)}{(3x + 5)^2} = \frac{-3}{(3x + 5)^2}$$ ■

Example 5 If $y = \dfrac{x}{x^2 + 1}$, find $\dfrac{dy}{dx}$.

Solution Taking $f(x) = x$ and $g(x) = x^2 + 1$ in Formula [3], we obtain

$$\frac{dy}{dx} = \frac{\dfrac{d}{dx}(x) \cdot (x^2 + 1) - x \cdot \dfrac{d}{dx}(x^2 + 1)}{(x^2 + 1)^2}$$

$$= \frac{1 \cdot (x^2 + 1) - x(2x)}{(x^2 + 1)^2} = \frac{x^2 + 1 - 2x^2}{(x^2 + 1)^2}$$

$$= \frac{1 - x^2}{(x^2 + 1)^2}$$ ■

Example 6 In the lumber industry, timber is harvested and replanted cyclically. The growth function $w(t)$ is known from observations (see Figure 3.1). Here, $w(t) =$ the quantity of usable lumber obtainable from a tree t years old. As

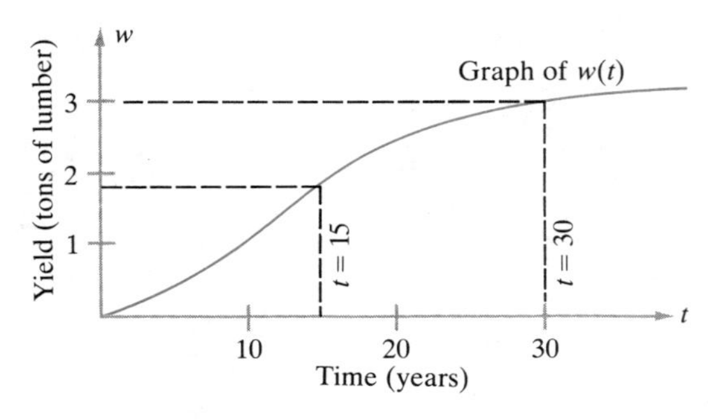

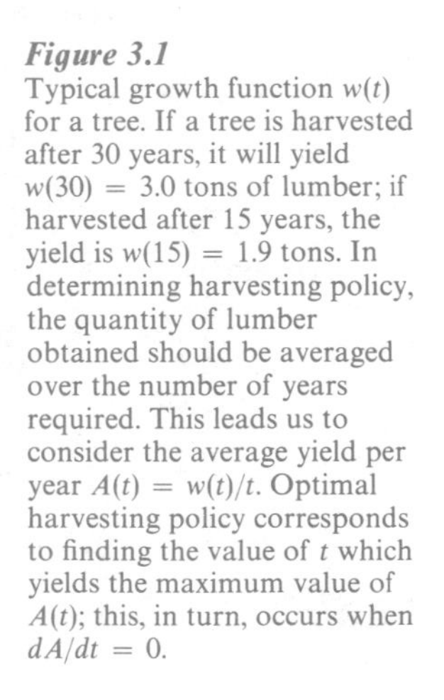

Figure 3.1
Typical growth function $w(t)$ for a tree. If a tree is harvested after 30 years, it will yield $w(30) = 3.0$ tons of lumber; if harvested after 15 years, the yield is $w(15) = 1.9$ tons. In determining harvesting policy, the quantity of lumber obtained should be averaged over the number of years required. This leads us to consider the average yield per year $A(t) = w(t)/t$. Optimal harvesting policy corresponds to finding the value of t which yields the maximum value of $A(t)$; this, in turn, occurs when $dA/dt = 0$.

explained in the caption to Figure 3.1, the function

$$A(t) = \frac{w(t)}{t} \quad \text{for} \quad t > 0$$

is of primary importance in determining the optimal harvesting policy. Find an expression for dA/dt in terms of $t, w(t)$, and dw/dt. When is $dA/dt = 0$?

Solution Because $A(t)$ is a quotient $w(t)/t$, we can use Formula [3] to obtain

$$\frac{dA}{dt} = \frac{\frac{dw}{dt} \cdot t - w(t) \cdot \frac{d}{dt}(t)}{t^2} = \frac{t\frac{dw}{dt} - w(t)}{t^2}$$

To determine where the derivative is zero, $dA/dt = 0$, we set

$$\frac{t\frac{dw}{dt} - w(t)}{t^2} = 0$$

For the quotient to be zero, the numerator must be zero; thus

$$t\frac{dw}{dt} - w(t) = 0$$

$$t\frac{dw}{dt} = w(t)$$

or

$$\frac{dw}{dt} = \frac{w(t)}{t} \qquad [4]$$

■

Exercises 3.1

1. Find dy/dx for the expression $y = (x^2 + x + 1)(x^3 - x^2 + 2)$ by each of the following methods. Reconcile the results.
   (i) Multiply out the product and differentiate the resulting polynomial.
   (ii) Apply the product rule directly.

In Exercises 2–13, differentiate the given function using the product rule.

2. $(x^2 + 1)(x^2 - 1)$

3. $x^4(1 - x - x^2)$

4. $(x^4 + x^2 + 1)(14x^3 - 3x + 2)$

5. $x^{-3}(x^2 + 1)(x^2 - 1)$

6. $\left(x + \frac{1}{x}\right)(\sqrt{x} + 1)$

7. $x(\ln x)$

8. $x^{1/3}(x^4 + 3x^2 + 1)$

9. $x^{-3}(x^2 - 1)$

10. t^2e^t

11. e^t/t

12. $(1 + x + x^2)e^x$

13. $e^x \cdot \ln x$

In Exercises 14–25, calculate the derivative using the quotient rule.

14. $\frac{1}{x}$

15. $\frac{1}{1 + x}$

16. $\frac{1 + x}{1 - x}$

17. $\frac{-4}{x^2 + 1}$

18. $\frac{1}{100 - 45x}$

19. $\frac{x + 2}{1 - x} + \frac{1}{x}$

20. $\frac{x^2 + 3x + 2}{x^2 - x + 1}$

21. $\frac{1 - 2x}{x^2 + x}$

22. $\frac{1 + \sqrt{x}}{1 - \sqrt{x}}$

23. $\frac{e^x}{1 - x}$

24. $\frac{x}{x^4 + 5x + 1}$

25. $\frac{x^2 + x + 2}{e^x + 1}$

26. Find the derivative $f'(1)$ if $f(x) = \dfrac{2 - 3x}{2x^2 + x + 3}$.

27. Find the derivative $f'(4)$ if $f(x) = \dfrac{1 - \sqrt{x}}{1 + \sqrt{x}}$.

28. Find $f'(1)$ if $f(x) = \dfrac{1}{1 + x^2}$.

29. Find the equation of the tangent line to the graph of $y = 2x/(x^2 + 1)$ at the point $(1, 1)$ and likewise at the point $(2, \frac{4}{5})$.

30. Find the slope of the tangent line and its equation at the point $(1, 0)$ on the graph of $y = (x - 1)/(x + 1)$.

31. Recall that the revenue of a manufacturing operation is given by $R(x) = x \cdot p(x)$ where x is the production level, and $p(x)$ is the demand function, which gives the price p at which all x units will be sold. Explain why

$$\frac{dR}{dx} = p(x) + x\frac{dp}{dx}$$

32. Referring to Exercise 31, explain why the condition

$$\frac{dp}{dx} = -\frac{p(x)}{x}$$

implies that the marginal revenue dR/dx is zero.

33. The *average cost per unit* in a manufacturing operation is defined to be

$$A(x) = \frac{\text{total cost}}{\text{number of units}} = \frac{C(x)}{x} \quad \text{for} \quad x > 0$$

Assuming that the cost function $C(x)$ is differentiable, express the rate of change dA/dx in terms of x, $C(x)$ and the marginal cost dC/dx. Show that $dA/dx = 0$ when the average cost equals the marginal cost, $A = dC/dx$.

34. If $f(x)$ is differentiable and $g(x) = x \cdot f(x)$, show that $g'(0) = f(0)$.

3.2 Composite Functions and the Chain Rule

Suppose we inflate a balloon in such a way that its radius r increases with time, say $r = 3t$. The volume V depends on the radius in the usual way, $V = \frac{4}{3}\pi r^3$. Thus

$$V = \frac{4}{3}\pi r^3 \quad \text{and} \quad r = 3t$$

The volume may be expressed as a function of t by substituting $r = 3t$ into $V = \frac{4}{3}\pi r^3$

$$V = V(t) = \frac{4}{3}\pi(3t)^3 = 36\pi t^3$$

Replacing the independent variable in one function by a function of some new variable is an example of **composition** of functions. In the preceding example we composed $V = (\frac{4}{3})\pi r^3$ with $r = 3t$ to obtain the **composite function** $V = 36\pi t^3$. In this section we shall give an important rule for differentiating composite functions.

Suppose the independent variable u in one function $y = f(u)$ depends on some other variable x according to some formula $u = g(x)$. Substituting $g(x)$ for u everywhere u appears in $f(u)$, we express y as a function of x

$$y = f(g(x)) \qquad [5]$$

This substitution is sometimes denoted by the symbol

$$y = \left[f(u) \Big|_{u=g(x)} \right] \qquad [6]$$

Here the vertical bar indicates that $u = g(x)$ should be substituted.

Example 7 If $f(u) = u^2 + 1$ and $u = g(x) = 3x + 2$, find the composite function $y = f(g(x))$.

Solution We have

$$\begin{aligned} y = f(g(x)) &= \left[u^2 + 1 \Big|_{u=3x+2} \right] \\ &= (3x + 2)^2 + 1 \\ &= 9x^2 + 12x + 5 \end{aligned}$$

■

Example 8 Find the composites of the following functions

(i) $f(u) = \dfrac{1}{u}$ and $u = g(x) = x^2 + 1$

(ii) $f(u) = u^2 + 1$ and $u = g(x) = \dfrac{1}{x}$

(iii) $y = \sqrt{u}$ and $u = 4x^2 + 3$

(iv) $y = e^{-2u}$ and $u = \dfrac{x^2}{2} + 1$

Solution

$$\text{(i)}\ y = \left[\frac{1}{u}\bigg|_{u=x^2+1}\right] = \frac{1}{x^2+1}$$

$$\text{(ii)}\ y = \left[u^2 + 1\bigg|_{u=\frac{1}{x}}\right] = \frac{1}{x^2} + 1$$

$$\text{(iii)}\ y = \left[\sqrt{u}\bigg|_{u=4x^2+3}\right] = \sqrt{4x^2+3}$$

$$\text{(iv)}\ y = \left[e^{-2u}\bigg|_{u=\frac{1}{2}x^2+1}\right] = e^{-x^2-2}$$

■

Often we must recognize when a complicated function of x is a composite of simpler functions.

Example 9 For each function of x in (i) and (ii), find $y = f(u)$ and $u = g(x)$, such that the given function is the composite $f(g(x))$.

$$\text{(i)}\ \frac{1}{(x-2)^{10}} \qquad \text{(ii)}\ \sqrt{1-x^2}$$

Solution In (i) it seems natural to try $u = x - 2$. Then we have

$$y = \frac{1}{(x-2)^{10}} = \frac{1}{u^{10}} = u^{-10}$$

Thus

$$y = \left[u^{-10}\bigg|_{u=x-2}\right]$$

and we can take $f(u) = u^{-10}$, $g(x) = x - 2$. In (ii) the natural choice is $u = 1 - x^2$. Then

$$y = \sqrt{1-x^2} = u^{1/2}$$

so that $y = f(g(x))$ if we take $f(u) = u^{1/2}$ and $u = 1 - x^2$. ■

The **chain rule** for differentiating composite functions tells us how to compute dy/dx from the derivatives dy/du and du/dx of the component functions $y = f(u)$ and $u = g(x)$.

CHAIN RULE If $y = f(u)$ and $u = g(x)$ are differentiable functions, their composite $y = f(g(x))$ has derivative

$$\frac{dy}{dx} = \frac{dy}{du} \cdot \frac{du}{dx} \qquad [7]$$

Here, $dy/du = f'(u)$ is a function of u, and dy/dx and du/dx are functions of x. To make sense of [7] we must write u in terms of x, substituting $u = g(x)$ everywhere u appears so that all functions involve the same variable x. Therefore, a more precise statement of Formula [7] is

$$\frac{dy}{dx} = \left[\left.\frac{dy}{du}\right|_{u=g(x)}\right] \cdot \frac{du}{dx} \qquad [8]$$

Notice that [7] and [8] say the same thing with slightly different notation. A proof of the chain rule, based on the approximation principle, is outlined at the end of this section.

We use the chain rule to differentiate a complicated function. To use it we search for simpler functions $y = f(u)$ and $u = g(x)$, such that the given function is the composite $y = f(g(x))$. If we know dy/du and du/dx, then we can find dy/dx.

Example 10 Differentiate the function $y = (x^2 + 1)^{20}$.

Solution We could expand $(x^2 + 1)^{20}$ by tedious multiplication to obtain a polynomial with 21 terms, then differentiate this polynomial. But the required differentiation is easily accomplished using the chain rule, taking $f(u) = u^{20}$ and $u = x^2 + 1$. Now

$$y = \left[\left. u^{20} \right|_{u=x^2+1}\right] = (x^2 + 1)^{20}$$

and we have

$$y = u^{20} \qquad u = x^2 + 1$$

$$\frac{dy}{du} = 20u^{19} \qquad \frac{du}{dx} = 2x$$

Thus

$$\frac{dy}{dx} = \frac{dy}{du} \cdot \frac{du}{dx} = (20u^{19}) \cdot (2x) = 20(x^2 + 1)^{19} \cdot (2x) = 40x(x^2 + 1)^{19} \quad ■$$

By using the same ideas, we may prove a general rule—a helpful special case of the chain rule.

Extended Power Law If $g(x)$ is a differentiable function and r a real number, consider the function

$$y = (g(x))^r = \left[u^r \Big|_{u=g(x)} \right]$$

obtained by composing $y = u^r$ and $u = g(x)$. Then its derivative is

$$\frac{dy}{dx} = ru^{r-1}\frac{du}{dx} = r(g(x))^{r-1}\frac{dg}{dx} \qquad [9]$$

The extended power law suffices for the previous example, in which $g(x) = x^2 + 1$, and $r = 20$. Here is another example of its use.

Example 11 Find the derivative of the function $y = (1 - x + x^2)^{1/2} = \sqrt{1 - x + x^2}$.

Solution Apply the extended power law, taking $g(x) = 1 - x + x^2$ and $r = \frac{1}{2}$:

$$\frac{dy}{dx} = \frac{1}{2}(1 - x + x^2)^{-1/2} \cdot \frac{d}{dx}(1 - x + x^2)$$

$$= \frac{1}{2\sqrt{1 - x + x^2}} \cdot (-1 + 2x)$$

$$= \frac{2x - 1}{2\sqrt{1 - x + x^2}}$$

Example 12 Differentiate $y = 1/(3 - x)^2$.

Solution We could use the quotient rule, but it is easier to use [9], taking $g(x) = 3 - x$ and $r = -2$. Then $y = (3 - x)^{-2}$, and

$$\frac{dy}{dx} = -2(3 - x)^{-3}\frac{d}{dx}(3 - x) = -2(3 - x)^{-3}(-1)$$

$$= \frac{2}{(3 - x)^3}$$

The extended power law will *not* work in the next two situations. They require the general chain rule given in [7] and [8].

Example 13 Find the derivative of $y = e^{-x^2/2}$.

Solution This function is the composite of two simpler functions

$$y = e^u \quad \text{and} \quad u = -\frac{x^2}{2}.$$

We know that $dy/du = e^u$. Thus

$$y = e^u \qquad u = -\frac{1}{2}x^2$$

$$\frac{dy}{du} = e^u \qquad \frac{du}{dx} = -x$$

and Formula [7] yields

$$\frac{dy}{dx} = \frac{dy}{du} \cdot \frac{du}{dx} = e^u \cdot (-x) = -xe^{-x^2/2}$$ ■

We remark parenthetically that the function $y = e^{-x^2/2}$ plays a central role in probability and statistics. Multiplied by a suitable constant, it gives the well-known normal, or "bell-shaped," distribution.

The same ideas used in Example 13 also yield the differentiation law for functions of the form $y = e^{kx}$, where k is a constant; just take $y = e^u$ and $u = kx$. We summarize the result for future reference.

DIFFERENTIATION OF $y = e^{kx}$ If k is a constant, the derivative of $y = e^{kx}$ is given by

$$\frac{d}{dx}(e^{kx}) = ke^{kx} \qquad [10]$$

When $k = 1$, this reduces to the law for differentiating $y = e^x$.

Example 14 If \$5000 is invested at interest rate q, continuously compounded, then in 5 years the amount in the account will be $A = 5000e^{5q}$. Find the rate of change dA/dq of investment yield A versus interest rate q.

Solution Apply Equation [10], taking $k = 5$:

$$\frac{dA}{dq} = \frac{d}{dq}(5000e^{5q})$$

$$= 5000\frac{d}{dq}(e^{5q})$$

$$= 5000 \cdot 5e^{5q} = 25{,}000e^{5q}$$ ■

If interest is not compounded continuously, the growth of money is given by a different formula, which can also be handled by the chain rule.

Example 15 If \$5000 is invested at interest rate q, compounded annually, then in 5 years the amount in the account will be $A = 5000(1 + q)^5$. Find the rate of change dA/dq of investment yield A versus interest rate q.

Solution Apply the extended power law, taking $g(q) = 1 + q$ and $r = 5$:

$$\begin{aligned}\frac{dA}{dq} &= \frac{d}{dq}(5000(1 + q)^5)\\ &= 5000\frac{d}{dq}((1 + q)^5)\\ &= 5000 \cdot 5(1 + q)^4 \frac{d}{dq}(1 + q)\\ &= 25{,}000(1 + q)^4\end{aligned}$$

■

We conclude this section with an (optional) discussion of the proof of the chain rule.

SKETCH PROOF OF THE CHAIN RULE If we give x an increment $\Delta x \neq 0$, the corresponding increment Δu in $u = g(x)$ is $\Delta u = g(x + \Delta x) - g(x)$. The increment in u in turn gives an increment Δy in $y = f(u)$ at the point $u = g(x)$:

$$\Delta y = f(g(x + \Delta x)) - f(g(x)) = f(u + \Delta u) - f(u) \tag{11}$$

Using the approximation principle twice (at each $\approx$ symbol), we obtain

$$\begin{aligned}f(u + \Delta u) - f(u) &\approx f'(u) \cdot \Delta u\\ &= f'(u)[g(x + \Delta x) - g(x)]\\ &\approx f'(u) \cdot [g'(x) \cdot \Delta x]\end{aligned}$$

so that

$$\frac{\Delta y}{\Delta x} \approx f'(u) \cdot g'(x) \cdot \frac{\Delta x}{\Delta x} = f'(u) \cdot g'(x)$$

The degree of approximation gets better and better as Δx approaches zero. In the limit we find that

$$\frac{dy}{dx} = \lim_{\Delta x \to 0} \frac{\Delta y}{\Delta x} = f'(u) \cdot g'(x) = f'(g(x)) \cdot g'(x)$$

as required. ■

Exercises 3.2

In Exercises 1–6, compute the composite function $f(g(x)) = [f(u)|_{u=g(x)}]$ for the given f and g.

1. $f(u) = u^3, \quad g(x) = 1 - x$

2. $f(u) = \dfrac{1}{u}, \quad g(x) = x^2 + 1$

3. $f(u) = u^2 + 1, \quad g(x) = \dfrac{1}{x}$

4. $f(u) = \sqrt{u}, \quad g(x) = \dfrac{1 + x}{1 - x}$

5. $f(u) = \dfrac{1 + u}{1 - u}, \quad g(x) = \sqrt{x}$

6. $f(u) = u^2 + u + 1, \quad g(x) = \sqrt{x^2 + 1}$

In Exercises 7–15, express the given function as a composite $y = f(g(x))$ of simpler functions $y = f(u)$ and $u = g(x)$.

7. $(x^2 - x + 1)^{45}$ **8.** $(7 - x)^5$ **9.** $\sqrt{1 - x^2}$

10. $\sqrt[3]{x^2 - x + 2}$ **11.** $\dfrac{1}{\sqrt{x^2 + x + 3}}$ **12.** $\sqrt{1 + \sqrt{x}}$

13. e^{-x} **14.** $\log(1 - x^2)$ **15.** $\dfrac{1}{(5 - 4x)^3}$

16. Use the chain rule to differentiate the functions given in Exercises 7–15.

In Exercises 17–22, use Formula [9] to differentiate each function.

17. $(x + 1)^3$ **18.** $\sqrt{x^2 + 1}$ **19.** $(1 + 2x)^5$

20. $\dfrac{1}{(1 + 2x)^5}$ **21.** $\left(1 + 2x - \dfrac{1}{2}x^2\right)^{3/2}$ **22.** $\dfrac{1}{\sqrt{x^2 + 1}}$

In Exercises 23–32, differentiate each function by any combination of methods.

23. $\sqrt{1 - x^2}$ **24.** $x\sqrt{1 - x^2}$ **25.** $\dfrac{2x}{\sqrt{1 - x}}$

26. $\dfrac{1 - x^2}{\sqrt{x^2 + x + 1}}$ **27.** $\left(\dfrac{x}{x + 1}\right)^{10}$ **28.** e^{5x}

29. xe^{-x} **30.** $\sqrt{\dfrac{1 - x}{1 + x}}$ **31.** $(3 + 4x)^{0.6}$

32. $\dfrac{x\sqrt{1 + x^2}}{x + 1}$

33. Use the chain rule and the fact that $d/du(e^u) = e^u$ to verify the differentiation rule

$$\frac{d}{dx}(e^{kx}) = k \cdot e^{kx} \quad (k \text{ is a constant})$$

34. If \$100 is invested at an interest rate r compounded annually, the amount on hand after 6 years will be $A = 100(1 + r)^6$. Find dA/dr, the rate at which the yield A varies with interest rate r in this scheme.

35. If \$100 is invested at an interest rate of 12% compounded continuously, the amount on hand after t years will be $A = 100e^{0.12t}$ dollars. Find the growth rate dA/dt. (*Hint*: $e^{0.12t}$ is of the form e^{kt}, where $k = 0.12$.)

36. A tracer sample of radioactive iodine ^{131}I injected into the bloodstream rapidly decays into inert products. After t days the amount remaining is

$$A(t) = e^{-0.0886t} \quad \text{micrograms}$$

if 1 microgram is injected at time $t = 0$. How fast is the amount of ^{131}I changing after t days? Initially, when $t = 0$? After 1 week? (*Note*: The constant $k = -0.0886$ in the exponential is related to the "half-life" of ^{131}I, which is 8.0 days.)

Physiological response to a certain drug depends on its blood concentration u (milligrams per liter) according to the formula

$$R(u) = 4u - 0.01u^2 \quad \text{response units}$$

After an injection, the concentration is known to vary with time

$$u = u(t) = 100e^{-0.5t} \quad \text{milligrams per liter}$$

after t hours. Answer the following questions:

37. Find the response level R as a function $R = R(t)$ of time.

38. What is the response level after 3 hours? After 5 hours?

39. Use the chain rule to calculate the rate of change dR/dt of response with respect to time t.

40. How rapidly is the response level R changing after 3 hours? Is it increasing or decreasing at this time?

3.3 Optimization Problems—A Case Study from Economics

Here are some typical items that might be found in any newspaper.

> *News item*: "The rate of inflation last month reached an all-time peak."
>
> *Advertisement*: "The NRG Motor Company offers the lowest prices in town on used cars."

Much of our everyday life is concerned with maximizing or minimizing certain variables. In this section we will show how derivatives can be used to solve such "optimization problems," illustrating the idea with a case study from economics. The approximation principle (Section 2.6) provides the key for these applications.

ILLUSTRATION A manufacturer can freely choose the level of production x for his operation. Suppose that his weekly profit is described by the function

$$P(x) = -1000 + 20x - 0.01x^2 \quad \text{dollars}$$

If he chooses his production level at random, say $x = 100$ or $x = 1500$ units per week, he might not be operating at maximum profit. The approximation principle may be applied to determine the optimum level of production.

Discussion We determine the derivative dP/dx using the standard differentiation rules

$$\begin{aligned}\frac{dP}{dx} &= \frac{d}{dx}(-1000 + 20x - 0.01x^2)\\ &= -1000(0) + 20(1) - 0.01(2x)\\ &= 20 - 0.02x\end{aligned}$$

The derivative $P'(x)$ is just the marginal profit at production level x, discussed in Section 2.5. By the marginal principle, which is just a variant of the more general approximation principle, we know that small increments Δx in production starting from some base level $x = x_0$ will cause the profit to change by an amount

$$\Delta P \approx P'(x_0) \cdot \Delta x \qquad [12]$$

Now suppose the production level is set at $x_0 = 100$, where the marginal profit is $P'(100) = 20 - 0.02(100) = 18$. Is it more profitable to change the production level slightly? By [12], if we make a small change Δx in production level to either side of the base level $x_0 = 100$ we obtain

$$\Delta P \approx 18(\Delta x) \quad \text{dollars} \qquad [13]$$

Obviously a rational strategist will choose to increase profits if possible; that is, to make $\Delta P > 0$. But in the present situation [13]

ΔP is positive (profit increases) if $\Delta x > 0$ (production increased)
ΔP is negative (profit decreases) if $\Delta x < 0$ (production decreased)

Thus, *at the particular production level* $x_0 = 100$, we can increase profits by increasing production somewhat.

Suppose we operate at a high level of production, say $x_0 = 1500$. Should we maintain this level, increase it, or decrease it? Again, Formula [12] and the sign of the marginal profit $P'(1500)$ at this production level tell us what to do. At this x_0 we see that $P'(1500) = 20 - 30 = -10$ dollars per unit. Since the marginal profit is now *negative*, the approximation formula

$$\Delta P \approx P'(1500) \cdot \Delta x = -10(\Delta x) \quad \text{dollars} \qquad [14]$$

tells us that we should *decrease* production (take $\Delta x < 0$) to increase profit. If $x_0 = 1500$, the production level is so high that substantial diseconomies of size occur, and it pays to cut back production.

Ultimately we want to pick a production level that *maximizes* profits, so that P is as large as it can get. If we randomly pick a production level x_0, the preceding remarks tell us which way to move to increase P. They also tell us something else of great importance: If $P'(x_0) \neq 0$, there is *some* direction we can move x to achieve an even higher profit. By default:

The only production levels at which P can possibly be a maximum are those for which $dP/dx = 0$.

There are not many values of x for which $dP/dx = 0$. In fact

$$0 = \frac{dP}{dx} = 20 - 0.02x$$

only if $x = 1000$. At this production level the profit is $P(1000) = \$9000$. That this is in fact the "maximizing" choice of x can also be seen from the graph of the profit function shown in Figure 3.2. But we located this maximum without ever taking the trouble to sketch the graph! ■

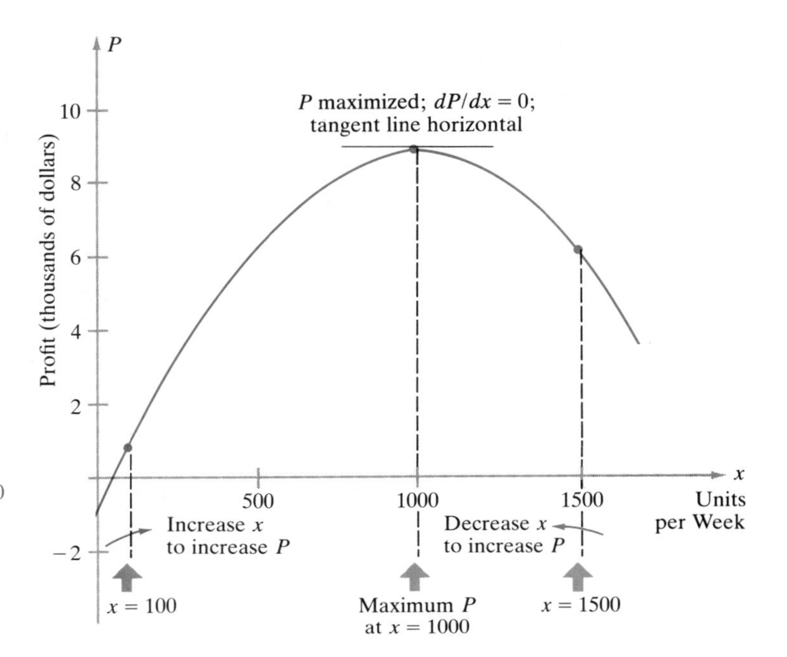

Figure 3.2
Graph of the profit function $P = -1000 + 20x - 0.01x^2$. At $x = 100$, marginal profit is positive ($dP/dx > 0$), so P *increases* if x is increased. If $x = 1500$, then $dP/dx < 0$, so P *decreases* if x is increased. Profit reaches a maximum at $x = 1000$. Then, $dP/dx = 0$ (horizontal tangent line). By solving the equation $dP/dx = 0$ we may find this point using algebra. This is an improvement over the earlier method of carefully plotting the graph and locating the maximum by eye.

We have deliberately kept the preceding discussion simple by avoiding any mention of the feasible set of x values, taking $P(x)$ to be defined for all x. In more complicated problems, the feasible set of x might be some interval such as $0 \leq x \leq 2000$. Then there is more to the story, as we will explain in the next section.

Meanwhile, we wish to point out some implications of the preceding discussion. That discussion could be applied verbatim to *any* differentiable function $y = f(x)$ defined for all x (or on an interval without endpoints, of the form $a < x < b$). Let us consider such a function at a base point $x = x_0$ where the derivative $f'(x_0)$ is nonzero. Exactly as in the case study

$$\Delta y \approx f'(x_0) \cdot \Delta x \tag{15}$$

for all small increments Δx away from this base point. If $f'(x_0) \neq 0$, Equation [15] tells us that we can always move away from x_0 in some direction so as to increase the value of f. (Moving in the opposite direction will decrease f; the sign of $f'(x_0)$ tells us which way to move.) In particular, if $f'(x_0) \neq 0$ the function cannot have a maximum or minimum value at $x = x_0$. If we wish to maximize (or minimize) f, our attention is therefore directed to those points where $df/dx = 0$. This establishes a basic result.

> **THE OPTIMIZATION PRINCIPLE** Let $y = f(x)$ be differentiable on an interval $a < x < b$. The points (if any) where $f'(x) = 0$ are the only ones at which the function can achieve a maximum or minimum value.

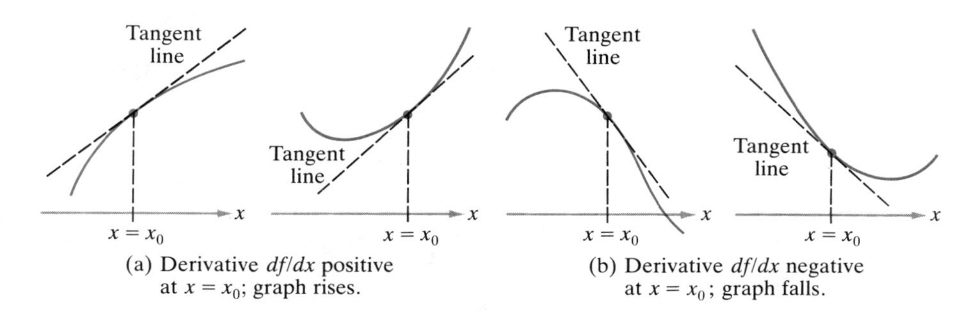

(a) Derivative df/dx positive at $x = x_0$; graph rises.
(b) Derivative df/dx negative at $x = x_0$; graph falls.

Figure 3.3
In (a) we show some graphs that are rising when $x = x_0$: the derivative is positive, $f'(x_0) > 0$. In (b) the graphs are falling at x_0 and $f'(x_0) < 0$. In general, the sign of the derivative determines whether the graph is rising or falling.

We were not interested in minimum values when we studied profit functions, but there are many applications in which we seek a minimum (studying costs, for instance).

This discussion also shows that whenever the derivative df/dx is nonzero, its sign determines whether $f(x)$ is increasing or decreasing as x moves to the right in the number line; in other words, whether the graph of $y = f(x)$ is rising or falling (see Figure 3.3).

> **GEOMETRIC SIGNIFICANCE OF THE SIGN OF df/dx** If $y = f(x)$ is differentiable then
>
> (i) $f(x)$ is increasing (graph rises) if $df/dx > 0$.
> (ii) $f(x)$ is decreasing (graph falls) if $df/dx < 0$.

Sign of dy/dx — $dy/dx = 0$ here

$+ + + + + + \tfrac{1}{2} - - - - -$ x

←Rising→←Falling→

(a) Behavior of graph as x increases.

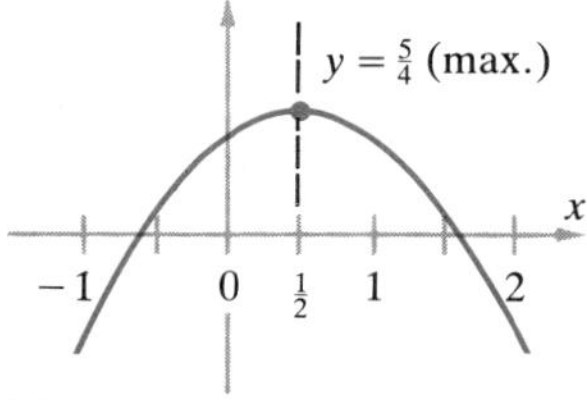

(b) General shape of the graph of $y = 1 + x - x^2$

Figure 3.4
A "sign diagram," showing where the derivative $dy/dx = 1 - 2x$ is positive or negative, is given in (a). In (b) we show the general shape the graph of the original function $y = 1 + x - x^2$ must have in view of the information in the sign diagram (a).

Example 16 Determine where the function $y = 1 + x - x^2$ is increasing and where it is decreasing. Use this information to find the maximum value of y.

Solution The derivative of this polynomial is

$$\frac{dy}{dx} = \frac{d}{dx}(1) + \frac{d}{dx}(x) - \frac{d}{dx}(x^2)$$

$$= 1 - 2x$$

The derivative is zero when $1 - 2x = 0$, or $x = \frac{1}{2}$. By the optimization principle, this is the only point at which a maximum or minimum could occur. To see that we actually have a maximum there, we determine where the function is increasing or decreasing. The derivative is positive, $dy/dx = 1 - 2x > 0$, if

$$2x < 1 \quad \text{or} \quad x < \frac{1}{2}$$

Thus, dy/dx is positive, and y is increasing (graph rises) for all x to the left of $x = \frac{1}{2}$. Similarly, dy/dx is negative, and y is decreasing (graph falls) to the right of $x = \frac{1}{2}$. As x moves from left to right, the graph rises until we reach $x = \frac{1}{2}$; thereafter, the graph falls steadily, as shown in Figure 3.4b. Thus, the largest value is attained at $x = \frac{1}{2}$, where the value of y is $\frac{5}{4}$. ■

Exercises 3.3

1. A manufacturer's profit is described by $P = -10{,}000 + 75x - 0.03x^2$. Calculate the marginal profit function. If the present level of production is $x = 1000$, will a slight increase in x cause profit to increase or decrease? What if $x = 2000$?

2. A retail gasoline station has profit function $P = -4320 + 0.12x - 0.000001x^2$ dollars per month. If current monthly sales are 40,000 gallons, does it pay to try to increase sales? What if $x = 70{,}000$?

3. The function $y = 1 - x^4$, defined for all x, has a maximum value. Find it. (*Hint*: Apply the optimization principle.)

4. Find all points in the interval $0 < x < \frac{4}{3}$ where the function $y = x^3 - 2x^2 + x + 1$ might have a maximum or minimum.

In Exercises 5–10 find *all* points where the derivative is zero.

5. $f(x) = 40 - 13x$

6. $f(x) = 1 + x + x^2$

7. $f(x) = 2x^3 - 3x^2 - 12x + 1$

8. $y = 1 + 3x + 3x^2 + x^3$

9. $f(r) = 1 + r^3$

10. $P(q) = -10{,}000 + 153q - 0.003q^2$

11. Calculate the derivative of $y = x^3 - 3x$. Is this function increasing or decreasing at $x = -2$? At $x = 0$? At $x = 0.5$? At $x = 3$?

12. Find the derivative of $f(x) = x^3 - 2x^2 + x - 1$. Is f increasing or decreasing at $x = -1$? At $x = 0$? At $x = 2$?

13. Find the derivative of $y = x - 3x^2 + 1$. For which x is y increasing? Decreasing?

14. If $P(q) = 5q - 0.004q^2 - 1000$, calculate the marginal profit dP/dq. For which production levels is the profit increasing with q? For which levels is it decreasing? Find the maximum profit and corresponding production level q.

In Exercises 15–22 make a sketch of the number line showing where the function is increasing and decreasing.

15. $f(x) = 3x - 2$

16. $f(x) = 1 + x + x^2$

17. $f(x) = 3 + 6x - x^2$

18. $y = x^2$

19. $y = 1 - x^2$

20. $y = x^3$

21. $y = 1 + r^3$

22. $P = -10{,}000 + 153q - 0.003q^2$

23. The profit function in Example 19, Section 1.3 was

$$P(x) = -5{,}000{,}000 + 61x - 0.000098x^2$$

Where is the marginal profit equal to zero? Make a sketch of the number line showing where P is increasing ($dP/dx > 0$) and where it is decreasing ($dP/dx < 0$). Where does P have the largest possible value?

24. Suppose a manufacturer's costs are $C(x) = 5500 + 47x + 0.1x^2$ at production level x. Then the average cost per unit at production level x is defined to be

$$A(x) = \frac{\text{total cost}}{\text{number of units}} = \frac{C(x)}{x}$$

so that, in the present situation

$$A(x) = \frac{5500}{x} + 47 + 0.1x \quad \text{for any } x > 0$$

For which $x > 0$ is the average cost increasing with x? For which is it decreasing? Where does the transition from decreasing to increasing take place?

3.4 Maxima and Minima of a Function

Having introduced some basic ideas in Section 3.3, we are ready to begin detailed study of optimization problems. We defer full discussion of applied "word problems" until Section 3.7 because they involve the additional difficulty of converting a verbal description into mathematical form. In this and the following sections we develop techniques for solving optimization problems, keeping a low profile on the subject of word problems. First we define some terms.

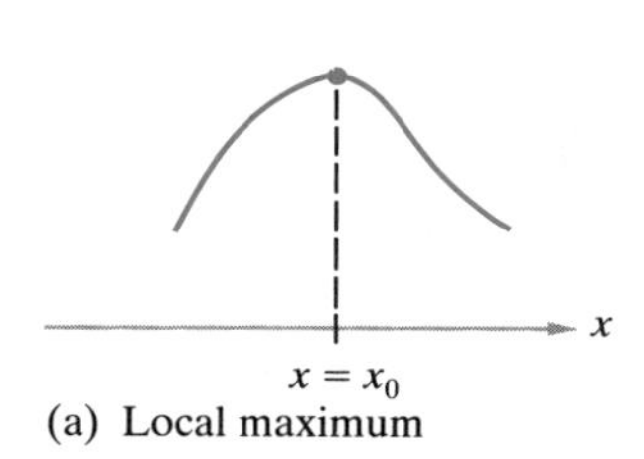

(a) Local maximum

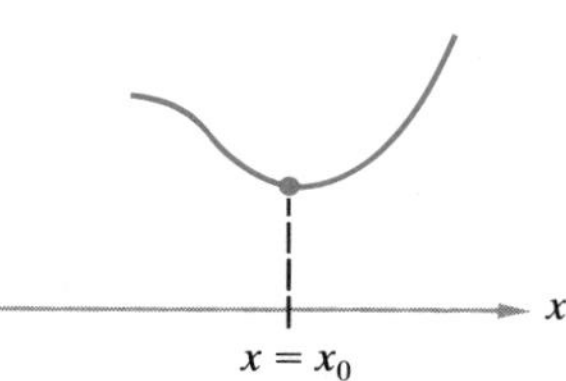

(b) Local minimum

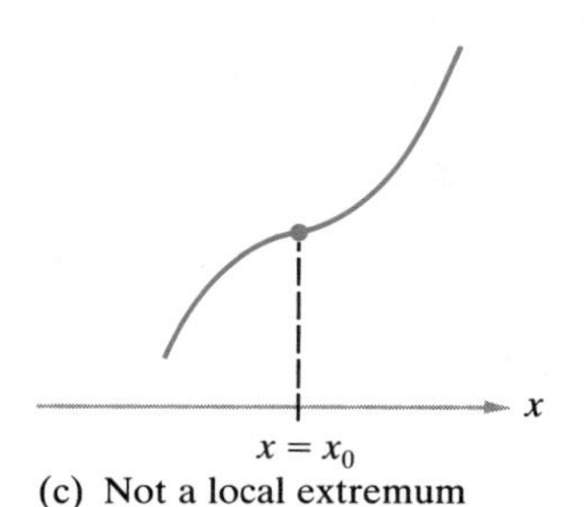

(c) Not a local extremum

Figure 3.5
We can decide if $f(x)$ has a local maximum or minimum at x_0 by comparing the value $f(x_0)$ at x_0 with the values $f(x)$ for points close to x_0. Some of the possibilities are shown here.

> DEFINITION If a function $y = f(x)$ is defined at $x = x_0$, we say that f has
>
> (i) a **local maximum** at x_0 if $f(x) \leq f(x_0)$ for all points x in the feasible set that lie near x_0. That is, the value of f is at least as large at x_0 as it is at any *nearby* point where f is defined.
>
> (ii) an **absolute maximum** at x_0 if $f(x) \leq f(x_0)$ for all points x in the feasible set, whether they lie close to x_0 or not.

Local (and **absolute**) **minima** are defined similarly: For minima $f(x) \geq f(x_0)$ for all nearby points x (for all points x) in the feasible set. If there is either a maximum or minimum at $x = x_0$, we sometimes combine these two possibilities by saying f has an **extremum** at x_0. Notice: once we know the place (value of x) where the largest or smallest value of f occurs, the value $y = f(x)$ is easy to calculate. Calculus methods will be used to locate the appropriate x.

The general shape of the graph of $y = f(x)$ near a local maximum or minimum is shown in Figure 3.5. You might reread the definition with these pictures in mind. In Figure 3.6, we show a more general graph. On it we have indicated all the local and absolute extrema. Notice two things:

(i) There are a number of local extrema that are not absolute extrema.
(ii) An extremum can occur at an endpoint of the feasible set.

An **optimization problem** is one in which we seek the absolute maximum (or minimum) of some function $y = f(x)$. We have already seen that the places where the derivative is equal to zero are important, so the following definition will prove useful.

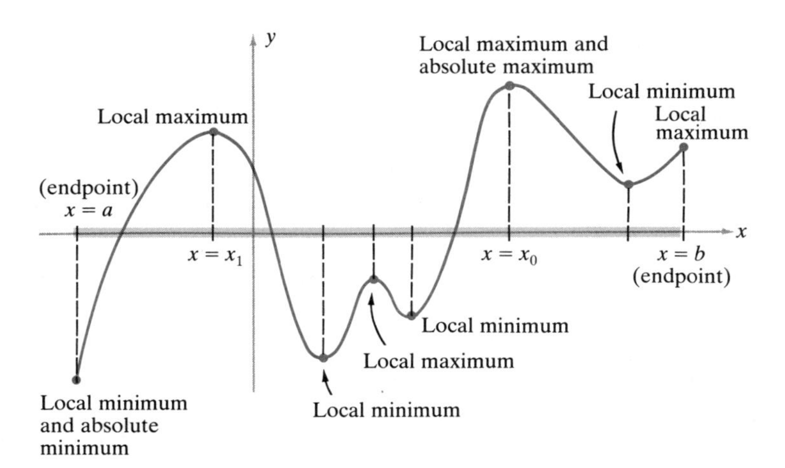

Figure 3.6
On the graph of $y = f(x)$, local maxima correspond to high points (peaks) and local minima to low points (valleys). If f is defined on the interval $a \leq x \leq b$ (shaded), the largest value is achieved at x_0; f has an absolute maximum there. A local maximum occurs at x_1, but this is not an absolute maximum, because $f(x)$ has an even higher value at x_0, some distance away. The various absolute and local extrema are labeled. Note carefully that a maximum or minimum may occur at either endpoint (if they are feasible). Here the absolute minimum occurs at endpoint $x = a$. There is a local maximum at the other endpoint $x = b$.

> **Definition** A point x in the feasible set of a differentiable function $f(x)$ is a **critical point** for f if $df/dx = 0$ there.†

The critical points are instrumental in locating the extrema of f, but if the feasible set has endpoints, these too must be examined. This brings us to the first kind of optimization problem we shall solve.

Problem (Optimization on a Bounded Interval) Locate the absolute maximum and minimum of a function $f(x)$ that is defined and differentiable on a *closed bounded interval* $a \leq x \leq b$, one that is bounded and includes both endpoints.

In many practical situations the domain of definition of f is bounded from above and below, hence our interest in this special problem. It is surprising that a very simple procedure solves the problem.

† We will consider differentiable functions in this book. In more general discussions, a critical point is taken to be any point where $df/dx = 0$ or the derivative is undefined.

> **LOCATING ABSOLUTE EXTREMA ON A BOUNDED INTERVAL** Suppose $f(x)$ is differentiable in a closed, bounded interval $a \le x \le b$. Then
>
> *Step 1* Locate all critical points $x_1, \dots, x_k$ within the interval.
> *Step 2* Calculate the values $y = f(x)$ at these critical points and at the endpoints $x = a$ and $x = b$.
> *Step 3* Compare the values $f(a), f(x_1), \dots, f(x_k), f(b)$. The largest of these is the absolute maximum for f on the interval $a \le x \le b$; the smallest is the absolute minimum.

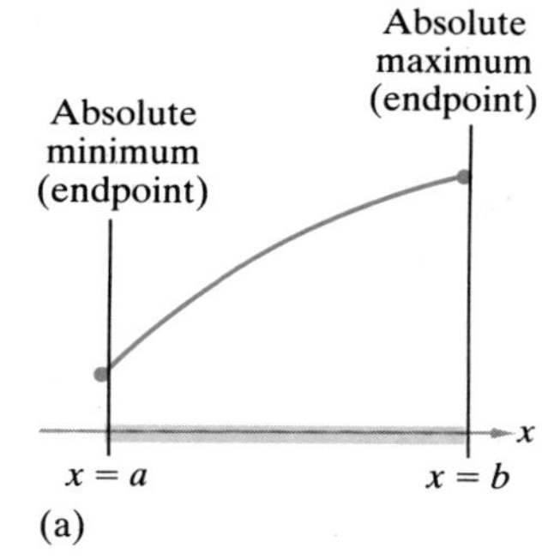

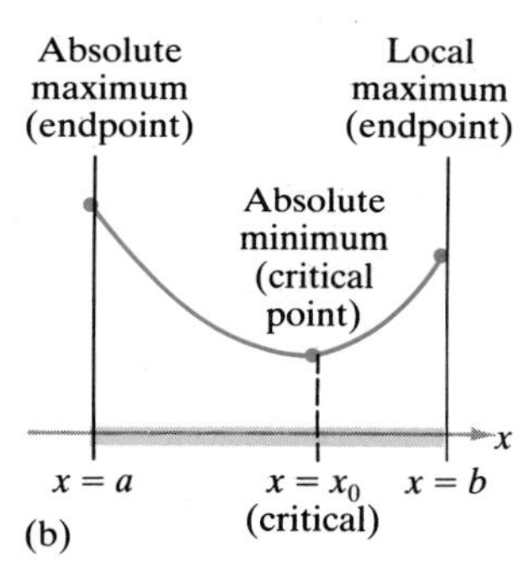

Figure 3.7
Graphs of two differentiable functions defined on an interval $a \le x \le b$, including the endpoints. Critical points occur where the tangent line is horizontal (slope $dy/dx = 0$). No critical points occur in (a) and just one occurs in (b), at x_0. Endpoint extrema are labeled.

To see why this is so, suppose x_0 is not a critical point or an endpoint. Then f is defined on both sides of x_0, f is differentiable at x_0, and $f'(x_0)$ is nonzero. By the approximation principle, $\Delta f \approx f'(x_0) \cdot \Delta x$ for all small increments Δx away from x_0, so we can increase $f(x)$ by moving in one direction and decrease $f(x)$ by moving in the opposite direction, starting from x_0. Thus an extremum cannot occur at x_0. By default, an extremum can occur only at a critical point or an endpoint.

To complete the justification of the procedure stated above we invoke a result that is simple enough to state, but whose proof must be left to more advanced courses: *If $f(x)$ is continuous on a closed bounded interval $a \le x \le b$, then absolute maxima and absolute minima actually exist somewhere in the interval.* (Simple examples show that absolute extrema might not exist at all if $f(x)$ is defined on an unbounded interval or one that fails to include both endpoints, see Exercises 26 and 27.) Once the existence of absolute extrema is assured, the rest of the argument is easy. An absolute maximum occurs somewhere, say at $x = x_0$. By the preceding observations, x_0 must appear in our list of points $a, x_1, \dots, x_k, b$ from Steps 1 and 2. There is an absolute maximum at x_0, so $f(x_0) \ge f(x)$ for all x, and in particular $f(x_0) \ge f(a), f(x_0) \ge f(x_1), \dots, f(x_0) \ge f(x_k), f(x_0) \ge f(b)$. That is, x_0 must be the point in our list at which f has the largest value, as stated in Step 3. The same reasoning applies to finding absolute minima.

The nature of "endpoint extrema" deserves more comment. Extrema in the *interior* $a < x < b$ of the domain of f signal their presence by occurring only at critical points. Endpoint extrema are different and can occur even if the endpoint involved is not critical. In Figure 3.7 we show some of the ways endpoint extrema can occur. In Figure 3.7a, there are no critical points at all in the domain of f. (Remember: At critical points $f'(x_0) = 0$, and the tangent line is horizontal.) In Figure 3.7b, the absolute minimum occurs in the interior at the critical point $x = x_1$, but the absolute maximum is at the left endpoint.

Example 17 The following functions are defined and differentiable (hence continuous) on the interval $-1 \le x \le 1$:

(i) $f(x) = 3x - 2$ (ii) $g(x) = x^2 + 1$

Find the critical points and absolute extrema for each function. What is the maximum value of each function on this interval?

Solution In (i) we have $df/dx = 3$ (constant function). Because df/dx is never zero, there are no critical points. Next, calculate the endpoint values of f

$$f(-1) = 3 \cdot (-1) - 2 = -5 \qquad f(1) = 3(1) - 2 = 1$$

to complete Step 2. Finally, the absolute maximum value is achieved at the right-hand endpoint $x = 1$, where $f(1) = 1$. (The absolute minimum occurs at $x = -1$.)

In (ii) we get $g'(x) = 2x$ for all x. Therefore, $g'(x) = 0$ at $x = 0$; $x = 0$ is the only critical point. The values $y = g(x)$ at this critical point and the endpoints $x = -1, x = 1$ are

$$g(-1) = 2 \qquad g(0) = 1 \qquad g(1) = 2$$

As in Step 3, the absolute maximum occurs at $x = -1$ and also at $x = 1$, where $g(-1) = g(1) = 2$. The absolute minimum occurs at $x = 0$. ■

Example 18 Find the absolute maximum value of $f(x) = -x^4 + 4x^3 + 8x^2 + 3$ on the interval $-3 \le x \le 3$.

Solution To find the critical points we must solve the equation

$$0 = \frac{df}{dx} = -4x^3 + 12x^2 + 16x$$

We do this by factoring the polynomial to obtain

$$0 = -4x(x^2 - 3x - 4) = -4x(x - 4)(x + 1)$$

The only way the product can be zero is for one of the factors to be zero. Thus the solutions of the equation $df/dx = 0$ are

$$x_1 = -1 \qquad x_2 = 0 \qquad x_3 = 4 \tag{16}$$

Because x_3 does not lie in the feasible set $-3 \le x \le 3$, we do not consider it any further; the *feasible* critical points to be listed in Step 1 are $x_1 = -1$ and $x_2 = 0$.

Next, compute $y = f(x)$ at the feasible critical points and at the endpoints $a = -3, b = 3$:

$$f(a) = f(-3) = -114 \qquad f(x_2) = f(0) = 3$$
$$f(x_1) = f(-1) = 6 \qquad f(b) = f(3) = 102$$

The largest of these values is $y = 102$, at the right endpoint $x = 3$. This is the absolute maximum value of f on $-3 \le x \le 3$. ■

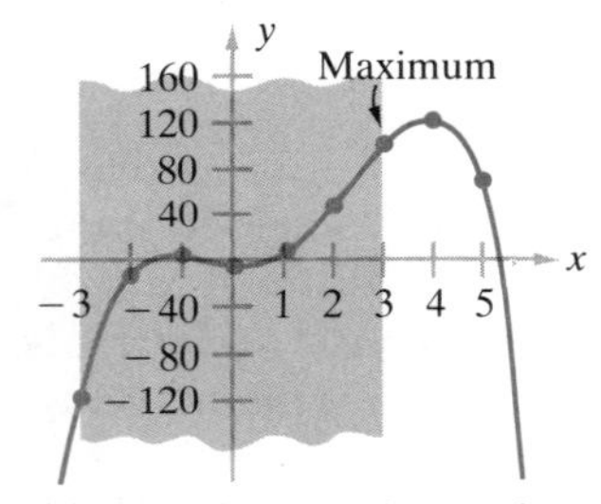

(a) Feasible set: $-3 \le x \le 3$

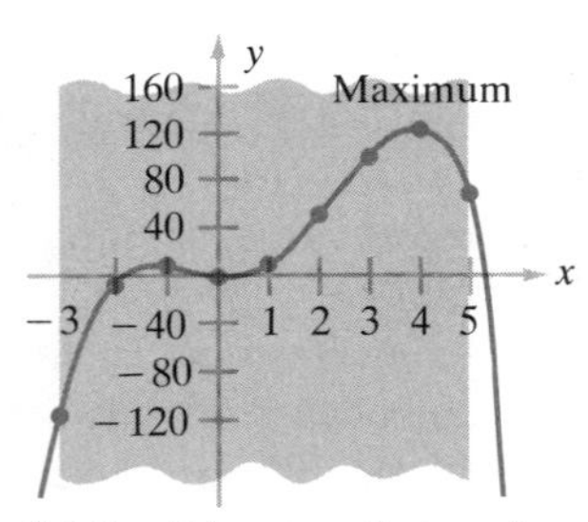

(b) Feasible set: $-3 \le x \le 5$

Figure 3.8
Graph of $y = -x^4 + 4x^3 + 8x^2 + 3$. In (a) the feasible set is $-3 \le x \le 3$; in (b) it is $-3 \le x \le 5$. Feasible sets correspond to the vertical shaded strips. Though the function is the same in both cases, the nature of the feasible set strongly affects the maximum and minimum values of y.

We could vary this problem a bit by trying to maximize $f(x)$ on a different interval, say $-3 \le x \le 5$. When we get to Step 1, all solutions [16] of the equation $df/dx = 0$ are now feasible, and the final answer is different; the absolute maximum now occurs at $x = 4$, where $y = 131$ (see Exercise 12 for details). Although we did not need a graph of $y = -x^4 + 4x^3 + 8x^2 + 3$ to solve either problem, we may find it interesting to refer to the graph in Figure 3.8 to see why the answers turn out as they do for the different intervals $-3 \le x \le 3$ and $-3 \le x \le 5$.

Example 19 The weekly revenue R for the publisher of the stock market newsletter mentioned in Example 18, Section 2.3 rises with the amount x spent each week on advertising according to the formula

$$R(x) = 25{,}000 + 150\sqrt{x} \quad \text{dollars}$$

His costs for a fixed weekly printing run plus advertising are $C(x) = 10{,}000 + x$, so his weekly profit is

$$P(x) = R - C = 15{,}000 + 150\sqrt{x} - x \qquad [17]$$

If he is willing to spend as much as $5000 on advertising, how much should he spend to maximize his profit?

Solution The feasible set is $0 \leq x \leq 5000$. The critical points are determined by solving the equation

$$0 = \frac{dP}{dx} = \frac{75}{\sqrt{x}} - 1$$

The only solution occurs when $75/\sqrt{x} = 1$ or $\sqrt{x} = 75$, that is, when $x = (75)^2 = \$5625$. This solution is *not* in the feasible set; there are no feasible critical points, and the maximum for $P(x)$ can only occur at one of the endpoints $x = 0$ or $x = 5000$. Because

$$P(0) = \$15{,}000$$
$$P(5000) = 15{,}000 + 150\sqrt{5000} - 5000 = \$20{,}606.60$$

the maximum is at $x = 5000$. He should go the limit and invest the full $5000 in advertising. ■

If he were willing to invest up to $8000 in advertising, optimum results would no longer be obtained by spending the full amount (see Exercise 30). For a clearer understanding of what is going on we show the graph of the profit function [17] in Figure 3.9.

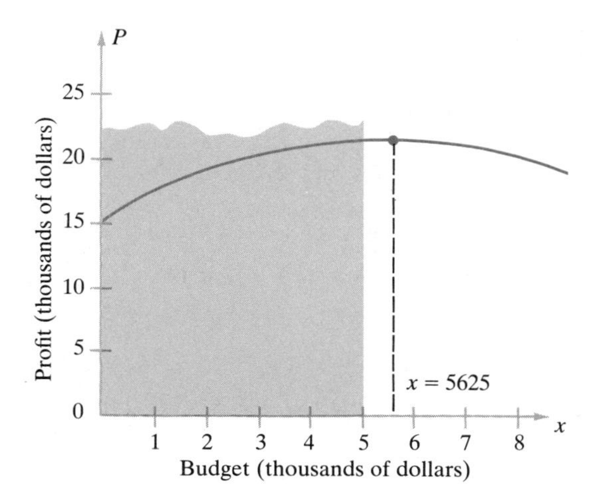

Figure 3.9
The publisher's profit function $P(x) = 15{,}000 + 150\sqrt{x} - x$ in Example 19. The feasible set $0 \leq x \leq 5000$ in that problem corresponds to the shaded strip. The critical point for $P(x)$ is at $x = 5625$ (not feasible). Absolute maximum occurs at the endpoint $x = 5000$.

We must emphasize that the optimization procedure we have been using in this section is valid only for functions defined on closed, bounded intervals, including endpoints. For other types of feasible set, absolute maxima and minima might not exist; and even if they do exist, more study is required to find them. These more general optimization problems will be discussed in Section 3.6.

Exercises 3.4

In Exercises 1–9 find the critical points for each function.

1. $y = -3x + 4$

2. $y = 4 + 7x - 3x^2$

3. $y = \frac{1}{3}x^2 - \frac{1}{4}x + 2$

4. $y = 3x - 3x^2 + x^3$

5. $y = x^3 - 12x + 7$

6. $y = 3x^4 - 16x^3 + 18x^2 - 5$

7. $y = \dfrac{1}{x} - 4x^2$

8. $y = \dfrac{x}{x^2 + 1}$

9. $y = x\sqrt{x^2 + 1}$

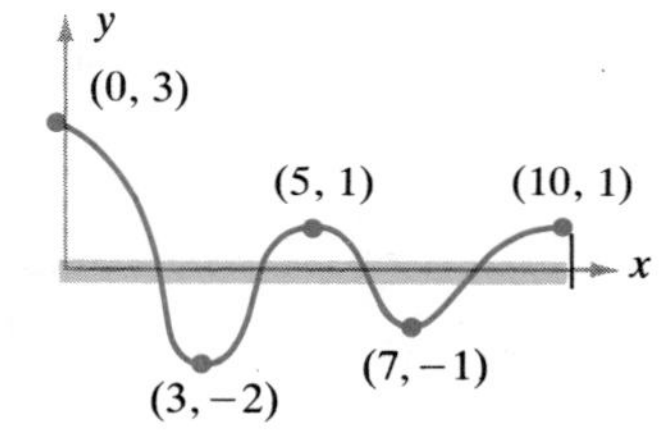

Figure 3.10
The graph in Exercise 10. The feasible set is shaded.

10. The function whose graph is shown in Figure 3.10 is defined for $0 \le x \le 10$. The coordinates of various interesting points on the graph are indicated. List all values of x where the function has
(i) a local maximum
(ii) a local minimum
(iii) an absolute maximum
(iv) an absolute minimum
What are the absolute maximum and minimum values of $f(x)$?

11. Find the absolute maximum and minimum values of the following functions on the interval $-1 \le x \le 2$.
(i) $y = -3x + 4$ (ii) $y = 4 + 7x - 3x^2$ (iii) $y = x^3 - 12x + 7$

12. Find the absolute maximum and minimum values achieved by $y = -x^4 + 4x^3 + 8x^2 + 3$ on the bounded interval $-3 \le x \le +5$.

In Exercises 13–25 use the procedure given in this section to locate the absolute maxima and minima for each function on the interval indicated.

13. $y = 5 - 2x$ on $-1 \le x \le 2$

14. $y = 1 + x - \frac{1}{2}x^2$ on $-2 \le x \le 2$

15. $y = x^2 - 6x + 5$ on $0 \le x \le 2$

16. $y = x + \dfrac{1}{x}$ on $\frac{1}{3} \le x \le 2$

17. $y = \frac{1}{3}x^3 + x^2 - 3x + 1$ on $-5 \le x \le 4$

18. $y = x^3 - x^2 + x$ on $-2 \le x \le 0$

19. $y = x(x - 1)$ on $0 \le x \le 1$

20. $y = 2x - 3$ on $2 \le x \le 5$

21. $y = x^2(x - 6)$ on $-2 \le x \le 5$

22. $y = x^2 + \dfrac{16}{x}$ on $1 \le x \le 4$

23. $C = 4320 + 0.60q - 0.000005q^2$ on $0 \leq q \leq 40{,}000$

24. $s = \dfrac{4t}{t^2 + 1}$ on $0 \leq t \leq 10$

25. $w = \dfrac{1 - r}{1 + r}$ on $0 \leq r \leq 100$

26. Sketch the graph of each of the following functions. In each case take the entire number line as the domain of definition.
(i) $y = x + 1$ (linear function)
(ii) $y = x^2$ (parabola)
(iii) $y = x^3$ (cubic curve)
By examining the graphs, decide whether there are any *absolute extrema*. Pay particular attention to what happens as x becomes large positive or large negative.

27. Repeat Exercise 26, this time taking the bounded interval $-1 \leq x \leq 2$ as the domain in each case.

28. The profit function in Example 19, Section 1.3 was

$$P(q) = -5{,}000{,}000 + 61q - 0.000098q^2$$

for $0 \leq q \leq 750{,}000$. Find the absolute maximum and the corresponding value of q. Repeat the problem, assuming the domain of P to be $0 \leq q \leq 250{,}000$.

29. At production level q, suppose that the cost function for a factory is

$$C(q) = 1000 + 100q - q^2$$

the demand function is

$$p(q) = 300 - 5q$$

and the feasible set of production levels is $0 \leq q \leq 40$.
(i) Find the revenue function $R(q)$.
(ii) Find the profit function $P(q)$.
(iii) Find the value of q that maximizes profit.

30. Repeat Example 19, assuming that the publisher is willing to spend any amount up to \$8000 per week on advertising.

31. A person expels air from the lungs through a circular tube-like organ called the trachea, or windpipe. Coughing serves the useful purpose of clearing congestion from the windpipe. The mechanism of a cough involves contraction of the trachea and simultaneous build-up of air pressure in the lungs, which forces air to rush at high velocity through the trachea (the cough). The greater the air speed, the more effective the coughing process. Experimental and theoretical studies show that the air speed v through the trachea during the cough is given by

$$v = k(0.45 - r)r^2$$

where $0.45 =$ normal trachea radius (in inches), and $r =$ smaller radius of trachea during the cough, and k is a constant. Find the radius r for which v is a maximum. (*Hint*: r lies between $r = 0$ (collapsed windpipe) and $r = 0.45$.)

3.5 Higher-Order Derivatives and Their Meaning

If a function $y = f(x)$ is differentiable, its derivative is another function $f'(x)$. If $f'(x)$ is itself differentiable, we can apply the differentiation process to it. The resulting function, called the **second derivative**, is denoted by any one of the following interchangeable symbols:

$$\frac{d^2y}{dx^2} = \frac{d}{dx}\left(\frac{dy}{dx}\right) \qquad \frac{d^2f}{dx^2} \qquad y'' \qquad f'' \qquad f^{(2)}$$

Example 20 Find d^2y/dx^2 if $y = x^3 - 4x^2 + 2x - 3$.

Solution Our rule for differentiating polynomials yields

$$\frac{dy}{dx} = 3x^2 - 8x + 2$$

We then differentiate the function dy/dx to obtain the second derivative

$$\frac{d^2y}{dx^2} = 6x - 8$$

■

There is no reason why this process cannot be continued, provided that we obtain differentiable functions at each step. Differentiating the second derivative function, we obtain a new function called the **third derivative** d^3y/dx^3; differentiating the third derivative yields the **fourth derivative** d^4y/dx^4, and so on.

Example 21 Find d^ny/dx^n for all $n = 1, 2, \ldots$ if $y = x^3 - 4x^2 + 2x - 3$.

Solution We have found the first derivative dy/dx and the second derivative d^2y/dx^2 in Example 20. Differentiating the function $d^2y/dx^2 = 6x - 8$, we obtain

$$\frac{d^3y}{dx^3} = 6 \quad \text{(a constant function)}$$

Differentiating again, we obtain

$$\frac{d^4y}{dx^4} = 0$$

Now any further differentiation will yield the zero function, so that

$$d^ny/dx^n = 0 \quad \text{for all } n \geq 4$$

■

Example 22 .If $f(x) = (x^2 + 1)^{3/2}$, find $f''(x)$.

Solution We first calculate $f'(x)$ using Formula [9], with $r = 3/2$.

$$f'(x) = \frac{3}{2}(x^2 + 1)^{1/2} \cdot (2x) = 3x(x^2 + 1)^{1/2}$$

Differentiating once more, we obtain

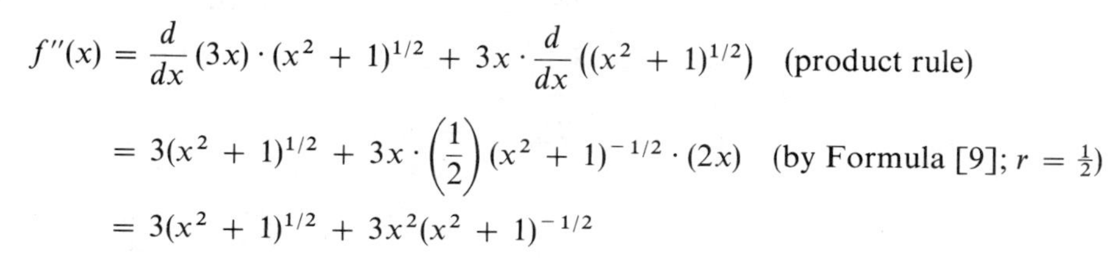

$$f''(x) = \frac{d}{dx}(3x)\cdot(x^2+1)^{1/2} + 3x\cdot\frac{d}{dx}\left((x^2+1)^{1/2}\right) \quad \text{(product rule)}$$

$$= 3(x^2+1)^{1/2} + 3x\cdot\left(\frac{1}{2}\right)(x^2+1)^{-1/2}\cdot(2x) \quad \text{(by Formula [9]; } r = \tfrac{1}{2}\text{)}$$

$$= 3(x^2+1)^{1/2} + 3x^2(x^2+1)^{-1/2}$$

Simplifying, we obtain

$$f''(x) = 3(x^2+1)^{-1/2}[(x^2+1)+x^2]$$

$$= 3(x^2+1)^{-1/2}(2x^2+1)$$

$$= \frac{3(2x^2+1)}{\sqrt{x^2+1}}$$

■

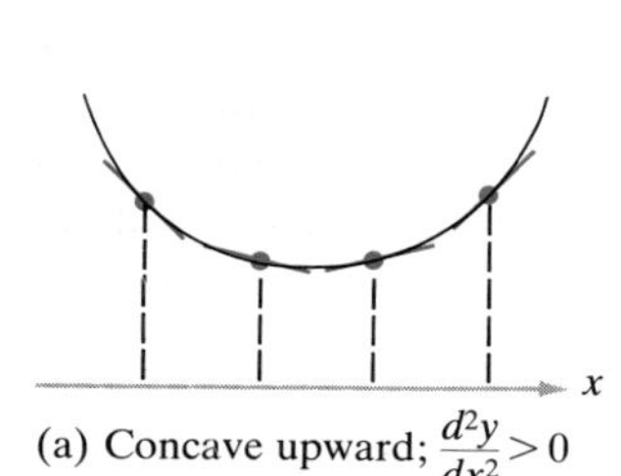

(a) Concave upward; $\frac{d^2y}{dx^2} > 0$

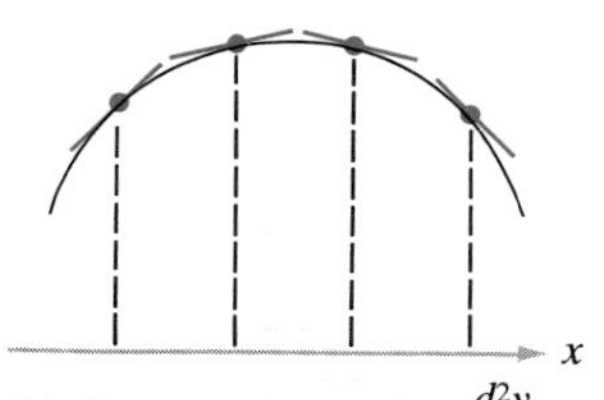

(b) Concave downward; $\frac{d^2y}{dx^2} < 0$

Figure 3.11
The concept of concavity and its relation to the second derivative d^2y/dx^2. In (a) the graph is concave upward. The slope of the tangent line steadily increases as x moves to the right, so that d^2y/dx^2 = *rate of change of slope* is positive. In (b) the graph is concave downward; the slope steadily decreases (starting positive and finally becoming more and more negative), and $d^2y/dx^2 < 0$.

There is a useful geometric interpretation of the second derivative. Recall that a function is increasing if its derivative is positive and decreasing if it is negative (Section 3.3). Now apply this remark to the function dy/dx and its derivative d^2y/dx^2 to conclude that

$$\frac{dy}{dx} \text{ is } \textit{increasing} \text{ if its derivative } \frac{d^2y}{dx^2} \text{ is } \textit{positive}$$

$$\frac{dy}{dx} \text{ is } \textit{decreasing} \text{ if its derivative } \frac{d^2y}{dx^2} \text{ is } \textit{negative}$$

But dy/dx is the slope of the graph of $y = f(x)$. If dy/dx increases as x increases, then the slope increases and the graph must "bend up" as in Figure 3.11a. A curve with this shape is said to be **concave upward**. Similarly, if the slope dy/dx decreases as x increases, then the graph "bends down" as in Figure 3.11b and is said to be **concave downward**. We summarize these observations in the following way.

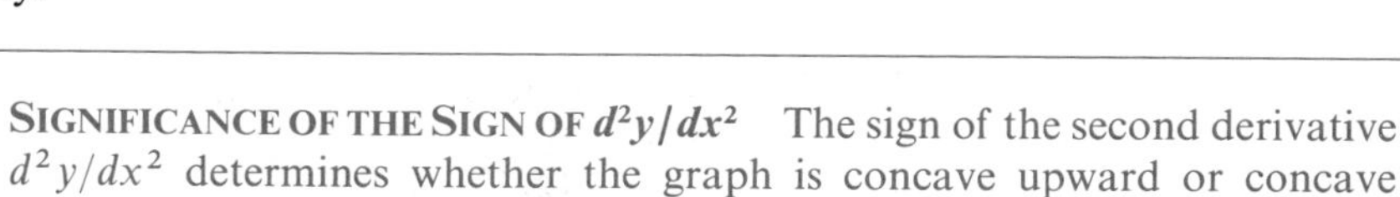

SIGNIFICANCE OF THE SIGN OF d^2y/dx^2 The sign of the second derivative d^2y/dx^2 determines whether the graph is concave upward or concave downward.

(i) The graph is concave up if $\frac{d^2y}{dx^2} > 0$ (see Figure 3.11a).

(ii) The graph is concave down if $\frac{d^2y}{dx^2} < 0$ (see Figure 3.11b).

At points where $d^2y/dx^2 = 0$, we can say nothing definite about the concavity of the graph. Often, the concavity changes from upward to downward, or vice versa, at such points.

Example 23 Where is the graph of the function $y = x^3 - 4x^2 + 2x - 3$ concave upward? Concave downward?

Solution In Example 20 we calculated the second derivative of this function, $d^2y/dx^2 = 6x - 8$. It is positive where $6x - 8 > 0$ or $x > 4/3$, and is negative if $x < 4/3$. Therefore, the graph is concave upward when $x > 4/3$, as indicated in Figure 3.12, and is concave downward when $x < 4/3$. At $x = 4/3$ we have $d^2y/dx^2 = 0$. There, the curve actually changes from concave downward to concave upward. Such a transition point where the concavity changes is called an **inflection point** of the curve. The presence of an inflection point is usually marked by a zero value for d^2y/dx^2. ■

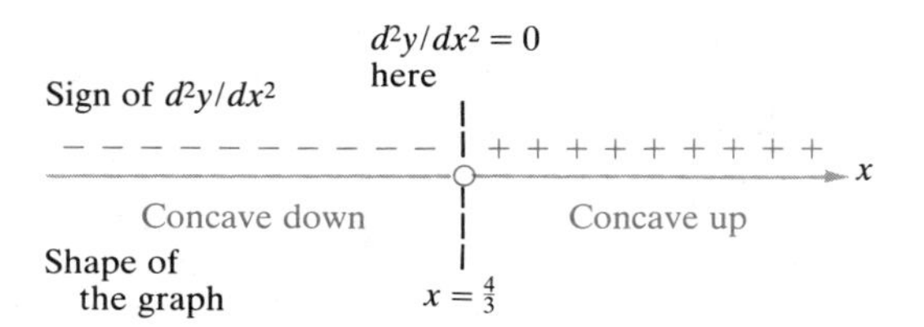

Figure 3.12
A sign diagram for the second derivative of $y = x^3 - 4x^2 + 2x - 3$, showing where the graph is concave upward or downward.

The second derivative of a function $f(x)$ is useful both for what it tells us about the function and also for applications. For example, what is it that causes the queasy feeling you sometimes experience in a high-speed elevator? It does not occur when you travel at constant velocity, even though your velocity may be very great. You get this feeling when the elevator starts and stops; in other words, it is the result of a *change* in velocity. In both common parlance and in physics, change in velocity is called acceleration. That is, the **acceleration** of a moving object is the instantaneous rate of change of its velocity $v(t)$. If the position of the object is given as a function of time by $s = s(t)$, the velocity is (by definition) the first derivative

$$v(t) = s'(t) = \frac{ds}{dt}$$

Thus, the acceleration must be the second derivative of $s(t)$,

$$\text{acceleration} = \frac{d}{dt}v(t) = \frac{d}{dt}\left(\frac{ds}{dt}\right) = \frac{d^2s}{dt^2}$$

We shall illustrate the concept of acceleration in the next two examples, the first describing free fall and the second describing the motion of an elevator.

Example 24 A stone falls a distance $s = 16t^2$ feet in t seconds after it has been released. Find the acceleration as a function of time.

Solution The velocity (measured in feet per second) is $ds/dt = 32t$, and the acceleration is therefore $d^2s/dt^2 = 32$ (measured in *feet per second*^{2} or *feet per second per second*). The acceleration is therefore *constant*, equal to 32 feet per second^2. This acceleration is a physical constant, characteristic of all dense objects falling freely under the influence of gravity at the Earth's surface. (On the surface of the Moon a different constant would apply, roughly 5.33 feet per second^2, because $s = 2.66\,t^2$.) ■

Example 25 An elevator travels for 4 seconds between stops. Suppose the distance travelled s (in feet) in t seconds is given by

$$s = \frac{32}{5}t - \frac{2}{25}(t-2)^5 - \frac{64}{25} \quad \text{for} \quad 0 \le t \le 4$$

Find the velocity and acceleration of the elevator: (i) for all t, and (ii) for $t = 1$.

Solution The velocity $v(t)$ is given by

$$v = \frac{ds}{dt} = \frac{32}{5} - \frac{2}{5}(t-2)^4$$

and the acceleration $a(t)$ by

$$a = \frac{dv}{dt} = \frac{d^2s}{dt^2} = -\frac{8}{5}(t-2)^3$$

When $t = 1$, $v(1) = 30/5 = 6$ feet per second, and $a(1) = 8/5 = 1.6$ feet per second^2. ■

Graphs of $s(t)$, $v(t)$, and $a(t)$ are shown in Figure 3.13. The acceleration is large when $t = 0$ and $t = 4$. Large accelerations correspond to rapid changes in velocity, and our bodies are sensitive to such changes. Hence our discomfort at the beginning and end, but not the middle of the trip.

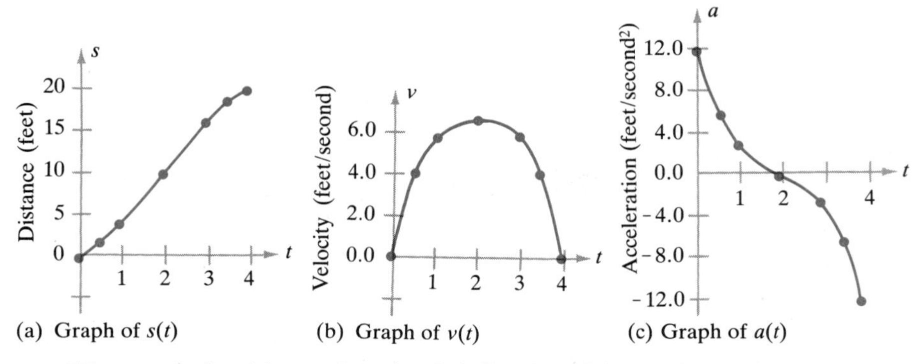

Figure 3.13 Graphs of the distance, velocity, and acceleration functions in Example 25. Because $v(t) = ds/dt$, the slope of the curve in (a) is always equal to the height of the velocity curve shown in (b). Similarly, the slope of the curve in (b) equals the height of the acceleration curve shown in (c). Acceleration has its largest magnitude at the beginning and end of the trip, when $t = 0$ or $t = 4$.

We conclude this section by briefly describing other uses of second derivatives. In economic theory the utility U of a commodity is a function $U = U(q)$ of the quantity consumed; it is a measure of the benefits derived by the consumer at various levels of consumption. As usual in economics, the rate of change dU/dq is called the *marginal utility* (recall Section 2.5). The *Weber–Fechner law* of decreasing marginal effect states that as q increases, the marginal

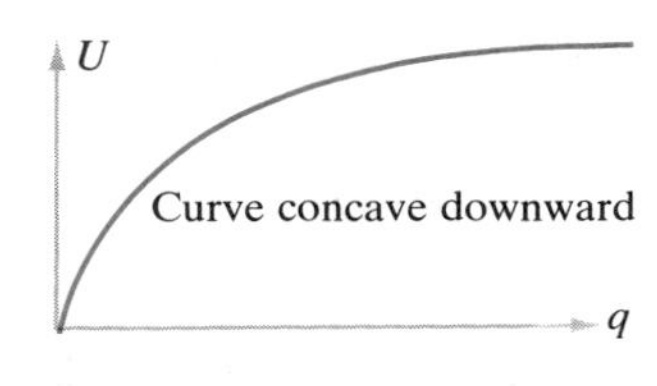

Figure 3.14
Graph of a typical utility function $U(q)$, with $d^2U/dq^2 < 0$. The slope dU/dq of the curve is always positive (curve is rising), but decreases steadily with q (curve flattens out).

utility decreases. That is, in mathematical terms

$$\frac{d^2U}{dq^2} < 0 \qquad [18]$$

This means that the graph of the utility function $U = U(q)$ is concave downward. A typical graph with property [18] is shown in Figure 3.14. In practical terms we might interpret this law as follows: If your base salary is \$10,000 per year, an increment of \$1000 per year is quite important; if you earn \$70,000 per year, the same increment will have much less effect on your life-style. Analogs of this "law of decreasing marginal effect" occur in many other fields, such as psychology. For example, the marginal effect of such stimuli as heat or pain tends to decrease as the total amount of stimulus increases.

Here is another case, a biological problem, in which the second derivative is useful. We could chart the progress of an epidemic by plotting a graph that shows, as in Figure 3.15, the cumulative number $N(t)$ of reported cases as a function of time t (days from the start of the outbreak). The rate of change dN/dt gives the *infection rate*—the number of new cases per day. At first the infection rate increases, but finally it reaches its peak and thereafter decreases slowly toward zero. The transition point, shown as $t = t_0$ in Figure 3.15 is of great importance. It signals the "break" in the epidemic and is directly related to the second derivative d^2N/dt^2. In fact, t_0 is the point at which the second derivative changes from

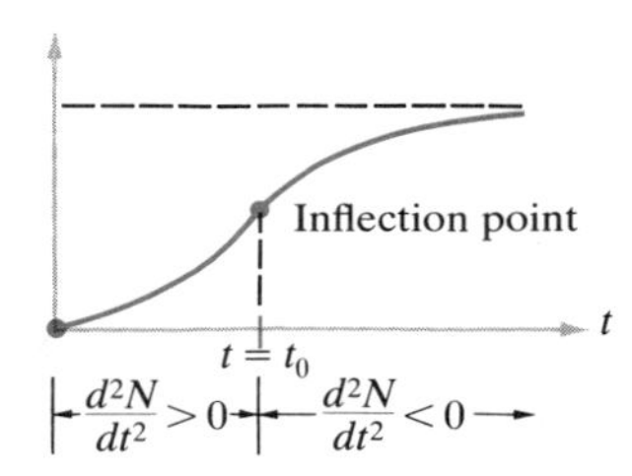

Figure 3.15
Cumulative number $N(t)$ of reported cases during an epidemic.

$$\frac{d^2N}{dt^2} > 0 \quad \left(\text{infection rate } \frac{dN}{dt} \textit{ increasing}; \text{ graph concave } \textit{upward}\right)$$

to

$$\frac{d^2N}{dt^2} < 0 \quad \left(\text{infection rate } \frac{dN}{dt} \textit{ decreasing}; \text{ graph concave } \textit{downward}\right)$$

At $t = t_0$ there is an inflection point, with $d^2N/dt^2 = 0$.

Exercises 3.5

1. Which graphs shown in Figure 3.16 are everywhere concave upward? Which are concave downward? Which graphs have points of inflection?

In Exercises 2–9 calculate d^2y/dx^2 for each function.

2. $\frac{1}{2}x^2 + 3x$

3. $\frac{1}{6}x^3 - x^2 + 1$

4. $x^4 - 15x^2 - 1400$

5. $\frac{x+1}{x-1}$

6. $\sqrt{x} - \frac{1}{\sqrt{x}}$

7. $\frac{1}{x^2 - 1}$

8. xe^{-x}

9. $-3 \ln x$

10. Find $f''(x)$ if
(i) $f(x) = \sqrt{x}$
(ii) $f(x) = \sqrt{1 - x^2}$
(iii) $f(x) = x\sqrt{x^2 + 1}$

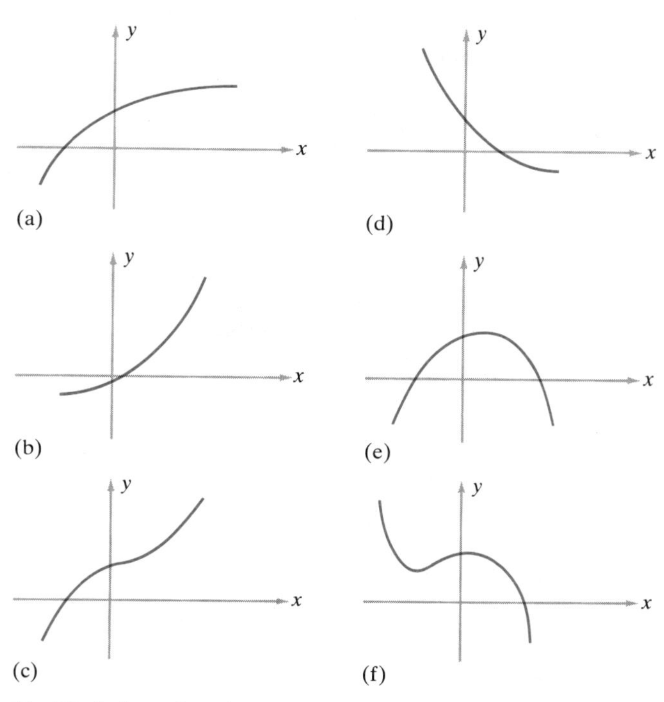

Figure 3.16
Graphs for Exercise 1.

11. Find the value $f''(0)$ of the second derivative at $x = 0$ for each of the functions
(i) $f(x) = x^2$
(ii) $f(x) = \sqrt{1 + x}$
(iii) $f(x) = x^3 + 7x^2 - 2x + 1$

12. Find $f(1)$, $f'(1)$, and $f''(1)$ for the functions
(i) $f(x) = 2x^2 - 3x + 7$
(ii) $f(x) = \sqrt{x}$
(iii) $f(x) = x^3 + 4x^2 - 7x$

In Exercises 13–16, make a sign diagram showing where the function graph is concave upward or downward. Find all inflection points (if any).

13. $y = 1 - \frac{1}{2}x - \frac{1}{3}x^2$

14. $y = x^3 - 12x + 2$

15. $y = x^4 + 2x^3 - 12x^2$

16. $y = xe^{-x}$ (*Hint*: $e^{-x} > 0$ for all x)

17. An object moving along a straight line has distance to the right of the origin

$$s = t^3 - 2t^2 + 1 \quad \text{for} \quad t \geq 0$$

Find its speed and acceleration as functions of time t. For which values of t is d^2s/dt^2 positive (acceleration to the right) and negative (acceleration to the left)?

18. If the cumulative number of reported flu cases during a small-town epidemic is given by

$$N(t) = \frac{1500t^2}{1 + t^2} \quad (t \text{ in weeks})$$

calculate: (i) the infection rate dN/dt; (ii) the second derivative d^2N/dt^2. When does the epidemic reach its peak ($d^2N/dt^2 = 0$)?

19. In a certain study, the cumulative number of flu cases reported in Lake County within t weeks after November 1 was described by an exponential formula

$$N(t) = 10{,}000\,(1 - (1 + t)e^{-t})$$

According to this mathematical model, when did the epidemic reach its peak ($d^2N/dt^2 = 0$)? (*Hint*: The value of e^{-t} is never zero: $e^{-t} = 1/e^t > 0$ for all t; simplify formulas using the fact that $e^a \cdot e^b = e^{a+b}$ for any a, b.)

20. Calculate d^3y/dx^3 for the functions

(i) $y = x^{10} + 7x^6 - x^3 + 13x + 5$

(ii) $y = x - \dfrac{1}{x}$

(iii) $y = xe^{-x}$

3.6 The Basic Methods of Optimization

(a) Graph concave downward; local maximum

x_0
Critical point

(b) Graph concave upward; local minimum

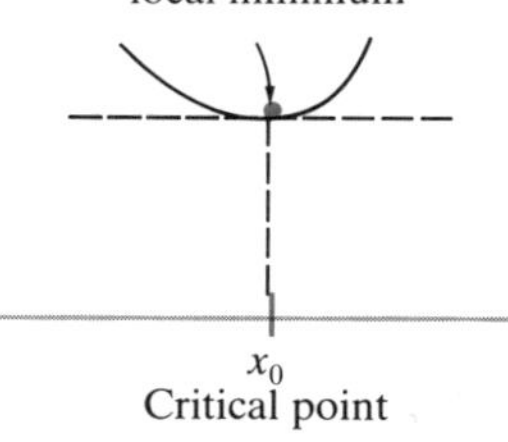

Figure 3.17
Behavior of the graph of $f(x)$ near a critical point x_0 is determined by the sign of d^2f/dx^2 at x_0. If $d^2f/dx^2 = 0$ there, the test is inconclusive.

We have already explained how to locate maxima and minima of a differentiable function $y = f(x)$ defined on a closed, bounded interval $a \leq x \leq b$, including endpoints. Further analysis is needed if the endpoints are not part of the feasible set or if the feasible set is an unbounded interval, such as $0 < x < +\infty$ or the whole number line. We now present methods for dealing with these more complicated optimization problems. All are based on examining the derivatives of the function $f(x)$, keeping in mind what they tell us about the shape of the graph.

In optimization problems the critical points are important because the absolute extrema we seek can only occur there or at feasible endpoints (if any). It is quite helpful to know the shape of the graph near each critical point, so we can quickly eliminate those that cannot possibly be solutions of our problem. This is sometimes done by the **second derivative test**.

> SECOND DERIVATIVE TEST At each feasible critical point x_0, where $df/dx = 0$, compute the value of the second derivative d^2f/dx^2. Then
>
> (i) If $d^2f/dx^2 < 0$, the graph is concave downward near x_0, and a local maximum occurs at x_0, as in Figure 3.17a.
>
> (ii) If $d^2f/dx^2 > 0$, the graph is concave upward near x_0, and a local minimum occurs at x_0, as in Figure 3.17b.
>
> If $d^2f/dx^2 = 0$ at x_0, the test is inconclusive and we must resort to a more detailed analysis.

Suppose we seek an absolute maximum. We may immediately ignore all critical points where $d^2f/dx^2 > 0$, focusing our attention solely on those that are local maxima. Combining the results of the second derivative test with our intuition about the problem, we can often find the absolute maximum without further effort. If not, we employ more forceful methods to be discussed later in this section.

Example 26 For each of the following functions find the critical points, calculate the second derivative at each, and determine the nature of the local extremum (if any).

$$\text{(i) } y = 1 - x^2 \qquad \text{(ii) } y = x^3 - 3x \qquad \text{(iii) } y = x^3$$

Solution We locate critical points by solving the equation $dy/dx = 0$. In (i)

$$\frac{dy}{dx} = \frac{d}{dx}(1 - x^2) = -2x$$

is zero at $x = 0$. This is the only critical point. There, the second derivative $d^2y/dx^2 = -2$ is negative, so a local maximum occurs at $x = 0$. In (ii)

$$\frac{dy}{dx} = 3x^2 - 3 = 3(x^2 - 1) = 3(x - 1)(x + 1)$$

This expression is zero only where one of the factors $(x - 1)$ or $(x + 1)$ is zero; thus, the only critical points are $x = 1$ and $x = -1$. The second derivative

$$\frac{d^2y}{dx^2} = \frac{d}{dx}(3x^2 - 3) = 6x$$

has the value $+6$ at $x = 1$; by the second derivative test, $x = 1$ is a local minimum. At $x = -1$, the second derivative has a negative value of -6; a local maximum occurs at $x = -1$.

In (iii) the only critical point of $y = x^3$ is $x = 0$. Because $d^2y/dx^2 = 6x$ is zero at $x = 0$, the second derivative test is inconclusive. (Actually, it turns out that there is neither a local maximum nor a local minimum at $x = 0$.) ■

Example 27 Apply the second derivative test to find all local maxima of the function $f(x) = -x^4 + 4x^3 + 8x^2 + 3$.

Solution First find the critical points by solving

$$\frac{df}{dx} = -4x^3 + 12x^2 + 16x = 0$$

We do this by factoring the polynomial to obtain the equation

$$-4x^3 + 12x^2 + 16x = -4x(x^2 - 3x - 4) = -4x(x - 4)(x + 1) = 0$$

The only roots occur where one of the factors $-4x$, $(x - 4)$, or $(x + 1)$ is zero, so the critical points are $x = 0$, $x = 4$, and $x = -1$.

Next, calculate the value of the second derivative $f''(x) = -12x^2 + 24x + 16$ at each critical point and apply the test:

$$f''(0) = 16 > 0 \text{ at } x = 0; \quad \text{local minimum at } x = 0$$
$$f''(4) = -80 < 0 \text{ at } x = 4; \quad \text{local maximum at } x = 4$$
$$f''(-1) = -20 < 0 \text{ at } x = -1; \quad \text{local maximum at } x = -1$$

Maxima occur only at $x = -1$ and $x = 4$. Observe how easy it is to apply the test. The graph of the function, which is not so easy to sketch, was shown in Figure 3.8; it confirms our results. ■

Example 28 A pharmaceutical manufacturer is planning a warehouse to store sensitive raw materials. He wants to minimize his annual stockholding costs (delivery plus storage). His dilemma is this: If the warehouse is small, he must take frequent deliveries, paying round-trip costs each time. It would be cheaper to have a larger warehouse, taking a few large orders per year rather than many small ones. But maintaining a lot of climatized warehouse space becomes very expensive. His analysts inform him that his annual stockholding costs C will depend on warehouse capacity x according to the formula

$$C(x) = 5 + 4x + \frac{1600}{x} \quad \text{thousand dollars per year} \qquad [19]$$

where x is measured in thousands of cubic feet. How should he choose the capacity x to minimize these costs?

Solution For various reasons, costs get high when x is very small or very large (see Figure 3.18). It is intuitively clear that there will be an optimum choice somewhere in between. The feasible values are given by $x > 0$, negative warehouse capacity being meaningless.† The derivative of $C(x)$ is

$$\frac{dC}{dx} = \frac{d}{dx}(5) + 4\frac{d}{dx}(x) + 1600\frac{d}{dx}\left(\frac{1}{x}\right)$$
$$= 4 - \frac{1600}{x^2} = \frac{4x^2 - 1600}{x^2}$$

To find the feasible critical points we note that $(4x^2 - 1600)/x^2$ is zero only where the numerator $(4x^2 - 1600)$ is zero, so that $dC/dx = 0$ where

$$4x^2 - 1600 = 0 \quad \text{or} \quad x^2 = 400$$

The critical points are $x = 20$ and $x = -20$; only $x = 20$ is feasible.

This problem has no feasible endpoints, so the critical point $x = 20$ is the only possible candidate for an absolute minimum. Because we are intuitively sure that there *is* an optimum choice of x, this must be it! If we are less willing to rely on intuition, we may reassure ourselves by applying the second derivative test to

† The endpoint $x = 0$ is not feasible. Formula [19] is undefined when $x = 0$, because it would involve division by zero.

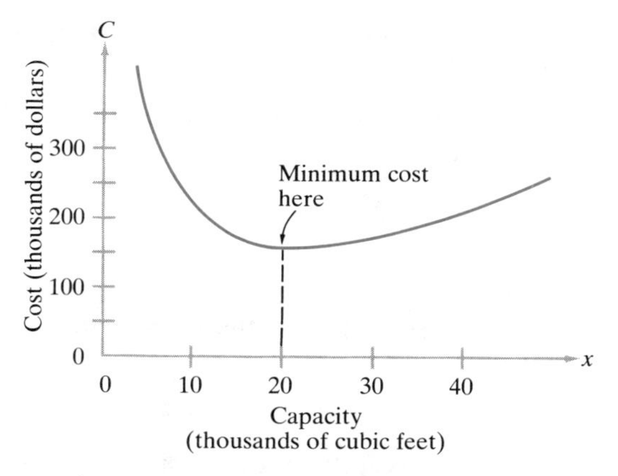

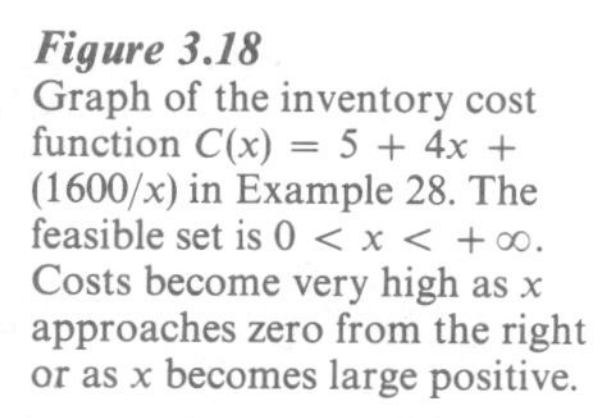

Figure 3.18
Graph of the inventory cost function $C(x) = 5 + 4x + (1600/x)$ in Example 28. The feasible set is $0 < x < +\infty$. Costs become very high as x approaches zero from the right or as x becomes large positive.

see that there is indeed a local minimum at $x = 20$. The second derivative

$$\frac{d^2C}{dx^2} = \frac{d}{dx}\left(4 - \frac{1600}{x^2}\right) = \frac{3200}{x^3}$$

is positive at $x = 20$, as required.

To sum up, minimum stocking costs are achieved by building a warehouse with capacity $x = 20$ thousand cubic feet. The minimized annual stockholding cost is then $C(20) = 165$ thousand dollars. ■

A more robust optimization method is to make a "sign diagram," showing where the function is increasing or decreasing on the feasible set. This information is often sufficient to find the absolute extrema—even if they occur at endpoints—and does not rely upon intuition. As explained in Section 3.3, $f(x)$ is increasing where $df/dx > 0$ and decreasing where $df/dx < 0$; so we need only determine the sign of df/dx in various parts of the feasible set.

FIRST DERIVATIVE SIGN TEST Suppose $f(x)$ is differentiable throughout an interval, bounded or unbounded.

Step 1 Make a sketch of the number line, marking in the feasible set.

Step 2 Calculate the first derivative df/dx and determine where it is positive, negative, or zero. This tells us where f is increasing, decreasing, or has a critical point.

Step 3 On your diagram, mark in the critical points and indicate where df/dx is positive or negative (graph rising or falling). Note that the sign of df/dx can change only as x passes a critical point.

Step 4 Examine the resulting sign diagram, keeping in mind what it tells us about the graph of f. Often we will have enough information to find the absolute extrema. For example, if the graph rises until we reach the point x_0 and then falls as we continue to the right, an absolute maximum occurs at x_0.

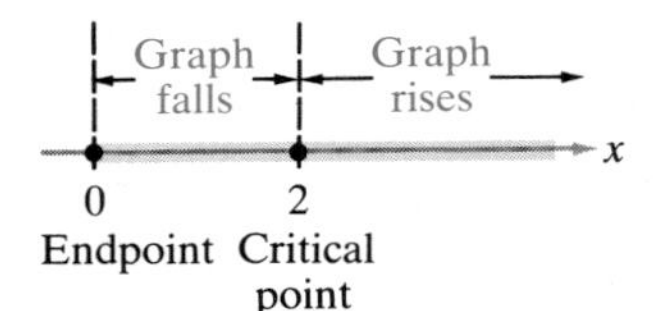

(a) Feasible set (shaded) and critical point, together with information extracted from the sign diagram for dy/dx (Figure 3.20).

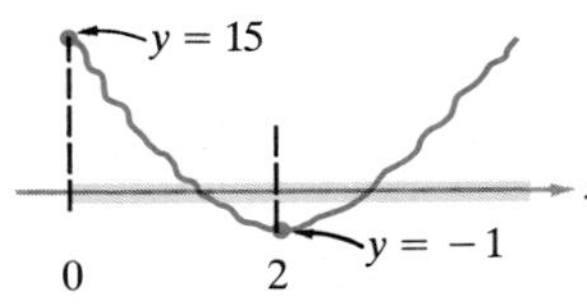

(b) General shape of the graph of $y = x^3 - 12x + 15$ over the feasible set, as deduced from the data in Figure 3.19a.

Figure 3.19

Example 29 Find the absolute minimum of $y = x^3 - 12x + 15$ on the feasible set $x \geq 0$.

Solution The feasible set $x \geq 0$ is shaded in Figure 3.19a. The derivative

$$\frac{dy}{dx} = 3x^2 - 12 = 3(x^2 - 4) = 3(x - 2)(x + 2)$$

is zero at $x = 2$ and $x = -2$, so $x = +2$ is the only feasible critical point. To see where dy/dx is positive or negative, we determine the sign of $dy/dx = 3(x - 2)(x + 2)$ from the signs of the factors $(x - 2)$ and $(x + 2)$ by applying the rule of signs:

$$(-)\cdot(-) = (+) \qquad (-)\cdot(+) = (-) \quad \text{and so on}$$

The easiest way to keep track of the signs is to set up a "sign diagram," as in Figure 3.20. For example, the signs for $(x - 2)$ are shown in the first line of the figure; $x - 2 > 0$ when $x > 2$ (x to the right of $+2$) and $x - 2 < 0$ when $x < 2$ (x to the left of $+2$); similarly, for the signs of $(x + 2)$. The signs for the product $dy/dx = 3(x - 2)(x + 2)$ are determined by comparing the signs of the factors.

Within the feasible set we see that the graph falls as x moves from 0 to $+2$ and thereafter rises steadily. Obviously, an absolute minimum occurs at $x = 2$. The minimum value of y achieved there is $y = f(2) = -1$. ■

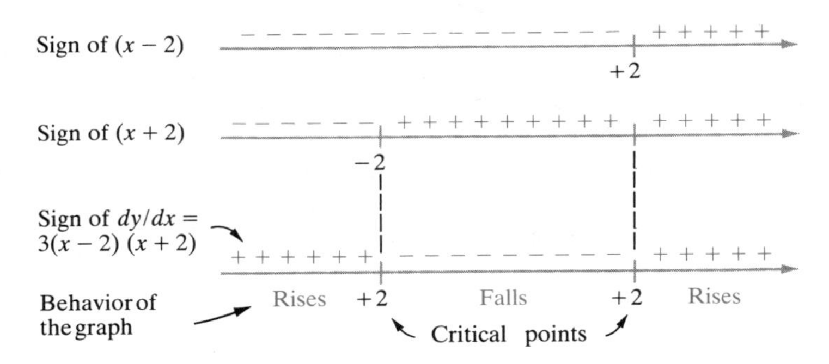

Figure 3.20
Sign diagram for $dy/dx = 3(x - 2)(x + 2)$. We first determine the signs of the individual factors $(x - 2)$ and $(x + 2)$. The sign of dy/dx is determined by comparing the signs of the factors.

Exercises 3.6

1. In Figure 3.21 we show several curves, graphs of functions $y = f(x)$. For which curves is $dy/dx > 0$ for all x? For which is $d^2y/dx^2 < 0$ for all x?

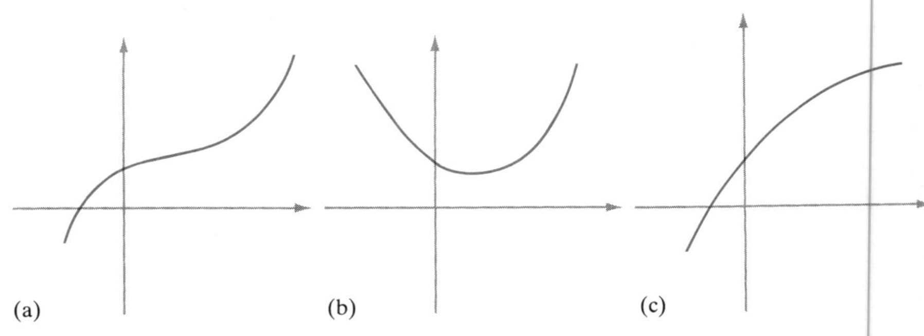

Figure 3.21
Graphs for Exercise 1.

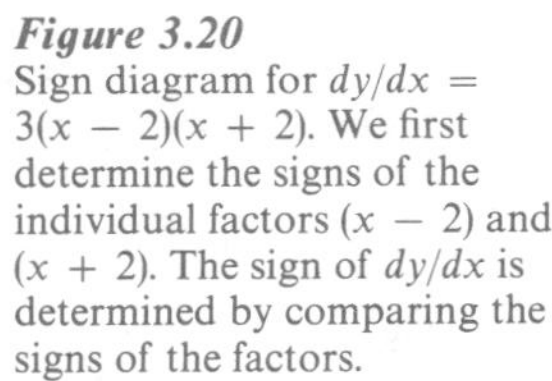

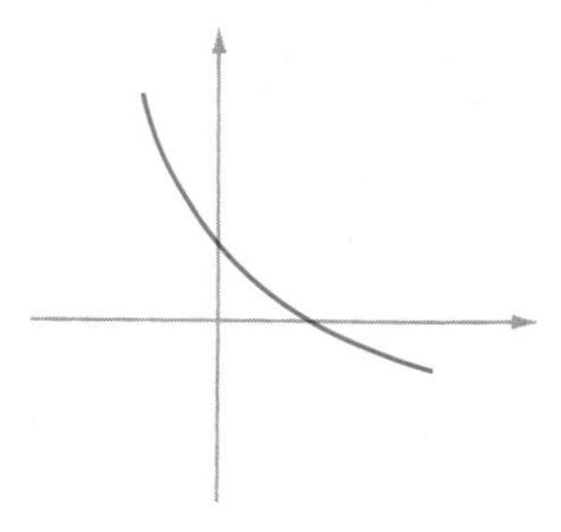

Figure 3.22
Graph for Exercise 2.

2. For the graph shown in Figure 3.22, what are the signs of dy/dx and d^2y/dx^2? Is there a value of x for which $dy/dx = 0$?

In Exercises 3–9 find all critical points of the given function, calculate the second derivative at each critical point, and determine the nature of the local extremum.

3. $y = x^2$
4. $y = 4x - 11$
5. $y = 8 + 4x - 3x^2$
6. $y = 3 + 6x^2 + x^3$
7. $y = 1 + 9x + 3x^2 - x^3$
8. $y = x^4 - 2x^2$
9. $y = x^5 - 5x + 7$

In Exercises 10–17 apply the second derivative test to find all local extrema (if any).

10. $f(x) = x^2 - 14x + 17$
11. $f(x) = 7 + 5x - 3x^2$
12. $f(x) = 3x^2 - x^3$
13. $f(x) = 1 + x + x^3$
14. $f(x) = 1 - x + x^3$
15. $f(x) = x + \dfrac{1}{x}$ (for $x \neq 0$)
16. $f(x) = \dfrac{2x^3 + 1}{x^2}$ (for $x \neq 0$)
17. $f(x) = xe^{-x}$

(*Hint*: In Exercise 17, $e^{-x} > 0$ for all x.)

In Exercises 18–23 set up a sign diagram showing where the following expressions are positive or negative.

18. $(x + 1)(x + 4)$
19. $x^2 - 5x$
20. $x^2 - 2x - 15$
21. $x^3 - x$
22. $\dfrac{4 - 5x}{x^2 + 1}$
23. x^3

(*Hint*: In Exercise 22 we have $(x^2 + 1) > 0$ for all x. Why?)

In Exercises 24–28, use the first derivative sign test to find the extremum indicated.

24. $y = x^3 - 3x^2 - 9x - 9$; absolute minimum for $x \geq 1$
25. $y = 4x^3 - x^4$; absolute maximum for all x
26. $y = x^2 + \dfrac{1}{x^2}$; absolute minimum for $x > 0$
27. $y = \dfrac{1}{4}x^4 - \dfrac{1}{2}x^2 + 3$; absolute minimum for $x \geq 0$
28. $y = xe^{-x}$; absolute maximum for all x

(*Hint*: In Exercise 28, we have $e^{-x} > 0$ for all x.)

29. The following functions are defined for *all* x. For each, work out a sign diagram for dy/dx. Are there any absolute maxima, absolute minima, or critical points?

(i) $y = x^3$ (ii) $y = 1 + x + x^3$ (iii) $y = 1 - x^4$

30. Show that the second derivative test is inconclusive when applied to the function $f(x) = x^4$ defined for all x. Then apply the first derivative sign test to show that f has an absolute minimum at $x = 0$.

31. On a fair day, the number N of people using the Park Circle tennis courts is

$$N = \frac{200}{(5 - 0.4T + 0.01T^2)}$$

where T is the temperature in degrees Celsius. Find the temperature that brings out the most tennis players.

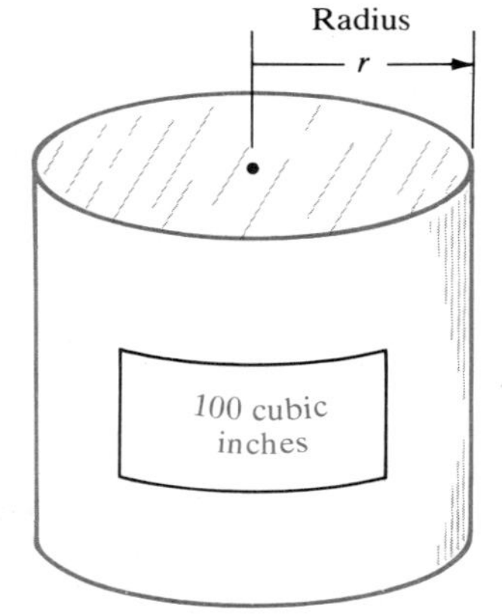

Figure 3.23
The 100-cubic-inch can in Exercise 33; r is its radius. Surface area includes side, top, and bottom.

32. For a certain electric motor the power P depends on the resistance R according to the formula

$$P = \frac{10R}{(1 + R)^2} \quad \text{for} \quad R \geq 0$$

What value of R gives the maximum power?

33. An industrial designer wants to minimize the cost of materials needed to make a 100 cubic inch can (Figure 3.23). He must find the radius $r > 0$ that minimizes the surface area of the can, which is given by the formula

$$A = 6.283r^2 + \frac{200}{r} \quad \text{square inches}$$

for any $r > 0$. What value of r should he choose?

34. The manufacturer of Brand X Cola has found, through market research, that his monthly profit P depends on the selling price per can x according to the formula

$$P = (10x - 130)e^{-x/30} \quad (x = \text{cents per can})$$

If he is unwilling to drop the price below 30¢ a can, what is the maximum feasible profit? (*Hint*: $e^a > 0$ for all real numbers a; thus $e^{-x/30} > 0$ for all x. For a rough idea of the situation, see Figure 3.24.)

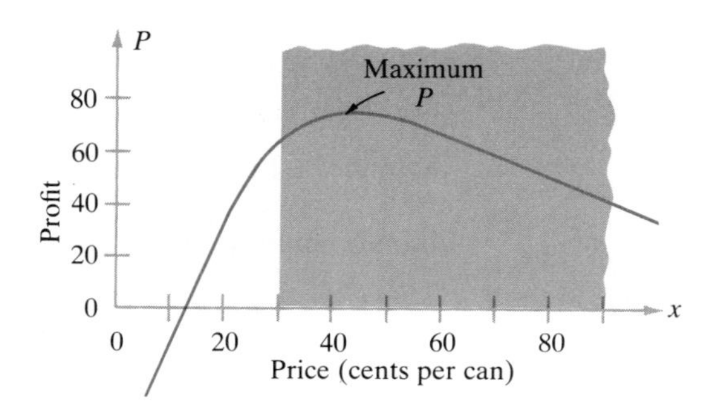

Figure 3.24
Profit on Brand X Cola in Exercise 34. What price $x \geq 30$ yields maximum profit?

3.7 Applied Optimization Problems (Word Problems)

We have now developed efficient methods of maximizing or minimizing a function. A major difficulty encountered in applied problems is translating this theory into practice. Usually, the situation is described as a "word problem"; we must translate this verbal description into a suitable mathematical description. There are some hints to help you accomplish this translation. Of course, with practice the task becomes easier.

PROCEDURE FOR HANDLING VERBAL PROBLEMS

Step 1 Read the problem carefully. Make a sketch, if appropriate.
Step 2 Label all the important variables in the problem, representing each variable by a different symbol.
Step 3 Identify the variable to be maximized or minimized. This will be the dependent variable in the final optimization problem.
Step 4 Express the relationship between this dependent variable and the other variables in the problem by an equation.
Step 5 Use the information in the problem to eliminate all independent variables but one from the equation you have found. In this way the dependent variable is expressed as a function of *just one* independent variable. (If you cannot do this, you have not extracted all the information from the original problem.)
Step 6 Determine the feasible set for the independent variable chosen in Step 5. This gives the domain of definition of the function to be maximized.
Step 7 Apply the techniques of Sections 3.4 through 3.6 that are most appropriate.

In solving a problem, we should use the simplest technique for the job:

1. If the domain of definition is bounded, of the form $a \le x \le b$, then the simple method of Section 3.4 suffices.
2. For all other domains, the methods of Section 3.6 apply. If we know from intuition that the problem has an answer, we should use the second derivative test to sort out the feasible critical points that are local maxima (or minima, as the case may be). The largest (smallest) value of the function at these points is the absolute extremum we seek.
3. If our intuition fails, we should apply the first derivative sign test to see where the function is increasing or decreasing. This information will usually allow us to decide whether the problem has a solution and to locate the solution if it exists.

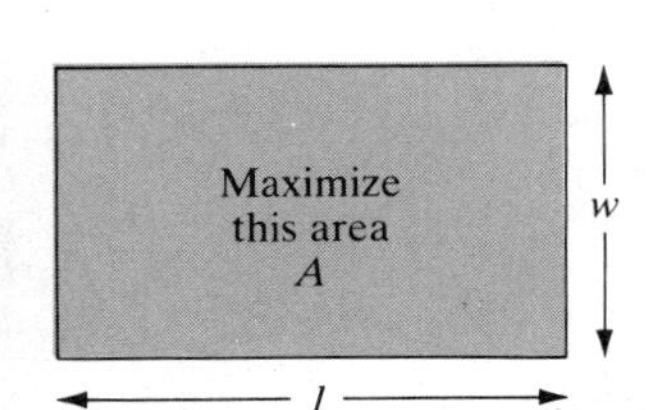

Figure 3.25
The rectangle in Example 30 has a perimeter of fixed length $2l + 2w = 12$ inches. Dimensions l and w are to be chosen that maximize the enclosed area A.

Example 30 Find the rectangle of largest area whose perimeter (sum of lengths of four sides) is 12 inches.

Solution The word "largest" tell us at once that it is a maximization problem. The situation is shown in Figure 3.25, where we have labeled the relevant variables l = length, w = width, and A = area of the rectangle. What variable is to be maximized? Of course it is the area A. Upon what variables does A depend?

Because the area of a rectangle is the product of its length and its width, we obtain

$$A = lw \tag{20}$$

We must next express A as a function of just one variable. To do this, we use the condition that the perimeter is 12 inches. Two sides of the rectangle have length l, and the other two have length w, so that

$$\text{perimeter} = 2l + 2w = 12$$
$$l + w = 6$$

We may use this relation to eliminate one variable, say w, by writing it in terms of the other:

$$w = 6 - l \tag{21}$$

Using equation [21] we may express A as a function of the variable l alone:

$$A = l \cdot w = l(6 - l) \tag{22}$$

Does the problem impose any restriction on the independent variable l? In fact, the sides of the rectangle cannot have negative length, so that $l \geq 0$ and $w \geq 0$. The latter inequality, combined with [21], yields another condition on l,

$$w = 6 - l \geq 0 \quad \text{or} \quad l \leq 6$$

Thus l is restricted to a closed, bounded interval $0 \leq l \leq 6$. We have now recast the original word problem in mathematical form: Find the absolute maximum value of the function $A(l) = l(6 - l)$ defined on the interval $0 \leq l \leq 6$. We may apply the optimization techniques of Section 3.4 directly to this problem. First, locate the critical points of

$$A(l) = l(6 - l) = 6l - l^2$$

by setting dA/dl equal to zero

$$\frac{dA}{dl} = 6 - 2l = 0$$

There is just one critical point, namely $l = 3$. Then compare the values of $A(l)$ at the critical point $l = 3$ and at the endpoints $l = 0$ and $l = 6$:

$$\text{If } l = 0,\ A = A(0) = 0(6 - 0) = 0$$
$$\text{If } l = 3,\ A = A(3) = 3(6 - 3) = 9$$
$$\text{If } l = 6,\ A = A(6) = 6(6 - 6) = 0$$

The largest value, $A = 9$, is the absolute maximum value achieved by $A(l)$; it occurs at $l = 3$. Note that, from [21], $w = 6 - l = 3$ if $l = 3$, so that the rectangle is a *square*. The same analysis can be used to show that of all rectangles with a given perimeter, the one with largest area is always a square. ■

Example 31 The sum of two numbers is equal to 6. For which choice(s) of these numbers is their product a maximum?

Solution This example is almost identical to Example 30. We shall emphasize their mathematical similarity by using the same symbols. Let l and w be the two numbers, and let $A = lw$ be their product. Then $l + w = 6$, and we may express A as a function of l alone:

$$A = A(l) = l(6 - l) = 6l - l^2$$

There is one important difference between this example and the preceding one: In this problem, l is not restricted; it can take on any real value, positive or negative. Thus we seek the absolute maximum value of $A = A(l) = 6l - l^2$, with no restrictions on the independent variable l.

Because l is not restricted to a bounded interval, the method of Section 3.4 does not apply. Instead, we solve the problem by examining the first derivative dA/dl. The only critical point for $A(l)$ is $l = 3$, as before. Where is $A(l)$ increasing or decreasing? Because $dA/dl = 6 - 2l = 2(3 - l)$, we have

$$\frac{dA}{dl} = 2(3 - l) > 0 \quad \text{for} \quad l < 3$$

so A is increasing for $l < 3$. Similarly, A is decreasing for $l > 3$ as shown in Figure 3.26 since

$$\frac{dA}{dl} = 2(3 - l) < 0 \quad \text{for} \quad l > 3$$

Since $A(l)$ increases steadily until we reach $l = 3$ and decreases thereafter, the largest possible value of $A(l)$ occurs at $l = 3$. The two numbers we seek are therefore $l = 3$ and $w = 6 - l = 3$. ■

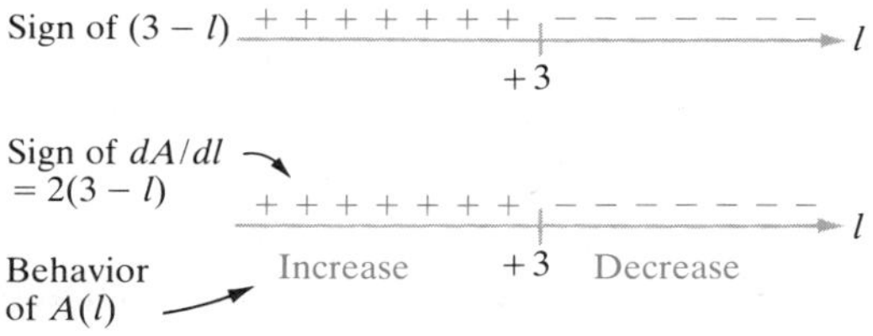

Sign of $(3 - l)$ + + + + + + + | − − − − − − − l
+3

Sign of $dA/dl = 2(3 - l)$ + + + + + + + | − − − − − − − l

Behavior of $A(l)$ Increase +3 Decrease

Figure 3.26 Analysis of the sign of $dA/dl = 2(3 - l)$ in Example 31.

Example 32 The Random Walk Moving Company is going to make a 1200-mile trip. Expenses are 15¢ per mile for the truck, \$12.50 per hour for driver's wages, plus diesel-fuel costs. Fuel costs run $v/288$ dollars per mile if the trip is made at a steady speed of v miles per hour. Keeping in mind that freeway speed limits for trucks are 40 mph minimum and 55 mph maximum, at what speed v should the trip be made to minimize total expenses?

Solution Obviously, v is the independent variable; its feasible values are $40 \le v \le 55$. Some items, such as the driver's wages, depend on the time t required to make the trip; but t is easily expressed in terms of the variable v,

$$t = \frac{\text{(distance)}}{\text{(speed)}} = \frac{1200}{v} \quad \text{hours}$$

Then the total expenses for the trip are

$$\begin{aligned} C(v) &= \text{(mileage costs)} + \text{(driver's wages)} + \text{(fuel costs)} \\ &= 0.15(1200) + 12.50t + 1200\left(\frac{v}{288}\right) \\ &= 180 + 12.50\left(\frac{1200}{v}\right) + 4.1667\,v \\ &= 180 + \frac{15{,}000}{v} + 4.1667\,v \end{aligned}$$

The derivative

$$\frac{dC}{dv} = -\frac{15{,}000}{v^2} + 4.1667$$

is zero for $v^2 = 15{,}000/4.1667 = 3600$, or $v = \sqrt{3600} = 60$ mph. Only values $40 \le v \le 55$ are feasible, so there are *no* feasible critical points. Because we are dealing with a closed, bounded interval, the procedure of Section 3.4 tells us that the minimum cost must occur at one of the endpoints. But

$$C = \$721.67 \quad \text{if } v = 40$$
$$C = \$681.89 \quad \text{if } v = 55$$

so the minimum cost $C = \$681.89$ occurs if we drive at the upper speed limit, $v = 55$ mph. ■

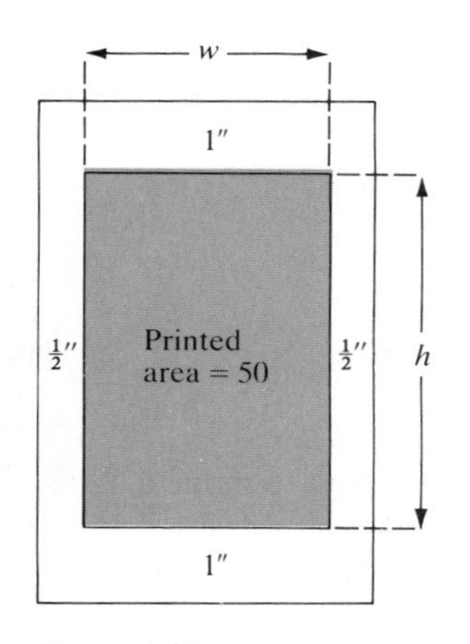

Figure 3.27
The printed page in Example 33, showing the meaning of the variables h, w, and P. The printed area P is shaded; the rest of the page is taken up by margins. The total printed area is $hw = 50$ square inches, and the total area of the whole page $A = (h + 2)(w + 1)$ is to be minimized.

Example 33 A publisher decides to print the pages of a book with 1-inch margins at top and bottom, and $\frac{1}{2}$-inch side margins. He is willing to vary the page dimensions, subject to the condition that the *printed area* of the page is 50 square inches. What dimensions will minimize the total area of the page (thereby minimizing the paper cost for the book)?

Solution For most of us, intuition assures us that there is an optimal solution to the problem, but fails to reveal what the solution is. Is a square page optimal? (How many books have you read with square pages?) Obviously, we should try to transform this into a mathematical optimization problem.

Let the height and width of the *printed area P* be h and w respectively (see Figure 3.27). We want to minimize the total page area A. Because the page is a rectangle with height $(h + 2)$ and width $(w + 1)$, the page area is

$$A = (h + 2)(w + 1) \tag{23}$$

We next express A as a function of just one independent variable. We may eliminate the variable h, writing it in terms of w, by recalling that the printed area is a rectangle with dimensions $h \cdot w$ so that

$$h \cdot w = 50 \quad \text{or} \quad h = \frac{50}{w} \tag{24}$$

Substituting [24] into [23], we get A expressed as a function of w,

$$A = \left(\frac{50}{w} + 2\right)(w + 1)$$

$$= 50 + 2w + \frac{50}{w} + 2$$

$$= 52 + 2w + 50w^{-1}.$$

The dimensions of the printed area must be positive, so the only restriction on the independent variable is $w > 0$ (unbounded feasible set). Now consider the first derivative

$$\frac{dA}{dw} = 2 - 50w^{-2} = 2 - \frac{50}{w^2} \quad \text{for} \quad w > 0$$

We find the critical points by setting $dA/dw = 0$,

$$0 = \frac{dA}{dw} = 2 - \frac{50}{w^2}$$

Thus

$$2 = \frac{50}{w^2} \quad \text{or} \quad 2w^2 = 50 \quad \text{or} \quad w^2 = 25$$

The solutions of $w^2 = 25$ are $w = -5$ and $w = 5$. Because we consider only $w > 0$, $w = +5$ is the only feasible critical point. Because there is only one candidate for a solution, this must be the location of the absolute minimum of A—if our intuition is to be believed. If we desire firmer evidence, we can apply the second derivative test to this critical point. The second derivative

$$\frac{d^2A}{dw^2} = \frac{d}{dw}\left(2 - \frac{50}{w^2}\right) = +\frac{100}{w^3}$$

is positive at $w = 5$, so there is a minimum there. If we have no faith at all in intuition, we can apply the first derivative sign test to dA/dw to see that A is decreasing in the interval $0 < w < 5$ and is increasing for $w > 5$. This is ironclad evidence that A has an absolute minimum at $w = 5$.

If $w = 5$, the corresponding value of h is $h = 50/w = 50/5 = 10$ inches, by [24]. In the optimal page, the printed matter will have dimensions $h = 10$ inches and $w = 5$ inches, and the whole page will be 12 inches by 6 inches. ■

Example 34 The cost of constructing an apartment building x stories high is

$$C(x) = 1{,}800{,}000 + 200{,}000x + 2000x^2$$

The term $C(0) = \$1{,}800{,}000$ represents the cost of buying land and leasing equipment, incurred before construction begins. Each floor costs about \$200,000, but the higher the floor the greater the labor costs for that floor; hence the term $2000x^2$. Find the building height at which the average cost per floor $A(x) = C(x)/x$ is a minimum.

Solution The function

$$A(x) = \frac{1{,}800{,}000}{x} + 200{,}000 + 2000x$$

is defined for $x > 0$. Intuitively, we know the problem has an answer; the minimum exists. This minimum must occur among the critical points. The derivative

$$\frac{dA}{dx} = \frac{-1{,}800{,}000}{x^2} + 2000$$

is zero when $2000 = 1{,}800{,}000/x^2$ or $x^2 = 900$, so that $x = +30$ or $x = -30$. The only *feasible* ($x > 0$) critical point is $x = 30$. This critical point is the only possible candidate for a solution, so the minimum must occur when $x = 30$ stories. The (minimal) average cost per floor is $A(30) = \$320{,}000$. ■

Example 35 In a manufacturing operation, suppose that the quantity sold q is related to the selling price (dollars per unit) p by the demand function $q = Q(p)$. As a measure of how q responds to changes in p, economists use the important concept of **elasticity of demand**, which is defined to be

$$E = -\frac{p}{q} \cdot \frac{dq}{dp} \qquad [25]$$

This quantity E may be interpreted as follows. For small increases in price Δp, the change in demand Δq is negative because demand decreases. Thus $-\Delta q$ is positive. The percentage changes are

$$100\,\frac{\Delta p}{p} = \text{(percentage increase in price)}$$

$$100\,\frac{-\Delta q}{q} = \text{(percentage decrease in demand)}$$

(both positive numbers due to our use of $-\Delta q$) and their ratio is

$$\left(\frac{-\Delta q}{q}\right) \Big/ \left(\frac{\Delta p}{p}\right) = -\frac{p}{q} \cdot \frac{\Delta q}{\Delta p}$$

For small price increases Δp, this ratio is approximately equal to $(-p/q) \cdot dq/dp$. Thus elasticity E may be interpreted by noting that

$$\frac{\text{(percentage decrease in demand)}}{\text{(percentage increase in price)}} \approx E(p)$$

or

$$\text{(percentage decrease in demand)} \approx E(p) \cdot \text{(percentage increase in price)}$$

for small changes in p.

Economists use the following basic rule:

$$E = 1 \quad \text{when revenue is maximized} \qquad [26]$$

Justify this rule mathematically.

Solution In this example we seek to maximize the revenue R. As usual, revenue is the product of selling price p and the number q of units sold, $R = p \cdot q$. Substituting the demand function $q = Q(p)$ into this relation, we may express R as a function of p alone:

$$R = R(p) = p \cdot Q(p) \qquad [27]$$

The maximum value for R must occur at a critical point for $R = p \cdot Q(p)$, where we have $dR/dp = 0$.[†] Applying the product rule for derivatives [1], we first find dR/dp:

$$\frac{dR}{dp} = \frac{dp}{dp} \cdot Q + p \cdot \frac{dQ}{dp} = Q + p\frac{dQ}{dp} = q + p\frac{dq}{dp}$$

Setting dR/dq equal to zero, we obtain

$$0 = \frac{dR}{dp} = q + p\frac{dq}{dp} \quad \text{or} \quad -p\frac{dq}{dp} = q$$

so that

$$-\frac{p}{q} \cdot \frac{dq}{dp} = 1$$

Thus, at the maximum we must have $E = 1$. ■

Exercises 3.7

1. Among all rectangles whose area is 100 square inches, find the one with minimum perimeter.
2. Find two numbers whose sum is 10, such that the sum of their squares is as small as possible.
3. Find two numbers x and y whose sum is 30, such that $x^2 + y$ is a minimum.
4. Suppose you want to fence in a rectangular plot of land 1000 square feet in area to graze sheep. What should be the dimensions of your plot if you want to minimize fencing costs (minimize length of the perimeter)?
5. A prospector wants to stake a mining claim but has only 1000 yards of fencing with him. Obviously he wants to stake out the largest possible area. Government claim regulations require that the claim plot be rectangular. What should he take as the dimensions of his claim?

[†] Revenue is zero when $p = 0$, and there is a price so ridiculously high that there will be absolutely no demand for the product; R has a maximum somewhere in between.

6. If the price per ticket for a bus tour is set at x dollars, $1 \leq x \leq 5$, past experience indicates that $50 - 10x$ customers can be expected. Find the tour price x that maximizes the total revenue under these conditions.

7. Suppose that the cost of operating a truck is $(10 + (v/4))$ cents per mile when the truck runs at a steady speed of v miles per hour. Suppose the driver is paid $6.50 per hour. Find the most economical speed for a 1000-mile trip. (Total cost = driver's wages + operating costs is to be minimized.) (*Hint*: At speed v, how many hours will the trip take?)

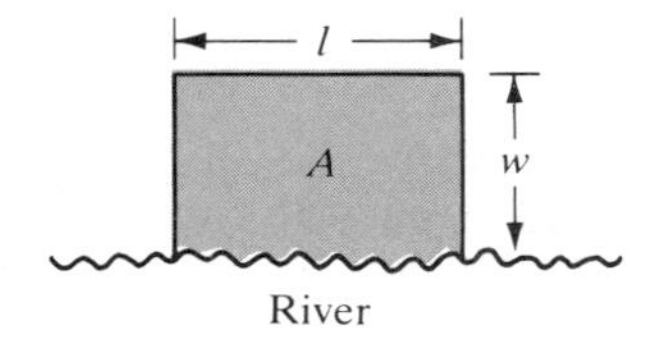

Figure 3.28
The farmer's field in Exercise 9.

8. A private jetliner travels at constant speed v miles per hour. Fuel charges are $0.01v^2$ dollars per hour. Salary for the crew and the cost of renting the jet amount to $1600 per hour. Find the speed v that minimizes the total cost (fuel plus salaries and rental) for a 3000-mile trip.

9. A farmer with a field adjacent to a river wishes to fence off a rectangular area for grazing, as shown in Figure 3.28. No fence is required along the river, and he has available 1600 yards of fencing. What choice of dimensions encloses the maximum area? What ratio l/w gives the maximum area?

10. A lot with river frontage is to be fenced in to enclose 5000 square feet. Fencing along the sides, perpendicular to the river, is wire fencing costing $2 per foot. But fencing along the back of the lot, parallel to the river, is to be picket fence costing $4 per foot. There is no fencing along the river edge. What dimensions for the lot will minimize *fencing costs*?

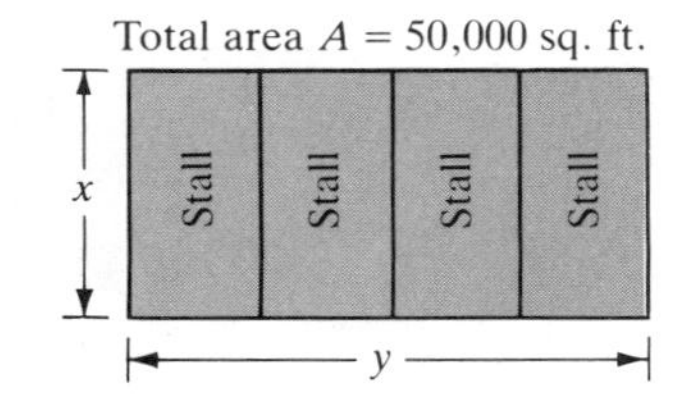

Figure 3.29
The cattle stalls in Exercise 11 all have the same dimensions.

11. A rancher wants to build four adjacent cattle stalls (Figure 3.29) to enclose a total area of 50,000 square feet. How should x and y be chosen to minimize the total amount of fencing required?

12. Rework Exercise 11, assuming that the rancher also wants to have $x \leq 125$ feet.

13. A store offering a certain brand of suit for p dollars can sell $240 - 2p$ of them. The wholesale price of these suits is $60 each. What price yields maximum profit?

14. Suppose a manufacturer's costs are given by

$$C(q) = 1600 + 20q + 0.01q^2$$

for $q > 0$. What production level q minimizes the *average cost per unit* $A(q) = C(q)/q$?

15. Repeat Example 33, taking the margins of the page to be 1 inch on one side and 2 inches on the other (to allow room for binding). The top and bottom margins are 1 inch, as before. If the printed area is to be 50 square inches, find the optimal dimensions. (*Hint*: Unbounded feasible set $w > 0$.)

16. A publisher has decided to publish a book using a format somewhat different from that discussed in Example 33. Each page is to have 1-inch margins at top and bottom and $\frac{1}{2}$-inch side margins. He is willing to vary the page dimensions, provided that the *total page area* is 50 square inches. What dimensions will maximize the printed area of the page? (*Hint*: If page dimensions are h and w, the feasible values of w are $1 \leq w \leq 25$. Why?)

17. A piece of wire 10 inches long is to be cut into two pieces. One piece will be formed into a square and the other into a circle. How should the wire be cut so that the combined

area of the circle and the square is a maximum? (*Hint*: Let x be the length used for the circle, $0 \le x \le 10$.)

18. If $P(q)$, $R(q)$, and $C(q)$ are the profit, revenue, and cost functions for a manufacturer, show that when the profit achieves a maximum the marginal revenue is equal to the marginal cost,

$$\frac{dR}{dq} = \frac{dC}{dq}$$

19. Show that when the average cost per unit $A(q) = C(q)/q$ for a manufacturer is minimized, the marginal cost is equal to the average cost.

20. A builder owns a lot in the form of a right triangle with legs of 300 and 400 feet, as shown in Figure 3.30. He plans to build a skyscraper on the lot with rectangular base dimensions x and y. How should x and y be chosen to maximize the base area of the skyscraper? (*Hint*: Total area of the triangle is 60,000 square feet; it is the sum of the areas of the three pieces that result when x and y are chosen. The areas of these pieces can be written in terms of x and y; doing so yields the relationship between x and y.)

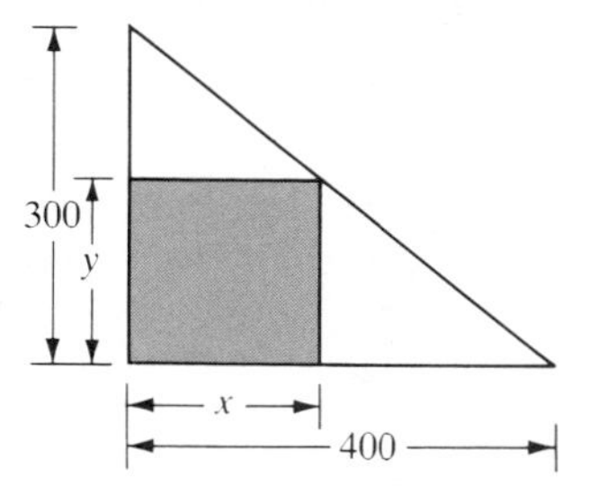

Figure 3.30
The builder's plot in Exercise 20. The total area 60,000 square feet equals the sum of areas of the three pieces obtained when the lot is subdivided as shown. Writing the area of each piece in terms of x and y yields the relationship between x and y, which should be used to eliminate one of these variables.

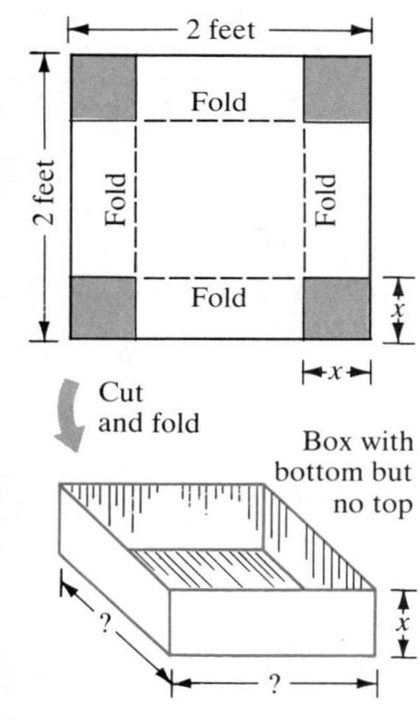

Figure 3.31
Making the box in Exercise 22. All cutout pieces are square and of the same size: x by x inches.

21. Potatoes can be harvested any time between August 1 and the first frost 2 months later. Unless harvested, their average weight grows from 0.3 to 1.0 pounds and is given by the formula

$$w = 0.3 + 0.0875t \quad (t \text{ in weeks elapsed from August 1})$$

Meanwhile, as more and more potatoes are marketed, the selling price p falls steadily from 40¢ per pound on August 1 to 10¢ per pound 8 weeks later:

$$p = 40 - \frac{30}{8}t \quad \text{cents per pound}$$

At what time t should the farmer harvest his crop to maximize his revenue

$$R = (\text{average weight}) \times (\text{price per pound})?$$

22. Given a sheet of tin 2 feet long and 2 feet wide, suppose we make a rectangular tray (no top!) by cutting a small square piece out of each corner and bending the sides to form a box. What is the maximum possible volume of such a box? (*Hint*: Let x be the depth of the cut; see Figure 3.31.)

23. A box with a square bottom and no top is to hold 5 cubic feet. Material for the sides costs 0.8¢ per square foot, but material for the bottom costs 1.0¢ per square foot. Find the dimensions for the box that minimize the *cost of materials* (see Figure 3.32).

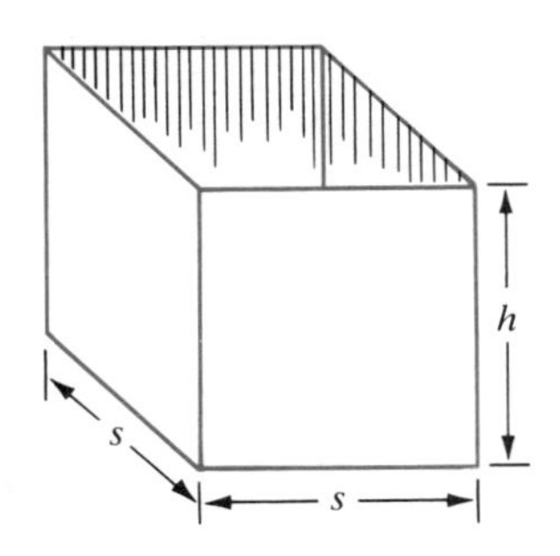

Figure 3.32
The box in Exercises 23–25. It has no top.

24. A rectangular box with an open top and square bottom is to be formed using just 36 square feet of cardboard. How should the height h and base length s be chosen to maximize the volume of the box. (*Hint*: See Figure 3.32.)

25. Repeat Exercise 24, imposing an additional constraint on the dimensions of the box: The bottom must measure at least 4 feet $\times$ 4 feet. When we impose this condition, what is the feasible set of values for the base length s?

26. Demand for copper is known to be 50,000 tons per month if copper is selling at 70¢ per pound. If the elasticity of demand at this price level is $E = 0.50$, by about how much will the demand change if the price of copper is raised to 71¢ per pound? Will demand increase or decrease?

27. An object moves along a straight line. Its distance to the right of the origin at time t is

$$s = \frac{1}{t+1} \quad \text{for} \quad t \geq 0 \quad \text{(unbounded feasible set)}$$

Determine the maximum distance from the origin. For which value of $t \geq 0$ is it achieved? (*Hint*: Does ds/dt ever change sign for $t \geq 0$?)

28. If a and b are fixed numbers, find the value of x that minimizes $(x - a)^2 + (x - b)^2$.

29. The volume of a cylindrical can with height h and radius r is $V = \pi r^2 h$. The surface area (sides, top, and bottom) is $S = 2\pi r^2 + 2\pi rh$. Suppose we want to make a can containing 16 cubic inches. If the cost of the can is proportional to the amount of sheet metal used (surface area), how should we choose the dimensions r and h to minimize the cost? What is the ratio height/diameter $= h/2r$ for the optimal can dimensions?

30. The demand for Brand X Cola, if offered at a price of p cents per can, is

$$q = 10e^{-p/15} \quad \text{million cans per month}$$

for $p \geq 20$. The cost of manufacturing a can of the stuff is 13¢. What selling price p would maximize the profit? (*Hint*: For any real number a, we have $e^a > 0$: Therefore, $e^{-p/15} > 0$ for any choice of p.)

31. Two points in the plane $P = (30, 0)$ and $Q = (0, 20)$ are shown in Figure 3.33. At time $t = 0$, P is moving *left* along the x axis at a constant speed of 7 units per minute, and Q moves *up* along the y axis at a constant speed of 5 units per minute.

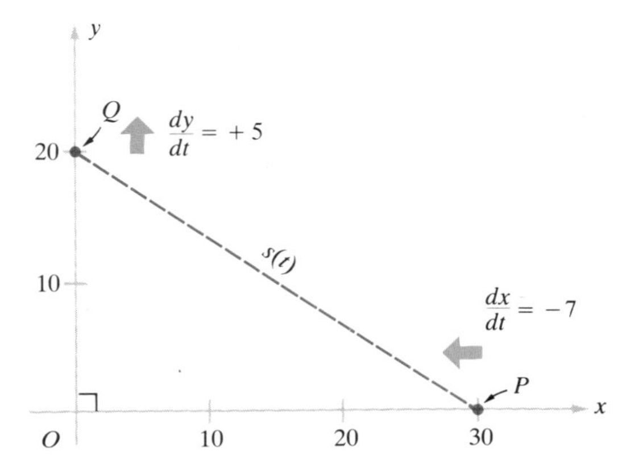

Figure 3.33
Distance between moving points P and S is the length $s(t)$ of the dashed straight line. The figure shows the situation when $t = 0$. Velocities of the points are constant, with values indicated. For P, dx/dt is negative because P moves left, and x is decreasing. The triangle $\triangle OPQ$ is a right triangle, so Pythagoras' theorem applies.

(i) Find the coordinates of P and of Q at time t minutes, assuming that the speeds remain constant.
(ii) Find the distance $s = s(t)$ between P and Q as a function of time.
(iii) At what time t is the distance s a minimum (for $t \geq 0$). What is the minimum distance?

3.8 Derivatives and Curve Sketching

Up to now we have relied heavily on a table of values when graphing a function. This is a basic but rather primitive technique. Now we have at our disposal rules for calculating first and second derivatives; we also know how to interpret these derivatives geometrically, as properties of the graph of the function. It is time for us to return to the subject of graphing, bringing to bear all our knowledge of derivatives.

Sign of $\frac{df}{dx}$	Information about the graph	Possible shape of the graph
$\frac{df}{dx} > 0$	$f(x)$ increases; graph rises	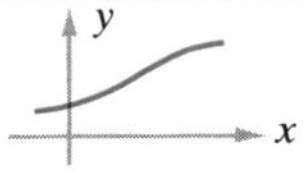
$\frac{df}{dx} < 0$	$f(x)$ decreases; graph falls	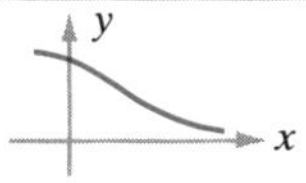

Figure 3.34
The sign of df/dx determines whether the graph is rising or falling as x increases.

Let us organize what we know about derivatives and graphs so we may apply this information in an orderly way. We begin with a differentiable function $y = f(x)$. The significance of its first derivative is indicated in Figure 3.34. As shown in Section 3.3, wherever the derivative is nonzero its sign determines whether $f(x)$ is increasing (graph rising) or decreasing (graph falling) as x moves to the right. At the critical points, where $df/dx = 0$, there are various possibilities as shown in Figure 3.35. The general shape of the graph near a critical point can often be determined using the second derivative test described in Section 3.5.

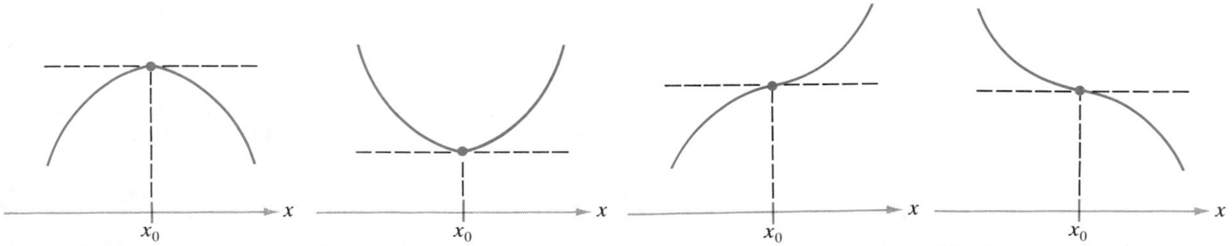

(a) Local maximum (b) Local minimum (c) No local extremum (d) No local extremum

Figure 3.35
Types of behavior of $y = f(x)$ near a critical point x_0. In each case, $f'(x_0) = 0$, so the tangent line is horizontal. Frequently, there is a transition from rising to falling, or vice versa, at a critical point, as in (a) and (b). But this is not always so, as we can see in (c) and (d).

Sign of $\frac{d^2f}{dx^2}$	Information about the graph	Possible shape of the graph
$\frac{d^2f}{dx^2} > 0$	Concave upward	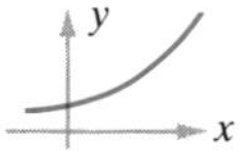 or 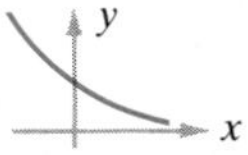
$\frac{d^2f}{dx^2} < 0$	Concave downward	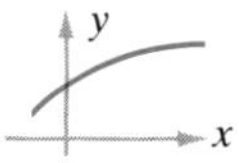 or 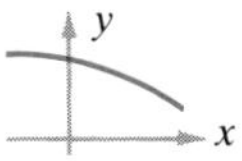

Figure 3.36
The sign of the second derivative d^2f/dx^2 determines whether the graph is concave upward or downward.

Figure 3.36 recalls the significance of the second derivative. Where d^2f/dx^2 is nonzero its sign determines whether the graph is concave upward or downward. A point x_0 marking a transition from concave upward to concave downward, or vice versa, is called an **inflection point**. At such points, $f'' = 0$. The graph of $y = f(x)$ may be sketched by carrying out the following steps.

PROCEDURE FOR CURVE SKETCHING

Step 1 Calculate the first derivative $df/dx = f'(x)$. Determine where $f'(x)$ is positive, negative, or zero. This tells us where f is increasing, decreasing, or has a critical point.

Step 2 Calculate the second derivative $d^2f/dx^2 = f''(x)$. Determine where $f''(x)$ is positive, negative, or zero. This tells us where the graph is concave upward or downward or has a possible inflection point.

Step 3 Plot the points on the graph corresponding to values of x such that $df/dx = 0$ (critical points) or $d^2f/dx^2 = 0$. These are important in making a sketch because the sign of df/dx or d^2f/dx^2 is likely to change at these points. If this is so, the nature of the graph also changes.

Step 4 Use the information from the preceding steps to sketch the graph. Plot additional points as needed to make a reasonably accurate sketch.

In practice, the information provided by the first derivative is of primary importance; the second derivative is useful for fine details, but is less important.

Example 36 Sketch the graph of $y = x^3 - 2x^2 - 4x + 4$ defined for all x.

Solution

Step 1 We observe that f has derivative $f'(x) = 3x^2 - 4x - 4 = (3x + 2)(x - 2)$. Thus

$$\frac{df}{dx} = 0 \quad \text{at} \quad x = +2 \quad \text{and at} \quad x = -\frac{2}{3}$$

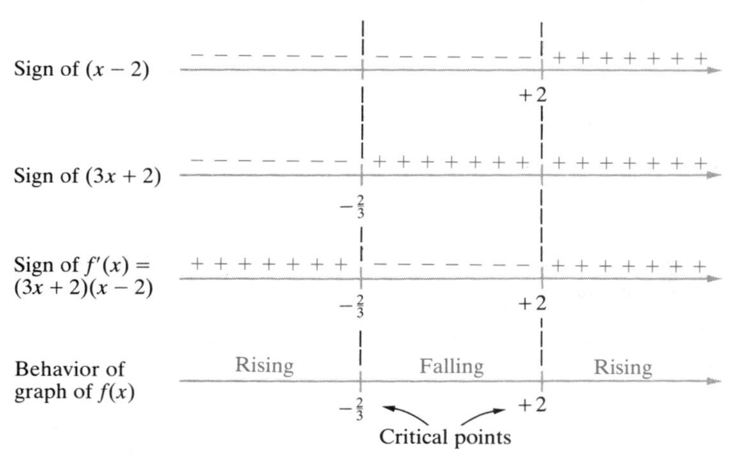

Figure 3.37 To determine the sign of $df/dx = (3x + 2)(x - 2)$, we first determine the signs of the factors $(3x + 2)$ and $(x - 2)$. For example, $x - 2 > 0$ when $x > 2$; similarly, $3x + 2 > 0$ when $x > -\frac{2}{3}$, as shown. Then we calculate the sign of their product df/dx by comparing signs. The information in the last line follows immediately.

The sign of $f'(x) = (3x + 2)(x - 2)$ is determined from the signs of the individual factors $(3x + 2)$ and $(x - 2)$ by applying the rule of signs to set up a sign diagram such as Figure 3.37.

Step 2 Because $f''(x) = 6x - 4 = 2(3x - 2)$, we see that $f''(x) = 0$ at $x = \frac{2}{3}$. Furthermore

$$f''(x) > 0 \quad \text{for} \quad x > \frac{2}{3} \qquad f''(x) < 0 \quad \text{for} \quad x < \frac{2}{3}$$

This information is summarized in Figure 3.38.

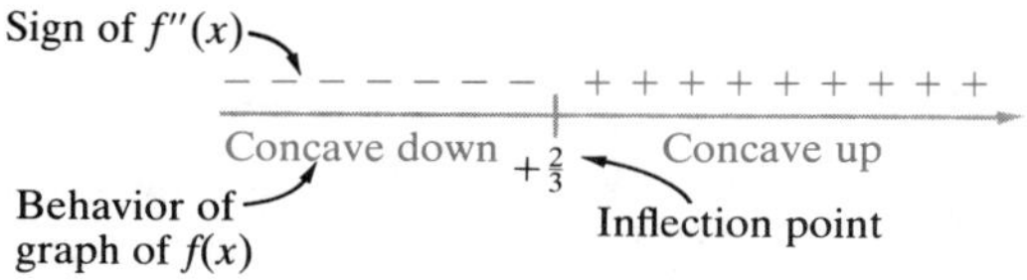

Sign of $f''(x)$

− − − − − − − + + + + + + + + +

Concave down $+\frac{2}{3}$ Concave up

Behavior of graph of $f(x)$

Inflection point

Figure 3.38 The second derivative $f''(x) = 2(3x - 2)$ is zero at $x = \frac{2}{3}$ (an inflection point) and has the signs shown to the right or left of this point. Notice that $3x - 2 > 0$ when $3x > 2$ or $x > \frac{2}{3}$. The concavity of the graph changes from downward to upward as x increases past $x = \frac{2}{3}$.

Step 3 Next we calculate the values $y = f(x)$ for the critical points $x = -\frac{2}{3}$ and $x = 2$, and for $x = \frac{2}{3}$, where $y'' = 0$.

$$f\left(-\frac{2}{3}\right) = \frac{148}{27} = 5.48 \qquad f\left(\frac{2}{3}\right) = \frac{20}{27} = 0.740 \qquad f(2) = -4$$

These points are shown as solid dots in Figure 3.39. The sign diagram in Figure 3.37 shows us that the graph changes from increasing to decreasing at $x = -\frac{2}{3}$, so a local maximum occurs at $x = -\frac{2}{3}$. Likewise, a local minimum occurs at $x = 2$, where the graph changes from decreasing to increasing. The sign diagram for d^2y/dx^2 shows that the graph changes from concave downward to concave upward as x passes the point $x = \frac{2}{3}$, where d^2y/dx^2 is zero, so an inflection point occurs at $x = \frac{2}{3}$.

Step 4 With Figures 3.37 and 3.38 in mind, we sketch the graph so it is increasing and decreasing and concave upward and downward in the right places. A local maximum occurs at $(-\frac{2}{3}, \frac{148}{27})$, a local minimum at $(2, -4)$, and an inflection point at $(\frac{2}{3}, \frac{20}{27})$. The final sketch is shown

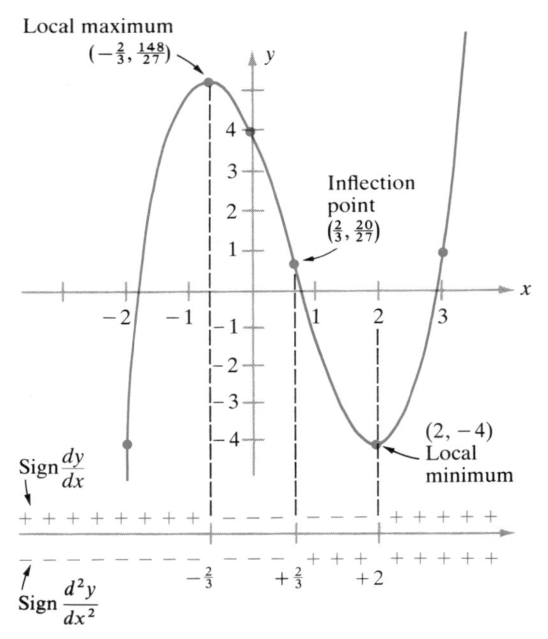

Figure 3.39 The graph of $y = x^3 - 2x^2 - 4x + 4$. The points corresponding to the critical points are a local maximum $(-\frac{2}{3}, \frac{148}{27})$ and a local minimum $(2, -4)$. The point on the graph corresponding to the inflection point $x = +\frac{2}{3}$ is $(\frac{2}{3}, \frac{20}{27})$. These are the points mentioned in Step 3. We know that the graph rises until x reaches $x = -\frac{2}{3}$, falls between $x = -\frac{2}{3}$ and $x = 2$, and rises again beyond $x = 2$. The graph is also concave downward to the left of $x = +\frac{2}{3}$ and concave upward to the right.

in Figure 3.39. To aid our rendering, we have plotted three additional points, corresponding to $x = -2$, $x = 0$, and $x = 3$. Notice that this function has no absolute maximum or minimum. ■

Example 37 Draw the graph of $f(x) = 3x^4 + 4x^3 - 6x^2 - 12x$.

Solution We shall only outline the solution, leaving the fine details as in Exercise 15. The first and second derivatives are easily factored. (For $f'(x)$ we must guess, by trial and error, that $x = 1$ is a root; recall Section 1.5.)

$$f'(x) = 12x^3 + 12x^2 - 12x - 12 = 12(x + 1)^2(x - 1)$$
$$f''(x) = 36x^2 + 24x - 12 = 12(3x - 1)(x + 1)$$

The signs of each factor are easily determined; by comparing signs of the factors

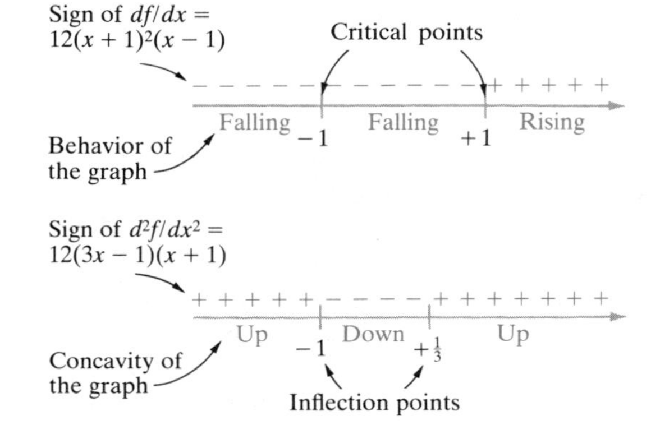

Figure 3.40 Signs of the derivatives of $y = 3x^4 + 4x^3 - 6x^2 - 12x$ and their geometric meaning. The derivatives have been factored and their signs determined from the signs of the factors. We leave the construction of detailed sign diagrams, as in Example 36, to the reader (see Exercise 15).

we arrive at the sign diagrams for $f'(x)$ and $f''(x)$ shown in Figure 3.40. There are two critical points: $f'(x) = 0$ at $x = 1$ and $x = -1$. From the sign diagram for $f'(x)$, we can see that a local minimum occurs at $x = 1$; but note carefully that there is no local extremum at $x = -1$, because the graph continues falling as x moves past $x = -1$. From the sign diagram for $f''(x)$, we see that there are two inflection points, at $x = -1$ and $x = \frac{1}{3}$.

Plotting the critical points, the inflection points, and a few additional graph points corresponding to $x = -2, x = 0, x = 2$, we can sketch an accurate graph (Figure 3.41). ■

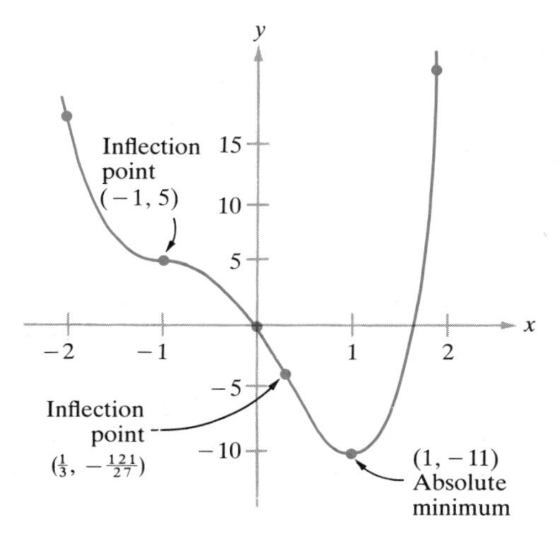

Figure 3.41
The graph of $y = 3x^4 + 4x^3 - 6x^2 - 12x$. A local minimum occurs at $(1, -11)$; points of inflection occur at $(-1, 5)$ and $(\frac{1}{3}, -\frac{121}{27})$. These points are indicated by solid dots. The graph falls and is concave upward until it reaches $x = -1$. At $x = -1$ the slope is zero (tangent line momentarily horizontal). In the interval $-1 < x < +\frac{1}{3}$ the graph falls and is concave downward. For $+\frac{1}{3} < x < +1$ the curve is still falling, but is concave upward. Finally, beyond $x = +1$ the graph rises and is concave upward.

The graph of a function is often badly behaved near any point where the formula used to define f has a zero in the denominator. At such points we should plot several nearby points and use the simple principles discussed at the end of Section 1.1 to obtain an accurate impression of the graph. These ideas are illustrated in the next example.

Example 38 Sketch the graph of $y = \dfrac{1}{x} + 4x$, defined for $x > 0$.

Solution The derivatives are

$$y' = -\frac{1}{x^2} + 4 \qquad y'' = \frac{2}{x^3}$$

The sign diagram for dy/dx is obtained by writing this function as a quotient

$$\frac{dy}{dx} = \frac{-1 + 4x^2}{x^2} = \frac{4\left(x^2 - \frac{1}{4}\right)}{x^2} = \frac{4\left(x - \frac{1}{2}\right)\left(x + \frac{1}{2}\right)}{x^2} \tag{28}$$

The denominator is positive $x^2 > 0$, so the sign of dy/dx is entirely determined by the sign of the numerator, which in turn is determined by the signs of its factors, as shown in Figure 3.42. The derivative is zero (critical point) at $x = \frac{1}{2}$.

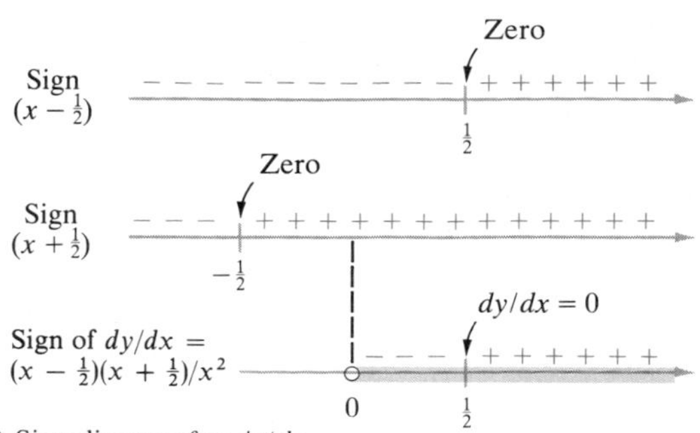

(a) Sign diagram for dy/dx

Sign of
$d^2y/dx^2 = 2/x^3$
0

(b) Sign diagram for d^2y/dx^2

Figure 3.42
Sign diagrams for dy/dx and d^2y/dx^2. In each case the feasible set is the shaded half-line where $x > 0$.

The graph falls for $0 < x < \frac{1}{2}$ and then rises steadily for $x > \frac{1}{2}$. An absolute minimum occurs at the critical point $x = \frac{1}{2}$.

The sign of the second derivative $d^2y/dx^2 = 2/x^3$ on the feasible set is easy to determine. If $x > 0$ then $x^3 > 0$ and $d^2y/dx^2 > 0$, so that the graph is concave upward throughout the feasible set. Plotting the critical point and several points near $x = 0$, we obtain the sketch shown in Figure 3.43. The general behavior of the graph as x approaches zero from the right is given by noting that for small positive x

$$y = \frac{1}{\text{(small positive)}} + 4(\text{small positive}) = (\text{large positive})$$ ■

x	y
0	Undefined
$\frac{1}{8}$	8.5
$\frac{1}{4}$	5.0
$\frac{1}{2}$	4.0
1	5.0
2	8.5

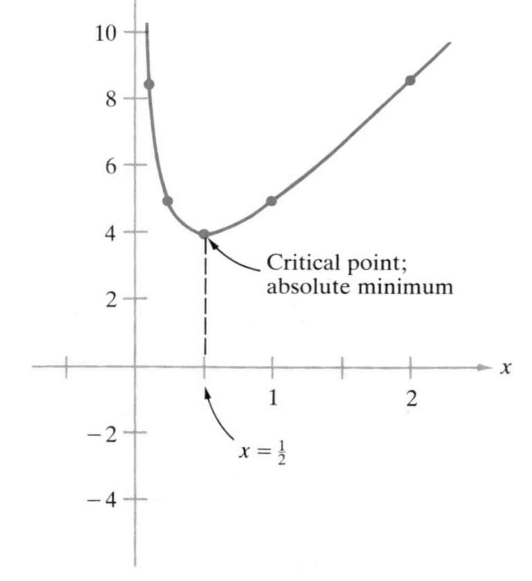

Figure 3.43
Graph of $y = (1/x) + 4x$, defined for $x > 0$. Tabulated values are plotted as solid dots; $x = \frac{1}{2}$ is the only feasible critical point. The formula breaks down at $x = 0$, and the behavior of the graph is somewhat peculiar there. As x approaches zero from the right, the value $y = (1/\text{small positive}) + (\text{small positive}) = (\text{large positive})$ gets larger and larger, as shown.

Exercises 3.9

In Exercises 1–14 find dy/dx and d^2y/dx^2 for each function. Determine all critical points and the parts of the number line where the graph is rising or falling. Determine where the graph is concave upward or downward, and find all inflection points. Then sketch the graph, and indicate all maxima and minima, if any.

1. $y = 7 - 4x + 2x^2$

2. $y = 2x - x^2$

3. $y = x^2 + x + 2$

4. $y = -2000 + 400x - 0.5x^2$

5. $y = x^3 - 3x$

6. $y = x^3 - 6x^2 + 20$

7. $y = x^3 - 9x^2 + 24x - 10$

8. $y = x^3 - 3x^2 - 9x + 13$

9. $y = \frac{1}{3}x^3 + x^2 - 3x + 1$

10. $y = x^4 + 4x$

11. $y = x^4 - 2x^2$

12. $y = 3x^4 - 4x^3 + 2$

13. $y = (x^4 - 18x^2)/27$

14. $y = -x^4 + 4x^3 + 8x^2 + 3$

15. Fill in the details in Example 37. In particular, work out the detailed sign diagram shown in Figure 3.40.

Each of the following functions is defined for $x > 0$. Sketch the graph, and decide if the function has an absolute maximum or absolute minimum.

16. $y = 2x + \dfrac{1}{x^2}$

17. $y = x^2 + \dfrac{1}{x}$

18. $y = \dfrac{4}{x} - \dfrac{16}{x^2}$

19. $y = \dfrac{4x - 16}{x}$

20. $y = \dfrac{4}{x} + \dfrac{1}{x^2}$

21. $y = \dfrac{4x^2 - 16}{x}$

22. Sketch the graph of $y = \dfrac{1}{x^2 + 3}$. (*Hint*: $x^2 + 3 > 0$ for all x; this will help determine the signs of dy/dx and d^2y/dx^2. Simplify your formulas for dy/dx and d^2y/dx^2 as much as you can before studying their signs.)

23. Sketch the graph of $y = \dfrac{1}{1 + x^2}$.

The following functions involve exponential functions. You should recall that $e^a > 0$ for all a. Sketch the graph of each function on the feasible set $x \geq 0$. (*Note*: The first two are "probability density functions" that turn up in statistics.)

24. $y = xe^{-x}$

25. $y = 2xe^{-x^2}$

26. $y = 1 - e^{-x}$

3.9 Implicit Differentiation and Related Rate Problems

The method of **implicit differentiation** is an outgrowth of the chain rule. Often a function $y = f(x)$ is described as the solution of an equation involving x and y, such as

$$x^2y + y = 1 \qquad [29]$$

or

$$x^3 - xy + y^3 = 1 \quad [30]$$

Sometimes we can solve the equation, rearranging it by algebraic manipulations to get y on one side and a formula involving x on the other. For example, we can solve [29] to obtain

$$y = \frac{1}{x^2 + 1}$$

Equation [30] is not so obliging; no simple formula expresses y directly in terms of x. Nevertheless, Equation [30] *implicitly* determines y as a function of x even though finding the actual formula may be beyond our capabilities. Using the chain rule, we may discuss the derivatives dy/dx of such functions without solving the equation that relates x and y.

The idea is to take the derivative d/dx of both sides of [30]. In doing so, *remember that y is regarded as a function of x*, $y = y(x)$. Thus an expression like $y^3 = y(x)^3$—or, more generally, $y^r = y(x)^r$ where r is a real exponent—must be handled using the chain rule. Let us recall the extended power rule, a simplified version of the chain rule discussed in Section 3.2. In the present situation it says

$$\frac{d}{dx}(y^r) = \frac{d}{dx}(y(x)^r) = ry(x)^{r-1} \cdot \frac{dy}{dx} = ry^{r-1} \cdot \frac{dy}{dx} \quad [31]$$

Applying it to such expressions as y^3, y^5, $y^{1/2}$, and so on, we obtain

$$\frac{d}{dx}(y^3) = 3y^2 \frac{dy}{dx} \qquad \frac{d}{dx}(y^5) = 5y^4 \frac{dy}{dx} \qquad \frac{d}{dx}(y^{1/2}) = \frac{1}{2} y^{-1/2} \frac{dy}{dx}$$

(*Note*: It is *not* correct to write $d/dx\,(y^r) = ry^{r-1}$; the expression y^r is not the same as x^r, which only involves the independent variable x.)

Returning to Equation [30], which implicitly defined y as a function of x, we take the derivative of the left side using the version of the chain rule just mentioned. We obtain

$$\frac{d}{dx}(x^3) = 3x^2 \quad \text{(usual power rule)}$$

$$\frac{d}{dx}(y^3) = \frac{d}{dx}(y(x)^3) = 3y(x)^2 \frac{dy}{dx} = 3y^2 \frac{dy}{dx} \quad \text{(chain rule)}$$

$$\frac{d}{dx}(xy) = \frac{d}{dx}(x \cdot y(x))$$

$$= \frac{d}{dx}(x) \cdot y + x \cdot \frac{dy}{dx} = y + x\frac{dy}{dx} \quad \text{(product formula)}$$

Putting these together, we find the derivative of the left side of [30]:

$$\frac{d}{dx}(x^3 - xy + y^3) = 3x^2 - y - x\frac{dy}{dx} + 3y^2 \frac{dy}{dx}$$

The derivative of the right side of [30] is $d/dx\,(1) = 0$. Comparing the results, we get an equation

$$3x^2 - y - x\frac{dy}{dx} + 3y^2\frac{dy}{dx} = 0 \qquad [32]$$

which can be solved for dy/dx.

$$3x^2 - y + (3y^2 - x)\frac{dy}{dx} = 0$$

$$(3y^2 - x)\frac{dy}{dx} = y - 3x^2$$

$$\frac{dy}{dx} = \frac{y - 3x^2}{3y^2 - x}$$

Though the derivative we have found is expressed in terms of both x and y, this is not much of a disadvantage. For example, it is easily checked that the point $P = (1, 0)$ is a solution of [30]. The implicitly defined solution curve $y = y(x)$ passing through P has derivative $y'(1)$ there. Because $x = 1$ and $y = 0$, we obtain

$$y'(1) = \frac{0 - 3(1)^2}{3(0)^2 - 1} = \frac{-3}{-1} = 3$$

Example 39 The equation of a circle with radius 1, centered at the origin, is $x^2 + y^2 = 1$. Use implicit differentiation to find the derivative dy/dx. What is the value of dy/dx at the point $P = (\frac{1}{2}, \frac{1}{2}\sqrt{3})$ on the curve?

Solution The part of the curve passing through P may be thought of as the graph of an implicitly determined function $y = y(x)$, satisfying

$$1 = x^2 + y^2 \qquad [33]$$

Differentiate with respect to x on both sides of [33] to get an equation involving dy/dx:

$$0 = \frac{d}{dx}(x^2) + \frac{d}{dx}(y^2) = 2x + 2y\frac{dy}{dx}$$

The chain rule was used here to obtain $d/dx\,(y^2) = 2y\,(dy/dx)$. Solving for dy/dx,

$$2y\frac{dy}{dx} = -2x \quad \text{or} \quad \frac{dy}{dx} = -\frac{x}{y}$$

At P we have $x = \frac{1}{2}$, $y = \frac{1}{2}\sqrt{3}$, so the derivative (= slope of the curve) there is

$$y'\left(\frac{1}{2}\right) = -\frac{\left(\frac{1}{2}\right)}{\left(\frac{\sqrt{3}}{2}\right)} = -\frac{1}{\sqrt{3}}$$

■

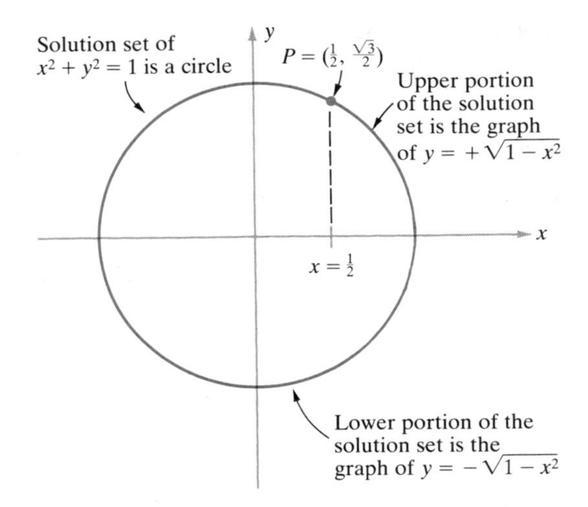

Figure 3.44
The solution set of the equation $x^2 + y^2 = 1$ is a circle with radius 1, centered at the origin. Solving for y we obtain $y = \pm\sqrt{1 - x^2}$. The upper half of the circle is the graph of $y_1(x) = +\sqrt{1 - x^2}$, because $y \geq 0$ there. The lower half is the graph of $y_2(x) = -\sqrt{1 - x^2}$. Both functions satisfy the equation $x^2 + y^2 = 1$, and their derivatives satisfy the equation $dy/dx = -x/y$ obtained in Example 39 by implicit differentiation.

The solution set for $x^2 + y^2 = 1$ and the point P are shown in Figure 3.44. In this problem we could have solved for y in terms of x, and found dy/dx directly; but in other problems this is not possible. The part of the curve passing through P is given by $y = \sqrt{1 - x^2}$ for $-1 \leq x \leq 1$. Its derivative is

$$\frac{dy}{dx} = \frac{d}{dx}((1 - x^2)^{1/2}) = \frac{1}{2}(1 - x^2)^{-1/2} \cdot (-2x) = \frac{-x}{\sqrt{1 - x^2}} = -\frac{x}{y}$$

in agreement with the result obtained by implicit differentiation.

Example 40 If $y^3 + xy = x^2 + 1$, find dy/dx by implicit differentiation. Then find the equation of the tangent line to the solution set of this equation at the point $P = (0, 1)$.

Solution Implicit differentiation yields

$$\frac{d}{dx}(y^3 + xy) = \frac{d}{dx}(x^2 + 1)$$

so that

$$\frac{d}{dx}(y^3) + \frac{d}{dx}(xy) = \frac{d}{dx}(x^2 + 1)$$

Using the chain rule to evaluate $d/dx\,(y^3)$, and the product formula for $d/dx\,(xy)$, we obtain

$$3y^2\frac{dy}{dx} + \frac{d}{dx}(x) \cdot y + x\frac{dy}{dx} = 2x$$

$$(3y^2 + x)\frac{dy}{dx} = 2x - y \qquad [34]$$

$$\frac{dy}{dx} = \frac{2x - y}{3y^2 + x}$$

It is easy to check that $P = (0, 1)$ satisfies our equation. We seek the tangent line to the solution curve $y = y(x)$ passing through P. The slope of this tangent line is the derivative $y'(0)$. Taking $x = 0$ and $y = 1$ in [34], we obtain

$$\text{slope} = y'(0) = \frac{2(0) - 1}{3(1)^2 + 0} = -\frac{1}{3}$$

Therefore, the equation of the tangent line is

$$y = y(0) + y'(0)(x - 0) = 1 - \frac{1}{3}x$$ ■

If two (or more) interrelated variables turn up in a problem, implicit differentiation can be used to understand how a change in one of the variables affects the other variables. Such problems are referred to as **related rate** problems.

Example 41 A ladder 7 feet long is propped between the ground and a wall, as shown in Figure 3.45. The bottom of the ladder is being pulled away from the wall at a steady speed of 3 feet per second. When the bottom of the ladder is 4 feet from the wall:

(i) How high is the top of the ladder above the ground?
(ii) How fast is the top of the ladder sliding downward?

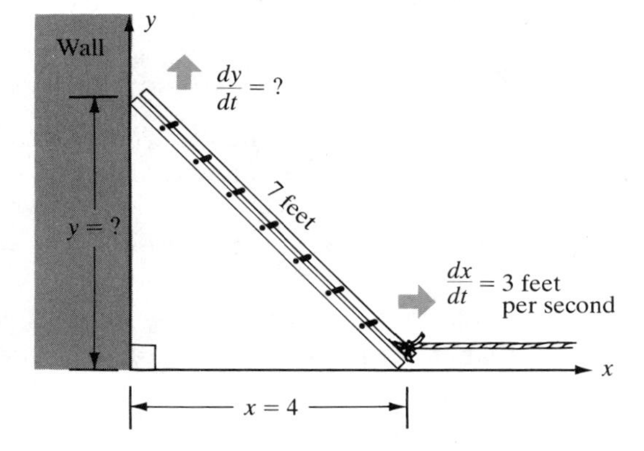

Figure 3.45
The sliding ladder in Example 41. The bottom is pulled steadily outward at 3 feet per second. We want to know what is happening at the moment when $x = 4$ feet. The ladder is 7 feet long.

Solution Let x and y be the distances shown in the figure. Here we have a right triangle, so by Pythagoras' theorem we obtain a relation between x and y that is valid at all times

$$x^2 + y^2 = 7^2 = 49 \qquad [35]$$

If $x = 4$ at a particular moment, then y is given by

$$4^2 + y^2 = 49 \qquad y^2 = 33 \qquad y = \sqrt{33} = 5.745 \quad \text{feet}$$

This answers (i). Both x and y vary with time, all the while related by [35]. Taking the derivative d/dt of both sides of [35], we obtain

$$2x\frac{dx}{dt} + 2y\frac{dy}{dt} = 0$$

$$\frac{dy}{dt} = -\frac{x}{y}\frac{dx}{dt}$$

Notice that the chain rule must be used to find both $d/dt\,(x^2)$ and $d/dt\,(y^2)$, because both x and y are functions of t! In our situation, $x = 4$, $dx/dt = 3$, and we have found that $y = \sqrt{33}$. Thus, at this moment

$$\frac{dy}{dt} = -\frac{4}{\sqrt{33}}(3) = -2.089 \quad \text{feet per second}$$

Because y is measured from the ground to the ladder top, the minus sign means that y is decreasing, and the ladder top slides downward. ■

Example 42 A pebble is dropped into a pond, causing a circular ripple to expand. The radius of the ripple is expanding at a rate of 2 inches per second when the radius is 7 inches. How fast is the area enclosed by the ripple increasing at that moment?

Solution Let r be the radius of the ripple, and A the enclosed area. Then both A and r are functions of the elapsed time t. But, there is also a relation between A and r (area formula for a circle)

$$A = \pi r^2 \qquad [36]$$

that is valid at all times. Regard A and r as functions of t and differentiate both sides of [36] with respect to t. Using the chain rule on the right side, we obtain

$$\frac{dA}{dt} = 2\pi r\frac{dr}{dt} \quad \text{because} \quad \frac{d}{dt}(r^2) = 2r\frac{dr}{dt}$$

We have been told that r is increasing by 2 inches per second when $r = 7$, so $dr/dt = 2$ at that moment. From this we can evaluate dA/dt:

$$\frac{dA}{dt} = 2\pi r\frac{dr}{dt} = 2\pi(7)(2) = 28\pi = 87.92 \quad \text{square inches per second.}$$ ■

Example 43 Suppose a car located at A in Figure 3.46 is 4 miles due south of an intersection at C, travelling north at 40 mph. At the same instant, suppose a car at B is 3 miles east of C, moving east at 50 mph. Is the direct distance s between cars increasing or decreasing at this moment? How fast is s changing?

Solution Referring to Figure 3.46, let y be the distance of car A south of C, let x be the distance of car B east of C, and let s be the straight-line distance between A

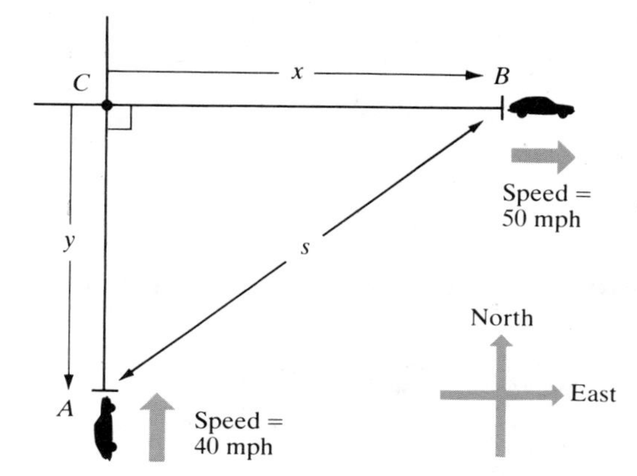

Figure 3.46
Two cars A and B in motion with respect to a highway intersection C. Distances x and y from the cars to the intersection (measured as shown) and the distance s between cars are each functions of time t.

and B. Then x, y, and s vary with time, and are therefore functions of the time t. The three variables x, y, and s are also related to each other; by Pythagoras' theorem on right triangles

$$s^2 = x^2 + y^2 \tag{37}$$

Differentiating both sides of [37] with respect to t, we obtain

$$2s\frac{ds}{dt} = 2x\frac{dx}{dt} + 2y\frac{dy}{dt}$$

or

$$s\frac{ds}{dt} = x\frac{dx}{dt} + y\frac{dy}{dt} \tag{38}$$

Now at the instant of time mentioned in the problem, $x = 3$ and $y = 4$ (miles). Therefore, $s^2 = 3^2 + 4^2 = 25$, and $s = 5$. Furthermore, $dy/dt = -40$ (because y is decreasing) and $dx/dt = +50$ (because x is increasing). Substituting these values into [38], we obtain

$$\begin{aligned} 5\frac{ds}{dt} &= 3(50) + 4(-40) \\ &= 150 - 160 = -10 \end{aligned}$$

so that

$$\frac{ds}{dt} = -2$$

Hence, s is decreasing, and the two cars are approaching one another at the rate of 2 mph. ■

Exercises 3.9

In Exercises 1–8 find dy/dx by implicit differentiation if:

1. $2x + 3y = 4$
2. $5x - 7y + 10 = 0$
3. $x^2 + y^2 = 16$
4. $x^2 - 4y^2 = 4$
5. $x^3 + y^3 = x + y$
6. $x + x^4 = y + y^3 + 2$
7. $xy = x^3 + y^3$
8. $x^3y^3 = 2x + 3y - 4$

In Exercises 9–12 find the slope and the equation of the tangent line at the indicated point on the curve.

9. $x^2 + xy - y^2 = 1$ at $P = (1, 1)$
10. $x^3 - y^3 = 4x - y$ at $P = (2, 0)$
11. $x^3 + y^3 = x - y$ at $P = (0, 0)$
12. $x^3 + y^3 = 2xy$ at $P = (1, 1)$

13. Suppose the cost function $C = C(x)$ of a manufacturer satisfies the equation

$$x^3C + C^3 = 1$$

Find an expression for the marginal cost dC/dx.

14. The surface area A of a sphere is related to its radius r by the well-known formula $A = 4\pi r^2$. Suppose we regard A as the independent variable and want to know how r varies as a function of A. Find dr/dA by implicit differentiation.

The following problems involve related rates.

15. A point moves along the curve $y^3 + x^2 + xy = 4$. At the moment that it passes the point $P = (-2, \sqrt{2})$ on the curve, it is observed that $dx/dt = 10$. Find dy/dt at that moment. (*Note*: As the point moves, both x and y vary with the elapsed time t.)

The area A and volume V of a sphere are given by the standard formulas

$$A = 4\pi r^2 \qquad V = \frac{4}{3}\pi r^3$$

where r is the radius. Answer the following questions in which A, V, and r are all varying with time t. Use implicit differentiation.

16. The volume of a spherical hot-air balloon is increasing at a rate of 10 cubic feet per minute (the rate at which hot air is being pumped in). How fast is the radius increasing when $r = 5$ feet?

17. As a hailstone falls through a cloud, its volume increases as more moisture freezes to it. If the stone is spherical and has radius $r = 1.2$ centimeters, how fast is its radius increasing if the surface area is increasing at a rate of $dA/dt = 0.5$ square centimeters per minute?

18. A low-flying jet aircraft covering a straight course is tracked by a radar station set 6 miles to one side of the flight path (Figure 3.47). Here, $x =$ (distance from a reference marker on the course), and $s =$ (direct distance from aircraft to radar unit). A radar unit can measure only the "range" s and the rate of change ds/dt. Suppose the observed values are $s = 10$ miles and $ds/dt = 800$ mph. Calculate the actual

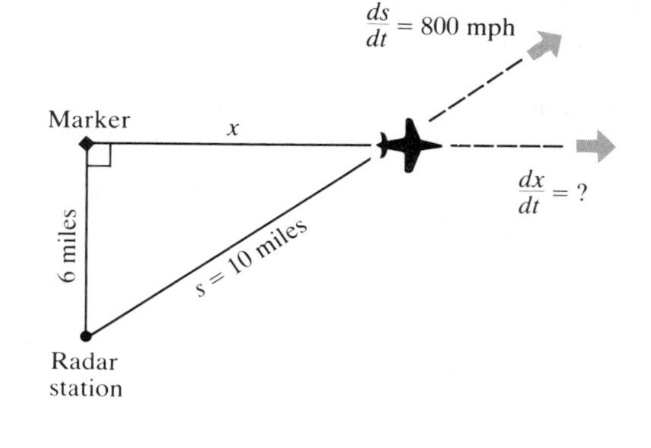

Figure 3.47
The situation in Exercise 18, as seen from directly overhead. The plane flies a straight course on the ground, indicated by the x axis.

speed dx/dt of the aircraft. (*Hint*: By Pythagoras' theorem, the side lengths of the right triangle in Figure 3.47 are related by $s^2 = x^2 + 6^2 = x^2 + 36$.)

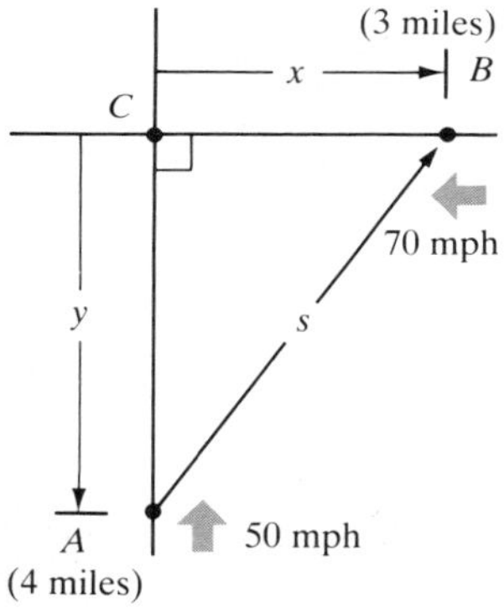

19. In Figure 3.48, the car A is 4 miles from an intersection C, moving toward it at 50 mph. Car B is 3 miles from the intersection, moving toward C at 70 mph. How fast is their direct distance s (shown in Figure 3.48) changing at this moment? What is the value of ds/dt?

(3 miles)
x
B
C
70 mph
y
s
A
50 mph
(4 miles)

Figure 3.48
The situation in Exercise 19. Two cars are approaching an intersection at C; s is the direct (straight-line) distance from car A to car B.

20. The population P of a certain parasite varies with the size H of the host population, according to the empirically determined law

$$P = 3H + 0.0001\, H^2$$

Both P and H depend on the elapsed time since the study began. Find the rate of increase dP/dt if the host population is presently 3000 and is increasing at the rate of 200 per year.

21. An airplane flies along a straight course at a constant speed of 500 mph, and constant altitude of 1 mile, as shown in Figure 3.49. The flight path passes directly over an observer on the ground. In the figure,

s = ground distance from observer to plane

r = direct (straight-line) distance from observer to plane.

When the plane is 2 miles downrange from the observer ($s = 2$), how fast is the direct distance r changing?

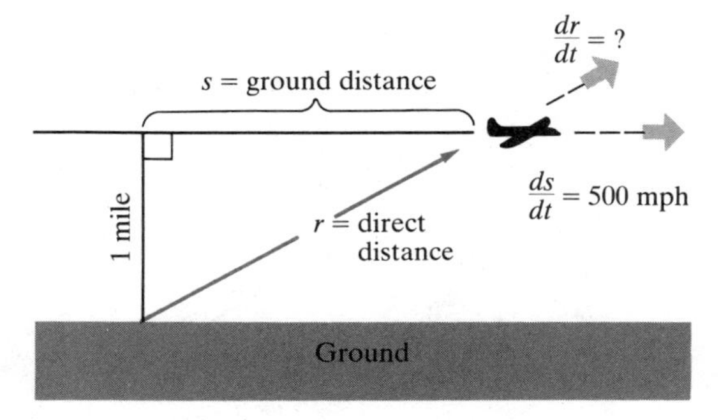

Figure 3.49
Exercise 21.

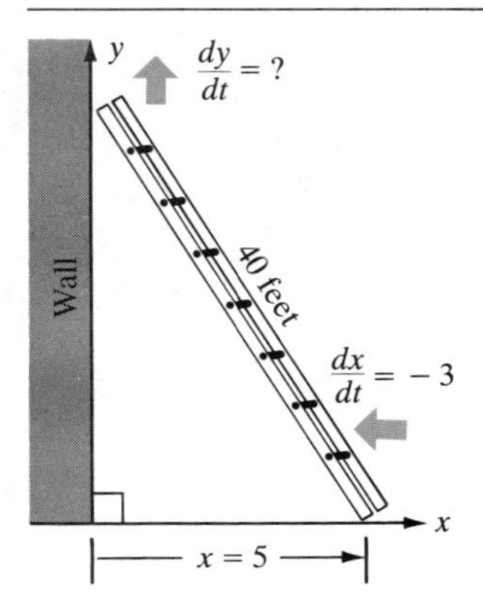

Figure 3.50
Exercise 22. Because x is measured from the base of the wall to the foot of the ladder and the ladder is being pushed toward the wall, x is decreasing with time, and $dx/dt < 0$. This is the source of the minus sign when we say that $dx/dt = -3$ feet per second.

22. A ladder is propped against a wall, as shown in Figure 3.50. The ladder is 40 feet long. Workers are pushing the base of the ladder toward the base of the wall at a steady speed of 3 feet per second ($dx/dt = -3$ if we take x as in the figure, because x is decreasing). How fast is the top of the ladder being pushed up the wall when its base is 5 feet from the wall?

23. A 6-foot-tall person walks at constant speed of 2 feet per second away from a 20-foot-tall lamp post. Let

$$x = \text{her distance from the lamp post}$$
$$s = \text{length of her shadow}$$

as in Figure 3.51. When the person is 15 feet from the post, how fast is the length of her shadow increasing? (*Hint*: The two right triangles in the figure are similar.)

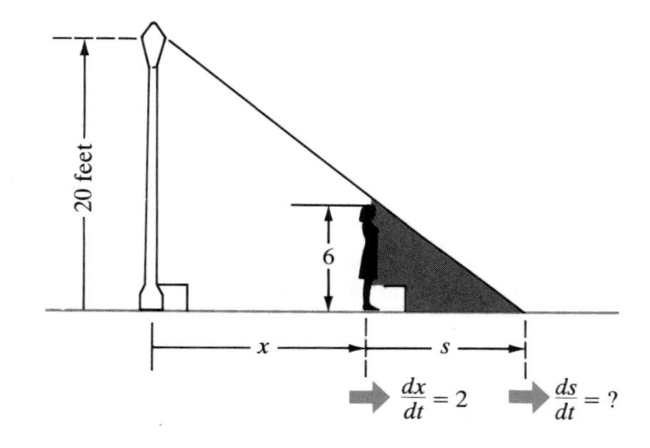

Figure 3.51
Exercise 23.

24. The cable of a suspension bridge has shape described by

$$y = 10 + \frac{x^2}{900} \quad \text{for} \quad -300 \le x \le 300$$

where y is the height of the cable above the roadway (in feet). A repair dolly climbing along the cable is at the point with coordinates (30, 11) and is moving with a horizontal speed $dx/dt = 6$ feet per second. Find the vertical speed dy/dt at this moment.

25. A ladder has been shoved over the top of a 10-foot-high fence, as shown in Figure 3.52. The ladder is 25 feet long. Its base is being pulled away from the base of the fence at a

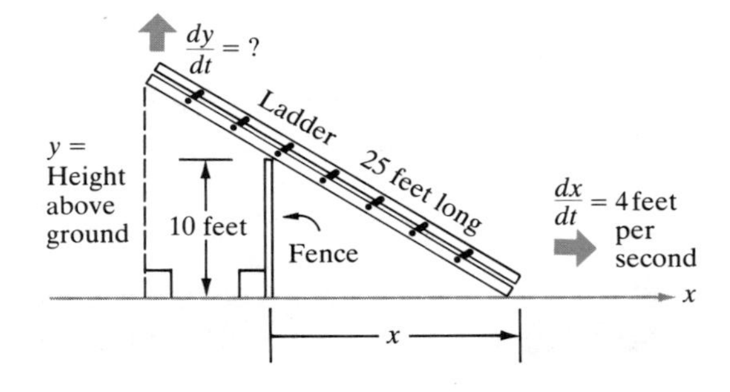

Figure 3.52
Exercise 25. The two triangles are similar. At the moment, $x = 10$.

steady rate of 4 feet per second ($dx/dt = 4$). At the moment when the base of the ladder is 10 feet from the base of the fence, how fast is the height of the ladder top above ground decreasing? (*Hint*: The two triangles in the figure are similar. Compare hypotenuse and vertical side in each to obtain a relation between x and y.)

Checklist of Key Topics

Product rule
Quotient rule
Composite functions
Chain rule
Extended power rule (special version of chain rule)
Optimization problems
Local maxima, minima, extrema
Absolute maxima, minima, extrema
Endpoint extrema
What the sign of dy/dx tells us about the graph
Critical points
Optimization on a bounded interval
Second derivative
Inflection points
Concavity of a graph
What the sign of d^2y/dx^2 tells us about the graph
Second derivative test at critical points
First derivative sign test
Word problems
Curve sketching
Implicit differentiation
Related rate problems

Derivative of $y = e^{kx}$:

$$\frac{d}{dx}(e^{kx}) = ke^{kx}$$

Chapter 3 Review Exercises

In Exercises 1–14 find the derivative of the function using any combination of methods.

1. $y = \dfrac{1}{(3 + 2x - x^3)^9}$

2. $y = \dfrac{1}{\sqrt{x^2 - x}}$

3. $f(x) = \dfrac{\sqrt{x + 4}}{3 + 2x + x^3}$

4. $f(x) = \sqrt{x + 4}\,(3 + 2x + x^2)^4$

5. $s = \dfrac{t^2 + 2t - 3}{4 - t}$

6. $s = \sqrt{4 - t}\,(t^2 + 2t - 3)$

7. $y = \dfrac{e^x}{(x^4 + 1)^7}$

8. $y = (1 - x^6)^8 \cdot \ln x$

9. $u = \dfrac{3v + 7}{4} \cdot e^{-v^2/2}$

10. $u = \dfrac{1}{5\ln(1 + v^2)}$

11. $f(x) = \left(1 - \dfrac{1 + x}{1 - x}\right)^4$

12. $f(x) = (x + \sqrt{1 - x^2})^{3/2}$

13. $y = \ln\left(\dfrac{x^2}{x^3 - 3x + 2}\right)$

14. $y = \dfrac{e^{-4x^2 + 1}}{1 + x^2}$

In Exercises 15–20, find the equation of the tangent line to the graph at the point indicated.

15. $y = \sqrt{1 + x^2}$ at $P = (0, 1)$

16. $y = \dfrac{x^2 - 1}{x^2 + 1}$ at $P = (1, 0)$

17. $y = x\sqrt{1 + x^2}$ at $P = (0, 0)$

18. $y = \sqrt{\dfrac{7 + x}{3 - x}}$ at $P = (-2, 1)$

19. The function $y = y(x)$ determined by $x^3 + 1 = xy + y^3$ at $P = (-1, 1)$

20. The function $y = y(x)$ determined by $xy + 1 = y^3e^x$ at $P = (0, 1)$

In Exercises 21–23 use the fact that the volume of a sphere of radius r is $\frac{4}{3}\pi r^3$ and the surface area is $A = 4\pi r^2$.

21. The volume of an expanding sphere is increasing at a steady rate of 5 cubic feet per minute. Find the rate dr/dt at which the radius is increasing at a moment when the radius is $r = 2$ feet.

22. In Exercise 21, how fast is the surface area increasing at that time?

23. A spherical balloon is deflating. At a given moment the radius is $r = 10$ feet and the volume is decreasing at a rate $dV/dt = -7$ cubic feet per minute. Find dr/dt.

24. Suppose a manufacturer's profit is given by $P = 48x - 0.1x^2 - 4350$, and the feasible production levels are $0 \leq x \leq 300$. Describe the production levels for which the profit is increasing with x. Find the maximum profit and the production level at which it occurs.

In Exercises 25–30 find the absolute maximum and absolute minimum for the function $y = f(x)$ on the interval indicated.

25. $f(x) = x^2 - x^3$ on $-1 \leq x \leq 3$

26. $f(x) = x^3 + 5x^2 - 8x + 13$ on $-1 \leq x \leq 2$

27. $f(x) = x^3 + x^2 - x + 1$ on $-2 \leq x \leq 1$

28. $f(x) = 2x + \dfrac{1}{2x - 1}$ on $\dfrac{2}{3} \leq x \leq \dfrac{3}{2}$

29. $f(x) = xe^{-x}$ on $0 \leq x \leq 3$

30. $f(x) = \dfrac{\ln x}{x}$ on $1 \leq x \leq 4$

In Exercises 31–36 find and classify all local extrema for the function $y = f(x)$.

31. $f(x) = x^2(x - 6)$

32. $f(x) = x^3 + 5x^2 - 8x + 5$

33. $f(x) = x^3 + x^2 - x + 4$

34. $f(x) = x^4 + 4x^3 - 16x + 10$

35. $f(x) = x^2(x - 3)^3$

36. $f(x) = xe^x$

37. Find two numbers a and b whose sum is 30, such that $a^2 + 10b$ is a minimum.

38. A rectangular lot containing 157,500 square feet is to be fenced in. Fencing along the front costs \$5 per linear foot, and fencing along the other three sides costs \$2 per linear foot. Find the dimensions of the lot that minimize the *total cost* of the fence.

39. State usury laws limit the interest charge on a loan to 18% annually. A loan company can lend all the money it can get its hands on at that maximum 18% rate. It gets the money to lend from investors who are willing to put up $100{,}000r$ dollars if they are paid interest on their investment at a rate of r% annually ($r \geq 6.0$). Find the value of r that maximizes the loan company's profit.

40. A farmer currently has 80 milk-producing cows. Past experience indicates that the milk production per cow will decrease linearly by 1% for every additional cow added to his herd. How large should his herd be to maximize total milk output?

41. An oil spill from a tanker mishap has formed a circular oil slick. If the radius is 200 feet and is increasing at the rate of 5 feet per minute, how fast is the area of the slick increasing?

42. A tracking radar is 30 miles from a space shuttle launch site. If the shuttle rises vertically, how fast is the distance from it to the radar unit increasing when the shuttle's altitude is 40 miles and its velocity is 2 miles per second?

In Exercises 43–49 sketch the graph of the function. Indicate the location of all local extrema and inflection points.

43. $y = x^3 - 3x^2 + 2$

44. $y = x^4 - 4x^3 + 10$

45. $y = \dfrac{x}{x^2 + 1}$

46. $y = x + \dfrac{4}{x}$ for $x > 0$

47. $y = x - \dfrac{4}{x}$ for $x > 0$

48. $y = \dfrac{x^2}{x + 1}$ for $x \neq -1$

49. $y = xe^x$

CHAPTER 4

Exponentials, Logarithms, and Applications

4.1 Simple and Compound Interest

If we invest \$1200 for 5 years at an annual rate of 6%, the annual interest is 6% of \$1200, or $(0.06) \times \$1200 = \$72$; the total interest for the five years is $5 \times (0.06) \times \$1200 = \$360$. This kind of return on invested capital is called **simple interest**. If we let r be the annual interest rate written in decimal form and let P stand for the amount invested, the amount of simple interest accumulated after t years is $t \cdot r \cdot P$ dollars,

$$\text{total interest } (t \text{ years}) = trP \qquad [1]$$

The total value of the investment, initial capital plus interest is given by the formula

$$P + trP = (1 + tr) \cdot P \qquad [2]$$

Example 1 If \$700 is invested for five years at simple interest of 6.5% annually, how much interest will be received by the end of five years?

Solution Writing the interest rate as a decimal, we have $r = 0.065$; here, $t = 5$ years. Thus, by Formula [1], the total interest paid is

$$trP = 5(.065)(700) = 227.50 \quad \text{dollars}$$

■

Suppose we deposit \$1200 in a bank account earning 6% annually, but that the bank pays interest quarterly. This does not mean that at the end of each quarter we receive 6% of the amount invested; 6% is the *annual* rate of interest. Actually, at the end of one quarter we receive $\frac{1}{4}$ of the total interest, $\frac{1}{4} \times 6\% = 1.5\%$ of the invested amount, or \$18. Interest paid monthly at a 6% *annual* rate is figured in the same way. After one month we receive $\frac{1}{12} \times 6\% = 0.5\%$ of the invested amount, or \$6.

Suppose we deposit money in a savings account and that the interest payments periodically credited to our account are left in the account untouched. Soon we will be earning interest on the accumulated interest, a process called **compounding** of interest. Consider this example: We deposit \$1200 on January 1 in an account paying 6% annual interest, compounded quarterly (paid in quarterly installments). How much will be in the account on the following January 1 or on January 1 several years later? As already explained, each quarter we are paid $(\frac{1}{4}) \times 6\% = 1.5\% = 0.015$ of the amount in the account, so that our investment grows as shown in Table 4.1. To better understand compounding, we will rewrite the entries in Table 4.1 in a way which emphasizes the underlying pattern. The initial amount (Line 1 in the table) is \$1200, and the quarterly interest rate is 0.015. The first interest payment is 0.015 × (previous balance) = 0.015(1200), raising the balance in the account to

$$\begin{aligned}
\text{(balance at end of first quarter)} &= \text{(previous balance)} + \text{(interest payment)} \\
&= 1200 + (0.015)(1200) \\
&= (1 + 0.015)(1200) \\
&= (1.015)(1200)
\end{aligned}$$

or \$1218, as shown in Line 2 of the table. The second interest payment is based on this *new* balance. The interest to be paid at the end of the second quarter is

$$0.015 \times \text{(previous balance)} = (0.015) \times (1.015)(1200)$$

This payment raises the balance in the account to

$$\begin{aligned}
\text{(balance at end of second quarter)} &= \text{(previous balance)} + \text{(interest payment)} \\
&= (1.015)(1200) + (0.015) \times (1.015)(1200) \\
&= (1 + 0.015)(1.015)(1200) \\
&= (1.015)^2(1200)
\end{aligned}$$

or \$1236.27. The next payment, at the end of the third quarter, is based on this balance. The interest to be paid is

$$0.015 \times \text{(previous balance)} = (0.015) \times (1.015)^2(1200)$$

raising the balance in the account to

$$\begin{aligned}
\text{(balance at end of third quarter)} &= \text{(previous balance)} + \text{(interest payment)} \\
&= (1.015)^2(1200) + (0.015) \times (1.015)^2(1200) \\
&= (1 + 0.015)(1.015)^2(1200) \\
&= (1.015)^3(1200)
\end{aligned}$$

Payment number	Interest payment	Accumulated amount
Start	——	1200.00
1	18.00	1218.00
2	18.27	1236.27
3	18.54	1254.81
4	18.82	1273.64

Table 4.1 Interest on $1200 compounded quarterly at 6% annual interest. At the end of each quarter we are paid $\frac{1}{4} \times 6\% = 1.5\%$ of the amount in the account.

or $1254.81. Notice the emerging pattern, which indicates that there is a simple formula for the total amount in the account after k interest payments:

$$(\text{Amount at the end of } k^{\text{th}} \text{ quarter}) = (1.015)^k(1200)$$

Similar reasoning shows that there is a formula like this, giving the amount accumulated after k interest payments in any compounding scheme. To describe the compounding scheme we must know

(i) The initial amount P, called the principal
(ii) The annual interest rate r (given in decimal form)
(iii) The number of times N that interest is paid per year. Thus, $N = 1, 2, 4$, and 12 for annual, semiannual, quarterly, and monthly compounding, respectively. There are N payment periods in a year.

After k payment periods, the accumulated amount is given by

$$A = P\left(1 + \frac{r}{N}\right)^k \qquad [3]$$

This is called the **compound interest formula**. In particular, if

$$P = 1200 \text{ (principal)}$$
$$r = 0.06 \text{ (annual interest rate)}$$
$$N = 4 \text{ (quarterly compounding)}$$

then [3] gives

$$A = P\left(1 + \frac{r}{N}\right)^k = 1200\left(1 + \frac{0.06}{4}\right)^k = 1200(1.015)^k$$

in agreement with the preceding example.

In many situations the account is held for t years. Then there will be $k = Nt$ payment periods, and [3] takes the form

$$A = P\left(1 + \frac{r}{N}\right)^{Nt} \tag{4}$$

which gives A in terms of t, rather than the number of payments.

Example 2 Find the amount accumulated after 7 years if $P = \$1200$ is compounded quarterly at 6% interest.

Solution Here, $r = 0.06$, $N = 4$, and there are $Nt = 4 \cdot 7 = 28$ payment periods. Substituting these values into [4], we obtain†

$$A = 1200\left(1 + \frac{0.06}{4}\right)^{28} = 1200(1.015)^{28} = 1200(1.51722) = 1820.67$$ ■

Example 3 Find the amount accumulated after 1 year if a principal of $1200 draws interest at an annual rate of 6% compounded monthly.

Solution Here, $P = 1200$, $r = 0.06$, $N = 12$, and there are $k = 12$ payment periods. Using [3] we obtain

$$A = 1200\left(1 + \frac{0.06}{12}\right)^{12} = 1200(1.005)^{12} = 1200(1.06167) = 1274.01$$ ■

Compare this with the amount 1273.63, which is accumulated if the same principal draws interest at 6% for the same time, but with quarterly compounding (Table 4.1). All other things being equal, more frequent compounding is advantageous to the depositor.

Example 4 Suppose we have two investment possibilities: We can lend $1000 to Company A with repayment of $2000 in 10 years, or we can lend $1000 to Company B for 10 years at an annual rate of 8% compounded annually. Assuming the companies are equally good risks, which investment should we choose?

Solution If we choose to deal with Company B, in 10 years our principal will grow to $\$1000(1 + 0.08)^{10} = \$2158.92$. This is greater than $2000, the amount Company A will repay in 10 years; therefore we should lend to Company B. ■

---

† Multiplying 1.015 by itself 28 times is so tedious that a pocket calculator should be used in these examples. To compute $(1.015)^{28}$ on most calculators we use the "y^x" key: Enter 1.015, hit the "y^x" key, enter the exponent 28, and finally hit the "=" key. The result will be 1.51722.... The exercises at the end of this section have been designed to avoid unbearably long computations, and can be done by hand.

In comparing compound interest schemes it is helpful to introduce the notion of **effective annual interest rate**. This is the interest rate r that, applied at simple interest, would produce the same yield after 1 year as the compounding scheme under consideration. Let us examine the yield on an initial investment of \$1, the final result being the same for any initial amount; each dollar grows by the same amount. After 1 year, the yield on simple interest is $(1 + r)$ dollars, so the effective rate r is obtained by solving for r in the equation

$$(\text{value of \$1 compounded one year}) = (1 + r)$$

Example 5 If 5% interest is compounded quarterly, what is the effective annual interest rate r?

Solution Starting with \$1, after one year the compounding scheme yields

$$P\left(1 + \frac{.05}{4}\right)^4 = 1(1.0125)^4 = 1.0509 \text{ dollars}$$

Simple interest at r percent for the same amount of time yields $(1 + r)$ dollars, so that $(1 + r) = 1.0509$, or $r = .0509$ (= 5.09%). ■

So far, our examples have involved realistic interest rates, at least for the United States. Our next illustration may seem exaggerated, but considering recent rates of inflation in certain countries, it may not be far from the mark.

ILLUSTRATION

In the country of Inflatia, things have reached a sorry state. Interest rates have climbed so rapidly that the government has declared a ceiling of 100% on the annual rate. The Inflatia National Bank desperately needs depositors. Realizing that competitors are offering a rate of 100% compounded quarterly, the bank advertises a rate of 100% compounded monthly. Competing banks respond with the same rate compounded weekly. Inflatia National takes the challenge and immediately offers a 100% annual rate compounded daily. Where is all this leading?

Discussion

More frequent compounding raises the "effective" rate of interest. To gauge this quantitatively, we can compare the final amount on a principal of $P = \$1$ for $t = 1$ year under different compounding schemes. With $r = 100\% = 1.00$ and $N = 4$, the final amount is $(1 + \frac{1}{4})^4 = 2.4414\ldots$ for quarterly compounding. For $N = 12$ (monthly compounding) the final amount is $(1 + \frac{1}{12})^{12} = 2.6130\ldots$. For $N = 52$ (weekly compounding), it is $(1 + \frac{1}{52})^{52} = 2.6926\ldots$. For any number N of annual payments, it is $(1 + 1/N)^N$. With more and more frequent compounding, the final amount $(1 + 1/N)^N$ increases, approaching a definite limiting value. That this is so is proved in more advanced texts; the limit value is universally denoted by the symbol e

$$e = \lim_{N \to \infty} \left(1 + \frac{1}{N}\right)^N \qquad [5]$$

Its decimal expansion is $e = 2.71828\ldots$

The banking situation is now clear. The mad rush to compound more frequently with the mandated 100% annual rate leads to final amounts that increase, approaching but never exceeding $e = 2.71828\ldots$ The limiting situation is called **continuous compounding**, for obvious reasons. We can get a pretty good idea of continuous compounding if we think of compounding interest every *microsecond* (1/1,000,000 of a second); this would be done $N = 31{,}536$ billion times per year. Then the final amount $(1 + (1/N))^N$ agrees with $e = 2.71828\ldots$ to many decimal places.

If we take $r = 1.00$ in Equation [4] and allow N to become larger and larger, then the amount accumulated after t years

$$A = P\left(1 + \frac{1}{N}\right)^{Nt} = P\left(\left(1 + \frac{1}{N}\right)^N\right)^t$$

is very nearly equal to Pe^t because $(1 + (1/N))^N$ is close to e. In the limiting situation of continuous compounding, based on an annual interest rate of $r = 1.00$, the value of the investment is

$$A = Pe^t \quad \text{dollars } (P = \text{initial amount})$$

after t years. ■

If the interest rate r is arbitrary instead of $r = 1.00$, similar calculations show that the amount A after t years is given by the following **continuous compounding formula**.

$$A = Pe^{rt} \quad (r = \text{annual interest rate expressed as a decimal}) \qquad [6]$$

A WORD ON THE USE OF THIS FORMULA WITH A CALCULATOR Most pocket calculators are equipped to compute powers of e needed in [6]. To obtain e^x on most calculators, just enter x, then hit the e^x key. We will assume a calculator is used in the following examples, though they could also be done with the tables of e^x given in Appendix 2. Sometimes it helps to simplify formulas using the laws of exponents (Appendix 1) before you grab your calculator; remember that $e^{x+y} = e^x \cdot e^y$, $e^{-x} = 1/e^x$, and $e^0 = 1$. ■

Example 6 Find the amount accumulated after 5 years on a principal of $1000 if interest is continuously compounded at the annual rate of 8%. Compare this with the amount accumulated when compounding is semiannual (two payments per year).

Solution In this example $P = 1000$, $r = 0.08$, and $t = 5$ years. If we substitute these values into [6], we obtain the amount after continuous compounding:

$$A = 1000e^{(0.08)(5)} = 1000e^{0.40} = 1491.82$$

If we compound semiannually, then we apply [4] with $N = 2$ to obtain

$$A = 1000\left(1 + \frac{0.08}{2}\right)^{2(5)} = 1000(1.04)^{10} = 1000(1.48024) = 1480.24$$ ■

Example 7 Compute the value of $100 after 50 years of continuous compounding at annual interest rates of 4%, 5%, and 6%.

Solution In every case, the amount after 50 years is $A = 100e^{50r}$, where $r = 0.04$, $r = 0.05$, and $r = 0.06$, respectively. Using a calculator to obtain the appropriate values of e^x, namely $e^{2.00} = 7.3891\ldots$, $e^{2.50} = 12.182\ldots$, and $e^{3.00} = 20.085\ldots$, we find the final amounts

$$100(e^{2.00}) = 100(7.3891) = \$738.91$$
$$100(e^{2.50}) = 100(12.1825) = \$1218.25$$
$$100(e^{3.00}) = 100(20.0855) = \$2008.55$$

■

It is interesting to compare this growth rate with what you know from common knowledge about the effects of inflation. Do you think that $100 a half century ago would buy as much as $738.91 today? As $1218.25 or $2008.55?

Example 8 The United States is a bit more than 200 years old. If you had invested $100 compounded annually at a rate of 4%, starting 200 years ago, how much would you have today? What if this interest were compounded continuously?

Solution In the compound interest formula [4], $P = 100$, $N = 1$, and $t = 200$; the final amount is

$$A = 100(1.04)^{200} = 100(2550.748) = \$255{,}074.98$$

Under continuous compounding at a rate of 4% annually, Formula [6] applies with $r = 0.04$. The final amount is

$$A = 100(e^{0.04(200)}) = 100(29809.579) = \$298{,}095.80$$

■

Exercises 4.1

1. If $150 is held at simple interest, how much interest will we receive at the end of 5 years if the annual interest rate is

(i) 14% (iii) 15.75%
(ii) 5% (iv) 6.80%

What will be the total value: (initial capital) + (interest) in each case?

2. Find the interest and the total value of $1300, drawing simple interest at 7.25% per year for a period of

(i) 1 year (ii) 2 years (iii) 10 years

3. Find the amount accumulated after 2 years if $100 is invested at an annual rate of 15% compounded

(i) annually (ii) quarterly (iii) continuously

How much would you have, had you started with $10? With $1000?

4. Find the amount accumulated after 2 years if a principal of $100 is invested at an annual rate of 17% compounded

(i) annually (iii) quarterly (v) continuously
(ii) semiannually (iv) monthly

5. A U.S. Treasury bond will mature in one year and pay \$10,000. Suppose the bond costs \$9300. Would it be better to buy the bond or to invest in a savings account paying 7% interest compounded quarterly?

6. A 10-year U.S. Treasury bond has a value of \$25 at maturity. What must the selling price be if the treasury pays
(i) 8.6% interest compounded annually
(ii) 8.6% interest compounded continuously
on funds invested in these bonds?

7. If you borrow \$600 through a credit card, you must pay back the original amount plus compound interest on the loan. Many credit cards charge about 18% annually on unpaid loans. How much must you repay if you borrow \$600 for

(i) 3 months (ii) 6 months (iii) 1 year

if interest is compounded quarterly?

8. Compute the yield on \$100 invested at 6% interest
(i) for one year, compounded annually
(ii) for one year, compounded continuously
The yield on any other compounding scheme (with 6% annual interest) falls between these extremes. What is the annual yield on \$1000 under each scheme? What is the effective annual interest rate in (ii); that is, what annual rate r, uncompounded, yields this amount on \$100 after 1 year?

9. What is the effective annual interest rate if interest is compounded continuously at the annual rate of 5%?

10. How much money should you invest in an account paying 8% compounded annually if you wish to have \$3000 in the account after 5 years?

11. How much money should you invest in an account paying 8% compounded continuously if you wish to have \$3000 in the account after 5 years?

12. A certain municipal bond promises to pay its face value of \$5000 when it matures 1.5 years hence. A bond salesman offers to sell us the bond now at a discount price of \$4500. Alternately, we could invest present funds at 8% interest, compounded quarterly. Should we buy the bond? (What will \$4500 bring us when the bond matures?) At what price P does the bond become a bargain?

13. A manufacturer assumes, for tax-accounting purposes, that a certain piece of plant equipment depreciates in value by $r = 15\%$ each year. (That is, if the value at the beginning of the year is P, then its value at the beginning of the next year is $P - 0.15P = (1 - r)P$.) Assuming its value is \$10,000 when new, compute its successive values $P(n)$ after n years, $n = 1, 2, \ldots$. If equipment is replaced when its value $P(n)$ falls below $\frac{1}{3}$ of its original value, in which year should this equipment be replaced?

14. A piece of manufacturing equipment becomes less valuable as time passes, because of wear and gradual obsolescence. This depreciation in value is often estimated by saying that the unit loses a certain percentage of its value each year. If it depreciates by r percent annually, and the value at the beginning of a year is P, then the value at the beginning of the next year is

$$P - rP = (1 - r)P \quad (r \text{ in decimal form})$$

Find its value $P(N)$ after $N = 1, 2,$ and 3 years. Show that these values agree with the following **depreciation formula** relating $P(N)$ to the initial value P:

$$P(N) = P \cdot (1 - r)^N \quad \text{for} \quad N = 0, 1, 2, \ldots \qquad [7]$$

(*Hint*: Refer to the discussion of the compound interest Formula [3]; the discussion of depreciation is much the same.)

15. Mathematics books become obsolete quickly. A young Ph.D with a technical library valued at \$8500 wants to estimate their value 4 years later, for tax purposes. If he assumes that they depreciate by 30% each year, what are they worth at the end of the fourth year? Use Formula [7].

16. For insurance purposes it is often assumed that an automobile depreciates by about 35% per year. If we buy a car for \$7000 at the beginning of 1983, what will it be worth at the beginning of 1985? What will it be worth at the beginning of 1988? Use Formula [7].

17. How much would \$1 be worth at the beginning of 1985 if it had been invested at the beginning of the year 1850 at
(i) 5% interest compounded annually?
(ii) 5% interest compounded quarterly?
(iii) 5% interest compounded continuously?
What would happen if we started with \$10 or \$100?

18. Legend has it that in 1626 Manhattan Island was sold to Peter Minuit by the Wappinger Indians for \$24. If the \$24 had been deposited immediately in a bank paying 3% compounded continuously, what would it be worth in 1976?

19. A large corporation offers to sell a bond issue to a broker. The bonds mature in 50 years to a face value of \$1,000,000. How much are they worth now if investors have a choice between buying the bonds or investing money at 8% interest compounded annually?

4.2 The Exponential Function $y = e^x$ and Logarithm Function $y = \ln(x)$

In Appendix 1 we review the meaning of a^x for any $a > 0$ and summarize the laws of exponents. Furthermore, we have seen that functions of the form $y = a^x$ describe certain growth phenomena: Recall Section 1.4, where we encountered functions such as $y = 2^x$ and $y = 2^{kx}$ in describing population growth. For various reasons, there is a particular choice of the base a in a^x that has overwhelming advantages, namely $a = e = 2.71828\ldots$. The function defined by

$$y = \exp(x) = e^x \quad \text{for all real} \quad x$$

is called the **exponential function**, and we will concentrate our attention on it. Besides, any exponential function $y = a^x$ to any base $a > 0$ can be written in terms of $y = e^x$, as is shown in more advanced courses.

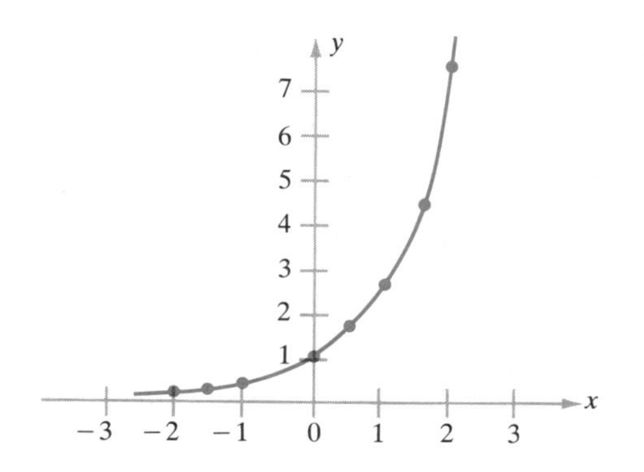

x	$y = e^x$
−3.0	0.0498
−2.0	0.1353
−1.5	0.2231
−1.0	0.3679
0.0	1.0000
0.5	1.6487
1.0	2.7183
1.5	4.4817
2.0	7.3891

Figure 4.1
Graph of the function $y = \exp(x) = e^x$. The tabulated values are plotted as solid dots. Because e^x is always positive, the graph lies above the x axis. It rises very rapidly as x moves to the right, and approaches the x axis as x moves to the left.

The function $y = e^x$ is defined for all x. Notice that the base e is fixed; the exponent x is the variable in this function. In Figure 4.1, we have computed a few values of $y = e^x$ using a calculator and sketched the graph of $y = e^x$. Its values are always positive, and the graph rises steadily as x increases. These geometric facts will be verified in Section 4.5. The exponent laws for real numbers force the exponential function $y = e^x$ to have the following algebraic properties.

ALGEBRAIC PROPERTIES OF $y = e^x$ If x_1, x_2, and x are arbitrary numbers, we have

(i) $e^{x_1 + x_2} = e^{x_1} \cdot e^{x_2}$
(ii) $e^{x_1 - x_2} = e^{x_1}/e^{x_2}$
(iii) $(e^{x_1})^{x_2} = e^{x_1 x_2}$
(iv) $e^{-x} = 1/e^x$
(v) $e^0 = 1$
(vi) $e^1 = e$ [8]

The exponential function $y = e^x$, based on the special number $e = 2.71828\ldots$ is important because it turns up in all kinds of continuous growth problems. We have just seen this in the law governing continuous compounding of interest [6], but it appears in much the same way in the laws of population growth and decline, the laws of radioactive decay in physics, and many others. These applications will be discussed later in Section 4.4.

Generally, if an operation or function is useful in mathematics, so is its inverse. For example, subtraction is the inverse of the operation of addition; likewise, division is the inverse of multiplication. Having defined the exponential function $y = e^x$, we may introduce its inverse, the **logarithm function** $y = \ln(x)$. This function arises when we ask such questions as:

$$\text{For which value of } y \quad \text{is} \quad e^y = 2?$$

Instead of starting with a number x and calculating $y = e^x$, we have reversed the roles of independent and dependent variables; given the value 2 we ask how the variable y should be chosen in order to obtain $e^y = 2$. The solution to this problem is the *logarithm of 2*, written ln(2). Taking any other number x in place of 2 we may ask the same question:

$$\text{For which value of} \quad y \quad \text{is} \quad e^y = x? \tag{9}$$

The solution y, if there is one, is the **logarithm of x**, written $\ln(x)$. Evidently,

$$y = \ln(x) \quad \text{means precisely that} \quad e^y = e^{\ln(x)} = x. \tag{10}$$

If we know how to compute values of e^x, how can we evaluate logarithms such as ln(2) from this definition? The value of e^y increases as y increases, as shown in Figure 4.1. If we try various choices of y, looking for one such that e^y is close to 2, we would find that

$$e^{0.69} = 1.9937 \quad \text{so that} \quad y = 0.69 \quad \text{is too low for} \quad \ln(2)$$
$$e^{0.70} = 2.0137 \quad \text{so that} \quad y = 0.70 \quad \text{is too high for} \quad \ln(2)$$

The correct value $y = \ln(2)$ lies between 0.69 and 0.70. From this we could quickly pin down the next decimal place to see that ln(2) lies between 0.693 and 0.694, and so on. The actual value is $\ln(2) = 0.693147\ldots$. Fortunately, almost all modern calculators have keys for computing both e^x and $\ln(x)$, so it is not necessary to conduct this kind of search. The following examples are done with a pocket calculator, although they could also be done with the table of logarithms provided in Appendix 2.

Now for the graph of $y = \ln(x)$. The number e^y is positive for every real number y (see Figure 4.1). This means that the equation [10] defining $\ln(x)$ has a solution only if $x > 0$; thus the function $y = \ln(x)$ is defined only for $x > 0$. Its graph, sketched in Figure 4.2, must therefore lie to the right of the vertical axis, above/below the domain $x > 0$. At the end of this chapter we will verify the basic geometric properties shown in this figure: As x increases, the graph rises steadily from large negative values to large positive values.

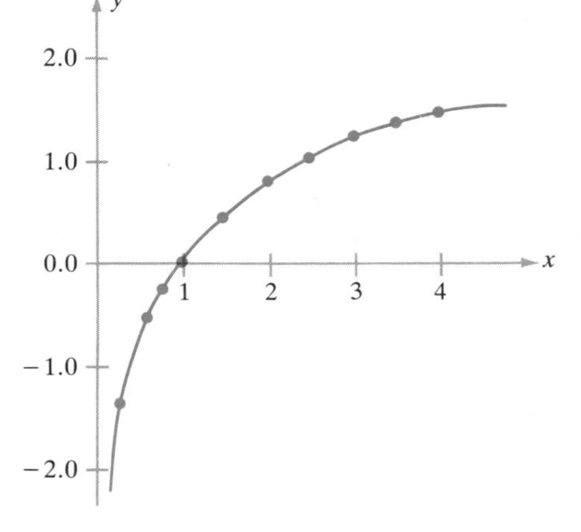

$x>0$	$y=\ln(x)$
0.25	−1.3863
0.50	−0.6931
0.75	−0.2877
1.00	0.0000
1.50	0.4055
2.00	0.6931
2.50	0.9163
3.00	1.0986
3.50	1.2528
4.00	1.3863

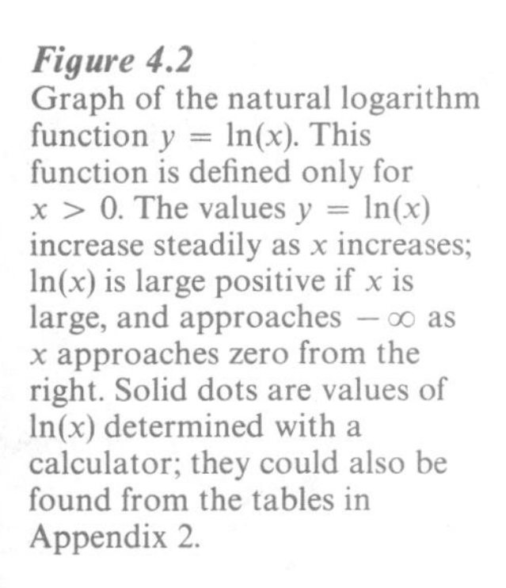

Figure 4.2
Graph of the natural logarithm function $y = \ln(x)$. This function is defined only for $x > 0$. The values $y = \ln(x)$ increase steadily as x increases; $\ln(x)$ is large positive if x is large, and approaches $-\infty$ as x approaches zero from the right. Solid dots are values of $\ln(x)$ determined with a calculator; they could also be found from the tables in Appendix 2.

Example 9 From the definition of $y = \ln(x)$, determine the values of

(i) $\ln(1)$ (iv) $\ln(\frac{1}{2})$, given that $\ln(2) = 0.693147\ldots$
(ii) $\ln(e)$ (v) $\ln(-2)$
(iii) $\ln(1/e^2)$ (vi) $\ln(0)$

Solution If $x = 1$, then $y = \ln(1)$ is defined to be the solution of the equation

$$e^y = 1$$

But we know that $e^0 = 1$, and there is at most one solution, so $\ln(1) = 0$. Similarly,

$$e^y = e \quad \text{has solution} \quad y = 1, \text{ so} \quad \ln(e) = 1$$

$$e^y = \frac{1}{e^2} = e^{-2} \quad \text{has solution} \quad y = -2, \text{ so} \quad \ln(1/e^2) = -2$$

In our discussion we mentioned that $\ln(2) = 0.693147\ldots$; to find $\ln(\frac{1}{2})$ we must solve

$$e^y = \frac{1}{2} \quad \text{or} \quad e^{-y} = 2$$

This means that $-y = \ln(2)$, so $\ln(\frac{1}{2}) = -\ln(2) = -0.693147\ldots$.

Finally, if we try to find $\ln(-2)$ or $\ln(0)$, we must solve

$$e^y = -2 \quad \text{or} \quad e^y = 0$$

But $e^y > 0$, so *there is no solution*; $\ln(-2)$ and $\ln(0)$ are not defined. ■

Our logarithm function $y = \ln(x)$ is sometimes called the **natural logarithm** or **logarithm to base *e***. It is possible to define logarithms to an arbitrary base $a > 0$, $a = 10$ being the most common choice other than $a = e$. This is done by solving for y in the equation $a^y = x$ instead of the equation $e^y = x$; then the inverse function is denoted by $y = \log_a(x)$ to distinguish it from the natural logarithm $\ln(x) = \log_e(x)$, which we discuss. Just as any exponential can be written in terms of the exponential function $y = e^x$, any problem involving logarithms $\log_a(x)$ can be rewritten in terms of the natural logarithm $\ln(x)$. We therefore leave most discussion of functions such as $y = a^x$ and $y = \log_a(x)$ to more advanced courses, though we do provide a brief introduction at the end of this section.

REMARKS ON USE OF CALCULATORS WITH $y = \ln(x)$ More elaborate calculators provide keys for $\log_{10}(x)$ as well as the natural logarithm $\ln(x)$. It is quite common to abbreviate $\log_{10}(x) = \log(x)$, reserving the symbol $\ln(x)$ for the natural logarithm. Be sure you know which key gives the natural logarithm before starting the exercises! ■

We have already pointed out that, by definition of $\ln(x)$

$$e^{\ln(x)} = x \quad \text{for all } x > 0 \qquad [11]$$

That is, if we successively apply the functions "ln" and "exp" to a number $x > 0$, then exp undoes the action of ln so that we end up with x. If instead we apply exp first and then ln , we again end up where we started

$$\ln(e^x) = x \quad \text{for all } x \qquad [12]$$

This is one reason we say that ln and exp are inverses of each other. The **inversion formulas** [11] and [12] are useful in solving algebraic problems.

Because exp and ln are so closely related, the algebraic properties of exp summarized in [8] lead to corresponding properties of ln that are extremely useful.

ALGEBRAIC PROPERTIES OF $y = \ln(x)$ If x_1, x_2, and x are positive numbers, then

(i) $\ln(x_1 \cdot x_2) = \ln(x_1) + \ln(x_2)$

(ii) $\ln\left(\dfrac{x_1}{x_2}\right) = \ln(x_1) - \ln(x_2)$

(iii) $\ln(x^r) = r \cdot \ln(x)$ for any real number r [13]

(iv) $\ln\left(\dfrac{1}{x}\right) = -\ln(x)$

(v) $\ln(1) = 0$

(vi) $\ln(e) = 1$

We will now use these rules and the inversion formulas in some applications. Logarithms were first invented because they convert complicated arithmetic operations into simpler ones, thereby speeding up hand calculations.

$$\ln(x_1 \cdot x_2) = \ln(x_1) + \ln(x_2) \qquad (\ln \text{ converts products to sums})$$
$$\ln(x^r) = r \cdot \ln(x) \qquad (\ln \text{ converts powers to products}) \qquad [14]$$

Although this application has been superseded by the use of calculators (it was still important 20 years ago!), we present one illustration to warm up to the use of logarithms.

Example 10 Use logarithms to calculate $1.5 \times (1.10)^{10}$.

Solution The idea is to find $x = 1.5 \times (1.10)^{10}$ by first computing $\ln(x)$, which is easy if we use the rules [13].

$$\begin{aligned} \ln(x) &= \ln(1.5 \times (1.10)^{10}) \\ &= \ln(1.50) + \ln((1.10)^{10}) \quad \text{by [13i]} \\ &= \ln(1.50) + 10 \cdot \ln(1.10) \quad \text{by [13iii]} \\ &= 0.4055 + 10(0.09531) \quad \text{by calculator or tables} \\ &= 1.3586 \end{aligned}$$

We recover x by taking exp of both sides:

$$x = e^{\ln(x)} = e^{1.3586} = 3.8907$$

■

Example 11 Compute using the rules [13]:

(i) $\ln(0.0000471)$ (ii) $\ln(6.03 \times 10^{23})$

Solution Write $0.0000471 = 4.71 \times 10^{-5}$. Then

$$\begin{aligned} \ln(4.71 \times 10^{-5}) &= \ln(4.71) + \ln(10^{-5}) \quad \text{by [13i]} \\ &= \ln(4.71) - 5 \cdot \ln(10) \quad \text{by [13iii]} \\ &= 1.5497 - 5(2.3025) \\ &= -9.9628 \end{aligned}$$

Similarly

$$\begin{aligned} \ln(6.03 \times 10^{23}) &= \ln(6.03) + 23 \cdot \ln(10) \\ &= 1.7967 + 23(2.3025) = 54.7562 \end{aligned}$$

■

The most important application of logarithms is to solve certain equations. Here, calculators are no help at all; they can do arithmetic, but they can't do algebra! The basic idea is this: To find an unknown x, take ln of both sides of the equation and solve. This is usually much easier than solving the original equation.

Example 12 How long will it take \$1000 to grow to \$2400 if it is invested at 8% annual interest, compounded quarterly?

Solution Here, $P = 1000$, $A = 2400$, $r = 0.08$, and $N = 4$ in the equation

$$A = P\left(1 + \frac{r}{N}\right)^{Nt}$$

and we want to find t. For our values of A, P, and N, we have $r/N = 0.02$, and the equation is

$$2400 = 1000(1.02)^{4t} \quad \text{or} \quad (1.02)^{4t} = 2.40$$

Taking logarithms of each side we obtain

$$4t \cdot \ln(1.02) = \ln(2.40)$$
$$4t(0.01980) = 0.87547$$
$$t = 11.052 \quad \text{years}$$

After $11\frac{1}{4}$ years (45 interest periods) the fund will grow to slightly more than \$2400—to \$2437.85 to be exact; after 11 years (44 interest periods) it has not quite reached our goal. ■

Example 13 Suppose an investment of \$1000 made 10 years ago in a mutual fund is now worth \$2000. At what interest rate, compounded continuously, would we have had to invest the principal to achieve the same yield?

Solution Let r be the interest rate we seek. Taking $A = 2000$, $P = 1000$, and $t = 10$ in the continuous compounding formula [6], we want to find r in the equation

$$2000 = 1000\,(e^{10r}) \quad \text{or} \quad 2 = e^{10r}$$

Taking logarithms on each side, we obtain

$$\ln(2) = 10r$$
$$r = \frac{\ln(2)}{10} = \frac{0.6932}{10} = 0.06932$$

or about 7%. ■

Exercises 4.2

Use a calculator or the exponential tables (Appendix 2) to find the following values of e^x.

1. e^0 **2.** $e^{0.75}$ **3.** $e^{2.14}$ **4.** $e^{1.827}$

5. $e^{0.023}$ **6.** $e^{1.0228}$ **7.** e^{10} **8.** e^{-10}

9. $e^{-1.742}$ **10.** $e^{-3.19}$ **11.** $e^{\sqrt{2}}$ **12.** $\sqrt{e} = e^{1/2}$

13. $\dfrac{1}{\sqrt{e}} = e^{-1/2}$ **14.** $1/e^2$ **15.** $\sqrt[3]{e}$ **16.** $e^{-4/5}$

Use the continuous compounding formula [6] to calculate the amount accumulated under the following continuous compounding schemes.

17. \$1000 at 6% for $4\frac{1}{2}$ years

18. \$10,000 at 6% for $4\frac{1}{2}$ years

19. \$500 at 5.5% for 100 years

20. \$1 at 5%, starting in 1875 and ending in 1975

21. \$1500 at 8.075% for 84 months

22. Calculate the values of the function $f(x) = e^{-x^2}$ for $x = 0.0, \pm 0.25, \pm 0.50, \pm 0.75, \pm 1.0, \pm 1.5$, and ± 2.0. Use these to plot the graph of this function, which is important in probability (the error function).

Find the following logarithms using a calculator or the table of logarithms in Appendix 2. Use the algebraic laws to simplify where this is convenient.

23. $\ln(3)$ **24.** $\ln(5.47)$ **25.** $\ln(0.0038)$

26. $\ln(3.14159)$ **27.** $\ln(10)$ **28.** $\ln(100)$

29. $\ln(1000)$ **30.** $\ln(\sqrt{2.732})$ **31.** $\ln(3^{1/4})$

32. $\ln(2^{1/15})$ **33.** $\ln((1.075)^{50})$ **34.** $\ln(34 \times 10^{22})$

35. $\ln(3.4 \times 10^{-16})$ **36.** $\ln\left(\frac{1}{143}\right)$ **37.** $\ln\left(\frac{1}{143 \times 10^8}\right)$

38. How much should we deposit in a savings account paying 6.25% annually, compounded continuously if we want to end up with a total of \$10,000 after 5 years?

39. Find the interest rate r that will double an investment in precisely 8 years under continuous compounding.

40. A gift of \$5000 to a young girl on her 5th birthday is to be held in trust for her, continuously compounded, until she turns 18. If the gift is to grow to \$15,000 at that time, what interest rate r would yield the desired growth?

41. The size of a pheasant population is modeled by the exponential function

$$N(t) = 8 \cdot 2^{1.60t} \quad t \text{ in years}$$

After how many years will the population triple, according to this formula?

42. A culture of 5000 cloned insulin-producing cells that doubles every 5.3 hours will grow according to the formula

$$N(t) = 5000 \cdot 2^{t/5.3} \quad t \text{ in hours}$$

How many hours must we maintain this culture before we have 10 billion cells (1×10^{10} of them)?

43. A demographic study of Country X has produced the following model for its population growth starting in 1970:

$$P(t) = 87{,}500{,}000 \cdot e^{t/15.596} = 87{,}500{,}000 \cdot e^{(0.064119)t}$$

where t = (years elapsed since 1970). Find the doubling time for this population: the time T after which $P(T) = 2 \cdot P(0)$.

44. Suppose the size of a colony of yeast cells is modeled by an exponential law of the form

$$N(t) = 1000 \cdot e^{kt} \quad t \text{ in hours}$$

where k is to be determined from experimental data.

(i) What is k if the population doubles in size in 5 hours?

(ii) What is k if the population doubles in size in T hours? (In this case, express k in terms of the doubling time T.)

45. One gram of a radioactive isotope with a half-life of T years decays with the passage of time. The amount left after t years is

$$A(t) = 2^{-t/T} \quad \text{grams}$$

Rewrite this formula in the form $A(t) = e^{kt}$ where k is a suitable constant. How is k related to the half-life T?

A demographic study has produced the following model for the growth in population of a certain country:

$$P(t) = 27{,}000{,}000 \cdot 2^{t/30} \quad t \text{ in years}$$

Answer the following questions based on this model.

46. After how many years will the population be twice its initial size of 27,000,000?

47. After how many years will the population triple?

48. After how many years will the population hit 100,000,000 according to this model?

4.3 Present Value, Geometric Series, and Applications

If you asked a banker how he would interpret the proverb "A bird in the hand is worth two in the bush," he might put it this way: A payment of \$1 right now may be more advantageous than an assured promise of \$2 in the year 2000. That may be so, but how about \$1 now versus an assured payment of \$2 three years from now—or even 10 years from now? The answer is no longer clear. Such questions are interest problems in disguise and are handled, psychological factors aside, by introducing the concept of **present value**.

The only way to compare a payment of P dollars now with payment of a different amount A dollars some years hence is to consider what we might do with the P dollars in the meantime. If a safe investment scheme is available that would yield more than the promised amount A dollars at the end of this time, we would take immediate payment. Otherwise, we would take the deferred payment. To make a decision, we must specify the investment scheme we would use to compare present and future funds. The **present value** of an amount A dollars to be paid t years from now is defined to be

(present value) = (the amount now that, invested according to the specified plan, would yield A dollars at the end of t years)

Example 14 Some regular savings accounts pay 5% annual interest, compounded continuously. What is the present value of \$2000 to be paid 5 years from now if the alternative is to place present funds in such a savings account?

Solution According to the continuous compounding formula [6], P dollars invested now and held for 5 years would yield

$$P \cdot e^{(0.05)t} = P \cdot e^{5(0.05)} = P \cdot e^{0.25} = P \cdot (1.2840) \quad \text{dollars}$$

The present value of \$2000 is the initial amount P such that $P \cdot (1.2840) = 2000$; therefore, P is given by

$$P = \frac{2000}{1.2840} = \$1557.63$$

With this investment scheme in mind, \$1557.63 now and a payment of \$2000 after 5 years are equally valuable, ignoring outside factors such as tax exemptions, inflation, and so on. (These factors could also be taken into account, but this would make the problem more complicated. We leave such adjustments to more advanced courses.)

Example 15 The State of Texas wants to sell us bonds worth \$1000 each at maturity 10 years hence. The selling price is \$525. But, we could also put our money in a secure Canadian savings bank that pays 7% compounded annually. What should we do?

Solution Buy the bonds if the present value of \$1000 exceeds the selling price. The present value P is the amount that will grow to \$1000 after 10 years. In view of the compound interest formula [4], this means that

$$1000 = P(1 + 0.07)^{10} = P(1.07)^{10} = P(1.96715)$$

or

$$P = \frac{1000}{1.96715} = \$508.35$$

The present value is less than the selling price, not a very favorable deal from our point of view.

It is not hard to work out general formulas for the present value P of an amount A dollars to be paid t years from now. Comparing things with compound interest (annual rate r expressed as a decimal) paid N times a year, we obtain

$$P\left(1 + \frac{r}{N}\right)^{Nt} = A \qquad [15]$$

Solving for P, we obtain the desired formula.

$$P = \frac{A}{\left(1 + \frac{r}{N}\right)^{Nt}} = A\left(1 + \frac{r}{N}\right)^{-Nt} \qquad [16]$$

Similarly, if amounts are compared relative to continuously compounded interest (annual rate r), we get $P \cdot e^{rt} = A$, or

$$P = \frac{A}{e^{rt}} = A(e^{-rt}) \qquad [17]$$

Geometric progressions arise naturally in evaluating investment schemes with periodic payments. A geometric progression is obtained by starting with a number x and listing its successive powers: $x^0 = 1$, $x^1 = x$, x^2, x^3, and so on.

$$1, x, x^2, x^3, \ldots, x^n, x^{n+1}, \ldots$$

If we take $x = \frac{1}{2}$, the particular progression we obtain looks like this:

$$1, \frac{1}{2}, \frac{1}{4}, \frac{1}{8}, \ldots, \left(\frac{1}{2}\right)^n, \left(\frac{1}{2}\right)^{n+1}, \ldots$$

We will soon be confronted with the task of adding up a block of successive terms in a geometric progression. As an example, we might have to evaluate the sum

$$1 + \frac{1}{2} + \frac{1}{4} + \frac{1}{8} + \cdots + \left(\frac{1}{2}\right)^{10} \quad \text{(11 terms in all)}$$

This could be tedious, involving ten additions of numbers of various sizes. Fortunately, a simple algebraic method yields the sum in a few steps. To evaluate a sum such as

$$1 + x + x^2 + \cdots + x^n$$

we multiply this expression by the quantity $(1 - x)$. Many terms cancel, and we get a remarkably simple result,

$$\begin{aligned}(1 - x) \cdot (1 + x + x^2 + \cdots + x^n) \\ = 1 + \cancel{x} + \cancel{x^2} + \cdots + \cancel{x^n} \\ - \cancel{x} - \cancel{x^2} - \cdots - \cancel{x^n} - x^{n+1} \\ = 1 - x^{n+1}\end{aligned}$$

To put it another way, the sum we are interested in may be rewritten as

$$1 + x + x^2 + \cdots + x^n = \frac{1 - x^{n+1}}{1 - x} \quad \text{for } x \neq 1 \qquad [18]$$

The expression on the right can be evaluated in just four steps, no matter how many terms there are ($n + 1$ of them) in the original sum! In our example, where $x = \frac{1}{2}$ and $n = 10$, we obtain

$$1 + \frac{1}{2} + \frac{1}{4} + \cdots + \left(\frac{1}{2}\right)^{10} = \frac{1 - \left(\frac{1}{2}\right)^{11}}{1 - \left(\frac{1}{2}\right)} = \frac{1 - \left(\frac{1}{2048}\right)}{1 - \left(\frac{1}{2}\right)}$$

$$= \frac{1 - (0.000488)}{0.5} = 1.99902$$

because $(1/2)^{11} = 1/(2^{11}) = 1/2048 = 0.000488$.

If instead we want to add up some other block of successive terms, say $x^r + x^{r+1} + \cdots + x^s$, just remove the common factor x^r from each term and apply [18]; we obtain

$$x^r + x^{r+1} + \cdots + x^s = x^r \cdot (1 + x + x^2 + \cdots + x^{s-r}) \qquad [19]$$
$$= x^r \cdot \left(\frac{1 - x^{s-r+1}}{1 - x}\right)$$

Example 16 Evaluate $(\frac{1}{2})^5 + \cdots + (\frac{1}{2})^{10}$

Solution Separating off the common factor $(\frac{1}{2})^5 = \frac{1}{32}$, we may write the sum as

$$\left(\frac{1}{2}\right)^5 \cdot \left(1 + \frac{1}{2} + \cdots + \left(\frac{1}{2}\right)^5\right)$$
$$= \left(\frac{1}{2}\right)^5 \cdot \frac{1 - \left(\frac{1}{2}\right)^6}{1 - \left(\frac{1}{2}\right)} = \left(\frac{1}{32}\right)\frac{1 - \left(\frac{1}{64}\right)}{1 - \left(\frac{1}{2}\right)}$$
$$= \left(\frac{1}{32}\right)\frac{1 - 0.01562}{0.5} = 0.06152$$

■

Sums such as [18] or [19] are called **geometric sums**. Here is an application in which they arise.

Example 17 A wealthy benefactor plans to set aside a fixed amount of money each year for 20 years, so that at the end of 20 years he may provide his university with a scholarship endowment fund of \$100,000. If his annual payments are placed in a savings account paying 6% compounded annually, what amount A must he deposit each year, starting now, so that the account, including interest, will total \$100,000 at the end of the 20th year?

Solution From the compound interest formula [4] we may compute the future value, at the end of the 20th year, of each payment (see Table 4.2). These future values must add up to \$100,000,

$$100{,}000 = A(1.06) + A(1.06)^2 + \cdots + A(1.06)^{20} \qquad [20]$$
$$= A(1.06)(1 + (1.06) + (1.06)^2 + \cdots + (1.06)^{19})$$

The geometric sum on the right may be evaluated using [18],

$$100{,}000 = A(1.06)\frac{1 - (1.06)^{20}}{1 - (1.06)} \qquad [21]$$
$$= A(1.06)\frac{1 - 3.20713}{-0.06} = A(36.7586)$$

or $100{,}000 = A(36.7856)$. Solving for A we get $A = \$2718.45$. ■

Start of year number	Payment	Number of years in account	Future values
1	A	20	$A(1.06)^{20} = A(3.2071)$
2	A	19	$A(1.06)^{19} = A(3.0256)$
$\vdots$	$\vdots$	$\vdots$	$\vdots$
20	A	1	$A(1.06)^{1} = A(1.0600)$

Table 4.2 Schedule of payments in Example 17 and their future values at the end of 20 years. Payments are made at the start of each year, starting with Year 1 and ending with Year 20. Funds are drawn out at the end of Year 20.

This is an example of an **annuity problem**. These problems have many variants, as the next example shows.

Example 18 (Cost-benefit analysis) A state senator favors a government hydroelectric project at an initial cost of $40 million, on the grounds that once built it will yield $2 million in benefits each year and will "pay itself off after 20 years." Is his comparison of costs and benefits realistic? Does the total present value of these benefits exceed the initial cost (assuming funds grow at 6% compounded annually)?

Solution Because we (as taxpayers) are being asked to put up the initial $40 million *now*, while benefits will appear only in the future, the only realistic way to make a judgment is to compare the present values of the costs and benefits. Costs are incurred immediately, so their present value is $40 million. Present values of the benefits, received at the end of each year from the 1st through the 20th, are given by formula [16]. The $2 million benefit at the end of the nth year has present value

$$2(1.06)^{-n} = 2\left(\frac{1}{1.06}\right)^n = 2(0.9434)^n \quad \text{million dollars}$$

These values are indicated in Table 4.3. Total present value of the benefits is evaluated using the formula for geometric sums

$$\begin{aligned}
(\text{present value}) &= 2(0.9434) + 2(0.9434)^2 + \cdots + 2(0.9434)^{20} \\
&= 2(0.9434)(1 + (0.9434) + \cdots + (0.9434)^{19}) \\
&= 2(0.9434)\frac{1-(0.9434)^{20}}{1-(0.9434)} \qquad [22] \\
&= 2(0.9434)\frac{1-0.3118}{1-0.9434} \\
&= 22.9406 \quad \text{million dollars}
\end{aligned}$$

This is a lot less than $40 million. ■

End of year number	Benefit	Cost	Present values
Start	–	40	-40
1	2	–	$2(0.9434)^1 = 1.8868$
2	2	–	$2(0.9434)^2 = 1.7800$
3	2	–	$2(0.9434)^3 = 1.6793$
⋮	⋮	⋮	⋮
19	2	–	$2(0.9434)^{19} = 0.6611$
20	2	–	$2(0.9434)^{20} = 0.6237$

Table 4.3 Present values of the costs and benefits in Example 18. All values are in millions of dollars. Minus sign indicates a cost, plus sign a benefit. Sum of the benefits is $22.9406 million.

Suppose we accumulate benefits for a longer period, say $n = 100$ years. The calculation of present value of all the benefits goes the same way, yielding

$$2(0.9434)\frac{1-(0.9434)^{100}}{1-(0.9434)} = 33.2374 \quad \text{million dollars}$$

still no breakeven! In fact, if we wait n years, the cumulative present value of the benefits is just

$$\begin{aligned} &2(0.9434) + 2(0.9434)^2 + \cdots + 2(0.9434)^n \\ &= 2(0.9434)\frac{1-(0.9434)^n}{1-(0.9434)} \quad \text{million dollars} \end{aligned} \tag{23}$$

But if x is any real number between zero and one, $0 < x < 1$, its powers x^n get smaller and smaller, approaching zero as n gets larger and larger. Thus, as n increases $(0.9434)^n$ gets small, the expression $1 - (0.9434)^n$ approaches the value 1, and the cumulative benefit [23] gets closer and closer to the "limiting value."

$$2(0.9434)\frac{1}{1-(0.9434)} = 33.3357 \quad \text{million dollars}$$

Even if we accumulated benefits *forever* their total present value would never reach $40 million! The proposed installation is a very bad investment.

Remark We have used the fact that if $|x| < 1$, the powers x^n become smaller and smaller and approach zero as n increases.

Exercises 4.3

1. A 10-year U.S. Government bond yields $100 at maturity 10 years from now. If interest is 6% compounded continuously, what is the present value of the bond? What if interest is 6% compounded semiannually?
2. Find the present value of $10,000 to be paid 8 years from now, if the alternative is to place present funds in a savings account paying (i) 7% annual simple interest, (ii) 7%

annual interest compounded semiannually, and (iii) 7% annual interest compounded continuously.

3. A stockbroker is trying to sell us some bonds with face value \$10,000 that mature 2.5 years from now. To get us interested, he offers them at a discount price of \$8400. If the alternative is to place funds in a term savings account yielding 8.18% compounded continuously, is the deal worth pursuing?

4. A 55-year-old employee is laid off after many years of service with a company. The company agrees to severance payments of \$10,000 now, \$7000 in one year, and \$4000 in two years. What is the present value of the severance package if money earns interest at 7% compounded continuously?

5. Legally, the maximum rate of interest that can be charged is 20% compounded continuously; a higher rate is illegal and is called *usury*. Which of the following lending schemes are legal? (The scheme is legal if the present value of all payments, computed using the maximum legal interest rate, is less than or equal to the initial loan of \$1000.)
(i) A loan shark lends \$1000 to be repaid in three installments of \$400 each, due in 2, 4, and 6 months.
(ii) The loan shark lends his favorite cousin \$1000 to be repaid in three installments of \$350 each, due in 2, 4, and 6 months.
(iii) The loan shark lends his mother \$1000 to be repaid in three installments of \$333.33, due in 2, 4, and 6 *years*.

6. Evaluate $1 + 3 + 3^2 + \cdots + 3^9$ by multiplying out each term and adding, and then by using the geometric sum formula.

Evaluate the following geometric sums

7. $1 + (1.06) + (1.06)^2 + \cdots + (1.06)^{10}$

8. $2 + 4 + 8 + \cdots + 2^8$

9. $1 + \left(\frac{3}{2}\right) + \left(\frac{3}{2}\right)^2 + \left(\frac{3}{2}\right)^3 + \left(\frac{3}{2}\right)^4$

10. $\left(\frac{3}{4}\right) + \left(\frac{3}{4}\right)^2 + \cdots + \left(\frac{3}{4}\right)^6$

11. $(1.08)^{10} + (1.08)^{11} + \cdots + (1.08)^{30}$

12. $(e^{-0.65})^2 + (e^{-0.65})^3 + \cdots + (e^{-0.65})^{40}$

Sum the following geometric progressions

13. $0.9 + (0.9)^2 + \cdots + (0.9)^{10}$

14. $(0.9)^3 + (0.9)^4 + \cdots + (0.9)^{12}$

15. $1 + (0.9) + (0.9)^2 + \cdots + (0.9)^{100}$

16. $(0.9)^5 + (0.9)^6 + \cdots + (0.9)^{50}$

17. $(0.9) + (0.9)^2 + (0.9)^3 + \cdots$ (forever) . . .

18. Rework Example 17, assuming that savings are held at 6% interest compounded continuously.

19. A simplified college trust fund is organized as follows. Each year a certain amount is to be set aside in an account paying 6% interest compounded annually, starting when a child turns 5 and ending with a payment at age 16. When the child turns 17 the funds are withdrawn. What amount A should be set aside yearly if the fund is to be worth \$20,000 at age 17?

20. Repeat Exercise 19, assuming that interest is compounded quarterly instead of annually.

21. A \$5000 coupon bond pays \$160 semiannually beginning 6 months from now. At the end of 10 years, the last coupon will be redeemed and the face value of \$5000 will come due. What is the present value of the bond (including coupon payments) if the money earns

(i) 10% compounded annually?
(ii) 5% compounded quarterly?
(iii) 6.4% compounded continuously?

22. How much should you donate now to establish a prize fund at your local high school that will award \$100 annually. Assume that the first award is 1 year hence and the last is 20 years hence, and that your donation will be put in a bank account paying 7% per year compounded continuously.

23. Suppose you want to establish a "perpetual" scholarship prize fund yielding \$100 each year forever. What amount of money, placed in a savings account yielding 7% interest compounded continuously, will yield \$100 each year? (Each year the interest is removed and used for the prize; the capital is never depleted.) (*Note*: This problem can be done simply, without resorting to present value calculations.)

24. Suppose we want to build up a scholastic award fund whose total value is \$1380. We want to stretch out our payments over a period of 10 years, paying in equal amounts at the beginning of Years 1 through 10 to a savings account that pays 7% interest compounded continuously. The desired total \$1380 is to be achieved at the beginning of the 11th year. What amount A must we set aside yearly?

25. If we used a lower interest rate to compare costs and benefits in Example 18, say 4%, would this make the project sound more profitable or less profitable? What if we used a higher interest rate, say 7%? Answer without making detailed calculations.

26. Compare the costs and benefits in Example 18 if we assume that the "cost of money," the interest rate at which present and future funds are compared, is 3.25% instead of 6%. Does this lower interest rate make the project sound more, or less, feasible than the 6% rate? (*Note*: Presently, 3.25% is an unrealistic value for the cost of money. If a bank were willing to lend us \$100,000 at this rate, we would be foolish not to take the loan. Yet in a recent case[†] the U.S. Bureau of Reclamation tried to justify an Idaho dam project, using precisely this 3.25% rate to "prove" that benefits exceeded costs. Why do you think the Bureau would use such an unrealistic rate in its calculations?)

27. A college trust fund is to be set up. A fixed amount will be invested each year at 6% interest compounded quarterly. Payments begin when a child turns 5 and end with a final payment at age 16. Then \$5000 is to be drawn out each year for the next four

[†] The 1975 Teton Dam Project in Idaho. Incidentally, the dam collapsed disastrously in 1975, shortly after its completion. For details, see the *New York Times Magazine*, 18 September 1975.

years, at ages 17 through 20. The fund terminates with the last payment. What amount A must be set aside each year?

What happens to the following geometric sums as n gets large? Use this fact: If $|x| < 1$, then x^n becomes very small as n increases.

28. $1 + \frac{1}{2} + \frac{1}{4} + \cdots + \left(\frac{1}{2}\right)^n$

29. $\frac{1}{2} + \frac{1}{4} + \frac{1}{8} + \cdots + \left(\frac{1}{2}\right)^n$

30. $0.9 + (0.9)^2 + \cdots + (0.9)^n$

31. $0.95 + (0.95)^2 + \cdots + (0.95)^n$

32. $1 + \frac{1}{4} + \frac{1}{16} + \cdots + \left(\frac{1}{4}\right)^n$

33. (This problem concerns economic multipliers.) Suppose the average person, on receiving \$1 in income, saves $\frac{1}{10}$ of it and spends the remaining $\frac{9}{10}$. The money spent becomes income for someone else, and so on, leading to the pattern of transactions shown in Figure 4.3. By the time the Nth person is reached, how much cash has changed hands from person to person? (Include the initial dollar, write down a geometric sum, and simplify.) As N gets larger and larger, what is the limiting value of the amount of cash that has changed hands? (*Note*: This limiting value is called the **multiplier** in economics. It measures the total amount of cash circulation in the entire economy caused by a typical dollar of income.)

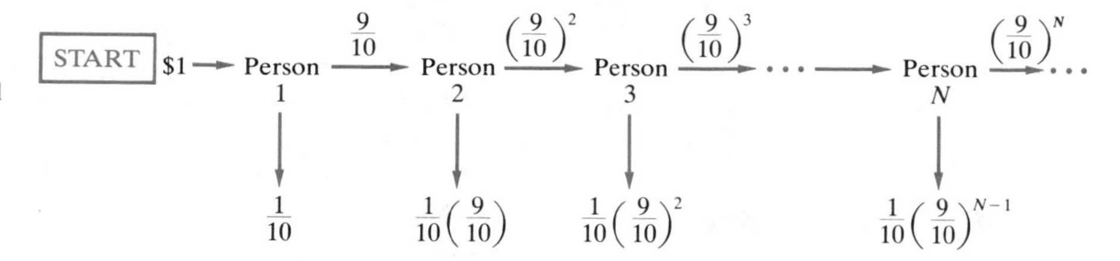

Figure 4.3
The transactions discussed in Exercises 33 and 34. Horizontal arrows indicate amounts of cash changing hands between individuals; downward arrows indicate amounts put into savings, out of circulation.

34. Rework Exercise 33, assuming that the fraction of income diverted to savings is $\frac{1}{20}$ (5% of income) instead of $\frac{1}{10}$.

4.4 Application: Growth and Decay Phenomena

Growth and decay are opposites. In this section we will explain how both may be described by a single all-encompassing theory. Exponentials and logarithms are the key to this theory. As we shall see, the growth or decay of a quantity Q with time is often governed by an exponential equation

$$Q(t) = Ce^{kt} \quad (C \text{ and } k \text{ suitably chosen constants}) \qquad [24]$$

Let us start with a particular problem, the decay of radioactive elements, before describing the general theory.

Since 1900 the concept of radioactivity has slowly become part of the public consciousness. No doubt you have heard about radioactive fallout from weapons tests, and its steady but slow decay into inert products. Or you may have heard about carbon dating of objects and archeological sites, which has revolutionized our understanding of history. This dating method involves careful measurement of the decay of radioactive C^{14} (carbon-14), which is produced by the action of cosmic rays on nonradioactive atmospheric nitrogen (see Example 21).

In any object the amount of radioactive material steadily decreases as unstable radioactive atoms decay into stable inert products. To describe this decrease we consider the quantity Q of radioactive material remaining at time t. Then, $Q = Q(t)$ is a function of time t, whose value decreases steadily toward zero. The crucial fact about this phenomenon is that:

The rate of change dQ/dt at any moment is proportional to the amount of material present at that moment.

That is, there is a constant of proportionality k such that

$$\frac{dQ}{dt} = kQ(t) \quad \text{for all } t \qquad [25]$$

In a decay process material is lost, so Q decreases, dQ/dt is negative, and k is a negative number; in a growth process, $k > 0$. This proportionality [25] can be justified by a "thought experiment." Suppose we take away part of the sample, leaving, say, $\frac{1}{2}$ (or $\frac{1}{3}$ or $\frac{15}{27}$) of the original sample. Then the number dQ/dt of atoms decaying per second in the remaining sample is obviously $\frac{1}{2}$ (or $\frac{1}{3}$ or $\frac{15}{27}$) the number for the full sample. That is, the decay rate per unit time dQ/dt is proportional to the amount of radioactive material in the sample. The value of k corresponds to the rate at which atoms decay; it depends on the material studied and must be determined from experimental data.

The equation [25] is called a **differential equation** because it involves an unknown function *and its derivatives.* Differential equations will be discussed in more detail in Section 6.4. For the particular equation [25] it can be shown that the only solutions $Q(t)$ are functions having the form

$$Q(t) = Ce^{kt} \quad \text{where } C \text{ is some constant}$$

Once we know the nature of the solutions [24], it remains only to see how the constants C and k are determined from observations. The constant C out front has a simple interpretation:

$$C \text{ is the initial value of } Q\text{, the value } Q(0) \text{ when } t = 0 \qquad [26]$$

In fact, setting $t = 0$ in [24] we get

$$Q(0) = Ce^{k \cdot 0} = Ce^0 = C \cdot 1 = C$$

The other constant k is determined once we know the amount remaining at some later time $t = t_0$, because

$$Q(t_0) = Ce^{kt_0} = Q(0)e^{kt_0}$$

$$e^{kt_0} = \left[\frac{Q(t_0)}{Q(0)}\right]$$

Taking logarithms of both sides we obtain

$$kt_0 = \ln(e^{kt_0}) = \ln\left[\frac{Q(t_0)}{Q(0)}\right]$$

Thus

$$k = \frac{1}{t_0}\ln\left[\frac{Q(t_0)}{Q(0)}\right] \quad \text{where } Q(t_0) \text{ is the amount at time } t_0 \qquad [27]$$

Here are some problems where we determine the constants C and k and then use the formula $Q = Ce^{kt}$ to make predictions.

Example 19 For the radioactive element radium the constant k in [24] determined by experiment is $k = -4.27 \times 10^{-4}$ if t is measured in years. In how many years will 1 ounce decay to the point that only half the original amount remains? (*Note*: In any decay process, the time T it takes for half of the original material to disintegrate is characteristic of the material and is called the **half-life**.)

Solution If $Q(t)$ is the amount remaining after t years, we know that

$$Q(t) = Ce^{(-4.27 \times 10^{-4})t}$$

for a suitable choice of the constant C. Initially, when $t = 0$, there is 1 ounce of radium, so $C = 1$. This completely determines the function $Q(t)$

$$Q(t) = e^{(-4.27 \times 10^{-4})t} \quad \text{ounces} \qquad [28]$$

After a certain amount of time T (the half-life of radium) only $\frac{1}{2}$ ounce will remain:

$$\frac{1}{2} = Q(T) = e^{(-4.27 \times 10^{-4})T}$$

Solve for T by taking logarithms of both sides:

$$\ln\left(\frac{1}{2}\right) = (-4.27 \times 10^{-4})T$$

$$T = \frac{\ln\left(\frac{1}{2}\right)}{-4.27 \times 10^{-4}} = \frac{-0.69315}{-4.27 \times 10^{-4}} = 1623 \quad \text{years}$$ ■

If we start with any initial amount of radium other than 1 ounce, half of that amount will remain after $T = 1623$ years; the half-life is independent of the amount we start with. There is no sense in talking about the "full" lifetime of radium: Because $Q(t) = e^{(-4.27 \times 10^{-4})t}$ never reaches zero, there is always some radium left no matter how much time has passed.

Example 20 The half-life of C^{14}, the radioactive form of carbon, is 5720 years: Half of any sample will decay in this period of time. Determine the constants C and k in [24] and write out the formula for the amount of C^{14} remaining after t years if there are 5 grams of pure material initially. Calculate the amount remaining after 10,000 years. How much remains after 3 half-lives?

Solution In Equation [24], $C = 5$ is the initial amount. From the definition of half-life we know that

$$Q(5720) = 5e^{5720k} = 2.5 \quad \text{grams}$$

or

$$\frac{1}{2} = e^{5720k}$$

Taking logarithms, we obtain k:

$$5720k = \ln(e^{5720k}) = \ln\left(\frac{1}{2}\right) = -\ln(2) = -0.69315$$

$$k = \frac{-0.69315}{5720} = -1.212 \times 10^{-4} = -0.0001212$$

Thus the amount after t years is

$$Q(t) = 5e^{(-1.212 \times 10^{-4})t} \quad \text{grams} \qquad [29]$$

If $t = 10{,}000$ we get

$$Q = 5e^{(-1.212 \times 10^{-4}) \times 10^4} = 5e^{-1.212} = 5(0.2976) = 1.488 \quad \text{grams}$$

or 29.8% of the original amount. After 3 half-lives $t = 3(5720) = 17{,}160$ years; entering this into [29] we obtain $Q = 0.6250$ grams. ■

After each successive half-life the amount remaining is reduced by half, so without any computation at all we can see that

$$Q = 5\left(\frac{1}{2}\right) \quad \text{after 1 half-life } (t = 5720 \text{ years})$$

$$Q = 5\left(\frac{1}{4}\right) \quad \text{after 2 half-lives } (t = 11{,}400 \text{ years})$$

$$Q = 5\left(\frac{1}{8}\right) \quad \text{after 3 half-lives } (t = 17{,}160 \text{ years})$$

$$\vdots$$

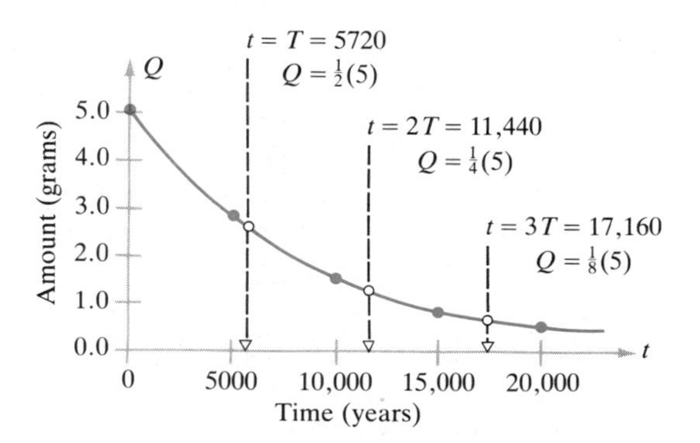

Figure 4.4
Graph of the exponential function $Q(t) = 5e^{kt}$, taking $k = -1.212 \times 10^{-4} = -0.0001212$. Initially, $Q(0) = 5$ grams; thereafter Q values decrease steadily toward zero. This function gives the amount remaining after t years from a 5-gram initial sample of C^{14}. The half-life of C^{14} is $T = 5720$ years. Note the values of Q after successive half-lives have elapsed (open dots).

For the reader's convenience, we have sketched the graph of $Q(t)$ in Figure 4.4, plotting the points corresponding to successive half-lives $t = T, 2T, 3T$, and so on.

The next example shows how these facts can be used in radiocarbon dating (see Exercises 3–6). The carbon in the atmosphere is mostly the stable isotope C^{12}, with a small percentage of the radioactive isotope C^{14}. During the life of any plant or animal the ratio of weights C^{14}/C^{12} in its body is almost identical to the corresponding ratio for carbon in the air; organic carbon is ultimately drawn from the air by plant photosynthesis, and biochemical processes cannot distinguish between C^{12} and C^{14}. After death, there is no further exchange of carbon between the air and the remains, so the radioactive C^{14} decomposes, while the inert C^{12} remains unaltered. The ratio of C^{14} to C^{12} steadily decreases toward zero. By comparing the (known) C^{14}/C^{12} ratio at the time of death with the ratio observed in the remains, we can determine their age.

Example 21 If bones discovered in an archaeological dig contain 25% of their original C^{14} content, find their age using Formula [29] worked out in Example 20.

Solution If C is the initial amount of C^{14}, then at any later time the amount remaining is given by Formula [29]

$$Q(t) = Ce^{(-1.212 \times 10^{-4})t} \quad (t = \text{years after death})$$

At present, t_0 years after death, the weight of C^{14} is only 25% of the original amount, so $Q(t_0) = 0.25C$ and

$$0.25C = Ce^{(-1.212 \times 10^{-4})t_0}$$

The (undetermined) constant C cancels out, leaving

$$0.25 = e^{(-1.212 \times 10^{-4})t_0}$$

Taking the logarithm of both sides, we find t_0:

$$-1.3862 = \ln(0.25) = (-1.212 \times 10^{-4})t_0$$

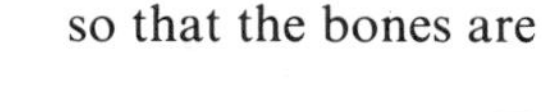

so that the bones are

$$t_0 = \frac{-1.3862}{-1.212 \times 10^{-4}} = \frac{1.3862}{1.212} \times 10^4 = 11{,}437 \quad \text{years old}$$

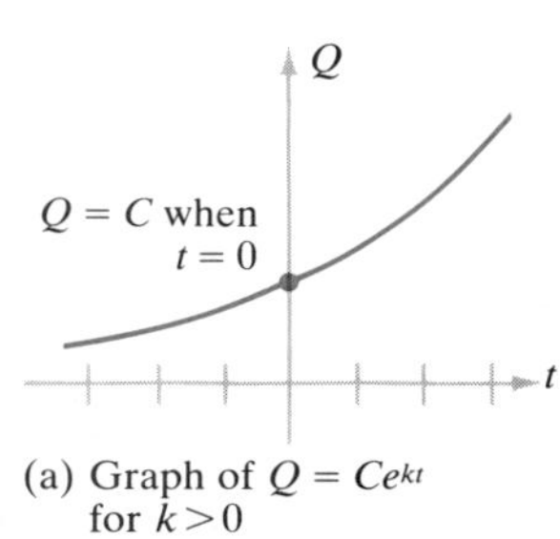

(a) Graph of $Q = Ce^{kt}$ for $k > 0$

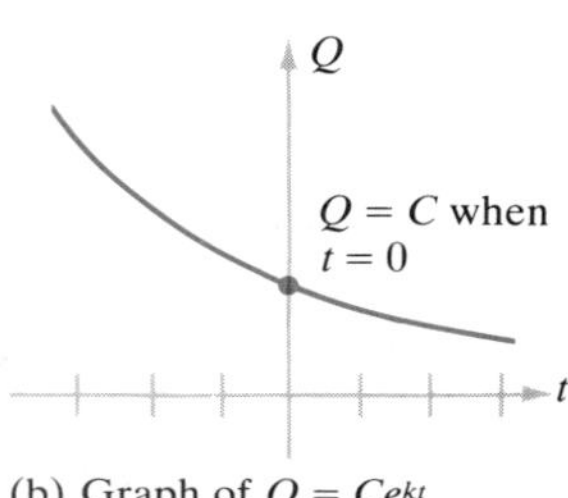

(b) Graph of $Q = Ce^{kt}$ for $k < 0$

Figure 4.5
The graph of an exponential $Q = Ce^{kt}$ rises (a) or falls (b) according to whether the constant k is positive or negative. If $k = 0$, then $Q = Ce^{kt} = e^0 = 1$ for all t (a constant function). The graphs are very steep if k is large positive or large negative. They pass through $(0,C)$ no matter what the value of k.

The same sort of analysis applies to many other problems involving growth or decay. In *any* situation where the rate at which some quantity Q changes is proportional to the size of Q

$$\frac{dQ}{dt} = k \cdot Q(t) \quad \text{where } k \text{ is some constant of proportionality} \qquad [30]$$

then Q will depend on the elapsed time according to the formula

$$Q(t) = Ce^{kt} \qquad [31]$$

This sort of proportionality arises in simple population growth, continuous compounding of interest, environmental degradation of pesticide residues, and many other situations. For example, in a colony of bacteria supplied with abundant nutrients the number of new bacteria produced per hour (the rate of change in population size) is clearly proportional to the population size at any time. Thus the population size $Q = Q(t)$ will be governed by an equation of the form $Q = Ce^{kt}$. Similar remarks apply to human populations; as long as the net birthrate remains constant and food supplies are abundant, a population will grow exponentially. You might recall that the growth of capital under continuous compounding is also governed by such a law: $A = Pe^{rt}$. Why? Because the proportionality condition [30] holds; the rate at which your investment grows is at all times proportional to the size of the investment at the time—that is, each dollar grows at the same rate.

The behavior of the function $Q(t) = Ce^{kt}$ depends on the sign of the constant k: $k > 0$ makes Q grow as time passes, and $k < 0$ makes Q decrease toward zero (Figure 4.5). In either case, the constant C out front is the initial size of Q, the value when $t = 0$.

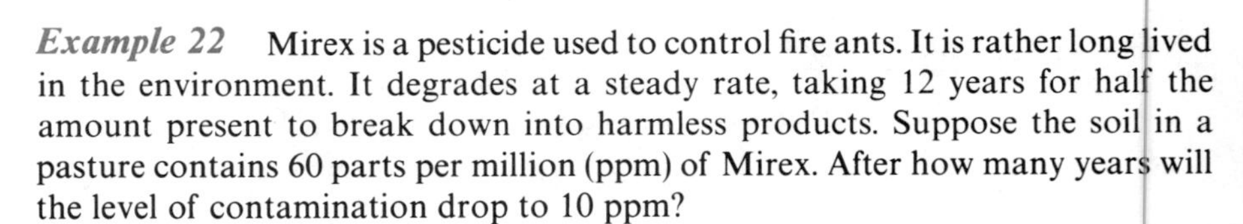

Example 22 Mirex is a pesticide used to control fire ants. It is rather long lived in the environment. It degrades at a steady rate, taking 12 years for half the amount present to break down into harmless products. Suppose the soil in a pasture contains 60 parts per million (ppm) of Mirex. After how many years will the level of contamination drop to 10 ppm?

Solution The rate at which Mirex breaks down is proportional to the amount present—as in radioactive decay problems—so the amount $Q(t)$ after t years is described by an exponential function

$$Q(t) = Ce^{kt} \quad \text{where } C \text{ and } k \text{ are constants}$$

First we determine C and k. Because $Q = 60$ when $t = 0$, $C = 60$ in this formula. To find k use the fact that $Q = 30$ when $t = 12$; then, in our formula

$$30 = Q(12) = 60e^{12k}$$

or

$$\frac{1}{2} = e^{12k}$$

Thus

$$-0.69315 = \ln\left(\frac{1}{2}\right) = 12k$$

$$k = \frac{-0.69315}{12} = -0.05776$$

and in our situation the amount of residue after t years is

$$Q(t) = 60e^{-(0.05776)t} \quad \text{parts per million} \tag{32}$$

To find the time $t = t_0$ when $Q(t_0) = 10$, we solve

$$10 = Q(t_0) = 60e^{-(0.05776)t_0}$$

$$\frac{1}{6} = e^{-(0.05776)t_0}$$

Taking logarithms, we obtain

$$-1.79176 = \ln\left(\frac{1}{6}\right) = -(0.05776)t_0$$

$$t_0 = \frac{-1.79176}{-0.05776} = 31.02 \quad \text{years}$$ ■

To give you some idea what is going on, Figure 4.6 shows the graph of $Q(t)$ determined in [32]. The arrow indicates the time when the residues have dropped to a safe level of 10 ppm.

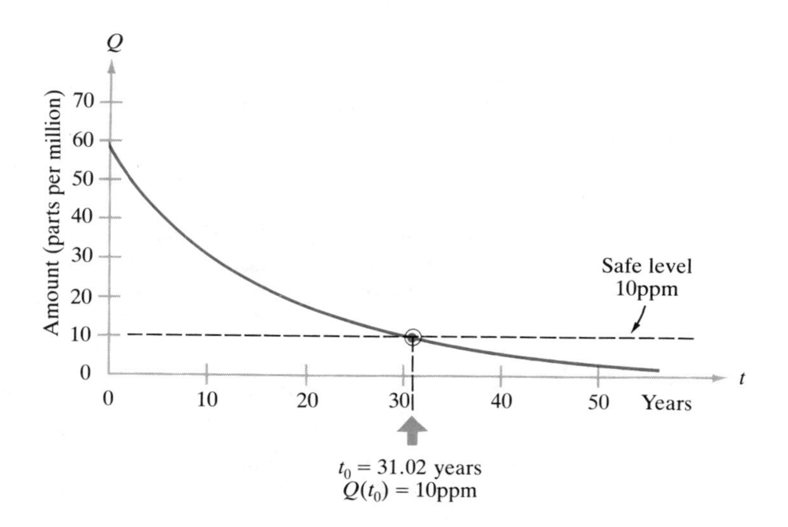

Figure 4.6
Graph of the function $Q(t) = 60\,e^{-(0.05776)t}$, giving the amount of Mirex residues in the soil after t years, as in Example 22. After $t_0 = 31.02$ years the amount has fallen to a safe level of 10 parts per million.

Example 23 Country X, with a high birth rate, has a population that grows 3% annually. Estimate the size of the population in 50 years if it is now 10 million.

Solution The population after t years is given by an equation of the form

$$N(t) = Ce^{kt} \quad (t \text{ in years})$$

Since $N(0) = 10{,}000{,}000$, $C = C \cdot e^0 = 10{,}000{,}000$; our main task is to find k. To say that the population increases by 3% in one year means that

$$\frac{N(1) - N(0)}{N(0)} = 0.03 \quad (=3\% \text{ growth from initial value } N(0) \text{ in 1 year})$$

But $N(1) = Ce^{k \cdot 1} = N(0) \cdot e^k$; thus the value $N(0) = 10{,}000{,}000$ cancels out in the formula

$$0.03 = \frac{N(0) \cdot e^k - N(0)}{N(0)} = e^k - 1$$

Adding 1 to both sides, we obtain

$$1.03 = e^k \quad \text{or} \quad k = \ln(1.03) = 0.02956$$

Consequently, the growth of this population is governed by the formula

$$N(t) = 10{,}000{,}000e^{(0.02956)t} \quad (t \text{ in years}) \qquad [33]$$

In $t = 50$ years the predicted population is

$$\begin{aligned} N(50) &= 10{,}000{,}000e^{(0.02956)\cdot 50} \\ &= 10{,}000{,}000e^{1.478} \\ &= 43.8 \quad \text{million} \end{aligned}$$

■

With Formula [33] in hand we could go on to answer many other questions. For instance, how long does it take this population to double in size?

Exercises 4.4

1. Sketch the graphs of the functions

   (i) $Q(t) = e^{0.5t}$ (ii) $Q(t) = e^{-0.5t}$ (iii) $Q(t) = 4e^{-0.27t}$

   for $t \geq 0$.

2. Sketch the graph of $N(t) = 10{,}000{,}000e^{(0.02956)t}$, the equation of the population growth in Example 23. Choose a suitable scale of units on each axis so your graph fits on a single piece of graph paper. You should at least show what happens for $0 \leq t \leq 50$.

3. Carbon in the cloth wrapper of an Egyptian mummy has been found to contain 71% of its original C^{14} content. Presumably, the wrappings and the mummy have nearly equal ages. What is the age of the entombed luminary? (*Hint*: Refer to Example 21.)

4. Physical chemists have determined that the uranium isotope U^{238} is radioactive, with a very long half-life of 4.5×10^9 years. If we start with 1 gram of pure uranium

isotope when $t = 0$, show that the amount left after t years is

$$Q(t) = e^{-(0.154 \times 10^{-9})t} \quad \text{grams} \qquad [34]$$

How much remains after 1 billion years? After 10 billion years? After 100 billion years? (1 billion $= 1 \times 10^9$.)

5. The uranium isotope U^{238}, with half-life of 4.5×10^9 years, decays into a form of lead that does not occur naturally. Thus, chemical tests of a rock specimen can tell us what fraction of the uranium isotope in the rock has decayed into lead since the rock solidified. Suppose the fraction remaining in three specimens of granite (from different locations) is

(i) 78% (ii) 85% (iii) 92%

Determine the age of each rock. What conclusion can you draw about the age of the Earth? (*Hint*: You could use Formula [34] from Exercise 4, rather than working things out from scratch.)

6. Suppose that 0.157% of the original uranium U^{238} in a sample of Canadian pitchblende (a rich uranium ore) has decayed into lead; 99.843% remains. Estimate the number of years since the ore mass was formed. (*Hint*: Use the formula [34] given with Exercise 4.)

7. The radioactive isotope of iodine I^{131} is used to treat thyroid disorders. Starting with a 1-gram sample of pure isotope, 80% remains after 2.598 days. Find the half-life T of this isotope, the time after which only half the sample remains. (*Hint*: Find C and k in the equation $Q(t) = Ce^{kt}$; then determine T using the fact that $Q(T) = \frac{1}{2}$.)

8. DDT absorbed in lake-bottom mud is slowly degraded by bacterial action into harmless products. Careful measurements show that 10% of the initial amount is eliminated in 5 years. As in all decay processes, the concentration of DDT is described by an equation of the form $Q(t) = Ce^{kt}$, where C is the initial concentration. Determine k. What is the half-life of DDT in this environment (after how many years will half of the initial contamination be degraded)? After how many years will 95% be degraded?

9. The population of Mexico tripled from 1940 to 1975, increasing from 20 million to 60 million. Assuming the same birthrate persists, estimate the Mexican population in the year 2000. (*Hint*: Set up a formula $N = Ce^{kt}$ where $t =$ (years elapsed from the base year 1940).)

10. Countries W, X, Y, and Z have populations that grow by 1%, 2%, 3%, and 5% annually. Assuming that population growth is not suppressed by famine, disease, or a shift to smaller families, the population in each country will double in a certain amount of time. Calculate these "doubling times" using ideas from Example 23. (*Note*: Initial population is irrelevant.)

11. In Country X, population grows 3% annually, and the gross national product (GNP) grows at 2% annually. If the per capita income (GNP/population) is \$750, what will it be in

(i) 2 years? (ii) 10 years? (iii) t years?

12. An initial population of 10,000 bacteria increases to 30,000 after 3 hours in an incubating medium. The size of the population obeys a growth law of the form

$N(t) = Ce^{kt}$ for suitably chosen constants C and k, as long as nutrients are plentiful. Determine the constants, and write down an explicit formula for $N(t)$. After how many hours will the population double in size?

13. Country Y doubles in population every 37 years. Find the *percentage increase* in population

$$\frac{N(10) - N(0)}{N(0)} \times 100\%$$

in 10 years. (*Hint*: First find $N(t)$. Here $N(0)$ = initial population.)

14. Country A has a population of 6,000,000 that doubles every 40 years. The population of Country B is 4,000,000 and doubles every 25 years. In how many years will the countries have equal populations?

15. The intensity of sunlight reaching a depth x in a murky lake decreases exponentially with x according to the formula

$$I = I_0 e^{-0.5x} \quad (x \text{ measured in feet})$$

where I_0 is the intensity at the water surface. At what depths are 50%, 70%, and 90% of the sunlight absorbed?

16. Suppose that the growth of some quantity $Q = Q(t)$ is governed by an exponential formula

$$Q(t) = Ce^{kt} \quad (C \text{ and } k \text{ are positive constants})$$

Prove that there is a "doubling time" T such that Q doubles if we start at *any* time t and allow T to elapse:

$$Q(t + T) = 2 \cdot Q(t) \quad \text{for any starting time } t$$

How is T related to k and C?

4.5 The Calculus of Exponentials and Logarithms

We have been using the differentiation formulas

$$\frac{d}{dx}(e^x) = e^x \qquad [35]$$

$$\frac{d}{dx}(\ln x) = \frac{1}{x} \qquad [36]$$

for some time. In this section we shall prove these facts and use them to discuss the graphs of $y = e^x$ and $y = \ln x$. As a bonus we will see why the number $e = 2.71828\ldots$ is the natural base to use when discussing exponentials and logarithms. We start with some comments on the significance of these formulas. The formula

$$\frac{d}{dx}(e^x) = e^x$$

shows that the function $y = e^x$ has a truly remarkable property

It is equal to its own derivative.

Thus, it is a solution of the differential equation

$$\frac{dy}{dx} = y(x)$$

and it is essentially the *only* solution—the only other functions with this property being simple constant multiples Ce^x, where C is some constant. Once this is known, it follows immediately that the more general exponential functions

$$y = e^{kx} \quad (k \text{ some constant})$$

have equally interesting properties:

The derivative of e^{kx} is proportional to e^{kx}.

In fact, by the chain rule for derivatives

$$\frac{d}{dx}(e^{kx}) = k \cdot e^{kx}$$

This means that $y = e^{kx}$ is a solution of the differential equation

$$\frac{dy}{dx} = k \cdot y(x)$$

which is of supreme importance in growth and decay problems (see Section 4.4). It is this property that accounts for the appearance of these exponential functions in such applications.

The differentiation formula [35] is so simple because the base of the exponential function is e. Exponential functions $y = a^x$ with different bases, such as 10^x or 2^x, have similar derivatives of the form

$$\frac{d}{dx}(a^x) = (\text{constant}) \cdot a^x \qquad [37]$$

but a messy calculation is needed to evaluate the constant, which turns out to be $\ln(a)$.

> **THE DERIVATIVE OF $y = a^x$** For any base $a > 0$, the exponential function $y = a^x$ has derivative $d/dx\,(a^x) = \ln(a) \cdot a^x$.

For example

$$\frac{d}{dx}(2^x) = (0.6931) \cdot 2^x \qquad \frac{d}{dx}(10^x) = (2.3026) \cdot 10^x$$

for all x. Only when $a = e$ is the constant out front equal to 1, yielding the simplest possible differentiation formula. For this reason the particular exponential function $y = e^x$ is universally employed in calculus.

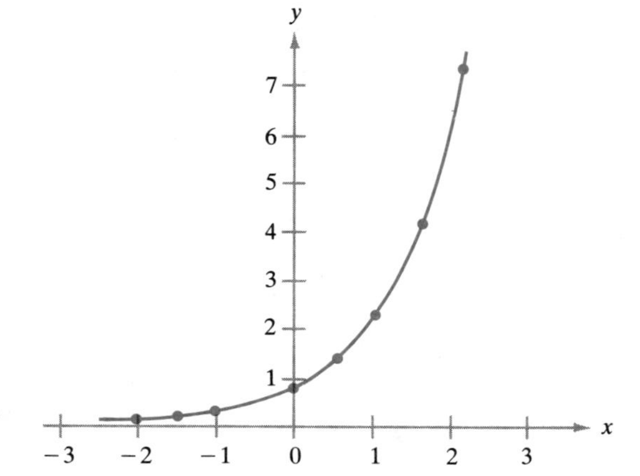

Figure 4.7
Graph of the function $y = \exp(x) = e^x$. Because e^x is always positive, the graph lies above the x axis. It rises very rapidly as x moves to the right, and approaches the x axis as x moves to the left.

Next we use these formulas to discuss the graphs and some optimization problems, deferring proof of the formulas until last. In sketching the graph of $y = e^x$, we know $e^x > 0$ for all x, so the graph stays above the horizontal axis. The derivatives have the properties

$$\frac{d}{dx}(e^x) = e^x \quad \text{is positive for all } x$$

$$\frac{d^2}{dx^2}(e^x) = \frac{d}{dx}(e^x) = e^x \quad \text{is positive for all } x$$

which tells us that the graph rises steadily as x increases and is concave upward, as shown in Figure 4.7. In addition

$$\begin{aligned} &e^x \quad \text{becomes large positive for large positive values of } x \\ &e^x \quad \text{becomes small positive, approaching zero, for large negative values of } x. \end{aligned} \qquad [38]$$

The second statement follows from the first in view of the exponent law

$$e^{-x} = \frac{1}{e^x}$$

which says that the value at $-x$ is the reciprocal of the value at x. The first statement follows from another exponent law:

$$e^{1+x} = e \cdot e^x = (2.71828) \cdot e^x \geq 2 \cdot e^x$$

If we move one unit right from x, the graph is more than twice as high at $1 + x$ as it was at x! Taking this observation together with the fact that the graph rises steadily, we get the first part of [38].

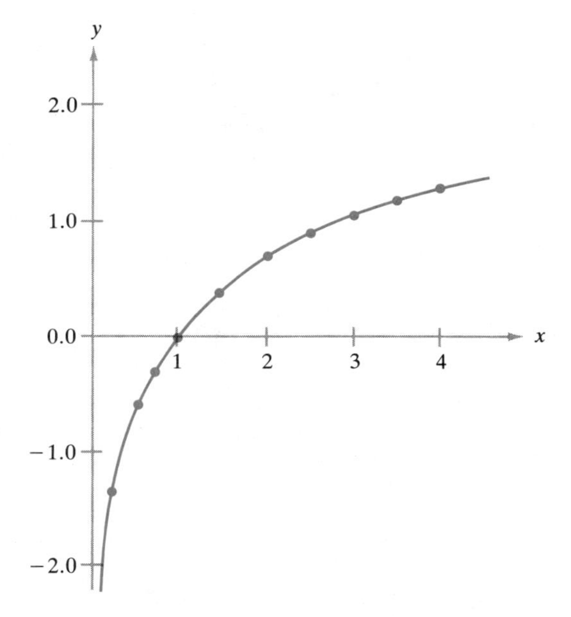

Figure 4.8
Graph of the natural logarithm function $y = \ln(x)$. This function is defined only for $x > 0$. The values of $y = \ln(x)$ increase steadily from large negative values when x is small positive to large positive values when x is large.

The graph of $y = \ln(x)$, shown in Figure 4.8, is defined for $x > 0$. It is rising and concave down because

$$\frac{d}{dx}(\ln(x)) = \frac{1}{x} \quad \text{is positive for all } x > 0$$

$$\frac{d^2}{dx^2}(\ln(x)) = \frac{d}{dx}\left(\frac{1}{x}\right) = -\frac{1}{x^2} \quad \text{is negative for all } x > 0$$

Furthermore

$\ln(x)$ becomes large positive for large positive values of x.
$\ln(x)$ becomes large negative for small positive values of x.

The latter facts could be proved directly, but they also follow from a more interesting fact about graphs of functions, such as $y = \ln(x)$ and $y = e^x$, which are inverses of each other:

The graph of $y = \ln(x)$ is obtained by reflecting the graph of $y = e^x$ across the 45° line $y = x$ in the coordinate plane.

The idea is shown in Figure 4.9: If we were to draw the graph of $y = e^x$ on transparent graph paper and then fold the paper along the 45° line $y = x$, we would obtain the graph of $y = \ln(x)$. In effect, this interchanges the roles of x and y, just as we did in defining $y = \ln(x)$ in terms of $y = e^x$.

Certain optimization problems involve the applications of exponential functions $y = e^{kt}$ discussed in Sections 4.1–4.4. Once we are familiar with these models of exponential growth, optimization can be done just as in Chapter 3.

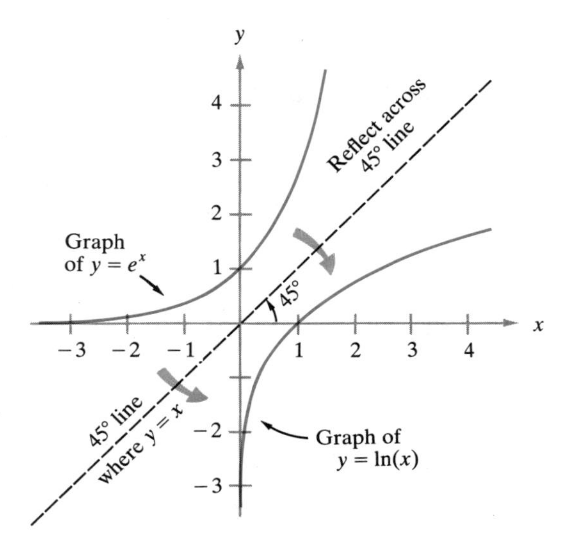

Figure 4.9
The graph of $y = \ln(x)$ is obtained from the graph of $y = e^x$ by reflecting across the (dashed) line $y = x$. Therefore, the shape of the graph of $y = \ln(x)$ can be determined from the graph of $y = e^x$, and vice versa. The same is true of any two functions $y = f(x)$ and $y = g(x)$ that are inverses in the sense that $f(g(x)) = x$ for all x and $g(f(x)) = x$ for all x.

Example 24 Mr. C. Devoon, an art dealer, has just purchased a Modigliani painting for \$35,000. He estimates that the painting will increase in value by \$3000 per year, so that in t years its dollar value will be $35{,}000 + 3000t$. He also estimates that the future rate of inflation will be 6% compounded continuously. Thus if he sold the painting in t years for $35{,}000 + 3000t$ dollars, he would actually receive an amount $(35{,}000 + 3000t) \cdot e^{-0.06t}$ when measured in present-day dollars (or *constant dollars*). When should Devoon sell the painting to maximize his "real return"—his return measured in constant dollars?

Solution Devoon seeks to maximize

$$A(t) = (35{,}000 + 3000t) \cdot e^{-0.06t}$$

The critical points of this function are obtained by setting $dA/dt = 0$; now

$$\begin{aligned}\frac{dA}{dt} &= 3000e^{-0.06t} + (35{,}000 + 3000t)(-0.06)e^{-0.06t} \\ &= (3000 + (35{,}000 + 3000t)(-0.06)) \cdot e^{-0.06t} \\ &= (900 - 180t)e^{-0.06t}\end{aligned}$$

Because $e^{-0.06t}$ is never zero, it follows that $dA/dt = 0$ when $(900 - 180t) = 0$, or $t = 5$ years. This is the only critical point. We can easily check that it is a local maximum if we wish. Thus the painting should be sold after 5 years. ■

This example is part of a larger picture. If an investment has value $A(t)$ after t years, it is useful to know the "present value" $P(t)$ of the investment. According to our discussion in Section 4.3, the present value is obtained by comparing future dollars (say after t years) with the value of a present dollar invested for the same length of time in a continuously compounded savings account. If the annual interest on savings is r (expressed as a decimal), and if this interest rate remains constant, the present value of an investment worth $A(t)$ dollars t years hence is

$$\text{present value } P(t) = e^{-rt}A(t) \qquad [39]$$

With this result in mind, we can analyze various investment schemes.

Example 25 If the interest rate on savings is a constant 8% annually, find the present value of an investment that grows according to the formula

$$A(t) = 1000(1 + t) \quad \text{after } t \text{ years}$$

When is the largest present value achieved? What is the maximum present value?

Solution Here, $r = 0.08$, and by Formula [39] the present value is

$$P(t) = 1000(1 + t)e^{-0.08t} \quad \text{for all } t \geq 0.$$

After 10 years, for example, the present value would be

$$P(10) = 1000(11)e^{-0.08(10)} = \$4942.62$$

and the actual value (in future dollars) is $A(10) = \$11{,}000$. In essence, \$4942.62 in hand right now is as good as having \$11,000 delivered on your doorstep 10 years from now.

The maximum value of $P(t)$ can occur only when $dP/dt = 0$:

$$0 = \frac{dP}{dt} = 1000e^{-0.08t} + 1000(1 + t)(-0.08)e^{-0.08t} \qquad \text{(product formula)}$$

$$= 1000e^{-0.08t}(0.92 - 0.08t)$$

Because $e^{-0.08t} > 0$ for all t, this means

$$0 = 0.92 - 0.08t \quad \text{or } t = 11.5 \text{ years}$$

Then we find the maximum possible present value:

$$P = P(11.5) = \$4981.49$$

Notice that the actual value at this time is $A = A(11.5) = \$12{,}500$. ■

The graphs of $A(t)$ and $P(t)$ in Example 25 are shown in Figure 4.10 so you can see the dramatic difference between the apparent value of this investment $A(t)$ and its present value. After a long time, you would be better off having your money in a savings account than in this investment scheme!

We conclude with a proof of the differentiation formulas for $\ln x$ and e^x.

SKETCH PROOF OF THE DIFFERENTIATION FORMULAS It is easiest to start with the logarithm function. By definition of the derivative,

$$\frac{d}{dx}\ln x = \lim_{\Delta x \to 0}\left(\frac{\ln(x + \Delta x) - \ln(x)}{\Delta x}\right)$$

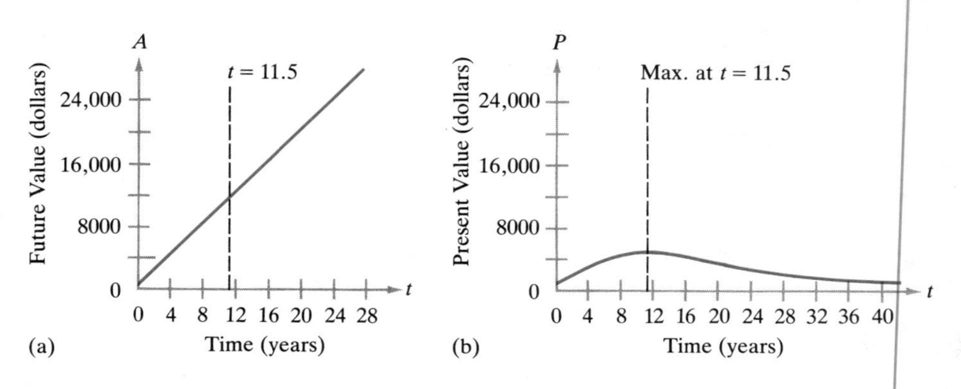

Figure 4.10
Graph (a) shows the actual value (future dollars) of the investment in Example 25. Graph (b) shows the present value, which is largest when $t = 11.5$ years. As t becomes very large, the present value declines toward zero!

We use the algebraic properties of $\ln x$ to simplify this expression

$$\begin{aligned}\frac{\ln(x + \Delta x) - \ln(x)}{\Delta x} &= \frac{1}{\Delta x}\Big(\ln(x + \Delta x) - \ln(x)\Big) \\ &= \frac{1}{\Delta x}\ln\left(\frac{x + \Delta x}{x}\right) \quad \text{by [13ii]} \\ &= \frac{1}{x}\cdot\frac{x}{\Delta x}\cdot\ln\left(1 + \frac{\Delta x}{x}\right) \\ &= \frac{1}{x}\cdot\ln\left[\left(1 + \frac{\Delta x}{x}\right)^{x/\Delta x}\right] \quad \text{by [13iii]}\end{aligned}$$

Now write N for the quantity $x/\Delta x$. As Δx gets small, N takes on larger and larger positive values. When we combine this observation with the formula that defines the number e,

$$e = \lim_{N\to\infty}\left(1 + \frac{1}{N}\right)^N = \text{limiting value of } \left(1 + \frac{1}{N}\right)^N \text{ as } N \to \infty$$

we conclude that

$$\left(1 + \frac{\Delta x}{x}\right)^{x/\Delta x} \quad \text{approaches} \quad e \quad \text{as } \Delta x \to 0$$

and

$$\ln\left[\left(1 + \frac{\Delta x}{x}\right)^{x/\Delta x}\right] \quad \text{approaches} \quad \ln(e) = 1 \quad \text{as } \Delta x \to 0$$

Consequently

$$\begin{aligned}\frac{dy}{dx} &= \frac{1}{x}\cdot\lim_{\Delta x\to 0}\ln\left[\left(1 + \frac{\Delta x}{x}\right)^{x/\Delta x}\right] \\ &= \frac{1}{x}\cdot\ln(e) = \frac{1}{x}\end{aligned}$$

To handle the exponential function, we first recall that the functions ln and exp are inverses of one another, so that

$$\ln(e^x) = x \quad \text{for all } x$$

Now differentiate the functions on each side of this identity, using the chain rule to differentiate the composite function $y = \ln(u)$, $u = e^x$ on the left:

$$1 = \frac{d}{dx}(x) = \frac{d}{du}(\ln u) \cdot \frac{du}{dx} = \frac{1}{u}\frac{du}{dx} = \frac{1}{e^x}\frac{d}{dx}(e^x)$$

Multiplying both sides by e^x, we obtain the differentiation formula for e^x:

$$\frac{d}{dx}(e^x) = e^x$$

Exercises 4.5

Find dy/dx for the following functions.

1. $1 - e^{-x}$ **2.** e^{3x} **3.** $e^{x^2/2}$

4. e^{2x+1} **5.** $\ln(2x)$ **6.** $\ln(1/x)$

7. $y = e^{-x}$ **8.** $y = e^{2x}$ **9.** $y = 4e^{-x^2}$

10. $y = \ln(4x)$ **11.** $y = \ln\sqrt{x}$ **12.** $y = 1 - e^{-2x}$

Find the derivatives of each function. What differential equation does y satisfy?

13. $y = 2^x$ **14.** $y = 10^x$ **15.** $y = \left(\frac{1}{2}\right)^x$

16. $y = 3^{-x}$ **17.** $y = e^{3.27x}$ **18.** $y = \ln x$

19. In Example 24, what should Mr. Devoon do if the annual inflation rate is

(i) 5%, (ii) 8%, (iii) 15%?

20. If we put $1000 in the bank, continuously compounded at an annual rate of 7%, the amount in our account after t years would be $A(t) = 1000e^{0.07t}$ dollars. Suppose the inflation rate is running at 10% annually. Then, just as in Example 25, the "constant dollar" value of our money in the bank would be

$$e^{-0.10t}A(t)$$

Give a formula for the net constant dollar value of our account at time t. What happens as t increases?

21. If the value of an investment is $A(t) = 5000(1 + 2t)$ dollars after t years and if we use $r = 0.08$ in computing its present value, what is the present value after

(i) 5 years, (ii) 10 years, (iii) 15 years?

22. Suppose an investment grows to $A(t) = 5000(1 + 2t)$ dollars after t years. Find the present value after 15 years if future and present funds are compared using an interest rate r of

(i) 5%, (ii) 8%, (iii) 12%

In each of the following investment schemes, find the present value after t years, assuming that $r = 0.07$ in Formula [39]. Then find the present value initially (when $t = 0$) and after 10 years.

23. A savings account that grows to $A(t) = 5000e^{0.075t}$ dollars after t years.

24. An investment worth $A(t) = 5000(1 + \frac{1}{20}t^2)$ dollars after t years.

25. You hide $5000 under your mattress for t years: $A(t) = 5000$ for all t.

26. If $r = 0.08$ and $A(t) = 5000(1 + 2t)$ dollars, when is the largest present value achieved? What is the maximum present value?

27. Repeat Exercise 26 using $r = 0.11$.

28. If $r = 0.08$ and $A(t) = 4000(1 + t + t^2)$, find the largest present value.

Checklist of Key Topics

Interest
Compound interest formula
Effective interest rate
Continuous compounding
Exponential function $e^x = \exp(x)$
Natural logarithm function $\ln(x)$
Algebraic properties of e^x and $\ln(x)$
Present value
Geometric progressions
Geometric sum formula
Annuity problems
Cost and benefit analysis
Growth and decay models
Half-life
Radioactive decay and carbon dating
Population growth models
Graphs of e^x and $\ln(x)$
Constant dollars

Chapter 4 Review Exercises

1. Find the annual interest rate at which $1000 grows to $1400 in 5 years under the following interest schemes:
(i) simple (uncompounded) interest
(ii) quarterly compounding
(iii) continuous compounding

2. If $7000 is invested for four years at simple interest of 8% annually, and then the total amount is reinvested for five years at 9% interest compounded semiannually, find the amount accumulated after the nine years.

3. In Exercise 2, suppose that the $7000 is invested for the first five years at 9% compounded semiannually and then reinvested for the next four years at 8% simple interest. Find the accumulated amount.

4. What is the effective annual interest rate if money is compounded continuously at the annual rate of 12%?

5. How much should we invest now in an account paying 10% compounded quarterly if we wish to have $1200 seven years from now?

6. How long will it take \$3000 to grow to \$10,800 if it is invested at 6% compounded continuously?

7. Express $\ln(x^2y^{1/3})$ in terms of $\ln x$ and $\ln y$. Do the same for $\ln(7y/\sqrt{x})$.

8. A study of recoverable oil reserves in a Texas county predicts a decline according to the formula

$$R = 10^7 \times 2^{-t/8} \quad (t \text{ in years, } R \text{ in barrels})$$

Write this formula in the form $R = 10^7 \times e^{kt}$ for a suitable value of the constant k. What is the appropriate value of k?

9. The population of a country is modeled by the formula

$$P(t) = (3 \times 10^6)e^{kt} \quad (t \text{ in years})$$

If the population will double in 30 years, find the value of the constant k. Then find the population in 10 years and in 40 years.

10. A state lottery advertises a first prize of \$5,000,000. Actually, it pays the winner \$250,000 in 20 yearly installments, beginning immediately. If money is worth 7% compounded continuously, find the present value of this "\$5,000,000" first prize.

11. The borough of Titheford pays the local earl £100 every January 1, to eternity. Assume a future rate of inflation of 5% per annum, compounded continuously. As of January 1, find the present "real value" of the sum of the payment made that day plus all future payments.

12. An entrepreneur offers us an annual interest rate of 10% for the next three years and 12% for the following four years, both compounded continuously. How much should we be willing to lend him now if we want to accumulate \$15,000 in seven years?

13. A \$5000 bond will make 16 semiannual payments of \$100 each, beginning 6 months from now. In eight years the principal of \$5000 will be due, in addition to the last \$100 payment. What should you pay for that bond if you want your money to earn 9% annual interest, compounded continuously?

14. The half-life of C^{14} is 5720 years. Logs found at an Indian site contain 82% of their original C^{14} content. What is the age of the site?

15. It is estimated that the population of Kenya will double in 21 years. How long will it take for the population to triple?

In Exercises 16–19 find the derivative $f'(x)$.

16. $f(x) = e^{-\frac{(x-3)^2}{20}}$

17. $f(x) = \ln(1 + x^4)$

18. $f(x) = e^{3\ln x}$

19. $f(x) = e^{-x} + 3x\ln x$

In Exercises 20–21, sketch the graph of the function $y = f(x)$.

20. $f(x) = \dfrac{e^x + e^{-x}}{2}$

21. $f(x) = \dfrac{e^x - e^{-x}}{2}$

CHAPTER 5

Integration

5.1 Indefinite Integrals

In this chapter we shall examine an important new operation: *integration.* In a sense, integration is the inverse of differentiation, so we start by saying a few words about reversing the process of differentiation.

An **indefinite integral** or **antiderivative** of a function $f(x)$ is any function $F(x)$ whose derivative is the original function

$$\frac{dF}{dx} = f(x)$$

For example, what is an indefinite integral for $f(x) = x^2$? We want to find a function $F(x)$ whose derivative is $f(x) = x^2$. This is a new and unfamiliar problem; but we do know a lot about derivatives, so it is not hard to make an informed guess. Recall that the derivative of any power x^r is a constant times a power of one lower degree, namely rx^{r-1}. If we want a function $F(x)$ such that $dF/dx = x^2$, this suggests that we consider something like x^3. Now

$$\frac{d}{dx}(x^3) = 3x^2$$

so $F(x) = x^3$ is not quite the correct choice. But if we choose $F(x) = \frac{1}{3}x^3$ then

$$\frac{dF}{dx} = \frac{d}{dx}\left(\frac{1}{3}x^3\right) = \frac{1}{3} \cdot \frac{d}{dx}(x^3) = \frac{1}{3}(3x^2) = x^2$$

Thus $F(x) = \frac{1}{3}x^3$ is an indefinite integral of $f(x) = x^2$. Similarly, an indefinite integral of $f(x) = x^7$ is given by $F(x) = \frac{1}{8}x^8$ because

$$\frac{dF}{dx} = \frac{d}{dx}\left(\frac{1}{8}x^8\right) = x^7$$

More generally, by reversing the differentiation formula

$$\frac{d}{dx}\left(\frac{1}{r+1}x^{r+1}\right) = x^r$$

we see that the function $f(x) = x^r$ has an indefinite integral of the form

$$F(x) = \frac{x^{r+1}}{r+1} \qquad [1]$$

except when $r = -1$. (Then the denominator $r + 1$ is zero, and the formula for $F(x)$ makes no sense.)

Example 1 Find an indefinite integral for each function:

$$\text{(i)}\quad f(x) = x^6 \qquad \text{(ii)}\quad g(x) = \frac{1}{x^3} \qquad \text{(iii)}\quad h(x) = \sqrt{x}$$

Solution We first express each function as a power of x,

$$f(x) = x^6 \qquad g(x) = x^{-3} \qquad h(x) = x^{1/2}$$

to apply Formula [1]. Taking $r = 6, -3, \frac{1}{2}$, respectively, we find that

(i) $f(x)$ has indefinite integral $F(x) = \dfrac{x^{6+1}}{6+1} = \dfrac{1}{7}x^7$

(ii) $g(x)$ has indefinite integral $G(x) = \dfrac{x^{-3+1}}{-3+1} = \dfrac{x^{-2}}{-2} = \dfrac{-1}{2x^2}$

(iii) $h(x)$ has indefinite integral $H(x) = \dfrac{x^{1/2+1}}{\frac{1}{2}+1} = \dfrac{x^{3/2}}{\frac{3}{2}} = \dfrac{2}{3}x^{3/2}$ ■

Because $g(x)$ and $h(x)$ have restricted domains of definition, we restrict $G(x)$ and $H(x)$ accordingly. In the rest of this chapter we will not concern ourselves with domains of definition.

A function $f(x)$ may have more than one indefinite integral. Indeed, if $F(x)$ is one indefinite integral and if c is a constant, we get another indefinite integral $G(x) = F(x) + c$ by adding a constant to $F(x)$, because

$$\frac{dG}{dx} = \frac{d}{dx}(F(x) + c) = \frac{dF}{dx} + \frac{d}{dx}(c) = \frac{dF}{dx} + 0 = f(x)$$

This leads us to ask: How can we find *all* indefinite integrals of a function $f(x)$?

Given one indefinite integral $F(x)$, there are others of the form

$$F(x) + \text{constant} \qquad [2]$$

Actually, these are all there are. In other words, if $f(x)$ has an indefinite integral at all this indefinite integral is "unique up to an added constant." For example, $f(x) = x^2$ has $F(x) = \frac{1}{3}x^3$ as an indefinite integral; all possible indefinite integrals are of the form

$$\frac{1}{3}x^3 + c \quad (c \text{ any constant})$$

This can be summed up as follows:

Uniqueness of the Indefinite Integral Suppose that $f(x)$ has an indefinite integral $F(x)$ on an interval. Then all other indefinite integrals of $f(x)$ have the form $F(x) + c$, where c is an arbitrary constant.

Sketch Proof Suppose $G(x)$ is any other indefinite integral for $f(x)$. Then

$$\frac{dF}{dx} = f(x) \quad \text{and} \quad \frac{dG}{dx} = f(x)$$

Their difference $H(x) = G(x) - F(x)$ has derivative zero, because

$$\frac{dH}{dx} = \frac{d}{dx}(G(x) - F(x)) = \frac{dG}{dx} - \frac{dF}{dx} = f(x) - f(x) = 0$$

That is, the rate of change of $H(x)$ is identically zero for all x. Intuitively, we can see that this forces $H(x)$ to be a constant function, $H(x) = c$ for all x (c some constant).† Because $G(x) - F(x) = c$, we obtain $G(x) = F(x) + c$. ■

Indefinite integrals of $f(x)$ are indicated by the special symbol

$$\int f(x)\,dx \qquad [3]$$

Thus

$$\int x^2\,dx$$

stands for any indefinite integral of $f(x) = x^2$. By [1] we see that

$$\int x^2\,dx = \frac{1}{3}x^3 + \text{const}$$

where "const" stands for the arbitrary added constant associated with an indefinite integral. Similarly, we may rewrite Formula [1] using this notation: If

† This fact is proved in advanced courses.

$r \neq -1$, the indefinite integral of x^r is

$$\int x^r \, dx = \frac{x^{r+1}}{r+1} + \text{const} \quad (r \neq -1) \qquad [4]$$

This is the first of several "integration formulas" we shall derive. To avoid errors, some special indefinite integrals are worth noting. The constant function $f(x) = x^0 = 1$ has indefinite integral

$$F(x) = \frac{x^1}{1} = x$$

Similarly, the constant function whose value is k, $f(x) = k$, has $F(x) = kx$ for an indefinite integral. This gives the integration formula

$$\int k \, dx = kx + \text{const} \quad (k \text{ any constant}) \qquad [5]$$

Taking $k = 1$ or $k = 0$ we get two commonly encountered special cases:

$$\int 1 \, dx = x + \text{const}$$
$$\int 0 \, dx = \text{const} \qquad [6]$$

Several differentiation rules can be reinterpreted as statements about indefinite integrals. The rule for differentiating a linear combination leads directly to the following rule for indefinite integrals.

Indefinite Integral of a Linear Combination If k_1 and k_2 are constants, then

$$\int (k_1 f(x) + k_2 g(x)) \, dx = k_1 \int f(x) \, dx + k_2 \int g(x) \, dx \qquad [7]$$

Example 2 Calculate the indefinite integral of $f(x) = x^3 + 6000x - \sqrt{2}$.

Solution By the formula for integrating powers [4], we obtain

$$\int x^3 \, dx = \frac{1}{4}x^4 + \text{const}$$

$$\int x \, dx = \frac{1}{2}x^2 + \text{const}$$

$$\int 1 \, dx = x + \text{const}$$

Now apply [7] and combine the undetermined constants.

$$\begin{aligned}\int (x^3 + 6000x - \sqrt{2})\, dx &= \int x^3\, dx + 6000 \int x\, dx - \sqrt{2} \int 1\, dx + \text{const} \\ &= \frac{1}{4}x^4 + 3000x^2 - \sqrt{2}x + \text{const}\end{aligned}$$

Do not be confused by the constants 6000 and $-\sqrt{2}$. They are brought outside of the integrals where they can be handled without difficulty. ■

In this and all other indefinite integration problems we can check our work by differentiating the result; we should get back the original function.

Example 3 Calculate the indefinite integral $\int f(t)\, dt$ if $f(t) = (3t^4 - 4)/t^2$.

Solution In this example the independent variable is labeled t instead of x; all of our integration formulas apply if we replace x by t throughout. So first we divide by t^2, writing $f(t)$ as a sum of powers of the variable t.

$$f(t) = 3t^2 - \frac{4}{t^2} = 3t^2 - 4t^{-2}$$

Taking $r = 2$ and $r = -2$ in Formula [4], we obtain

$$\int t^2\, dt = \frac{1}{3}t^3 + \text{const}$$

$$\int t^{-2}\, dt = \frac{t^{-2+1}}{-2+1} + \text{const} = \frac{t^{-1}}{-1} + \text{const} = -\frac{1}{t} + \text{const}$$

Therefore, by our rule for integrating linear combinations

$$\begin{aligned}\int \frac{3t^4 - 4}{t^2}\, dt &= \int (3t^2 - 4t^{-2})\, dt \\ &= 3\int t^2\, dt - 4\int t^{-2}\, dt \\ &= 3\left(\frac{1}{3}t^3\right) - 4\left(-\frac{1}{t}\right) + \text{const} \\ &= t^3 + \frac{4}{t} + \text{const}\end{aligned}$$ ■

We may calculate the indefinite integral of any polynomial or sum of powers x^r ($r \neq -1$) by these methods. The indefinite integral of the troublesome power $x^{-1} = 1/x$ can be found by examining the differentiation formula for the natural logarithm function $\ln(x)$,

$$\frac{d}{dx}(\ln(x)) = \frac{1}{x}$$

discussed in Sections 2.3 and 4.5. From this we see that the indefinite integral of $f(x) = 1/x = x^{-1}$ is *not* some power of x, but rather

$$\int \frac{1}{x}\, dx = \ln(x) + \text{const} \quad \text{for } x > 0 \tag{8}$$

Example 4 Calculate the indefinite integral $\int 1 - \frac{1}{x}\, dx$.

Solution The appropriate indefinite integrals are

$$\int 1\, dx = x + \text{const} \quad \text{and} \quad \int \frac{1}{x}\, dx = \ln(x) + \text{const}$$

Thus

$$\int 1 - \frac{1}{x}\, dx = \int 1\, dx - \int \frac{1}{x}\, dx = x - \ln(x) + \text{const}$$

By examining other differentiation formulas we can obtain several more integration formulas. For future reference, we compile a list (Table 5.1) of all the formulas that can be obtained this way. For example, from the differentiation rule

$$\frac{d}{dx}(e^x) = e^x$$

we obtain

$$\int e^x\, dx = e^x + \text{const} \tag{9}$$

Function	Indefinite Integral
x^r	$\int x^r\, dx = \frac{x^{r+1}}{r+1} + \text{const} \quad (r \neq -1)$
$\frac{1}{x}$	$\int \frac{1}{x}\, dx = \ln(x) + \text{const}$
e^{kx}	$\int e^{kx}\, dx = \frac{1}{k} e^{kx} + \text{const} \quad (k \neq 0)$

Table 5.1 A basic list of indefinite integrals.

and from the more general rule

$$\frac{d}{dx}\left(\frac{1}{k}e^{kx}\right) = \frac{1}{k}\frac{d}{dx}(e^{kx}) = \frac{1}{k}(ke^{kx}) = e^{kx}$$

worked out in Section 3.2, we get

$$\int e^{kx}\,dx = \frac{1}{k}e^{kx} + \text{const} \quad (k \text{ any constant} \neq 0). \qquad [10]$$

This accounts for the third entry in our table. When $k = 1$ we of course get the special case [9], which is worth remembering by itself.

Example 5 Calculate the indefinite integral of $f(s) = \sqrt{s} + \dfrac{4}{\sqrt{s}} + 3e^s$.

Solution Formula [4] is easier to apply if we write the first two terms in $f(s)$ as fractional powers of s,

$$f(s) = s^{1/2} + 4s^{-1/2} + 3e^s$$

Next, apply [4] to the first two terms and [9] to the last to get

$$\int s^{1/2}\,ds = \frac{s^{(1/2)+1}}{(\frac{1}{2}+1)} + \text{const} = \frac{s^{3/2}}{\frac{3}{2}} + \text{const} = \frac{2}{3}s^{3/2} + \text{const}$$

$$\int s^{-1/2}\,ds = \frac{s^{(-1/2)+1}}{(-\frac{1}{2}+1)} + \text{const} = \frac{s^{1/2}}{\frac{1}{2}} + \text{const} = 2s^{1/2} + \text{const}$$

$$\int e^s\,ds = e^s + \text{const}$$

Now put these together using the formula for integrating linear combinations,

$$\int\left(\sqrt{s} + \frac{4}{\sqrt{s}} + 3e^s\right)ds = \int s^{1/2}\,ds + 4\int s^{-1/2}\,ds + 3\int e^s\,ds$$

$$= \frac{2}{3}s^{3/2} + 8s^{1/2} + 3e^s + \text{const}$$

■

Example 6 Integrate $\int (e^{4x} - (x+1)^2)\,dx$.

Solution First, expand $(x+1)^2$ to obtain

$$(x+1)^2 = (x+1)(x+1) = x^2 + 2x + 1$$

so that

$$\int (e^{4x} - (x+1)^2)\,dx = \int e^{4x}\,dx - \int x^2\,dx - 2\int x\,dx - \int 1\,dx$$

Then use Formula [10] with $k = 4$ to obtain

$$\int e^{4x}\,dx = \frac{1}{4}e^{4x} + \text{const}$$

The final result is

$$\int (e^{4x} - (x+1)^2)\,dx = \frac{1}{4}e^{4x} - \frac{1}{3}x^3 - x^2 - x + \text{const}$$

Exercises 5.1

In Exercises 1–16, use Table 5.1 to find all indefinite integrals for each function. Check your answers by differentiation.

1. $f(x) = 1$ (constant function)

2. $f(x) = x^4$

3. $f(x) = x^{2.3}$

4. $f(x) = x^{-2}$

5. $f(x) = x^{1/3}$

6. $f(x) = x^{-4/5}$

7. $f(x) = x^{-1}$

8. $f(x) = \dfrac{1}{x^5}$

9. $f(x) = e^x$

10. $f(x) = e^{-x}$

11. $f(x) = e^{2x}$

12. $f(x) = (e^x)^3$

13. $f(s) = \dfrac{1}{s^3}$

14. $f(t) = e^{-t/2}$

15. $f(q) = \dfrac{1}{q}$

16. $f(x) = \ln(e^{x^2})$

In Exercises 17–28, calculate the indefinite integral $\int f(x)\,dx$ for each function.

17. $f(x) = 2$ (constant function)

18. $f(x) = 0$ (constant function)

19. $f(x) = -\dfrac{x^3}{4}$

20. $f(x) = \dfrac{1}{2x}$

21. $f(x) = \dfrac{2}{3}x + 6$

22. $f(x) = 3 - \dfrac{2}{\sqrt{3x}}$

23. $f(x) = 1 - 3x$

24. $f(x) = 1 + \dfrac{1}{x^2}$

25. $f(x) = x^7 - 4x^5 + \dfrac{1}{2}x^2 - 8$

26. $f(x) = -0.03x^2 - 10x + 4500$

27. $f(x) = 5\sqrt{x} + \dfrac{1}{\sqrt{x}} + 10$

28. $f(x) = \dfrac{4}{x} + 2e^{x/2}$

In Exercises 29–38, find the indefinite integral.

29. $\int 3\,dx$

30. $\int (5 - 2x)\,dx$

31. $\int (4x^3 + x^2 + 2x)\,dx$

32. $\int \frac{x^3 - 1}{x^2}\,dx$

33. $\int (1 - x + x^5)\,dx$

34. $\int (\sqrt{x} - 5\sqrt{3})\,dx$

35. $\int (0.9x^2 - 120x + 1600)\,dx$

36. $\int (x + e^x)\,dx$

37. $\int (e^3 - e^{3t})\,dt$

38. $\int x(3x^7 - 7x^2 - \frac{1}{2})\,dx$

39. Find all indefinite integrals of $f(x) = x$. Choose any three indefinite integrals of f and sketch their graphs on the same piece of graph paper. Geometrically, how do these graphs relate to one another? Find the particular indefinite integral F whose graph passes through the point (2, 0). How many indefinite integrals of f satisfy the condition $F(0) = 2$?

In Exercises 40–43, find all functions $f(x)$ with the given property.

40. $f'(x) = x^2 - 2$

41. $f'(x) = \frac{3}{\sqrt{x}}$

42. $f'(x) = x^3 - x^{3/2}$

43. $f'(x) = 3x^4 + 4x^3 - \sqrt{5x}$

44. Integration formulas more complicated than those listed in Table 5.1 are found in more advanced texts. Some were discovered by guesswork, others by systematic advanced methods. Verify the following integration formulas by differentiation.

(i) $$\int \frac{1}{x^2 - 1}\,dx = \frac{1}{2}\ln\left(\frac{x - 1}{x + 1}\right) + \text{const}$$

(ii) $$\int \frac{1}{a^2 - x^2}\,dx = \frac{1}{2a}\ln\left(\frac{a + x}{a - x}\right) + \text{const} \quad (a \neq 0, \text{constant})$$

(iii) $$\int xe^{-x}\,dx = -xe^{-x} - e^{-x} + \text{const}$$

45. Find the value of the constant k if

$$\int (3x - 2)^{10}\,dx = k \cdot (3x - 2)^{11} + \text{const}$$

46. Find $F'(x)$ if

(i) $F(x) = \int x^4\,dx$ (ii) $F(x) = \int \frac{e^x}{x}\,dx$

5.2 Applications of Indefinite Integrals

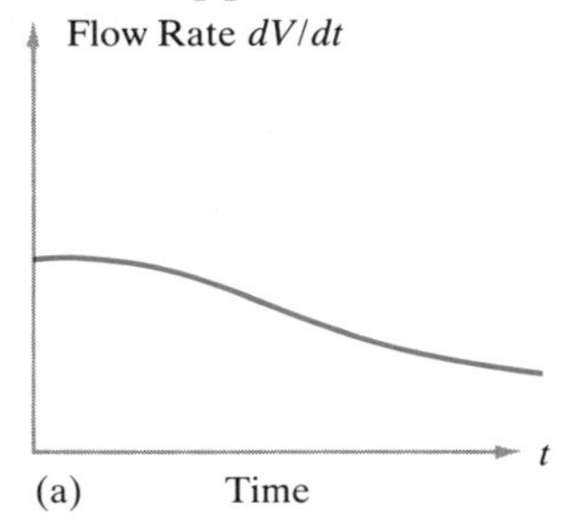

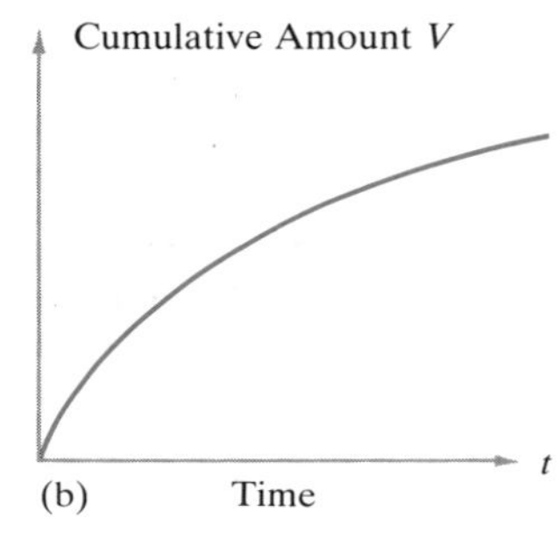

Figure 5.1
The graph (a) shows the flow rate of water through a dam, in gallons per minute. The graph (b) shows the total amount $V(t)$ of water released since the gates were opened at time $t = 0$; clearly, $V(0) = 0$. The rate the total amount increases, dV/dt, must equal the rate of water being released from the dam. The flow rate dV/dt is easily measured using a flow meter. The problem is to compute $V(t)$ once we know the flow rate data presented in (a).

Suppose we are interested in a function $f(x)$ but can only obtain information about its rate of change, the derivative df/dx. How can we reconstruct the original function $f(x)$? In this section we will see that indefinite integrals are the perfect tool for the job.

To illustrate the kind of problem we have in mind, consider a dam that starts releasing water into a stream at time $t = 0$. We would like to know the total volume $V(t)$ of water released up to time t. But this is not easily measured—after the water is released we cannot weigh it. Measuring the flow rate, the amount released per unit time, is quite easy; a simple flow meter will do the job. But the flow rate is just the derivative dV/dt:

$$\begin{aligned}(\text{flow rate}) &= (\text{volume released per minute})\\ &= (\text{rate of change of total volume released})\\ &= \frac{dV}{dt}.\end{aligned}$$

In this situation the problem is to reconstruct $V(t)$ from the meter readings, which give dV/dt. Figure 5.1 shows typical graphs for dV/dt and $V(t)$.

In general, suppose we are given the derivative $f'(x) = df/dx$ and want to find $f(x)$. The crucial observation is now almost obvious:

The desired function $f(x)$ is an indefinite integral of the given function df/dx.

This is just the definition of an indefinite integral. In our notation for indefinite integrals, this fact is expressed by the formula:

$$f(x) = \int \frac{df}{dx}\,dx + \text{const} \tag{11}$$

for a suitably chosen value of the constant. This observation is the basis of many applications. Thus if we are adept at finding indefinite integrals we can determine $f(x)$ up to an added constant by finding an indefinite integral of the given function df/dx. Usually, we can find the value of the added constant from supplementary information, such as the value of $f(x)$ when $x = 0$. We illustrate the idea in the following examples.

ILLUSTRATION To determine the heating requirements of a house in the design stage, a heating engineer analyzes the rate of heat loss over the heating season. She does this by taking into account such factors as the number of windows, size of the roof, and so on; combining the rates of heat loss of these various components, she arrives at an overall rate for the entire building. Once the rate of heat loss is known, she calculates the total heat loss by integration.

Suppose the heating season lasts for 180 days, and let x be the number of days since the start of the season. Suppose the rate of heat loss on day x is[†]

$$r(x) = 0.18x - 0.001x^2 \quad \text{therms per day } (0 \le x \le 180)$$

Find the cumulative heat loss (in therms) after x days. How much heat is lost over the entire season?

Discussion Let $h(x)$ be the cumulative amount of heat lost up to day x. Clearly the derivative dh/dx is the rate at which heat is being lost:

$$\frac{dh}{dx} = r(x)$$

Substituting our formula for $r(x)$ we obtain

$$\frac{dh}{dx} = 0.18x - 0.001x^2$$

Applying the basic principle [11], we see that

$$\begin{aligned} h(x) &= \int \frac{dh}{dx}\,dx \\ &= \int (0.18x - 0.001x^2)\,dx \\ &= 0.18 \int x\,dx - 0.001 \int x^2\,dx \\ &= 0.09x^2 - \frac{0.001}{3}x^3 + c \end{aligned} \tag{12}$$

where c is some constant. To evaluate c, we observe that the cumulative heat loss is zero ($h = 0$) at the start of the season when $x = 0$; in short, $h(0) = 0$. Using this fact in [12] we can determine c:

$$0 = h(0) = 0.09(0)^2 - \frac{0.001}{3}(0)^3 + c = c$$

$$c = 0$$

Now the heat loss function is completely determined,

$$h(x) = 0.09x^2 - \frac{0.001}{3}x^3 \quad \text{for } 0 \le x \le 180.$$

The loss over the entire season is the value of $h(x)$ on the last day, when $x = 180$.

[†] The rate of heat loss per day depends on x. In the middle of the heating season, when $x = 90$, the outdoor temperature is much lower than at the beginning or end, and the rate of heat loss per day is higher.

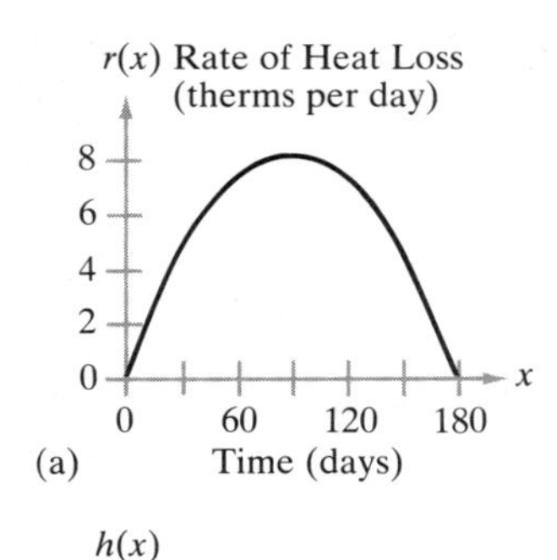

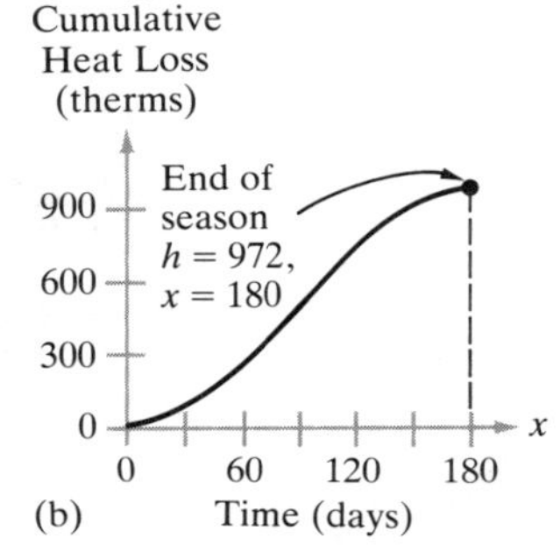

Figure 5.2
Graphs of the rate of heat loss $r(x) = 0.18x - .001x^2$ (see (a)) and the cumulative heat loss $h(x)$ after x days (see (b)) in Example 6. Because $r(x)$ is always positive, and $dh/dx = r(x)$, the graph of $h(x)$ rises steadily from zero when $x = 0$ to $h = 972$ when $x = 180$ (end of heating season). In this situation, we start with information about the rate of heat loss $r(x)$ and must calculate the cumulative heat loss $h(x)$.

Thus,

$$\begin{aligned}\text{(heat loss over entire season)} &= h(180)\\ &= 0.09(180)^2 - \frac{0.001}{3}(180)^3\\ &= 972 \quad \text{therms}\end{aligned}$$

In other words, the house will require 972 therms of energy each year for heating purposes. ■

To show what is going on in the last example, the graphs of the given function $r(x) = dh/dx$ and the desired cumulative heat loss $h(x)$ are sketched in Figure 5.2.

The next two examples involve economic applications.

Example 7 Given the following information, find the weekly profit function $P(x)$ of a small manufacturer.

(i) The marginal profit is $dP/dx = 70 - 0.6x$
(ii) $P(0) = -2200$ (that is, the fixed costs are \$2200 per week).

Solution Remember that the "marginal profit" is just another name for the derivative dP/dx. From it we reconstruct $P(x)$ up to an added constant c using the basic principle [11]:

$$P(x) = \int \frac{dP}{dx}\,dx = \int (70 - 0.6x)\,dx = 70x - 0.3x^2 + c$$

Next we use the fact that $P(0) = -2200$ to determine c:

$$-2200 = P(0) = 70(0) - 0.3(0)^2 + c = c$$
$$c = -2200$$

Now we have $P(x)$,

$$P(x) = 70x - 0.3x^2 - 2200$$ ■

Example 8 Between 1968 and 1973, world oil consumption grew exponentially. If t is the number of years since the start of 1968, the rate of oil consumption at time t was

$$r(t) = 14e^{0.08t} \quad \text{billion barrels per year } (0 \le t \le 5)$$

Find the total amount of oil used during the 5 years 1968 through 1973.

Solution Let $Q(t)$ be the cumulative amount of oil consumed (in billions of barrels) from the beginning of 1968, when $t = 0$, until time t. The rate of consumption $r(t)$ is just the derivative of $Q(t)$:

$$\frac{dQ}{dt} = r(t) = 14e^{0.08t}$$

Using the integration Formula [10] with $k = 0.08$, we determine $Q(t)$ up to an added constant c:

$$Q(t) = \int 14e^{0.08t}\,dt = \frac{14}{0.08}e^{0.08t} + c = 175e^{0.08t} + c$$

But $Q(0) = 0$, so we can determine the constant c[†]:

$$0 = Q(0) = 175e^{0.08(0)} + c = 175 + c$$
$$c = -175.$$

Thus we have found $Q(t)$,

$$Q(t) = 175e^{0.08t} - 175 = 175(e^{0.08t} - 1)$$

The total amount of oil used during the five years is $Q(5)$:

$$Q(5) = 175(e^{0.08(5)} - 1) = 86 \quad \text{billion barrels}$$

■

Next we give a motion problem that depends on the fact that velocity $v(t)$ of a moving object is just the derivative of the function $s(t)$ that gives the distance covered (recall Section 2.5).

Example 9 A recording speedometer in a race car indicates that the velocity $v(t)$ during an acceleration test on a straight track was given by

$$v(t) = 17.6(t - 0.003t^2) \quad \text{feet per second}$$

after t seconds. Calculate the length of track covered during the time intervals $0 \le t \le 5$ and $5 \le t \le 10$ seconds.

Solution Let $s(t)$ be the distance covered in t seconds, the distance between the starting point and the car at time t. Notice that $v(t) = ds/dt$; therefore, we may reconstruct $s(t)$ up to an added constant using indefinite integrals,

$$\begin{aligned} s(t) &= \int \frac{ds}{dt}\,dt = \int v(t)\,dt \\ &= \int 17.6(t - 0.003t^2)\,dt \\ &= 17.6\left(\frac{1}{2}t^2 - 0.001t^3\right) + c \end{aligned}$$

Because $s(0) = 0$ (at the starting time, no distance has been covered), $c = 0$, so the correct formula for $s(t)$ is

$$s(t) = 17.6\left(\frac{1}{2}t^2 - 0.001t^3\right) \quad \text{feet}$$

[†] Note that $e^{0.08(0)} = e^0 = 1$.

In particular, when $t = 5$ or $t = 10$, we have

$$s(5) = 17.6\left(\frac{1}{2}(5)^2 - 0.001(5)^3\right) = 217.8 \quad \text{feet}$$

$$s(10) = 17.6\left(\frac{1}{2}(10)^2 - 0.001(10)^3\right) = 862.4 \quad \text{feet}$$

and so the length of track covered between $t = 0$ and $t = 5$ is 217.8 feet; the length of track covered between $t = 5$ and $t = 10$ is

$$s(10) - s(5) = 862.4 - 217.8 = 644.6 \quad \text{feet}$$

The next example is a little more sophisticated: Sometimes more than one integration is needed to solve the problem at hand.

Example 10 (Inertial Navigation) A submarine must keep track of its position, or net distance travelled. Surfacing to obtain a location fix is subject to the vagaries of weather and may not be acceptable for other reasons. Modern navigation depends on inertial guidance systems. Without relying on external observations, these systems sense the acceleration of the boat and record its instantaneous acceleration as a function of time. From this information, the velocity and position of the boat can be calculated by integrating twice.

Consider a situation where the submerged submarine moves along a straight course, with acceleration given by

$$a(t) = 14 - 9t + t^2 \quad (t \text{ in hours}; 0 \leq t \leq 8)$$

Find the velocity function $v(t)$ and the position function $s(t)$, given that

$$v(0) = 6 \quad \text{(initial speed is 6 mph)}$$
$$s(0) = 0 \quad \text{(boat starts moving at time } t = 0)$$

Solution We discussed acceleration in Section 3.5 and noted that

$$a(t) = \frac{dv}{dt} \quad \text{where } v(t) \text{ is the velocity}$$

Therefore, we can determine the velocity by taking an indefinite integral

$$v(t) = \int \frac{dv}{dt}\,dt = \int a(t)\,dt$$
$$= \int (14 - 9t + t^2)\,dt = 14t - \frac{9}{2}t^2 + \frac{1}{3}t^3 + c$$

The condition $v(0) = 6$ allows us to find c:

$$6 = v(0) = 14(0) - \frac{9}{2}(0)^2 + \frac{1}{3}(0)^3 + c$$

$$c = 6$$

Thus the velocity after t hours is

$$v(t) = 14t - \frac{9}{2}t^2 + \frac{1}{3}t^3 + 6 \quad \text{mph}$$

Now that we know the velocity, we use the relation $v(t) = ds/dt$ to find $s(t)$ by integrating one more time.

$$\begin{aligned} s(t) &= \int \frac{ds}{dt}\,dt = \int v(t)\,dt \\ &= \int \left(14t - \frac{9}{2}t^2 + \frac{1}{3}t^3 + 6\right) dt \\ &= 7t^2 - \frac{3}{2}t^3 + \frac{1}{12}t^4 + 6t + k \end{aligned}$$

where k is some constant. The initial condition $s(0) = 0$ implies that $k = 0$. Thus we have found the distance covered after t hours:

$$s(t) = 7t^2 - \frac{3}{2}t^3 + \frac{1}{12}t^4 + 6t \quad \text{miles}$$

for $0 \le t \le 8$ hours. ■

Exercises 5.2

In Exercises 1–6, find the function satisfying the given conditions.

1. $f'(x) = 1 - 2x, \quad f(0) = 3$

2. $f'(x) = x^2 + 2, \quad f(1) = 2$

3. $P'(x) = 100 - x, \quad P(0) = -3000$

4. $\dfrac{dV}{dt} = 10t, \quad V(0) = 37$

5. $S'(t) = 88t - 12t^2, \quad S(1) = 0$

6. $\dfrac{df}{dx} = e^{3x}, \quad f(0) = 0$

7. Find the monthly cost function $C(x)$ of a manufacturer, given that the marginal cost is $C'(x) = 18 - 0.012x$ and the fixed costs are \$8500 per month. (*Hint*: Fixed costs $= C(0) = 8500$.)

8. A retailer finds that his marginal revenue function is $dR/dx = 15 - 0.04x$. Find the revenue function $R(x)$ using the fact that $R(0) = 0$. Why is $R(0) = 0$ a reasonable assumption?

9. Repeat Exercise 8, taking the marginal revenue function to be

$$\frac{dR}{dx} = 86 - 0.2x + 0.006x^2$$

10. Find the profit function $P(x)$ given that
(i) The marginal revenue is $R'(x) = 95 - 0.015x^2$
(ii) The marginal cost is $C'(x) = 30 + 0.04x$
(iii) When $x = 0$, we have $P(0) = -12{,}000$.

11. The heating season lasts for 240 days in Warrensburg. On day x of the season, the local mill loses heat at the rate $r(x) = 0.72x - 0.003x^2$ therms per day. Find the cumulative heat loss (in therms) up to day x of the heating season. How much heat is lost in the first third of the season? How much is lost over the entire season?

12. A dam starts to release water into a stream at time $t = 0$. Find the total volume $V(t)$ of water released by time t if the flow rate is given by $dV/dt = 1000 + 40t$.

13. Chestertown is just below a 9.6-million-gallon reservoir. Right now, at time $t =$ 0 hours, the reservoir is half full, but smaller creeks are feeding water into it at an increasing rate. The rate of inflow t hours from now is estimated to be $8000 + 800t$ gallons per hour. How many hours do the townspeople have to evacuate before the reservoir begins to overflow?

14. Kenisco Reservoir contains 15 million gallons of water. The outflow of water to meet consumer needs t hours from now is estimated to be $10{,}000 + 180t$ gallons per hour, while the inflow of feeder streams will be $7000 - 90t^2$ gallons per hour. Find the volume of water in Kenisco at time t hours.

15. A worker's efficiency decreases as the day wears on. Suppose that at time t the typical worker produces at the rate of $100 - 4t$ widgets per hour, where t is the number of hours the worker has been on the job. Find the number of widgets produced in an 8-hour day.

16. Repeat Exercise 15 taking the production rate to be $100 - 0.3t^2$ widgets per hour at time t.

17. After t days a rumor spreads at the rate of $100 + 70t$ new listeners per day. At $t = 0$, one hundred people have heard the rumor. According to this model, how many will have heard it after 14 days?

18. Thirty people in Morton Grove are down with the flu. It is estimated that x days from now there will be $5 + 4x$ new cases per day. Find a formula for the cumulative number of cases up to day x.

19. The rate of crude oil consumption in the United States between 1960 and 1970 was

$$r(t) = 3.2e^{0.036t} \quad \text{billion barrels per year}$$

with $t = 0$ corresponding to the beginning of 1960. Find a formula for the total U.S. oil consumption from $t = 0$ until time t, $0 \leq t \leq 10$. In all, how much crude was consumed between 1960 and 1970?

20. The rate of natural gas consumption in the United States between 1960 and 1970 was

$$r(t) = 2e^{0.06t} \quad \text{billion barrels per year (crude oil equivalent)}$$

with $t = 0$ corresponding to the start of 1960. How much gas was consumed between 1960 and 1967? Between 1960 and 1970? Between 1962 and 1968?

21. Between 1985 and the year 2000, it is estimated that the rate of U.S. energy consumption will be

$$r(t) = 13e^{0.014t} \quad \text{billion barrels per year (crude oil equivalent)}$$

with $t = 0$ corresponding to the start of 1985. Find a formula for the total U.S. energy consumption from $t = 0$ until time t, $0 \leq t \leq 15$.

22. The velocity of a car after t seconds is measured at $v(t) = 88 - 4t$ feet per second. Find the total distance covered between times $t = 0$ and $t = 5$, between $t = 0$ and $t = 10$, and between $t = 5$ and $t = 10$ seconds.

23. Repeat Exercise 22, taking the velocity to be $v(t) = 44 + 2t$ feet per second.

24. A rocket rises with constant acceleration of 15 feet per second^2; it starts with initial velocity $v = 0$ when $t = 0$ (standing start from the launch pad). Find the velocity 10 seconds and 100 seconds after launch. (*Hint*: Acceleration $= dv/dt$.)

25. In free fall, the Earth's gravity pulls on an object causing a constant downward acceleration

$$a(t) = -32 \quad \text{feet per second}^2$$

Find the velocity function $v(t)$ and the height function $h(t)$ if

$$\text{initial velocity is } v(0) = 22 \text{ feet per second (upward)}$$
$$\text{initial height is } h(0) = 16 \text{ feet.}$$

(*Hint*: Acceleration $a = dv/dt$ and, by definition of velocity, $v = dh/dt$.)

26. Repeat Exercise 25, taking initial velocity $v(0) = v_0$ feet per second and initial height $h(0) = h_0$ feet.

27. A single-stage rocket is launched vertically from rest at ground level. Its engines burn for 90 seconds, producing an acceleration

$$a(t) = 300 - 2t \quad \text{feet per second}^2$$

during the time interval $0 \leq t \leq 90$ seconds. Find the velocity v at burnout ($t = 90$). Find the height of the rocket at burnout. (*Hint*: Acceleration $a = dv/dt$ and, by definition of velocity, $v = dh/dt$. First find v as a function of t, and set $t = 90$. Then go on to find h as a function of t, and set $t = 90$.)

28. During the first 16 weeks after birth, the weight $w(t)$ of a certain type of white mouse increases at a rate

$$\frac{dw}{dt} = \frac{2 + t}{30} \quad \text{ounces per week}$$

If the weight at birth is 0.4 ounces, find a formula for the mouse's weight after t weeks. What is the weight after 16 weeks?

5.3 Definite Integrals and Areas; Fundamental Theorem of Calculus

In this section we define the definite integral of a function in terms of its indefinite integral and use it to calculate areas. Even in antiquity, the problem of calculating areas was important because of the need to survey land. Greek geometers regarded it as one of the central problems of plane geometry. Our methods will go far beyond any geometric technique for determining areas.

We start by defining the **definite integral** of a function $f(x)$ over the interval $a \leq x \leq b$. It is indicated by the symbol

$$\int_a^b f(x)\,dx$$

The number b is called the "upper limit," and a is the "lower limit" of the definite integral. In contrast with the indefinite integral, which is a function, the definite integral is a number calculated as follows.

RULE FOR CALCULATING THE DEFINITE INTEGRAL $\int_a^b f(x)\,dx$

Step 1 Find an indefinite integral $F(x)$ of $f(x)$ on the interval $a \leq x \leq b$.

Step 2 Compute the value of the indefinite integral $F(x)$ at the upper limit $x = b$; this is $F(b)$.

Step 3 Compute the value of $F(x)$ at the lower limit $x = a$; this is $F(a)$.

Step 4 Subtract the number in Step 3 from the number in Step 2; this is the definite integral

$$\int_a^b f(x)\,dx = F(b) - F(a) \qquad [13]$$

In carrying out this procedure it does not matter which indefinite integral $F(x)$ we use; the number obtained in Step 4 will always be the same. To see why, observe that if $G(x)$ is some other indefinite integral of $f(x)$, then $G(x) = F(x) + c$ for some constant c. Therefore,

$$\begin{aligned} G(b) - G(a) &= \big(F(b) + c\big) - \big(F(a) + c\big) \\ &= F(b) + c - F(a) - c = F(b) - F(a) \end{aligned}$$

and the added constant c cancels out of the final result.

Example 11 Evaluate the definite integral $\int_2^6 x\,dx$.

Solution Step 1: Using Formula [4] of Section 5.1, we see that $F(x) = \frac{1}{2}x^2$ is an indefinite integral of $f(x) = x$. In Steps 2 and 3 we have

$$\text{value of } F \text{ at the upper limit } b = 6\text{: } F(6) = \frac{1}{2}(6)^2 = 18$$

$$\text{value of } F \text{ at the lower limit } a = 2\text{: } F(2) = \frac{1}{2}(2)^2 = 2$$

In Step 4 we obtain the definite integral by taking the difference of these values,

$$\int_2^6 x\,dx = F(6) - F(2) = 18 - 2 = 16$$

■

With practice, the steps in this procedure can be combined by introducing a useful symbol used universally to indicate the difference of endpoint values of a function $F(x)$. If we define

$$F(x)\Big|_a^b = F(b) - F(a) \qquad [14]$$

the procedure for calculating the definite integral takes on a simple form

$$\int_a^b f(x)\,dx = F(x)\Big|_a^b \quad \text{where } F(x) \text{ is any indefinite integral of } f(x) \qquad [15]$$

Example 12 Evaluate $\int (3 - x^2)\,dx$ and $\int_{-1}^{2} (3 - x^2)\,dx$.

Solution By Formulas [4] and [7] of Section 5.1 we have

$$\int (3 - x^2)\,dx = 3x - \frac{1}{3}x^3 + \text{const}$$

To evaluate the definite integral, we may choose the simplest value for the constant in the indefinite integral—namely, zero. Then, using the indefinite integral $F(x) = 3x - \frac{1}{3}x^3$, we calculate the definite integral:

$$\begin{aligned}\int_{-1}^{2} (3 - x^2)\,dx &= \left(3x - \frac{1}{3}x^3\right)\Big|_{-1}^{2} \\ &= \left[3(2) - \frac{1}{3}(2)^3\right] - \left[3(-1) - \frac{1}{3}(-1)^3\right] \\ &= \left(6 - \frac{8}{3}\right) - \left(-3 - \frac{1}{3}(-1)\right) = \frac{10}{3} - \left(-\frac{8}{3}\right) = 6\end{aligned}$$ ■

Now we turn to the problem of area. We shall consider general regions in the plane such as the one shown in Figure 5.3. Such a region R is obtained by taking the part of the plane bounded by a continuous curve $y = f(x)$, the x axis, and the two vertical lines where $x = a$ and $x = b$. We always assume $a < b$, as in the figure, and restrict attention to the case where $f(x) \geq 0$; so the curve $y = f(x)$ stays above the horizontal axis. Then the area, Area(R), can be evaluated by applying a truly remarkable result connecting areas and integrals.

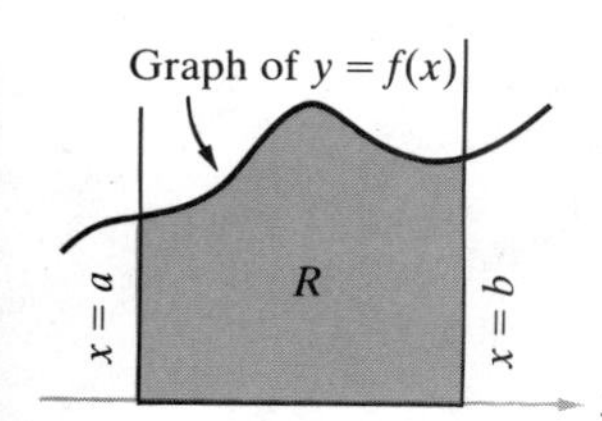

Figure 5.3
The shaded region R is bounded above by the graph of $y = f(x)$, below by the x axis, on the left by the vertical line $x = a$, and on the right by $x = b$. We shall assume that $f(x) \geq 0$ for $a \leq x \leq b$, so the graph does not cross the horizontal axis.

FUNDAMENTAL THEOREM OF CALCULUS For any plane region R of the form shown in Figure 5.3, bounded by the graph of a continuous function $y = f(x)$, the area of R is equal to the definite integral of $f(x)$ for $a \leq x \leq b$:

$$\text{Area}(R) = \int_a^b f(x)\,dx \qquad [16]$$

As part of the proof it is shown that every continuous function $y = f(x)$ actually *has* an indefinite integral $F(x)$, so it makes sense to talk about the definite integral $\int_a^b f(x)\,dx = F(x)|_a^b$. Unfortunately, it is not always easy to find an explicit algebraic formula for the indefinite integral, and the proof of the fundamental theorem does not provide one.

We shall give the proof at the end of this section. Right now we will concentrate on applying this result.

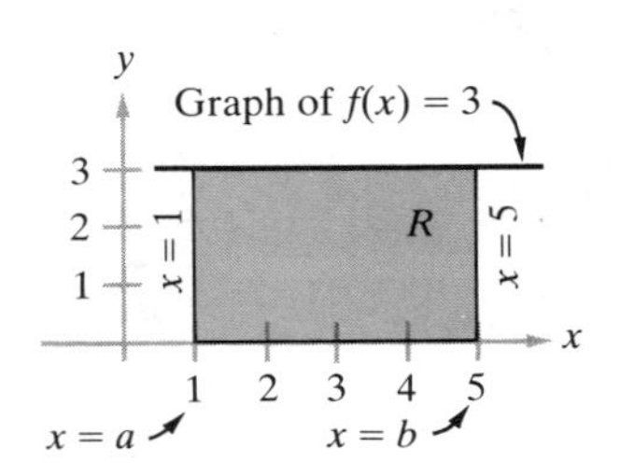

Figure 5.4
The rectangular region R in Example 13.

Example 13 Consider the constant function $f(x) = 3$ for $1 \leq x \leq 5$. The region R bounded by the graph of f is a rectangle (see Figure 5.4). Verify that the Fundamental Theorem of Calculus gives the usual area for R.

Solution The upper boundary of R is given by the constant function $f(x) = 3$, and the vertical lines are given by $a = 1$ and $b = 5$. Using the Fundamental Theorem of Calculus and the indefinite integral $F(x) = 3x$ we obtain

$$\text{Area}(R) = \int_a^b f(x)\,dx = \int_1^5 3\,dx$$

$$= 3x\Big|_1^5 = 15 - 3 = 12$$

This agrees with the usual area formula *Area* = *height* × *width*, because the height is 3, and the width is $5 - 1 = 4$. ■

Next we examine a region whose area cannot be calculated by the usual methods of plane geometry. Finding its area would be difficult without the Fundamental Theorem of Calculus.

Example 14 Calculate the area of the region lying above the interval $-1 \leq x \leq 2$ in the x axis and below the graph of $f(x) = x^2$ (see Figure 5.5).

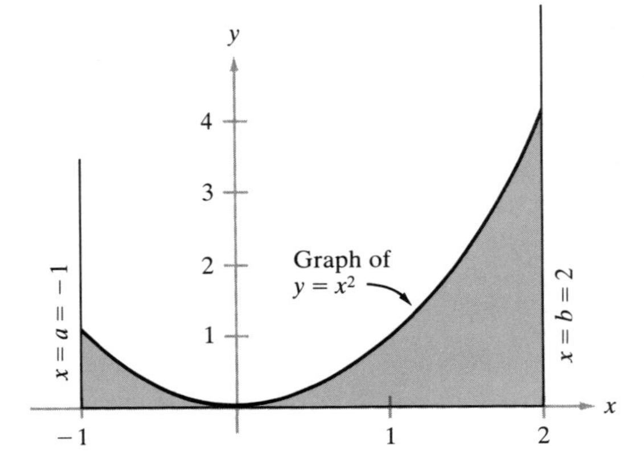

Figure 5.5
The region R lying under the graph of $y = x^2$ and over the interval $-1 \leq x \leq 2$ is shaded. Its area is computed in Example 14.

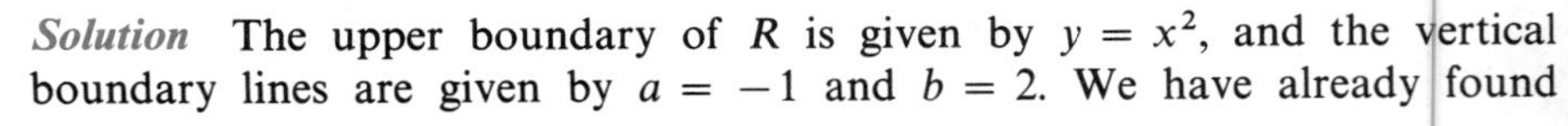

Solution The upper boundary of R is given by $y = x^2$, and the vertical boundary lines are given by $a = -1$ and $b = 2$. We have already found indefinite integrals for $f(x) = x^2$ in Section 5.1; we may use the particular indefinite integral $F(x) = \frac{1}{3}x^3$. Applying the Fundamental Theorem, we obtain

$$\begin{aligned}\text{Area}(R) &= \int_{-1}^{2} x^2\,dx \\ &= \frac{1}{3}x^3\Big|_{-1}^{2} \\ &= \left(\frac{1}{3}(2)^3\right) - \left(\frac{1}{3}(-1)^3\right) \\ &= \frac{8}{3} - \left(-\frac{1}{3}\right) = \frac{9}{3} = 3\end{aligned}$$

Example 15 Interpret the definite integral $\int_1^3 (1/x^2)\,dx$ as the area of a region R bounded by the curve $y = 1/x^2$. Sketch the region. Calculate the area by evaluating the definite integral.

Solution The graph of $y = 1/x^2$ is shown in Figure 5.6. Because it lies above the horizontal axis, the definite integral is equal to the area of the shaded region R bounded by vertical lines at $x = 1$ and $x = 3$.

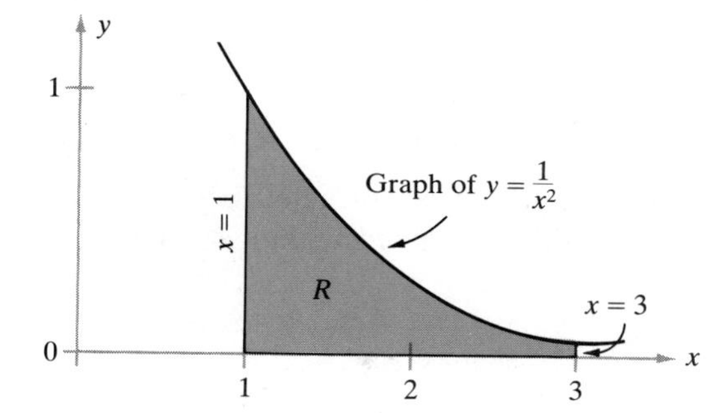

Figure 5.6
The shaded region R corresponds to the definite integral $\int_1^3 (1/x^2)\,dx$.

Using Formula [4] of Section 5.1, we find that $F(x) = -1/x$ is an indefinite integral of $f(x) = 1/x^2$. Therefore, by the Fundamental Theorem:

$$\begin{aligned}\text{Area}(R) &= \int_1^3 \frac{1}{x^2}\,dx = F(x)\Big|_1^3 = -\frac{1}{x}\Big|_1^3 \\ &= \left(-\frac{1}{3}\right) - \left(-\frac{1}{(-1)}\right) = -\frac{1}{3} + 1 = \frac{2}{3}\end{aligned}$$

Example 16 Find the area under the graph of $f(x) = e^{3x} + 6x^2 + 4$ above the interval from $x = 0$ to $x = 2$ on the x axis.

Solution For positive x, each term of $f(x)$ is positive, and therefore $f(x) = e^{3x} + 6x^2 + 4 \geq 0$. So, without bothering to sketch the graph (a laborious task), we can apply the Fundamental Theorem of Calculus:

$$\begin{aligned}\text{Area}(R) &= \int_0^2 f(x)\,dx = \int_0^2 (e^{3x} + 6x^2 + 4)\,dx \\ &= \frac{1}{3}e^{3x} + 2x^3 + 4x\Big|_0^2 \\ &= \left(\frac{1}{3}e^6 + 16 + 8\right) - \left(\frac{1}{3}\right) \\ &= \frac{e^6 + 71}{3} = 158.14\end{aligned}$$

■

We conclude this section with an explanation of why the Fundamental Theorem of Calculus is true. We start from certain simple facts about areas.

BASIC PROPERTIES OF AREAS For regions in the plane, areas have the following properties

(i) The area of a rectangle R is given by the usual formula

$$\text{Area}(R) = l \cdot w \quad (l = \text{length},\ w = \text{width})$$

(ii) If a region R is broken up into several nonoverlapping pieces $R_1, R_2, \ldots, R_n$, then the areas of the pieces must add up to that of R

$$\text{Area}(R) = \text{Area}(R_1) + \text{Area}(R_2) + \cdots + \text{Area}(R_n)$$

(iii) If a region R is bracketed between regions R_1 and R_2, so that R_1 is entirely contained within R, which in turn lies entirely within R_2, then

$$\text{Area}(R_1) \le \text{Area}(R) \le \text{Area}(R_2)$$

Now consider a region R such as the one shown in Figure 5.7a; R is bounded by a continuous nonnegative curve $y = f(x)$, vertical lines where $x = a$ and $x = b$, and the horizontal axis. How shall we calculate the area of R? The crucial idea is to study how the area *changes* as it is swept out from $x = a$ to $x = b$. For each x such that $a \le x \le b$, we let $R(x)$ be the region between the horizontal axis and the curve $y = f(x)$ lying over the interval $[a, x]$ (see Figure 5.7b). The function $A(x) = \text{Area}(R(x))$ is then defined for $a \le x \le b$. When $x = a$, the interval $[a, a]$ reduces to a point, and $A(a)$ is the area of a line segment (length $= f(a)$, width $= 0$), so that

$$A(a) = 0 \qquad [17]$$

When $x = b$, we get the area of the original region R,

$$A(b) = \text{Area}(R) \qquad [18]$$

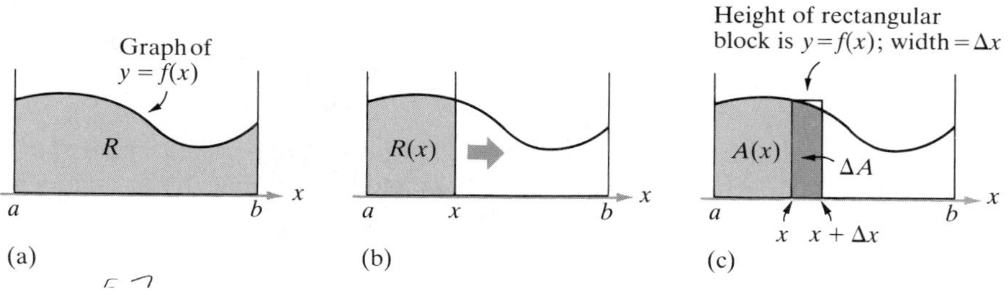

Figure 5.7
In (a) we show the region R bounded by the curve $y = f(x)$, the horizontal axis, and two vertical lines. In (b), the shaded subregion $R(x)$ is the part of R lying above the segment $[a, x]$ in the horizontal axis. As x moves from a to b, the region $R(x)$ covers more of R, and when $x = b$ we get $R(b) = R$. In (c), $A(x) = \text{Area}(R(x))$; this area function gives the area of the (lightly shaded) region $R(x)$ lying above the segment $[a, x]$. If x is increased slightly to $x + \Delta x$, then $A(x + \Delta x)$ is the area lying above the larger segment $[a, x + \Delta x]$. Then $\Delta A = A(x + \Delta x) - A(x)$ is the area of the (heavy shading) region above the small segment $[x, x + \Delta x]$. Because f is continuous, ΔA is approximately equal to $f(x) \cdot \Delta x$, the area of the rectangular block over $[x, x + \Delta x]$ with height $y = f(x)$.

Let us calculate the rate of change of area: $A'(x) = dA/dx$ as x increases from a to b. We first form the difference quotients, as in the delta process,

$$\frac{\Delta A}{\Delta x} = \frac{A(x + \Delta x) - A(x)}{\Delta x} \quad \text{for small displacements } \Delta x \neq 0$$

For simplicity we take $\Delta x > 0$ as shown in Figure 5.7c. Then, ΔA represents the area above the short line segment $[x, x + \Delta x]$ in the horizontal axis. Because f is continuous, the values of f above the interval $[x, x + \Delta x]$ stay very close to the value $f(x)$ at the left-hand endpoint of this interval. Thus the area ΔA, which corresponds to the heavily shaded strip in Figure 5.7c, is very nearly equal to that of a rectangle with base length Δx and height $f(x)$. That is, ΔA is approximately equal to $f(x) \cdot \Delta x$:

$$\Delta A \approx f(x) \cdot \Delta x \qquad [19]$$

Therefore, the difference quotient is

$$\frac{\Delta A}{\Delta x} \approx \frac{f(x) \cdot \Delta x}{\Delta x} = f(x)$$

This approximate formula becomes increasingly accurate as Δx is made smaller, so that

$$A'(x) = \lim_{\Delta x \to 0} \frac{\Delta A}{\Delta x} = f(x) \qquad [20]$$

Thus, if we differentiate the area function $A(x)$ we get back the original function $f(x)$ whose graph determines the region R. In other words, the area function $A(x)$

is an indefinite integral for $f(x)$!

Because $A(a) = 0$, we already know that

$$\text{Area}(R) = A(b) = A(b) - 0 = A(b) - A(a)$$

(see Equations [17] and [18]). Then, from the very definition of the definite integral $\int_a^b f(x)\,dx$ and the fact that $A(x)$ is an indefinite integral of $f(x)$, we obtain

$$\begin{aligned}\text{Area}(R) &= A(b) - A(a)\\ &= A(x)\Big|_a^b\\ &= \int_a^b f(x)\,dx\end{aligned}$$

This completes the proof of the Fundamental Theorem. ■

In this section we restricted attention to functions $y = f(x)$ whose graphs lie above the horizontal axis: $f(x) \geq 0$ for $a \leq x \leq b$. If $f(x)$ is sometimes positive and sometimes negative in the interval $a \leq x \leq b$, the definite integral $\int_a^b f(x)\,dx$ can still be given a geometric interpretation in terms of areas. Basically, regions above the horizontal axis are assigned positive area, and regions below the axis are assigned negative area. The sum of these areas, taking signs into account, is called the *signed area* of the region between the curve and the horizontal axis. In this situation the Fundamental Theorem says:

$$\text{Signed area}(R) = \int_a^b f(x)\,dx$$

We will not develop this idea here.

Exercises 5.3

In Exercises 1–4, evaluate $F(x)|_a^b = F(b) - F(a)$.

1. $F(x) = x;\quad a = 0, b = 4$

2. $F(x) = x^2;\quad a = -3, b = -1$

3. $F(x) = x^3 + x^2;\quad a = 1, b = 1.5$

4. $F(x) = e^x;\quad a = 0, b = 2$

In Exercises 5–22, evaluate the definite integral. (*Note*: In some cases we use a symbol other than x for the variable. The name applied to the variable makes no difference; just write out the indefinite integral using the same symbol for the variable.)

5. $\displaystyle\int_0^3 2\,dx$

6. $\displaystyle\int_{-1}^3 x\,dx$

7. $\displaystyle\int_{-3}^{-1} x^2\,dx$

8. $\displaystyle\int_1^8 x^{-1/3}\,dx$

9. $\displaystyle\int_1^{10} \frac{1}{x^2}\,dx$

10. $\displaystyle\int_1^{10} \frac{1}{t^2}\,dt$

11. $\displaystyle\int_0^4 -5x\,dx$

12. $\displaystyle\int_{-2}^2 \sqrt{3}\,dx$

13. $\int_1^4 \frac{3}{\sqrt{x}}\, dx$

14. $\int_{-1}^5 (4 - 2x)\, dx$

15. $\int_0^1 4e^x\, dx$

16. $\int_0^2 (5 - t + t^2)\, dt$

17. $\int_a^b \frac{1}{x^3}\, dx \quad (0 < a < b)$

18. $\int_{-2}^{-1} (3x + 1)\, dx$

19. $\int_0^2 (x^3 - 10x + 9)\, dx$

20. $\int_{-1}^1 (x^4 + x^3 - x + 5)\, dx$

21. $\int_1^3 \left(r^2 - \frac{4}{r}\right) dr$

22. $\int_1^2 \left(e^{x/2} - \frac{3}{x} + 4\right) dx$

23. Use the Fundamental Theorem of Calculus to find the area of the region lying above the interval $\frac{1}{2} \le x \le 3$ in the x axis and below the graph of $y = 2x - 1$. Check your answer by finding the area of this (triangular) region geometrically.

24. Make a rough sketch of the graph of $f(x) = 1 + x^2$ on graph paper. Draw a picture of the region under the graph whose area is given by the definite integral $\int_{-1}^2 (1 + x^2)\, dx$. Evaluate the integral.

25. Consider $y = 1/x^2$ on the interval $-4 \le x \le -1$. Sketch the part of the graph over this interval. Sketch the region whose area is represented by $\int_{-4}^{-1} (1/x^2)\, dx$, and evaluate this definite integral.

In Exercises 26–29, sketch a picture of the region whose area corresponds to the definite integral, and evaluate the definite integral.

26. $\int_{-1}^1 (4 - x^2)\, dx$

27. $\int_1^5 \frac{1}{x}\, dx$

28. $\int_0^4 \sqrt{x}\, dx$

29. $\int_{-2}^{-1} (t^2 + t + 4)\, dt$

In Exercises 30—36, use the Fundamental Theorem of Calculus to find the area of the region bounded by the curve $y = f(x)$, the x axis, and the vertical lines indicated.

30. $f(x) = 2$ (constant function) between $x = -1$ and $x = 3$

31. $f(x) = 3x$ between $x = 0$ and $x = 4$

32. $f(x) = x^4$ between $x = -1$ and $x = 0$

33. $f(x) = x^2 + 1$ between $x = -1$ and $x = 1$

34. $f(x) = \sqrt{x}$ between $x = 1$ and $x = 4$

35. $f(x) = \frac{1}{x}$ between $x = 1$ and $x = 5$

36. $f(x) = 8 - x^2$ between $x = 0$ and $x = 2$

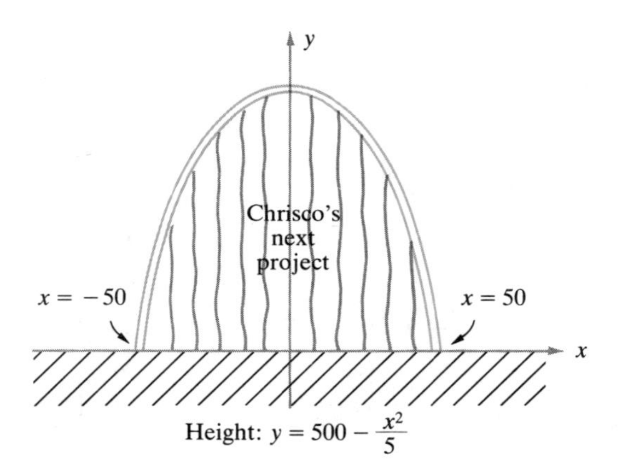

Figure 5.8
Mr. Chrisco's proposed curtain.

37. A famous midwestern memorial arch has a parabolic shape shown in Figure 5.8. In terms of the coordinates shown there, its height is given by $y = 500 - (x^2/5)$ feet. Mr. R. Chrisco, a conceptual artist famous for wrapping coastlines and other large objects in canvas, is planning to cover the archway with a hanging sheet of canvas. How many square feet will he need?

38. Explain this paradox: The graph of the function $f(x) = 1/x^2$ stays above the x axis; therefore the area A of the region under the graph and above the interval $-1 \leq x \leq 1$ is positive, $A > 0$. By the Fundamental Theorem of Calculus, the area is given by $\int_{-1}^{1} (1/x^2)\, dx$, so that

$$A = \int_{-1}^{1} \frac{1}{x^2}\, dx = \left(-\frac{1}{x}\right)\Bigg|_{-1}^{1} = (-1) - \left(-\frac{1}{-1}\right) = -2 < 0$$

Now the area cannot be both positive and negative at the same time! What is going on here? Have we made legitimate use of the Fundamental Theorem of Calculus?

5.4 Approximating Definite Integrals

On the surface, the area problem appears to be completely solved: Just apply the Fundamental Theorem of Calculus, and it reduces to evaluation of a definite integral. If the function at hand $y = f(x)$ is simple enough, so that we can find a formula for the indefinite integral $F(x)$, then indeed the definite integral of $f(x)$ is easily computed, and nothing more need be said. But, there are situations where we cannot find an explicit formula for $F(x)$, and consequently cannot evaluate the definite integral directly. Here are two ways this can occur:

1. We may only know $f(x)$ approximately, say by its graph. In this case we have no algebraic formula for $f(x)$ and cannot hope to find one for the indefinite integral $F(x)$.
2. We may have an algebraic formula for $f(x)$ but may not be able to find such a formula for the indefinite integral $F(x)$ by any available integration technique.

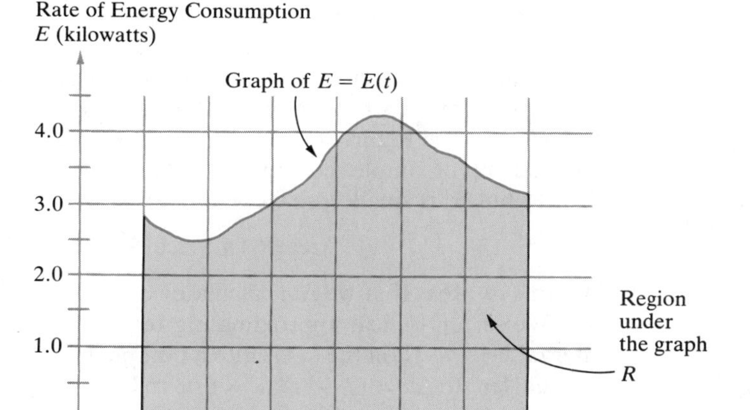

Figure 5.9
Rate of consumption of electrical energy (measured in kilowatts) is a function of time: $E = E(t)$. Time t is measured in hours starting with $t = 0$ at noon. The graph is plotted from meter data. The function $E(t)$ is not given by any simple algebraic formula. Total energy consumed between 1 PM and 7 PM (in kilowatt hours) is numerically equal to the area of the shaded region under the graph.

In either case we cannot evaluate the definite integral that appears in the Fundamental Theorem of Calculus. Then we need a new method for computing areas.

As an example of the first difficulty, consider the graph in Figure 5.9, which arises as follows. The *rate* at which a homeowner consumes electric energy is measured in units called kilowatts, and the *amount* of energy consumed in units called kilowatt-hours. A 500-watt light bulb consumes energy at the rate of 0.5 kilowatt; after, say, 3.5 hours of use, the amount of energy it consumes is

$$\begin{aligned}(\text{time}) \times (\text{rate of power consumption}) &= (3.5 \text{ hours}) \times (0.5 \text{ kilowatts}) \\ &= 1.75 \quad \text{kilowatt-hours}\end{aligned}$$

and so on. The term *kilowatt-hour* (kwh) should be familiar to anyone who has paid an electric bill and was charged for the number of kilowatt-hours used (at the rate of 16.5¢ per kwh in New York as of 1983). In Figure 5.9, we have recorded the rate of power consumption E (kilowatts) as a function of elapsed time t during a typical afternoon. Consumption varies irregularly, depending on which appliances are in use.

Suppose we want to compute from this information the total amount of energy (kwh) consumed, and thus the cost of electric service for this period. It turns out the total number of kilowatt-hours is numerically equal to the area under the graph of $E = E(t)$ between $t = 1$ and $t = 7$.† The function $E(t)$ is not

† To see this, let $A(t)$ be the amount of energy (kilowatt-hours) consumed up to time t. Because $E(t)$ is the *rate* at which energy is consumed, $E(t) = dA/dt$, $A(t)$ is an indefinite integral for $E(t)$, and by definition of the definite integral we have

$$(\text{kwh consumed from } t = 1 \text{ to } t = 7) = A(7) - A(1) = \int_1^7 E(t)\,dt$$

By the Fundamental Theorem of Calculus we know that this definite integral may be interpreted as the area under the graph of $E(t)$ from $t = 1$ to $t = 7$.

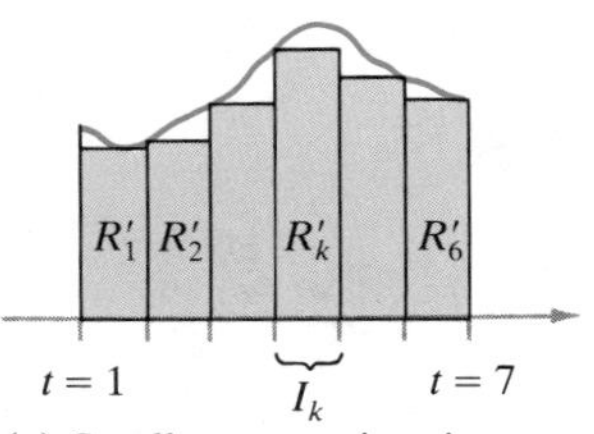

(a) Smaller approximating region R' is shaded.

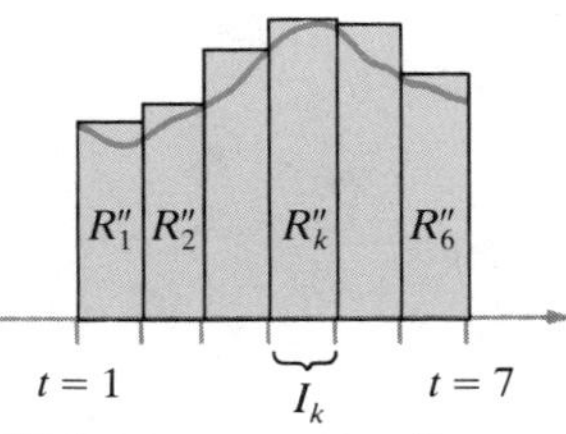

(b) Larger approximating region R'' is shaded.

Figure 5.10
The smaller and larger rectangular regions R' and R'' used to approximate the region R in Figure 5.9. Regions R' and R'' are shaded. Construction of the individual blocks R'_k and R''_k is described in the text; if the base interval $1 \leq t \leq 7$ is divided into n subintervals, each approximating region consists of n rectangular blocks.

given by an algebraic formula, so we cannot evaluate the area by the Fundamental Theorem of Calculus. Instead, we shall *estimate* the area by setting up certain regions made up of rectangles that approximate the given region R. The sort of "approximating regions" we have in mind are the shaded regions R' and R'' shown in Figure 5.10. The areas of R' and R'' can be calculated since they are made up of simple rectangular blocks. Furthermore, the region R'' is larger than R, but R' is smaller, so we obtain an estimate for Area(R)

$$\text{Area}(R') \leq \text{Area}(R) \leq \text{Area}(R'') \qquad [21]$$

in terms of areas that we *can* calculate.

We set up typical approximating regions R' and R'' by dividing the base interval $I = [1, 7]$ on the t axis into a number of subintervals $I_1, I_2, \ldots, I_n$, each of equal length $\Delta t = (7 - 1)/n = 6/n$. In Figure 5.10 we have divided I into $n = 6$ intervals, each of length $\Delta t = 6/6 = 1.0$ hour. (The discussion would be similar if we considered a much finer subdivision of I consisting of, say, $n = 100$ intervals. There would be more numbers to keep track of, but the idea would be the same.) The smaller region R' is obtained by setting up rectangular blocks $R'_1, R'_2, \ldots, R'_n$ over each of the intervals $I_1, I_2, \ldots, I_n$. The block R'_k over the interval I_k is made as high as possible, keeping the entire block under the graph. The appropriate block heights can be read from the graph without much trouble, and are tabulated in Table 5.2. Because Area(R'_k) = (height) × (width) we may also calculate the area of each block and the total area of the region R' (full shaded region in Figure 5.10a)

$$\begin{aligned}\text{Area}(R') &= \text{Area}(R'_1) + \text{Area}(R'_2) + \cdots + \text{Area}(R'_n)\\ &= 2.5 + 2.6 + \cdots + 3.2\\ &= 18.8\end{aligned}$$

Because the rectangular region R' lies inside the given region R we obtain an estimate Area(R) ≥ Area(R') = 18.8.

We construct the larger rectangular region R'' in much the same way. Over each interval I_k, set up a rectangular block R''_k just large enough to cover the graph. These blocks fill the shaded region R'' shown in Figure 5.10b. As in the preceding, the heights and areas of the blocks $R''_1, \ldots, R''_n$ may be determined by

k	Interval I_k	Height of R'_k	Area(R'_k)	Height of R''_k	Area(R''_k)
1	$1 \leq t \leq 2$	2.5	2.5	2.8	2.8
2	$2 \leq t \leq 3$	2.6	2.6	3.1	3.1
3	$3 \leq t \leq 4$	3.1	3.1	3.8	3.8
4	$4 \leq t \leq 5$	3.8	3.8	4.3	4.3
5	$5 \leq t \leq 6$	3.6	3.6	4.2	4.2
6	$6 \leq t \leq 7$	3.2	3.2	3.6	3.6

Table 5.2 Heights and areas of the rectangular blocks in Figure 5.10. Heights of blocks have been read from the graph in Figure 5.9.

referring to the graph of $E = E(t)$. The results are listed in Table 5.2. Clearly

$$\text{Area}(R'') = \text{Area}(R_1'') + \text{Area}(R_2'') + \cdots + \text{Area}(R_n'') = 21.8$$

Because R'' is larger than R we have $\text{Area}(R) \leq \text{Area}(R'') = 21.8$, giving a combined estimate for the desired area

$$18.8 = \text{Area}(R') \leq \text{Area}(R) \leq \text{Area}(R'') = 21.8$$

Thus we have estimated Area(R) by direct methods that do not involve the Fundamental Theorem of Calculus.

What if we divide $I = [1, 7]$ into many more pieces of smaller length? The regions R' and R'' will then give a much better approximation to R, and the estimate

$$\text{Area}(R') \leq \text{Area}(R) \leq \text{Area}(R'')$$

will also be much better. Of course more calculation will be necessary; but if a computer is programmed to do all the routine calculations, this will not be a serious difficulty. In fact, this is just what is done in practice. From this discussion we see that the exact area Area(R) is a limit of areas of rectangular approximating regions

$$\lim_{n\to\infty} \text{Area}(R') = \text{Area}(R) = \lim_{n\to\infty} \text{Area}(R'') \qquad [22]$$

as the number n of equal-length subintervals becomes larger and larger in the procedure described above.

The point of this discussion is that this procedure can be used to estimate the definite integral (area under the graph)

$$\int_a^b f(x)\,dx$$

of any reasonable function $f(x)$ over any interval $a \leq x \leq b$. For practical purposes an estimate is good enough, and this method works without having to find an explicit indefinite integral $F(x)$, which would be needed to compute the area using the Fundamental Theorem of Calculus. As a bonus we get an alternate interpretation of the definite integral as the limit of the areas of approximating rectangular regions R' and R'' as we divide the base interval $a \leq x \leq b$ into smaller and smaller pieces

$$\int_a^b f(x)\,dx = \lim_{n\to\infty} \text{Area}(R') = \lim_{n\to\infty} \text{Area}(R'')$$

We shall apply this technique to estimate

$$\int_0^1 e^{-x^2}\,dx$$

a definite integral important in statistics. We are forced to approximate in this example because we just cannot find an indefinite integral of $f(x) = e^{-x^2}$ in explicit algebraic form. (Don't try! By sophisticated reasoning well beyond the scope of this book, it can be shown to be impossible.) Without a formula in hand for an indefinite integral, the exact evaluation of the definite integral is short-circuited and we are forced to use an approximation procedure.

Example 17 Interpret $\int_0^1 e^{-x^2}\,dx$ as an area, and compute the approximate values Area(R') and Area(R'') by dividing the base interval $0 \le x \le 1$ into $n = 10$ subintervals of equal length.

Solution The graph of $y = e^{-x^2}$ is shown in Figure 5.11; it is closely related to the well-known "normal" (or bell-shaped) curve from statistics. Using the Fundamental Theorem of Calculus we see that

$$\int_0^1 e^{-x^2}\,dx = \text{Area under the graph from } x = 0 \text{ to } x = 1$$

The rectangular blocks R'_k are shown in the figure, and the areas of R'_k and R''_k are listed in Table 5.3. The heights are obtained by direct evaluation of $y = e^{-x^2}$ on a calculator. Because the function is decreasing for $0 \le x \le 1$, we need only the values at the endpoints of the subintervals to determine the heights of the blocks. All blocks have the same base length $\Delta x = 0.1$.

From the table we obtain the estimates

$$0.7155 \le \int_0^1 e^{-x^2}\,dx \le 0.7787$$

It is interesting to observe that the *average* of our lower and upper estimates

$$\frac{0.7155 + 0.7787}{2} = 0.7471$$

is an extremely accurate estimate for $\int_0^1 e^{-x^2}\,dx$. ■

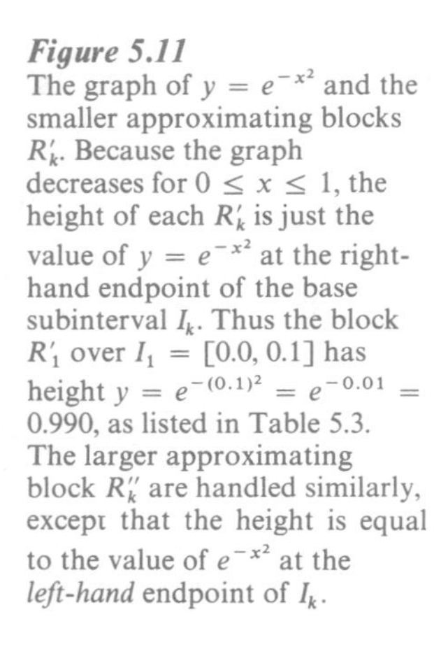

Figure 5.11
The graph of $y = e^{-x^2}$ and the smaller approximating blocks R'_k. Because the graph decreases for $0 \le x \le 1$, the height of each R'_k is just the value of $y = e^{-x^2}$ at the right-hand endpoint of the base subinterval I_k. Thus the block R'_1 over $I_1 = [0.0, 0.1]$ has height $y = e^{-(0.1)^2} = e^{-0.01} = 0.990$, as listed in Table 5.3. The larger approximating block R''_k are handled similarly, except that the height is equal to the value of e^{-x^2} at the *left-hand* endpoint of I_k.

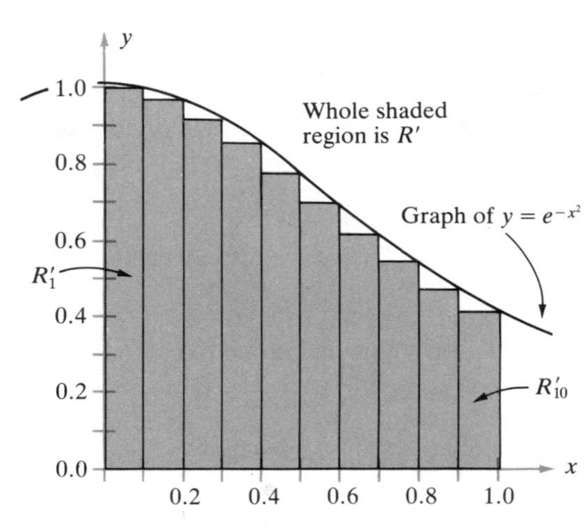

k	Height(R'_k)	Area(R'_k)	Height(R''_k)	Area(R''_k)
1	0.990	0.0990	1.000	0.1000
2	0.961	0.0961	0.990	0.0990
3	0.914	0.0914	0.961	0.0961
4	0.852	0.0852	0.914	0.0914
5	0.787	0.0787	0.852	0.0852
6	0.698	0.0698	0.787	0.0787
7	0.613	0.0613	0.698	0.0698
8	0.527	0.0527	0.613	0.0613
9	0.445	0.0445	0.527	0.0527
10	0.386	0.0386	0.445	0.0445
Totals:	Area(R') = 0.7155		Area(R'') = 0.7787	

Table 5.3 Heights and areas of the rectangular blocks R'_k and R''_k in Figure 5.11. All blocks have base length $\Delta x = 0.1$, so Area(R'_k) = height(R'_k) × (0.1) and similarly for the R''_k.

Exercises 5.4

1. Estimate the area of the region R in Figure 5.9 by dividing the base interval $I = [1, 7]$ into $n = 12$ equal subintervals and calculating Area(R') and Area(R''). Determine the heights of the rectangular blocks by inspecting the graph in the figure. Compare this estimate for Area(R) with those obtained in the text, which used a coarser subdivision into $n = 6$ subintervals.

In Exercises 2–5, draw the graph of $y = f(x)$ over the base interval $I = [a, b]$. Then divide I into n equal subintervals, and sketch the corresponding smaller rectangular region R'. Make a similar sketch for the larger rectangular region R''.

2. $f(x) = x \quad I = [0, 4] \quad n = 4$

3. $f(x) = 12 - 2x \quad I = [1, 4] \quad n = 9$

4. $f(x) = x^2 \quad I = [-1, 2] \quad n = 6$

5. $f(x) = \dfrac{1}{x} \quad I = [1, 5] \quad n = 8$

Exercises 6–13 refer to the region under the graph of $y = f(x)$ lying above the interval I in the x axis. Consider the smaller rectangular region R' and the larger rectangular region R'' obtained when I is divided into n equal subintervals. Calculate Area(R') and Area(R''). Then compare with the exact area calculated by the Fundamental Theorem of Calculus.

6. $f(x) = 3$ (constant function) $\quad I = [-2, 3] \quad n = 5$

7. $f(x) = \dfrac{x}{5} \quad I = [0, 3] \quad n = 9$

8. $f(x) = 6 - x \quad I = [-1, 2] \quad n = 15$

9. $f(x) = x^2 \quad I = [0, 2] \quad n = 6$

10. $f(x) = 4 - x^2 \quad I = [-2, 2] \quad n = 8$

11. $f(x) = 3 + 2x - x^2 \quad I = [-1, 2] \quad n = 6$

12. $f(x) = e^x \quad I = [-2, 1] \quad n = 9$

13. $f(x) = \dfrac{1}{x} \quad I = [1, 3] \quad n = 10$

14. Sketch the region bounded by the curve $y = 1 + 3x$, the x axis, and the vertical lines $x = 1$ and $x = 3$. Calculate the numerical value of Area(R') and Area(R'') when $I = [1, 3]$ is divided into $n = 4$ equal parts. Repeat, taking $n = 8$. (*Note*: The exact area of R (a trapezoid) is 14.)

15. Use the procedure of Example 17 to calculate the area under the normal curve $y = e^{-x^2}$ over the base intervals in the x axis listed below.
(i) From $x = 0$ to $x = 2$ ($n = 10$ subintervals)
(ii) From $x = 0$ to $x = 1.5$ ($n = 10$ subintervals)
(iii) From $x = 0$ to $x = 0.5$ ($n = 10$ subintervals)
Use the average of the upper and lower rectangular areas for your estimate.

16. Estimate the area under the normal curve $y = e^{-x^2}$ that corresponds to the integral $\int_0^1 e^{-x^2}\,dx$ using the methods of Example 17, but taking $n = 15$ subintervals. Is there much difference between your answer and the one obtained in the text using $n = 10$ subintervals?

17. The function

$$t_7(x) = 0.3850\left(1 + \frac{x^2}{7}\right)^{-4}$$

is one of a whole family of functions, called *t distributions*, which are important in statistics. It is necessary to determine areas of regions A_k lying under the graph of the kind shown in Figure 5.12. These areas are given by integrals:

$$A(k) = \text{Area}(A_k) = \int_0^k t_7(x)\,dx \quad \text{for each } k > 0$$

One way to calculate them is to make numerical estimates, as in Example 17. For each of the following values of k, estimate the area $A(k)$ by subdividing $I = [0, k]$ into $n = 10$ equal subintervals and taking the average of the upper and lower rectangular areas.
(i) $k = 1$
(ii) $k = 2$
(iii) $k = 0.75$

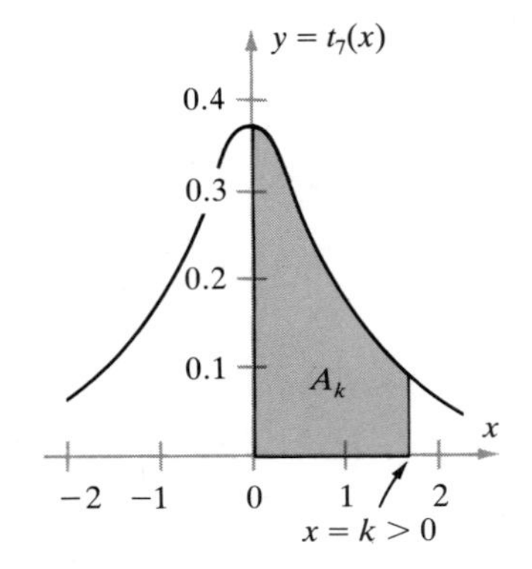

Figure 5.12
Graph of the t distribution function $y = t_7(x)$. The vertical line $x = k$ marks a cutoff; the area of the region A_k is to be determined. Total area under the graph is one. This function occurs in probability and statistics. It looks like the bell-shaped normal curve, but has a different formula and rather different properties for large values of x.

In Exercises 18–21, interpret $\int_a^b f(x)\,dx$ as an area, and compute the approximate values Area(R') and Area(R''), dividing the base interval $I = [a, b]$ into n equal subintervals. Do not try to apply the Fundamental Theorem of Calculus: The necessary indefinite integrals cannot be written in any simple form, so we must resort to approximations.

18. $\displaystyle\int_0^2 \sqrt{16 - x^3}\,dx \quad (n = 8)$

19. $\displaystyle\int_1^2 \frac{1}{\sqrt{16 - x^3}}\,dx \quad (n = 4)$

20. $\displaystyle\int_0^1 x\sqrt{1 + x^3}\,dx \quad (n = 4)$

21. $\displaystyle\int_1^2 \frac{e^x}{x}\,dx \quad (n = 5)$

If we are given the rate of change $r(x)$ of some quantity, the area under the graph of this function over $a \le x \le b$ is the total change in this quantity. Thus if $y = f(x)$ and $r(x) = df/dx$ is the rate of change, then

$$(\text{change in } f) = (\text{Area under the graph of } r(x))$$

or

$$f(b) - f(a) = \int_a^b r(x)\,dx$$

This follows from [15] because $r(x) = df/dx$, exactly as in our discussion of power consumption. Answer the following questions with this in mind.

22. Figure 5.13 gives the graph of the velocity of a car as a function of time, $v = v(t)$. Because $v = ds/dt$, the distance s travelled between 2 PM and 6 PM is numerically equal to the area under the graph over $2 \le t \le 6$. Estimate this area (and the distance) by dividing the interval $2 \le t \le 6$ into $n = 4$ equal pieces, calculating areas Area(R') and Area(R'') by inspecting the graph.

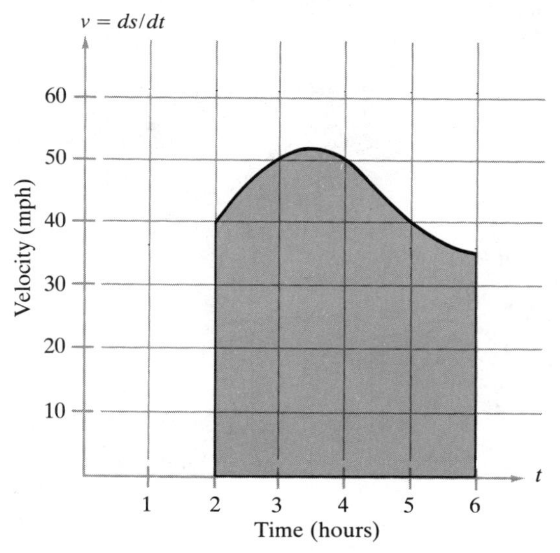

Figure 5.13
Velocity of the car in Exercise 22. The problem is to find the distance travelled (area under the graph).

23. The rate of oxygen consumption of an experimental subject in a 2-minute stress test is shown in Figure 5.14. How many liters of oxygen are consumed during the test? Estimate this by dividing the base interval $0 \le t \le 2$ into $n = 8$ equal pieces.

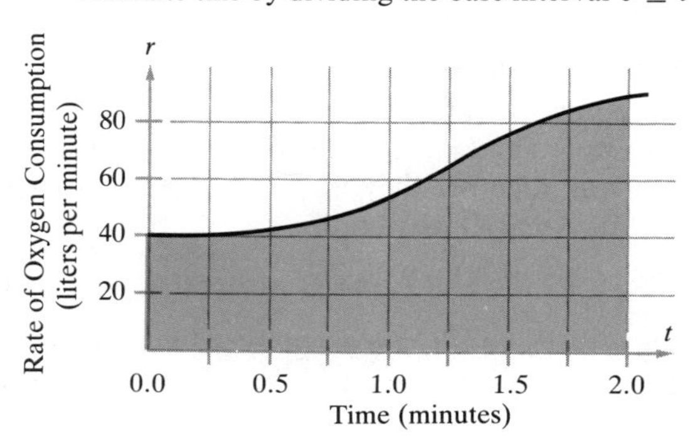

Figure 5.14
Rate of oxygen consumption (liters per minute) in Exercise 23.

5.5 Applications of Definite Integrals

The calculation of areas is by no means the only use for definite integrals. They play an important role in physics, geometry, statistics, economics, and systems analysis. When a function $y = f(x)$ appears in a practical problem, the definite integral $\int_a^b f(x)\,dx$ (area under its graph) takes on a meaning related to that of $f(x)$ in the problem at hand and often plays an important role in solving the problem. We hinted at this in the energy-consumption problem discussed in the last section. Here we discuss further applications of definite integrals; another far-reaching application will be given in Section 5.6.

ILLUSTRATION Areas of More Complicated Regions

Suppose a region R is bounded by vertical lines $x = a$ and $x = b$ and by two curves, as in Figure 5.15a. The upper boundary of R is given by $y = f(x)$, and the lower boundary by a different curve $y = g(x)$. Over each x the height of the region is given by the function $h(x) = f(x) - g(x)$, for $a \leq x \leq b$. We shall show that Area(R) is given by the definite integral of this height function

$$\text{Area}(R) = \int_a^b h(x)\,dx = \int_a^b (f(x) - g(x))\,dx \qquad [23]$$

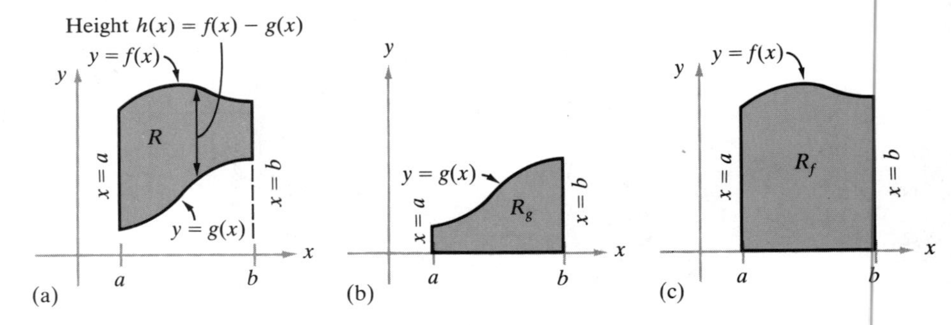

Figure 5.15 In (a) we show a region R bounded by two curves, $y = f(x)$ and $y = g(x)$, and by two vertical lines. The height of the region over x is $h(x) = f(x) - g(x)$. In (b) and (c) we show the regions R_g and R_f underneath $y = g(x)$ and $y = f(x)$, respectively; Area(R) + Area(R_g) = Area(R_f).

Discussion

For simplicity we assume that the lower boundary curve stays above the x axis for $a \leq x \leq b$. Then the region R lies completely above this axis. Write R_g for the region bounded by the curve $y = g(x)$, the x axis, and vertical lines $x = a$ and $x = b$; and write R_f for the region bounded by $y = f(x)$ and the x axis, between the lines $x = a$ and $x = b$, as in Figures 5.15b and 5.15c. Then R and R_g do not overlap, and together they fill up R_f. Hence Area(R) + Area(R_g) = Area(R_f). Solving for Area(R) we obtain

$$\text{Area}(R) = \text{Area}(R_f) - \text{Area}(R_g)$$

The areas of R_f and R_g are given by definite integrals. If $F(x)$, $G(x)$ are indefinite integrals of $f(x)$, $g(x)$, then $F - G$ is an indefinite integral for $h(x) = f(x) - g(x)$

(see Formula [7]). Thus, by the Fundamental Theorem of Calculus,

$$\text{Area}(R_f) = \int_a^b f(x)\,dx = F(b) - F(a)$$

$$\text{Area}(R_g) = \int_a^b g(x)\,dx = G(b) - G(a)$$

and by the definition of the definite integral of h

$$\begin{aligned}\int_a^b h(x)\,dx &= (F - G)(b) - (F - G)(a)\\ &= F(b) - G(b) - F(a) + G(a)\\ &= \big(F(b) - F(a)\big) - \big(G(b) - G(a)\big)\\ &= \text{Area}(R_f) - \text{Area}(R_g) = \text{Area}(R)\end{aligned}$$

as required. A more careful analysis shows that the positions of the curves $y = f(x)$ and $y = g(x)$ relative to the x axis do not matter. The formula is valid as long as the curve $y = f(x)$ stays above $y = g(x)$ for $a \le x \le b$. ■

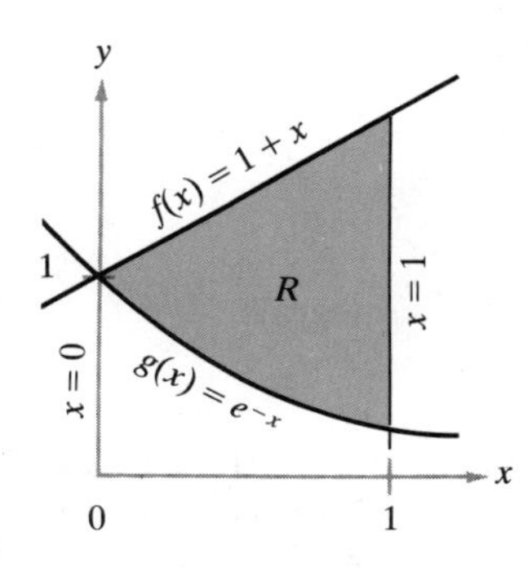

Figure 5.16
The region R in Example 18.

Example 18 Find the area of the shaded region in Figure 5.16, with upper boundary curve $f(x) = 1 + x$ and lower boundary curve $g(x) = e^{-x}$, over the interval $0 \le x \le 1$.

Solution For x between zero and one, we see that the height of R is

$$h(x) = f(x) - g(x) = 1 + x - e^{-x}$$

so the area is

$$\begin{aligned}\int_0^1 h(x)\,dx &= \int_0^1 (1 + x - e^{-x})\,dx\\ &= x + \frac{1}{2}x^2 + e^{-x}\Big|_0^1\\ &= \left(1 + \frac{1}{2} + e^{-1}\right) - \left(0 + \frac{1}{2}(0) + e^{-0}\right)\\ &= \left(\frac{3}{2} + \frac{1}{e}\right) - 1\\ &= \frac{1}{2} + \frac{1}{e} \approx 0.8679\end{aligned}$$

■

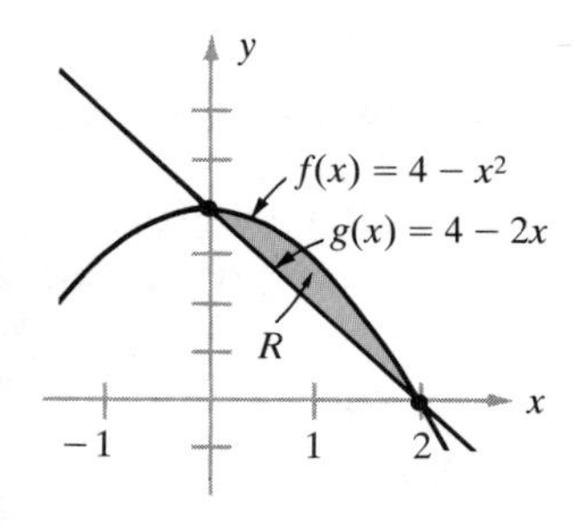

Figure 5.17
The shaded region R is bounded by the curves $f(x) = 4 - x^2$ (upper curve) and $g(x) = 4 - 2x$ (lower curve). The cross-over points for the curves are shown as solid dots. They occur at $x = 0$ and $x = 2$.

In the next example we are given two curves. Before applying [23] we must find the x values where the curves cross in order to describe the region R.

Example 19 Calculate the area of the region R (Figure 5.17) bounded by the curves $f(x) = 4 - x^2$ and $g(x) = 4 - 2x$.

Solution We find the x values where the curves cross by solving the equation $f(x) = g(x)$:

$$4 - x^2 = 4 - 2x \quad \text{or} \quad 2x - x^2 = 0$$

Crossing occurs at $x = 0$ and $x = 2$, and for $0 \leq x \leq 2$ the curve $y = f(x)$ lies above $y = g(x)$, as required in [23]. Thus the shaded region may be thought of as bounded by $y = f(x)$, $y = g(x)$, and the vertical lines $x = 0$ and $x = 2$. The height function for this region is $h(x) = f(x) - g(x) = (4 - x^2) - (4 - 2x) = 2x - x^2$ for $0 \leq x \leq 2$. Thus

$$\begin{aligned} \text{Area}(R) &= \int_0^2 h(x)\,dx = \int_0^2 (2x - x^2)\,dx \\ &= x^2 - \frac{1}{3}x^3 \Big|_0^2 \\ &= \left(4 - \frac{8}{3}\right) - \left(0 - \frac{1}{3}(0)\right) \\ &= \frac{4}{3} \end{aligned}$$

■

We deal next with an economic application of definite integrals.

ILLUSTRATION Consumers and Producers Surplus

As prices rise in a competitive market, manufacturers are willing to increase production; when prices fall they will cut back. The **supply function** gives the relationship between x, the number of units manufacturers are willing to produce (in some fixed short-term period), and the unit price p they can fetch for their product:

$$p = S(x) \quad \text{(supply function)} \qquad [24]$$

It should be clear that p increases with x, and the graph—the **supply curve**—rises as shown in Figure 5.18a. Consumer action produces opposite effects: As prices rise, buyers cut back their purchases, and when prices fall they increase consumption. The **demand function** expresses this relation between price p and

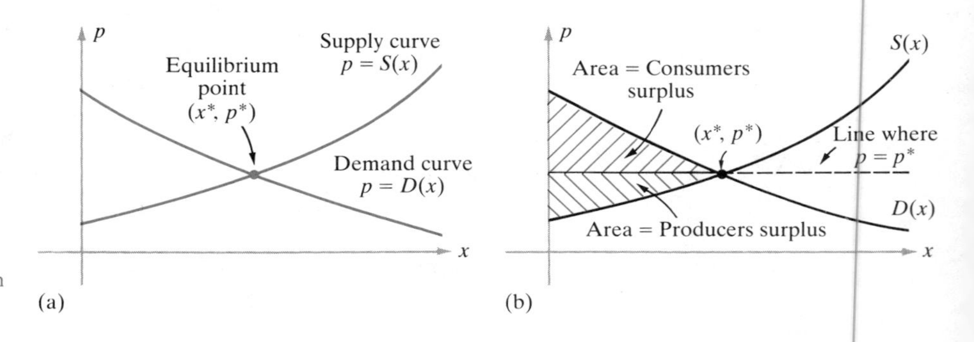

Figure 5.18 Typical supply and demand curves $p = S(x)$ and $p = D(x)$ are shown in (a). The graphs meet at the equilibrium point (x^*, p^*). In (b) the area of the upper shaded region is the consumers surplus, and the area of the lower shaded region is the producers surplus.

amount purchased x:

$$p = D(x) \quad \text{(demand function)} \qquad [25]$$

The demand function decreases, and its graph—the **demand curve**—falls as x increases.

The point where the supply and demand curves meet (when supply = demand) is the **equilibrium point** (x^*, p^*) shown in Figure 5.18a. The **equilibrium price** p^* is the actual competitive market price, and x^* is the actual number of units produced. The **consumers surplus** is defined as the area between the demand curve $p = D(x)$ and the horizontal line $p = p^*$, from $x = 0$ to $x = x^*$ (see Figure 5.18b). In terms of definite integrals

$$\text{Consumers surplus} = \int_0^{x^*} (D(x) - p^*)\, dx \qquad [26]$$

The **producers surplus** is the area between the supply curve and the line $p = p^*$, from $x = 0$ to $x = x^*$:

$$\text{Producers surplus} = \int_0^{x^*} (p^* - S(x))\, dx \qquad [27]$$

In economic terms, consumers surplus measures the total saving to consumers buying in an open market—in which all units are bought at the equilibrium price p^*—compared to what consumers would pay under a monopoly employing "price discrimination." A monopoly would auction off successive units to the highest bidder; prices would start high and gradually fall off toward price p^* as more units were sold. Similarly, producers surplus represents the total gain between the minimum price producers would be willing to charge and the price they actually can charge under open market conditions.

Example 20 Suppose the weekly demand for a certain brand of men's suit in Chicago is

$$p = D(x) = 200 - 5x - x^2 \quad \text{dollars per suit}$$

and the supply function is

$$p = S(x) = x^2 + 4x \quad \text{dollars per suit}$$

where x is in hundreds of suits per week. Find the consumers and producers surplus.

Solution We first find the equilibrium point by equating supply and demand: We have $S(x) = D(x)$ when

$$x^2 + 4x = 200 - 5x - x^2$$

or

$$2x^2 + 9x - 200 = 0, \quad \text{that is} \quad (x - 8)(2x + 25) = 0$$

The roots of the equation are $x = 8$ (first factor equal zero) and $x = -25/2$ (second factor equal zero). We reject the second root because feasible values of x must be positive. Therefore the equilibrium point occurs when $x^* = 8$. The equilibrium price is calculated by setting $x = x^*$ in the supply function (the demand function would do as well)

$$p^* = S(x^*) = S(8) = 96$$

Having found the equilibrium point $(x^*, p^*) = (8, 96)$ we use Equations [26] and [27] to calculate the consumers and producers surplus.

$$\begin{aligned}
(\text{consumers surplus}) &= \int_0^{x^*} (D(x) - p^*)\,dx \\
&= \int_0^8 ((200 - 5x - x^2) - 96)\,dx \\
&= \int_0^8 (104 - 5x - x^2)\,dx \\
&= 104\,x - \frac{5}{2}x^2 - \frac{1}{3}x^3 \Big|_0^8 = 501.33
\end{aligned}$$

and

$$\begin{aligned}
(\text{producers surplus}) &= \int_0^{x^*} (p^* - S(x))\,dx \\
&= \int_0^8 (96 - (x^2 + 4x))\,dx \\
&= \int_0^8 (96 - x^2 - 4x)\,dx \\
&= 96x - \frac{1}{3}x^3 - 2x^2 \Big|_0^8 = 469.33
\end{aligned}$$

If we want the two surpluses in dollars, we must multiply these figures by a factor of 100 (recall that x represented *hundreds* of suits sold per week). Hence, for a period of one week

consumers surplus = \$50,133
producers surplus = \$46,933 ■

Our next application shows how to express the average value of a function in terms of the definite integral.

ILLUSTRATION Average Value of a Function $f(x)$

Suppose $f(x)$ is a continuous function over $a \le x \le b$. What do we mean by its **average value** over this interval? The idea is certainly used intuitively when we say: "The average temperature in LA on Thursday was 67°F." Here $f(t) =$ (temperature in downtown Los Angeles), $t =$ (hours since midnight), and the average is taken over $0 \le t \le 24$. Another example of this would be a statement such as: "This beach extends 250 feet from the shore and has an average height above sea level of 12 feet," where $f(x)$ is the height of the sand at a distance x feet from the water's edge, and the average is taken over $0 \le x \le 250$.

Consider the beach example in more detail. In Figure 5.19a we show the height profile of a typical beach, with sand height varying irregularly as we move away from the water. Now imagine that the beach is completely leveled by bulldozers, as in Figure 5.19b. It is reasonable to take the constant height $\bar{h}$ of the sand after it has been leveled to be the average height of the original beach. Because bulldozers just move sand around and do not change the cross-sectional area, we have

$$\text{(cross-section area of leveled beach)} = \text{(cross-section area of original beach)}$$

or

$$\bar{h} \times 250 = \int_0^{250} f(x)\,dx$$

so that

$$\text{(average height)} = \bar{h} = \frac{1}{250}\int_0^{250} f(x)\,dx$$

Similar reasoning shows that if x varies from a to b, $a \le x \le b$, rather than from 0 to 250, $0 \le x \le 250$, we would obtain

$$\begin{aligned}(\text{average height over } a \le x \le b) &= (\text{average value of } f(x) \text{ over } a \le x \le b)\\ &= \frac{1}{b-a}\int_a^b f(x)\,dx \qquad [28]\end{aligned}$$

In general the average value of any function $f(x)$ may be thought of as the average height of the region under its graph. The previous discussion shows that this average height is given by an integral [28]. We therefore make the following

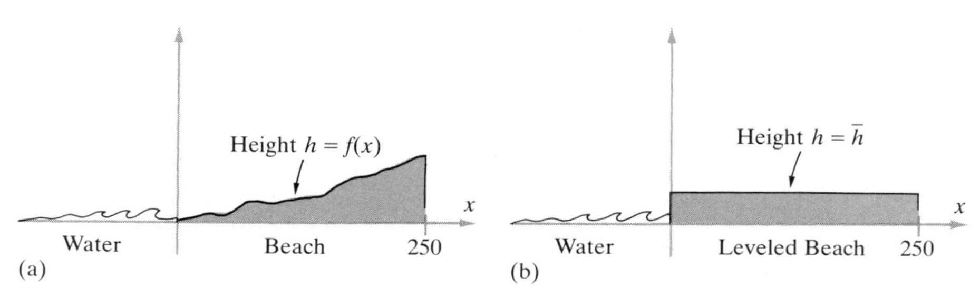

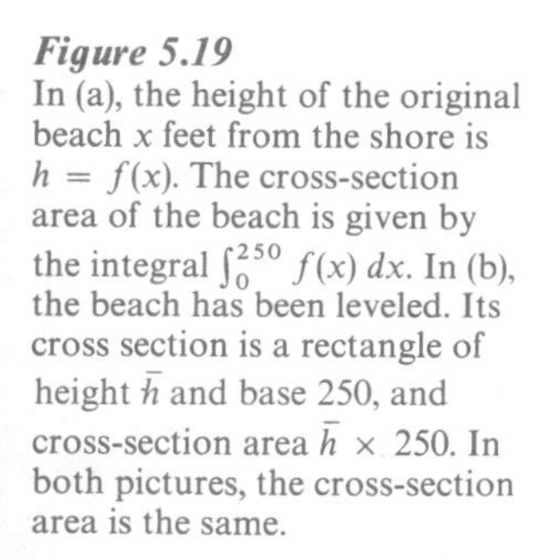

Figure 5.19
In (a), the height of the original beach x feet from the shore is $h = f(x)$. The cross-section area of the beach is given by the integral $\int_0^{250} f(x)\,dx$. In (b), the beach has been leveled. Its cross section is a rectangle of height $\bar{h}$ and base 250, and cross-section area $\bar{h} \times 250$. In both pictures, the cross-section area is the same.

definition:

$$\text{average value of } f(x) \text{ over } a \le x \le b \text{ is } \frac{1}{b-a}\int_a^b f(x)\,dx \qquad [29]$$

Thus, the statement that "the average temperature on Thursday was 67°" means

$$67 = \frac{1}{24}\int_0^{24} f(t)\,dt$$

where $f(t)$ is the temperature t hours into the day, $0 \le t \le 24$.

Here are three examples illustrating this averaging principle.

Example 21 Suppose the temperature during a 10-hour period was

$$T(t) = 72 - 2t + 0.08t^2 \quad (°\text{F})$$

where t = (hours elapsed). What was the average temperature during this time?

Solution Here $a = 0$ and $b = 10$, the whole time period consists of $b - a = 10$ hours, and the average is

$$\begin{aligned}\text{average} &= \frac{1}{10}\int_0^{10} (72 - 2t + 0.08t^2)\,dt \\ &= \frac{1}{10}\left(72t - t^2 + \frac{0.08}{3}t^3 \Big|_0^{10}\right) \\ &= \frac{1}{10}((720 - 100 + 26.7) - (0)) = 64.67°\text{F}\end{aligned}$$

■

Example 22 A beach is 100 feet wide. Its height $h(x)$ at a distance x feet from the water's edge is $h(x) = 0.6\sqrt{x}$ feet, $0 \le x \le 100$. Find the average height of this beach. Then find the average height of the strip of beach corresponding to $36 \le x \le 100$ feet.

Solution For the entire beach we have $a = 0, b = 100$ in Formula [29] and the average height is

$$\begin{aligned}\text{average} &= \frac{1}{100}\int_0^{100} 0.6\sqrt{x}\,dx \\ &= \frac{1}{100}\int_0^{100} 0.6x^{1/2}\,dx \\ &= \frac{1}{100}\left(0.4x^{3/2}\Big|_0^{100}\right) \\ &= \frac{1}{100}((0.4)(100)^{3/2} - (0)) = 4 \text{ feet}\end{aligned}$$

For $36 \leq x \leq 100$, the average height of this strip of beach is

$$\begin{aligned}\text{average} &= \frac{1}{100 - 36}\int_{36}^{100} 0.6\sqrt{x}\,dx \\ &= \frac{1}{64}\left(0.4x^{3/2}\Big|_{36}^{100}\right) \\ &= \frac{1}{64}(400 - 86.4) \\ &= 4.9 \text{ feet}\end{aligned}$$

Example 23 Suppose a country has a population of 100 million at the beginning of 1980 and that the population grows according to the formula

$$N(t) = 100e^{0.015t} \text{ million} \quad (t \text{ in years from 1980})$$

Planners concerned with resource allocation want to know the average population during the next 20 years. Find it.

Solution The average is

$$\begin{aligned}\text{average} &= \frac{1}{20}\int_{0}^{20} 100e^{0.015t}\,dt \\ &= \frac{1}{20}\left(\frac{100}{0.015}e^{0.015t}\Big|_{0}^{20}\right) \\ &= 333.3(e^{0.3} - e^{0}) \\ &= 116.6 \text{ million}\end{aligned}$$

Exercises 5.5

In Exercises 1–4, find the area of the region with upper boundary $y = f(x)$, lower boundary $y = g(x)$, over the interval $a \leq x \leq b$.

1. $f(x) = 7$ (constant function), $g(x) = 3$ (constant function), $a = -1, b = 3$

2. $f(x) = 7 + x$, $g(x) = 3 + x$, $a = 1, b = 3$

3. $f(x) = 7 + x^2$, $g(x) = 3 + x$, $a = 1, b = 3$

4. $f(x) = 7 + x^2$, $g(x) = 3 - x$, $a = -2, b = 2$

In Exercises 5–8, the region is bounded by the curves $y = f(x)$ and $y = g(x)$, from $x = a$ to $x = b$. Determine which curve is the upper boundary and which is the lower. Then find the area of the region.

5. $f(x) = 3 + x$, $g(x) = 7$ (constant function), $a = 0, b = 3$

6. $f(x) = 3 + x - x^2$, $g(x) = 3 + x^2 - 3x$, $a = 0, b = 2$

7. $f(x) = x^2 + 2$, $g(x) = 2x + 2$, $a = 0, b = 2$

8. $f(x) = x^3 + 3$, $g(x) = 3 + 2x - x^2$, $a = -1, b = 0$

In Exercises 9–12, sketch the region bounded by the curves $y = f(x)$ and $y = g(x)$. Then calculate its area. (*Hint*: First locate the crossover points.)

9. $f(x) = x^2, \quad g(x) = x^3$

10. $f(x) = x + 2, \quad g(x) = x^2$

11. $f(x) = \frac{1}{8}x^2, \quad g(x) = \sqrt{x}$

12. $f(x) = 4 - x^2, \quad g(x) = 3$ (constant function)

In Exercises 13–16, find the area of the region bounded by the following curves and vertical lines. If no vertical lines are specified, determine the crossover points for the curves, as in Example 19. (*Note*: In some of these problems the curves can extend below the x axis. Formula [23] applies as long as the upper and lower boundary curves are correctly identified.)

13. $f(x) = -x^2 + 3x - 3, \quad g(x) = x^2 - 2$

14. $f(x) = 2x - 3, \quad g(x) = 4x - x^2$

15. $f(x) = e^{-x}, \quad g(x) = x - 2$, between $x = -1$ and $x = 1$

16. $f(x) = 3x - 4, \quad g(x) = 1 - (2/x)$, between $x = 1$ and $x = 3$

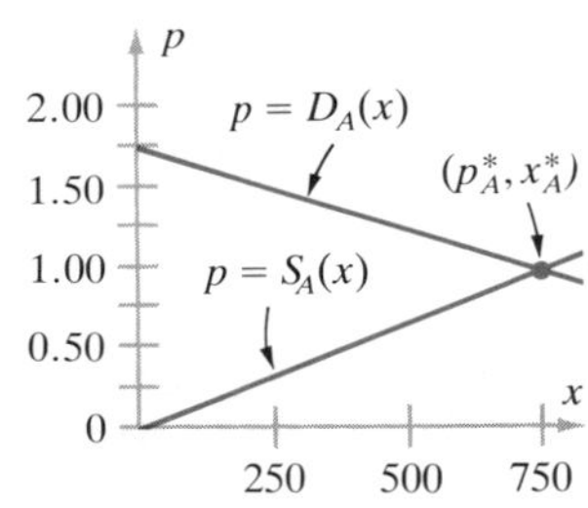

(a) August supply and demand

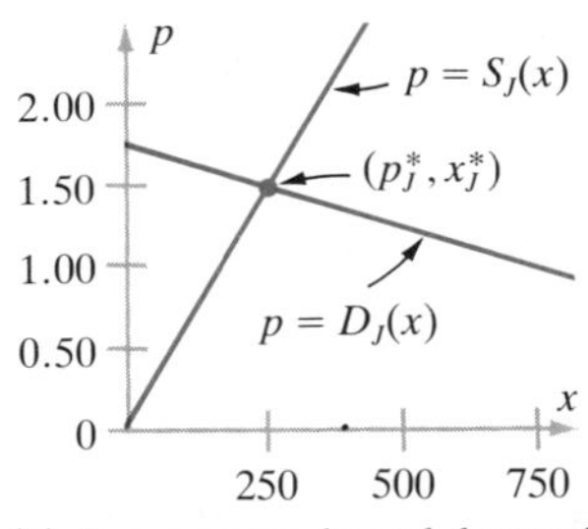

(b) January supply and demand

Figure 5.20
Supply and demand curves for peaches during two different months. The demand curve is the same in each case, but the supply curves are quite different. The equilibrium points shift accordingly (see Exercises 22–26).

In Exercises 17–21, find the coordinates (x^*, p^*) of the equilibrium point for the demand function $p = D(x)$ and the supply function $p = S(x)$.

17. $D(x) = 10 - \frac{x}{2}, \quad S(x) = \frac{x}{2}$

18. $D(x) = 10 - \frac{x}{2}, \quad S(x) = 3 + \frac{x}{5}$

19. $D(x) = 15 - x, \quad S(x) = 3 + \frac{x}{5}$

20. $D(x) = 9 - \sqrt{x}, \quad S(x) = 1 + \sqrt{x}$

21. $D(x) = 960 - \frac{1}{2}x^2, \quad S(x) = \frac{1}{10}x^2$

Exercises 22–26 deal with the price of peaches. Figure 5.20a shows the supply and demand curves for peaches in the markets of Salt Lake City during the month of August:

$$p = D_A(x) = -0.001x + 1.75 \qquad p = S_A(x) = \frac{1}{750}x$$

Here p = price (dollars per pound), x = quantity (thousands of pounds per month). In January, the demand is the same: $D_J(x) = -0.001x + 1.75$; but supplies are harder to come by, leading to a different supply curve shown in Figure 5.20b.

$$p = S_J(x) = \frac{3}{500}x$$

Naturally, the equilibrium points (x^*, p^*) are different in each month.

22. Intuitively, what do you expect to happen to the equilibrium price p^* and quantity sold x^* when you compare August to January? (Do they increase or decrease?) Don't calculate; use your common sense about market prices in August versus those in January.

23. Find the equilibrium price p_A^* and quantity x_A^* in August.

24. Find the equilibrium price p_J^* and quantity x_J^* in January.

25. Find the consumers surplus and producers surplus in August. (*Note*: Because x is measured in *thousands* of pounds, the definite integrals must be multiplied by 1000 to obtain the total dollar value of each surplus (see Example 20).)

26. Find the consumers surplus and producers surplus in January. (*Note*: Because x is measured in *thousands* of pounds, the definite integrals must be multiplied by 1000 to obtain the total dollar value of each surplus (see Example 20).)

27. Suppose the supply and demand curves for sirloin are given by

$$p = D(x) = \frac{1}{40}x - 3 \qquad p = S(x) = \frac{1}{60}x + \frac{1}{2}$$

where p = price (dollars per pound) and x = quantity (millions of pounds per week). Find the consumer surplus and producer surplus. (*Note*: Because x is measured in *millions* of pounds, the definite integrals must be multiplied by 1 million to obtain the dollar value of each surplus (see Example 20).)

28. Nuclear magnetic resonance (NMR) imaging machines are the latest noninvasive diagnostic devices used in hospitals. They are also very expensive. Suppose the yearly demand function for NMR machines is

$$p = D(x) = 1{,}400{,}000 - 100x^2$$

while the supply function is

$$p = S(x) = 740{,}000 + 5000x$$

where x is the number of machines sold in a year. Find the equilibrium point (x^*, p^*). Find the consumers surplus and producers surplus.

29. Repeat Exercise 28, using these demand and supply curves:

$$D(x) = 1{,}400{,}000 - 10{,}000x \qquad S(x) = 600{,}000 + 120x^2$$

30. The Dan–Eryn Motor Company produces custom-made sports cars. The monthly demand function for their cars is $p = D(x) = 100{,}000 - 500x$, and the supply function is $p = S(x) = 30{,}000 + 200x$, where x is the number produced per month. Find the consumers surplus and producers surplus.

31. Repeat Exercise 30 with $D(x) = 1000(45 - x)$ and $S(x) = 10(30 + x)^2$.

In Exercises 32–41, find the average value of $f(x)$ over the interval $a \le x \le b$.

32. $f(x) = 7$ (constant function), $a = -3, b = 12$

33. $f(x) = 4$ (constant function), $a = 0, b = 5$

34. $f(x) = 4 + 3x$, $a = -1, b = 2$

35. $f(x) = 12 - 2x, \quad a = 0, b = 5$

36. $f(x) = \dfrac{1}{x}, \quad a = 1, b = 3$

37. $f(x) = \dfrac{1}{x^2}, \quad a = 1, b = 3$

38. $f(x) = 10 - 2x + 6x^2, \quad a = 0, b = 1$

39. $f(x) = x^3 + 4x + 5, \quad a = -1, b = 1$

40. $f(x) = e^{-x}, \quad a = 0, b = 3$

41. $f(x) = e^{-1.25x}, \quad a = 0, b = 1$

42. The temperature in beautiful downtown Burbank over a 24-hour period was

$$T(t) = 80 + t - \frac{1}{24}t^2 \quad (t = \text{hours from start of the period})$$

Find the average temperature during that time.

43. Repeat Exercise 42, assuming that the temperature was $T(t) = 40 + 0.4t - 0.03t^2$.

44. The relative level of sulfur dioxide pollution in the air of Miami is estimated to be $P(x) = e^{0.07x}$, where x is the number of years from now. Find the average level of pollution over the next 5 years.

45. Repeat Exercise 44, assuming that $P(x) = e^{-0.03x}$.

46. The depth of the Atlantic Ocean x miles due east of Sandy Hook is

$$h(x) = 0.02x + 0.006x^2 \quad \text{(depth in miles)}$$

Find the average ocean depth over the first 10 miles.

47. Repeat Exercise 46, but now find the average depth over the first 15 miles from shore.

5.6 Describing Probability Distributions by Integrals

Another important application of integrals deals with probability distributions. In this section we shall give an elementary, self-contained discussion that does not presume any familiarity with probability or statistical methods.

Suppose we wish to describe, as efficiently as possible, the distribution of heights among adult U.S. citizens in the year 1980. Let N stand for the number of adult citizens, $N \approx 140{,}000{,}000$. A very crude approach would be to label all persons and list their heights. But compiling such a list would be very expensive, and a list with 140,000,000 entries would be unwieldy, to say the least. How then are we to give a concise description of the distribution of heights? We must stop for a moment and see what kind of questions we may hope to answer by referring to statistical data. Here are a few:

A designer of a compact car comfortable only for individuals less than 6 feet tall might ask: "How many people have height $0 \le h \le 6$ feet?" or "What fraction of the total population has height $0 \le h \le 6$?" Clearly the answer to the second question is $1/N$ times the answer to the first.

A buyer of raincoats for a clothing chain would be interested in the fraction of the total population with height in the range $r_1 \le h \le r_2$ for various choices of $r_1 < r_2$, say $5.5 \le h \le 5.75$ or $5.75 \le h \le 6.0$. This way, the number of garments purchased in each size will match the demand for that size.

For a large population we can summarize the statistical facts about height distribution by setting up a function $f(h)$, the **probability–density function** for heights, which allows us to determine the fraction of the population with height in any range $r_1 \le h \le r_2$ by calculating the definite integral $\int_{r_1}^{r_2} f(h)\,dh$.

DEFINING PROPERTY OF THE DENSITY FUNCTION $f(h)$ Given any range of heights $r_1 \le h \le r_2$ with $r_1 < r_2$, the fraction of the population with height in this range is given by

$$\int_{r_1}^{r_2} f(h)\,dh$$

Numerically, the fraction of the population with height $r_1 \le h \le r_2$ is equal to the area under the graph of $y = f(h)$ above the interval $I = [r_1, r_2]$.

In Figure 5.21 we show the graph of $f(h)$; the heavily shaded strip corresponds to the fraction of the population with height $5.0 \le h \le 5.5$. Because few heights lie outside the range $3 \le h \le 8$, we may take $f(h)$ to be defined for $3 \le h \le 8$ without excluding any significant portion of our population.

The description of height distribution by a probability density function applies just as well to any variable associated with a large population of objects

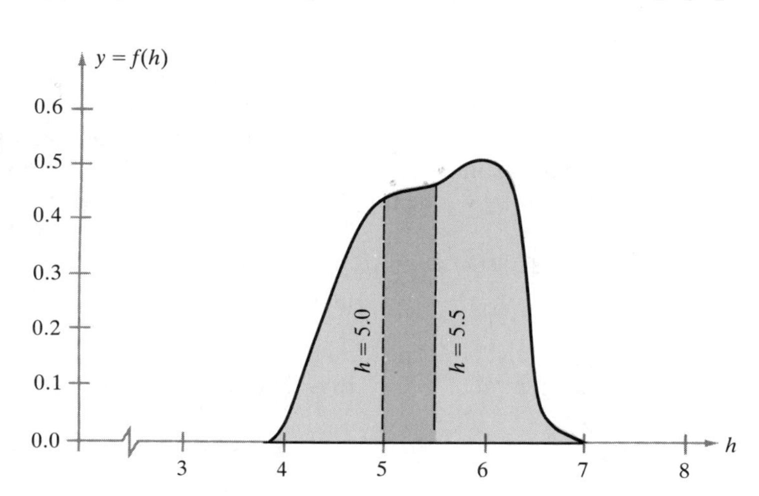

Figure 5.21
Probability density function $f(h)$ representing the distribution of heights among adult U.S. citizens in 1980. For all h, $f(h) \ge 0$. On an interval $r_1 \le h \le r_2$, the definite integral $\int_{r_1}^{r_2} f(h)\,dh$ gives the fraction of the population with height in this range. Because *all* individuals have heights $3 \le h \le 8$, we must have $\int_3^8 f(h)\,dh = 1$; the total area under the graph (entire shaded region) is equal to 1. The heavily shaded strip has area $\int_{5.0}^{5.5} f(h)\,dh$; numerically, this is just the fraction of the population with heights $5.0 \le h \le 5.5$ feet. The two bumps in the graph correspond to the different average heights for men and women.

or individuals. For example we might consider all family units in the United States for the year 1985 and the variable x = (gross family income). The density function $f(x)$ for this variable would be useful in formulating governmental economic and tax policy.

Suppose we have decided upon a variable x to be studied for a certain population of objects, and have chosen a range $a \leq x \leq b$ that includes essentially all values of x. Just as in the analysis of height distribution, there is a probability–density function $f(x)$ that has the property

$$\int_{r_1}^{r_2} f(x)\,dx = \text{fraction of the total population for which } x \text{ lies in the range } r_1 \leq x \leq r_2 \quad [30]$$

for every choice of $r_1 < r_2$. Two general properties of the density function follow directly from this definition:

(i) The density function must be non-negative; $f(x) \geq 0$ everywhere. For any range such as $r_1 \leq x \leq r_2$ the integral $\int_{r_1}^{r_2} f(x)\,dx$ is a positive fraction between 0 and 1. If $f(x)$ were negative, its integral over certain ranges would also be negative, which is impossible.

(ii) The total area $\int_a^b f(x)\,dx$ under the graph is 1. This integral is precisely the fraction of the population for which $a \leq x \leq b$. But *all* values of x fall within this range.

By forming appropriate integrals we may answer many statistical questions involving x, such as the ones posed for height distributions. For example, if r is given

$$\int_a^r f(x)\,dx = \text{fraction of the population such that } a \leq x \leq r$$

Because all objects have $a \leq x$, this is just the fraction of the population with $x \leq r$. Similarly

$$\int_r^b f(x)\,dx = \text{fraction of the population such that } x \geq r$$

because $x \leq b$ for all objects. If we wish to find the *total number* of objects such that x lies in a certain range $r_1 \leq x \leq r_2$, we simply multiply the appropriate fraction by N, the total number of objects in the population under study; thus

$$\begin{aligned} N\int_{r_1}^{r_2} f(x)\,dx &= (\text{total number of objects}) \times (\text{fraction with } r_1 \leq x \leq r_2) \\ &= \text{total number of objects such that } r_1 \leq x \leq r_2 \end{aligned}$$

Example 24 The "uniform-distribution" for a variable whose values fall in the interval $0 \leq x \leq 10$ is described by the probability density function

$$f(x) = \frac{1}{10} \quad (\text{constant}) \quad \text{for } 0 \leq x \leq 10$$

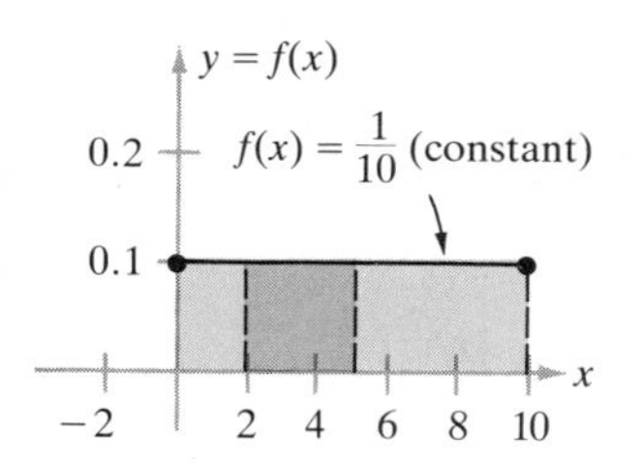

Figure 5.22
The "uniform" probability density function in Example 24. All observed values of x fall in the range $0 \le x \le 10$, so the density function is zero outside this interval. Within the interval, all x values are equally likely, so the density function is constant for $0 \le x \le 10$. Because the total area under the graph must equal 1, as for any probability density function, the constant value of $f(x)$ on this interval must be $1/10 = 1/(\text{length of interval})$. Heavily shaded area corresponds to the fraction of the population with x values in the range $2 \le x \le 5$.

as shown in Figure 5.22. (Constancy of the density function means that all values of x are equally likely to turn up if we measure x for randomly selected members of the population being studied.) For what fraction of the population does x lie in the range $2 \le x \le 5$? If a, b are such that $0 \le a \le b \le 10$, calculate the fraction of the population with x values in the range $a \le x \le b$.

Solution The fraction of the population for which x lies in the range $2 \le x \le 5$ is given by

$$\int_2^5 f(x)\,dx = \int_2^5 \frac{1}{10}\,dx = \frac{1}{10}x\Big|_2^5$$
$$= \left(\frac{5}{10}\right) - \left(\frac{2}{10}\right) = \frac{3}{10} = 0.30$$

Similarly, the fraction in the range $a \le x \le b$ is

$$\int_a^b f(x)\,dx = \int_a^b \frac{1}{10}\,dx = \frac{1}{10}x\Big|_a^b = \frac{b-a}{10}$$

In particular, if we set $a = 0$, $b = 10$ (thereby including all possible values of x for this population), the preceding formula gives the fraction $(b - a)/10 = 10/10 = 1$. This is in agreement with the property that the total area under the graph of a density function must equal 1. ■

The next example has a different probability density function defined for $0 \le x \le 10$. We will see that the fraction of the population having x values in the range $2 \le x \le 5$ changes accordingly.

Example 25 The graph of the probability density function

$$f(x) = (0.0012)(x^3 - 20x^2 + 100x) = 0.0012x^3 - 0.024x^2 + 0.12x$$

defined for $0 \le x \le 10$, is shown in Figure 5.23. Calculate the fraction of the population such that the variable x lies in the range $2 \le x \le 5$, corresponding to the shaded region under the graph. Verify that the total area under the graph is equal to 1.

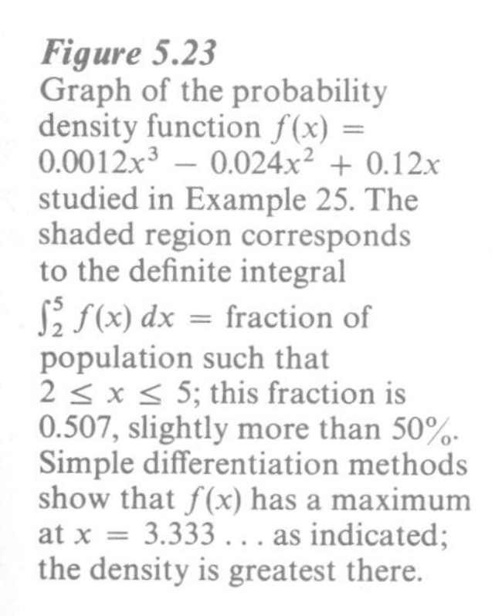
Figure 5.23
Graph of the probability density function $f(x) = 0.0012x^3 - 0.024x^2 + 0.12x$ studied in Example 25. The shaded region corresponds to the definite integral $\int_2^5 f(x)\,dx$ = fraction of population such that $2 \le x \le 5$; this fraction is 0.507, slightly more than 50%. Simple differentiation methods show that $f(x)$ has a maximum at $x = 3.333\ldots$ as indicated; the density is greatest there.

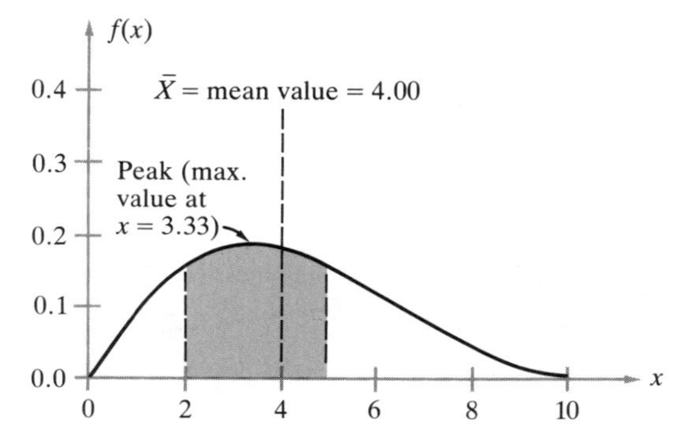

Solution The fraction of the population such that $2 \le x \le 5$ is given by

$$\int_2^5 f(x)\,dx = (0.0012)\int_2^5 (x^3 - 20x^2 + 100x)\,dx$$
$$= (0.0012)\left(\frac{1}{4}x^4 - \frac{20}{3}x^3 + 50x^2\Big|_2^5\right)$$
$$= (0.0012)\cdot(422.3) = 0.507$$

by definition of the probability density function. An easy calculation using the same indefinite integral shows that

$$\text{total area} = \int_0^{10} f(x)\,dx = 1$$

■

Example 26 Insurance companies are interested in how long an automobile will be on the road once it is purchased: its "lifetime." Suppose statisticians have determined that the lifetimes of a certain model manufactured in 1980 are governed by the probability density function

$$f(x) = \frac{1}{4.5}e^{-x/4.5} \quad \text{(lifetime } x \text{ in years from purchase)}$$

What fraction of these 1980 models will cease to be operational during the first 3 years? What fraction will be operational for at least 3 years?

Solution The graph of this probability density function is shown in Figure 5.24. The fraction that has lifetimes $0 \le x \le 3$ years (these are the ones that cease working during the first 3 years) is given by the shaded area in the figure:

$$\text{(fraction with lifetimes } 0 \le x \le 3 \text{ years)} = \int_0^3 f(x)\,dx = \int_0^3 \frac{1}{4.5}e^{-x/4.5}\,dx$$

An indefinite integral for $f(x) = (1/4.5)e^{-x/4.5}$ is $F(x) = -e^{-x/4.5}$, so the area is

$$\int_0^3 f(x)\,dx = F(x)\Big|_0^3 = -e^{-x/4.5}\Big|_0^3$$
$$= (-e^{-3/4.5}) - (-e^{-0/4.5})$$
$$= 1 - 0.5134 = 0.4866$$

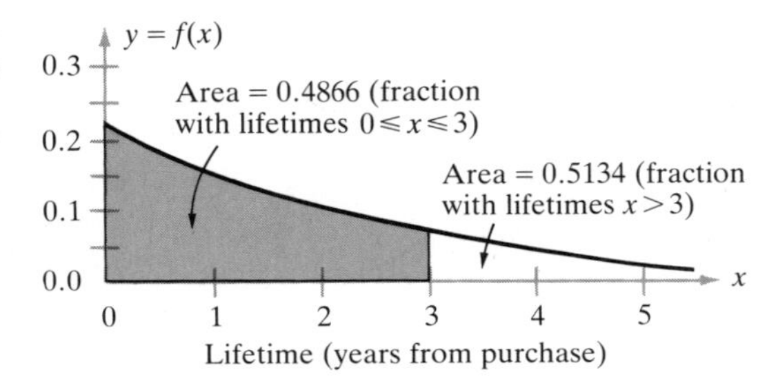

Figure 5.24
Probability density function for automobile lifetimes x = (years operating after purchase). Those with lifetimes $0 \le x \le 3$ (the cars that *fail* sometime during the first 3 years of operation) correspond to the shaded area. The rest, with lifetimes $x > 3$, correspond to the unshaded region. Total area under the graph is 1.

About 49% fail during the first 3 years. Because *every* car sold either fails during the first 3 years or has a lifetime of at least 3 years, we can answer the last question without further integration:

$$\begin{aligned}&(\text{fraction with lifetimes at least 3 years})\\&\quad= 1 - (\text{fraction with lifetimes } 0 \le x \le 3 \text{ years})\\&\quad= 1 - 0.4866 = 0.5134\end{aligned}$$

■

In statistics certain numbers are helpful in describing density functions. The most important of these is the **mean** or **mean value** of the distribution, and second in importance is the **standard deviation**. These numbers are obtained from definite integrals involving the probability density function $f(x)$.

THE MEAN VALUE $\bar{X}$ Intuitively, the mean value is the number about which the values x cluster. Suppose there are N objects and that the values of x have been compiled in a list $x_1, x_2, \ldots, x_N$. The **mean value** of x for this population of objects is defined by giving equal weight $1/N$ to each of the values, and forming the average

$$\bar{X} = \frac{1}{N}\cdot(x_1 + x_2 + \cdots + x_N) \qquad [31]$$

If N is very large this sum is difficult to calculate directly. For a large population with known density function $f(x)$ the mean value is given by the integral

$$\bar{X} = \int_a^b x \cdot f(x)\,dx \qquad [32]$$

THE STANDARD DEVIATION σ Once the mean value $\bar{X}$ has been calculated, the standard deviation is defined as

$$\sigma = \sqrt{\frac{(x_1 - \bar{X})^2 + (x_2 - \bar{X})^2 + \cdots + (x_N - \bar{X})^2}{N}} \qquad [33]$$

The standard deviation σ is a positive number that measures the scatter of the values x_i away from the mean value $\bar{X}$. For large populations with known density function $f(x)$, the standard deviation is given by the integral

$$\sigma = \sqrt{\int_a^b (x - \bar{X})^2 \cdot f(x)\,dx} \qquad [34]$$

The numbers $\bar{X}$ and σ tell us a lot about a population. The values of x cluster around the mean value $\bar{X}$. If σ is small they are concentrated close to $\bar{X}$, but if σ is large they are more spread out.

Example 27 Calculate the mean value $\bar{X}$ and standard deviation σ for the probability–density function given in Example 26.

Solution We calculate the mean value according to definition:

$$\begin{aligned}
\bar{X} &= \int_0^{10} x \cdot f(x)\,dx \\
&= (0.0012)\int_0^{10} (x^4 - 20x^3 + 100x^2)\,dx \\
&= (0.0012)\left(\frac{1}{5}x^5 - 5x^4 + \frac{100}{3}x^3\Big|_0^{10}\right) \\
&= 4.00
\end{aligned}$$

The location of $\bar{X}$ is shown in Figure 5.23.

In calculating the standard deviation σ, it helps to avoid entering the actual value $\bar{X} = 4.00$ until the very last step:

$$\begin{aligned}
\sigma^2 &= \int_0^{10} (x - \bar{X})^2 f(x)\,dx \\
&= (0.0012)\int_0^{10} (x^2 - 2\bar{X}x + \bar{X}^2)(x^3 - 20x^2 + 100x)\,dx \\
&= (0.0012)\int_0^{10} (x^5 - (20 + 2\bar{X})x^4 + (100 + 40\bar{X} + \bar{X}^2)x^3 \\
&\qquad - (200\bar{X} + 20\bar{X}^2)x^2 + (100\bar{X}^2)x)\,dx \\
&= (0.0012)\left(\frac{x^6}{6} - \frac{20 + 2\bar{X}}{5}x^5 + \frac{100 + 40\bar{X} + \bar{X}^2}{4}x^4\right. \\
&\qquad \left. - \frac{200\bar{X} + 20\bar{X}^2}{3}x^3 + \frac{100\bar{X}^2}{2}x^2\Big|_0^{10}\right)
\end{aligned}$$

Upon setting $\bar{X} = 4.00$ we get $\sigma^2 = 4.0$, so that $\sigma = 2.0$. ■

Exercises 5.6

In Exercises 1–6, the probability density functions all have common range $1 \le x \le 4$. For each density function $f(x)$, find the fraction of the population for which the variable x has values lying in the range $1 \le x \le 2$.

1. $f(x) = \frac{1}{3}$ (constant)

2. $f(x) = \dfrac{2x}{15}$

3. $f(x) = \dfrac{1}{21}x^2$

4. $f(x) = \left(\dfrac{1}{e^4 - e}\right) \cdot e^x$

5. $f(x) = \dfrac{1}{\ln 4} \cdot \dfrac{1}{x}$

6. $f(x) = \dfrac{2}{9}(x - 1)$

In Exercises 7–10, explain why the given function $f(x)$ over the indicated range of values for x is *not* a probability density function. (*Hint*: A density function must be non-negative over its range, and the area under its graph must equal 1.)

7. $f(x) = x \quad \text{for } 0 \le x \le 1$

8. $f(x) = \dfrac{3 - x}{4} \quad \text{for } 0 \le x \le 4$

9. $f(x) = x^2 \quad \text{for } -1 \le x \le 2$

10. $f(x) = x^3 \quad \text{for } -1 \le x \le 2$

In Exercises 11–18, a variable x is restricted to the interval $1 \le x \le 3$. Over that interval, its probability density function $f(x)$ has the form as follows, involving an undetermined constant k. Find the appropriate value of k. (*Hint*: The density function must satisfy the condition $\int_1^3 f(x)\,dx = 1$.)

11. $f(x) = k$ (constant function)

12. $f(x) = kx$

13. $f(x) = kx^2$

14. $f(x) = \dfrac{k}{x}$

15. $f(x) = ke^{-x}$

16. $f(x) = kx^3$

17. $f(x) = k(3 - x)$

18. $f(x) = \dfrac{x - 1}{6} + k$

Exercises 19–22 refer to the probability density function $f(x)$ discussed in Example 25,

$$f(x) = (0.0012)(x^3 - 20x^2 + 100x)$$

For what fraction of the population does the variable x lie in the range indicated?

19. $0 \le x \le 5$

20. $5 \le x \le 10$

21. $0 \le x \le 3$

22. $7 \le x \le 10$

23. A variable x measures a certain property of a large population of objects. The values of x for this population are found to lie between -1 and $+1$; the probability density function for this variable has been determined,

$$f(x) = \frac{3}{4}(1 - x^2) \quad \text{for } -1 \le x \le 1$$

Calculate the fraction of the population for which x lies in the range $-0.25 \le x \le 0.25$. If there are 100,000 objects in the population, how many of them have x values in this range?

24. The city of Evanston is considering a tax on all incomes above \$20,000 per year. Suppose that the annual income x (in thousands of dollars) of Evanston residents lies in the range $10 \le x \le 100$, with probability density function

$$f(x) = \frac{1}{4050}(100 - x) \quad \text{for } 10 \le x \le 100$$

Find the fraction of the population subject to the proposed tax (that is, the fraction for which $x \ge 20$).

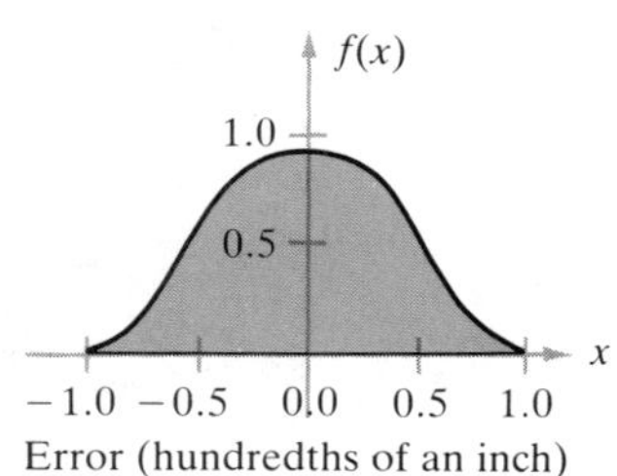
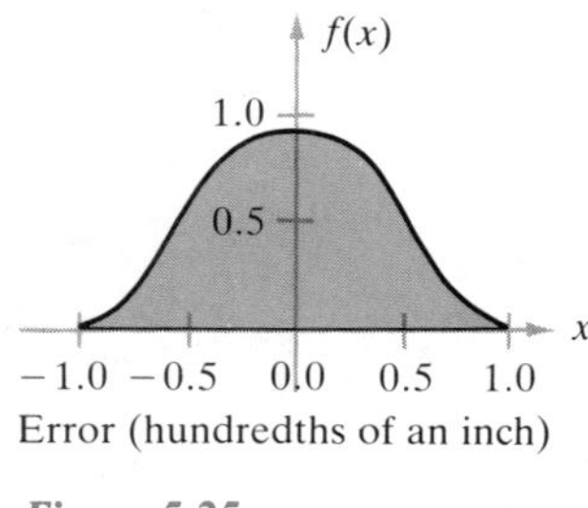

Figure 5.25
The probability density function $f(x) = (15/16)(x^2 - 1)^2$, defined for $-1 \le x \le 1$, as in Exercise 25. This function is equal to zero for x outside the range $-1 \le x \le 1$.

25. A precision bolt with diameter $D = 3.000$ inches is to be produced with a maximum allowable machining error of $\Delta D = \pm 0.005$ inches. From raw statistical data, the quality-control department has found that the variable

$$x = (\text{error (or deviation) from design diameter})$$

(measured in 1/100ths of an inch) has the probability–density function

$$f(x) = \begin{cases} \dfrac{15}{16}(x^2 - 1)^2 & \text{for } -1 \le x \le 1 \\ 0 & \text{elsewhere} \end{cases}$$

shown in Figure 5.25.
(i) Show that $\int_{-1}^{1} f(x)\,dx = 1$, as required of a density function.
(ii) What fraction of the bolts produced will have machining errors within the acceptable range $-0.5 \le x \le +0.5$ (hundredths of an inch)?

26. The function $y = e^{-x}$ occurs in a number of important density functions. Consider the density function $f(x) = 1.0067e^{-x}$, defined for $0 \le x \le 5$. Use the integration formula

$$\int xe^{-x}\,dx = -xe^{-x} - e^{-x} + \text{const}$$

to calculate the mean value $\bar{X}$ for this distribution. For what fraction of the population does x lie in the range $0 \le x \le 1$? (*Note*: The formula may be verified by direct differentiation.)

In Exercises 27–30, find the mean $\bar{X}$ and the standard deviation σ for the variable x whose probability density function is given.

27. $f(x) = 1$ (constant) for $0 \le x \le 1$

28. $f(x) = 1$ (constant) for $5 \le x \le 6$

29. $f(x) = \frac{1}{3}$ (constant) for $0 \le x \le 3$

30. $f(x) = 2x$ for $0 \le x \le 1$

31. A variable x associated with a large population has values in the interval $0 \le x \le 2$ and is described by a probability–density function of the form $f(x) = k \cdot x^3$, where k is a constant.
(i) Find k (to ensure that $\int_0^2 f(x)\,dx = 1.0$).
(ii) Find the fraction of the population with $0 \le x \le 1$.
(iii) Find the mean $\bar{X}$ for this population.
(iv) Find the standard deviation σ.

Figure 5.26
Graph of the probability density function $f(x) = 30(x^4 - x^5)$. Maximum value of $f(x)$ occurs at $x = X_0$.

32. Consider a variable x whose values lie in the interval $0 \le x \le 1$. Suppose the probability density function for this variable is

$$f(x) = 30(x^4 - x^5) \quad \text{(graph shown in Figure 5.26)}$$

Show that the total area under the graph is actually equal to one:

$$\int_0^1 f(x)\,dx = 1$$

Then find
(i) the mean value $\bar{X}$
(ii) the value $x = X_0$ at which $f(x)$ has its maximum value

(iii) the standard deviation σ
Do $\bar{X}$ and X_0 coincide?

33. Cornpone Cereal is machine packed in 1-pound boxes. Boxes with less than 16 ounces of cereal after filling are automatically rejected. Let the variable x represent the actual number of ounces in a box. Because machines are not perfect, we expect to find a spread of values for x during a day's run. Suppose x has probability density function

$$f(x) = \frac{3}{4}(1 - (16.4 - x)^2) \quad \text{for } 15.4 \leq x \leq 17.4$$

Find
(i) the fraction of boxes passing weight inspection ($x \geq 16$)
(ii) the mean weight $\bar{X}$
(iii) the standard deviation σ

34. Repeat Exercise 33, assuming that x has probability density function

$$f(x) = 6\left(\frac{1}{4} - (16.4 - x)^2\right) \quad \text{for } 15.9 \leq x \leq 16.9$$

(*Note*: Comparison of the results in Exercises 33 and 34 shows the advantage of reducing the value of σ even if the mean value $\bar{X}$ remains the same.)

Checklist of Key Topics

Indefinite integral $\int f(x)\,dx$
Application: reconstructing $f(x)$ given its rate of change
Definite integral $\int_a^b f(x)\,dx$
Areas and definite integrals
Fundamental Theorem of Calculus
Approximate calculation of areas and definite integrals
Application: area between two curves
Application: consumers and producers surplus
Application: average value of a function
Probability density functions
Mean value and standard deviation

Integration formulas:

$$\int x^r\,dx = \frac{x^{r+1}}{r+1} + c \quad (\text{if } r \neq -1)$$

$$\int \frac{1}{x}\,dx = \ln x + c$$

$$\int e^{kx}\,dx = \frac{1}{k}e^{kx} + c \quad (k \neq 0 \text{ any constant})$$

$$\int k_1 f(x) + k_2 g(x)\,dx = k_1 \int f(x)\,dx + k_2 \int g(x)\,dx$$

Chapter 5 Review Exercises

In Exercises 1–12, find the indefinite integral.

1. $\int (3x + 5)\, dx$

2. $\int \left(4 - \frac{x}{3}\right) dx$

3. $\int \left(2 - \frac{x^2}{3} + 12x^5\right) dx$

4. $\int \left(2 - \frac{x^2}{3}\right)^2 dx$

5. $\int \frac{2x - 1}{x}\, dx$

6. $\int \frac{2x - 1}{\sqrt{x}}\, dx$

7. $\int \frac{x^2 + 2x + 3}{x^3}\, dx$

8. $\int \frac{x^2 + 2x + 3}{\sqrt{x}}\, dx$

9. $\int \sqrt{x}\,(1 - x^2)\, dx$

10. $\int (1 - e^{-7t})\, dt$

11. $\int (e^{0.2t} - 2.5t)\, dt$

12. $\int 47.5e^{-1.257t}\, dt$

In Exercises 13–20, evaluate the definite integral.

13. $\int_{-1}^{2} 5\, dx$

14. $\int_{1}^{2} -5\, dx$

15. $\int_{2}^{4} (6 - 2x)\, dx$

16. $\int_{2}^{4} x(6 - 2x)\, dx$

17. $\int_{-1}^{1} (10s^4 - 6s + 7)\, ds$

18. $\int_{1}^{2} \frac{t - 2}{t^3}\, dt$

19. $\int_{-2}^{3} e^{-4x}\, dx$

20. $\int_{0}^{5} e^{4x}\, dx$

21. An oil field is depleting rapidly. It is estimated that the future yearly rate of production will be $r(t) = 2e^{-0.3t}$ million barrels per year, where $t = 0$ corresponds to the present time. Find the total amount of oil expected in the next decade ($0 \le t \le 10$).

22. The acceleration of a certain racing car starting from scratch is measured to be $a(t) = 12(1 - 0.02t)$ feet per second^2. Find the velocity after 10 seconds. Then find the distance covered in the first 10 seconds.

23. Find the weekly cost function $C(x)$ if you know that the marginal cost function is $C'(x) = 24 - 0.01x + 0.06x^2$ and the fixed costs are \$3600 per week.

In Exercises 24–29 find the area of the indicated region.

24. Region above the x axis, below the curve $y = x^{1/3}$, and between the vertical lines $x = 1$ and $x = 8$.

25. Region above the x axis, below the curve $y = 3 + (1/x^2)$, and between the vertical lines $x = 1$ and $x = 3$.

26. Region above the curve $y = 1/x$, below the curve $y = e^x$, and between the vertical lines $x = 1$ and $x = 5$.

27. Region above the curve $y = -\frac{1}{2}x$, below the curve $y = 1 + x^4$, and between the vertical lines $x = -2$ and $x = 1$.

28. Region bounded by the curves $y = x + 1$ and $y = x^2 + x$.

29. Region bounded by the curves $y = x^2 + x - 13$ and $y = 15 - 2x$.

In Exercises 30–31 interpret the integral as an area, making a simple sketch. Then compute the approximating areas of the smaller and larger rectangular regions corresponding to subdivision of the base interval into n equal parts.

30. $\int_1^3 \frac{e^x}{x^2}\,dx \quad (n = 8)$

31. $\int_0^1 \sqrt{1 - x^3}\,dx \quad (n = 5)$

32. The number of unemployed workers is predicted to be

$$U(t) = 9{,}000{,}000 - 800{,}000t + 70{,}000t^2$$

t years from now. Find the average number of unemployed in the next 8 years.

33. Repeat Exercise 32, taking $U(t) = 3{,}000{,}000e^{-t} + 6{,}000{,}000e^{0.05t}$.

34. Observers have found that the time t (in minutes) that a player spends on a certain video game is governed by the probability density function $f(t) = \frac{1}{15}e^{-t/15}$ for $t \geq 0$.
(i) Find the fraction of players that spend between 10 and 20 minutes ($10 \leq t \leq 20$) at the game.
(ii) Find the fraction that play more than 15 minutes ($t \geq 15$).

35. Suppose the distribution of values for a variable x associated with a large population is governed by the probability density function

$$f(x) = \frac{3}{4}(1 - (x - 3)^2) \quad 2 \leq x \leq 4$$

Calculate
(i) the fraction of the population for which x lies between 3 and 3.5
(ii) the mean value $\bar{X}$
(iii) the standard deviation σ

CHAPTER 6

Further Topics on Integration

6.1 Advanced Integration Techniques: Substitution

Almost every differentiation rule may be turned around and reinterpreted as a statement about indefinite integrals. This leads to new techniques for evaluating complicated integrals. We shall explain one of these, the **substitution method**, which is obtained by inverting the chain rule. In this method we reduce the problem of evaluating an indefinite integral $\int f(x)\,dx$ with variable x to the evaluation of a completely different integral $\int g(u)\,du$, with variable u, which may be easier to handle.

We start by introducing a formal notation. If u is a function of x, say $u = u(x)$, we let the symbol du stand for the expression $(du/dx)\,dx$, that is

$$du = u'(x)\,dx \qquad [1]$$

For example, if $u = x + 1$ then

$$u'(x) = 1 \quad \text{and} \quad du = u'(x)\,dx = dx$$

Similarly, if $u = 1 - x^2$, then

$$u'(x) = -2x \quad \text{and} \quad du = u'(x)\,dx = -2x\,dx$$

We use this notation in describing the "substitution" procedure.

THE SUBSTITUTION METHOD Given the integral $\int f(x)\,dx$, we look for a block of terms $u = u(x)$ and a function $g(u)$ such that

$$f(x)\,dx = g(u)\,du \qquad [2]$$

Here $du = u'(x)\,dx$ as in [1] and $g(u)$ stands for $g(u(x))$, the composite of $g(u)$ and $u(x)$. Once $f(x)\,dx$ has been written in this form we obtain

$$\int f(x)\,dx = \int g(u)\,du\Big|_{u=u(x)} \qquad [3]$$

That is, we obtain $\int f(x)\,dx$ by evaluating a new integral $\int g(u)\,du$ as a function of u and then replacing u with $u(x)$.

In practice, we first find a block of terms $u = u(x)$ that would make things look simpler if it were consolidated. With $u(x)$ in hand, we peel off $du = u'(x)\,dx$ from the expression $f(x)\,dx$. What remains has to be expressed in terms of u; this gives the function $g(u)$. By substituting u for x we convert $\int f(x)\,dx$ into a new integral $\int g(u)\,du$, which we may be able to integrate.

The next examples illustrate the procedure.

Example 1 Evaluate $\int (x+1)^8\,dx$.

Solution Here $f(x) = (x+1)^8$, and it seems reasonable to let $u = x + 1$. Then $du = dx$ and the expression $f(x)\,dx$ in [2] take the form

$$f(x)\,dx = (x+1)^8\,dx = (x+1)^8\,du$$

Peeling off du leaves the expression $(x+1)^8$, which in terms of u becomes u^8. Thus the choice $u = x + 1$, $g(u) = u^8$ satisfies condition [2]:

$$f(x)\,dx = (x+1)^8\,dx = u^8\,du = g(u)\,du$$

Now we apply [3] with $u = x + 1$, $g(u) = u^8$ to obtain

$$\begin{aligned}\int (x+1)^8\,dx &= \int u^8\,du\Big|_{u=x+1}\\ &= \frac{1}{9}u^9 + \text{const}\Big|_{u=x+1} \qquad \left(\text{because } \int u^8\,du = \frac{1}{9}u^9 + c\right)\\ &= \frac{1}{9}(x+1)^9 + \text{const}\end{aligned}$$

We can check our answer: Differentiating $\frac{1}{9}(x+1)^9$ by the chain rule we obtain $(x+1)^8$ as expected. ■

Example 2 Evaluate $\displaystyle\int \frac{x}{\sqrt{1-x^2}}\,dx$.

Solution The form of the integral suggests that we consolidate the block of terms $u = 1 - x^2$. Then

$$du = -2x\,dx \quad \text{or} \quad -\frac{1}{2}\,du = x\,dx$$

Substituting $u = 1 - x^2$ and $-\frac{1}{2}\,du = x\,dx$ in $f(x)\,dx$, and writing what is left in terms of u, we obtain

$$\frac{x}{\sqrt{1 - x^2}}\,dx = (1 - x^2)^{-1/2} \cdot x\,dx$$

$$= u^{-1/2} \cdot \left(-\frac{1}{2}\,du\right) = -\frac{1}{2}u^{-1/2}\,du$$

We now apply the substitution formula [3]:

$$\int \frac{x}{\sqrt{1 - x^2}}\,dx = \int -\frac{1}{2}u^{-1/2}\,du\Big|_{u=1-x^2}$$

$$= -u^{1/2} + \text{const}\Big|_{u=1-x^2}$$

$$= -\sqrt{1 - x^2} + \text{const}$$ ■

The next example shows how the method can be applied to definite integrals.

Example 3 Evaluate $\displaystyle\int_1^3 \frac{(\ln x)^2}{3x}\,dx$.

Solution We first determine the indefinite integral

$$\int \frac{(\ln x)^2}{3x}\,dx$$

It seems reasonable to try $u = \ln x$, especially because $du = u'(x)\,dx = (1/x)\,dx$ already appears in this integral. Then, substituting u for $\ln x$ and du for $(1/x)\,dx$, we find that

$$\int \frac{(\ln x)^2}{3x}\,dx = \int \frac{1}{3}u^2\,du\Big|_{u=\ln x}$$

$$= \frac{1}{9}u^3 + \text{const}\Big|_{u=\ln x}$$

$$= \frac{1}{9}(\ln x)^3 + \text{const}$$

Therefore, the definite integral we seek is

$$\int_1^3 \frac{(\ln x)^2}{3x}\,dx = \frac{1}{9}(\ln x)^3\Big|_1^3 = \frac{(\ln 3)^3}{9} - \frac{(\ln 1)^3}{9}$$
$$= 0.1473$$

Remember: $\ln 1 = 0$. ■

The key to the substitution method is the choice of the function $u = u(x)$ in the first place. Making the proper choice is a skill that improves with practice; the objective is to choose $u = u(x)$ so the new integral $\int g(u)\,du$ is simpler than the original integral $\int f(x)\,dx$. Although substitution is a technique that greatly extends our ability to integrate, and is well worth learning, it does not always help. Sometimes, no matter how reasonable the choice of $u = u(x)$, the new integral is no easier to evaluate than the original. For example, the substitution method is not helpful in evaluating $\int x \ln x\,dx$. You might try the substitution $u = \ln x$ to convince yourself of the difficulty. (For a different approach, which does evaluate this integral, we refer the reader to Section 6.2.)

We conclude this section with a brief justification of the substitution formula [3]. Suppose a substitution $u = u(x)$, $du = u'(x)\,dx$ converts $f(x)\,dx$ into $g(u)\,du$. This is equivalent to saying that

$$f(x) = g(u)\frac{du}{dx} = g(u(x))\frac{du}{dx} \qquad [4]$$

Now let

$$G(u) = \int g(u)\,du \qquad [5]$$

That is, $G(u)$ is an indefinite integral of $g(u)$, so that $dG/du = g(u)$. We want to show that formula [3]

$$\int f(x)\,dx = \int g(u)\,du\Big|_{u=u(x)} = G(u(x))$$

is valid. By definition of the indefinite integral $\int f(x)\,dx$, this amounts to saying that the derivative of the composite function $G(u(x))$ is equal to $f(x)$. Let's check it: Differentiating $G(u(x))$ by the chain rule, we obtain

$$\frac{d}{dx}(G(u(x))) = \frac{dG}{du}\cdot\frac{du}{dx} \quad \text{(by the chain rule)}$$
$$= g(u)\cdot\frac{du}{dx} \quad \text{(by definition of } G \text{ in [5])}$$
$$= f(x) \quad \text{(by [4])}$$

as required.

Exercises 6.1

1. If $f(x) = 2x(4 + x^2)^5$, show that $f(x)\,dx = u^5 du$ if we make the substitution $u = 4 + x^2$. Then apply the substitution formula to evaluate the indefinite integral

$$\int 2x(4 + x^2)^5\,dx$$

Check your answer by differentiation.

2. If

$$f(x) = \frac{x^2}{\sqrt{1 - x^3}}$$

show that $f(x)\,dx = -\frac{1}{3}u^{-1/2}\,du$ if we take $u = 1 - x^3$. Then apply the substitution formula to evaluate

$$\int \frac{x^2}{\sqrt{1 - x^3}}\,dx$$

In Exercises 3–10, determine the indefinite integral by the method of substitution, and check your answer by differentiation.

3. $\int (x + 2)^3\,dx$

4. $\int 2x\sqrt{x^2 + 4}\,dx$

5. $\int \sqrt{x + 2}\,dx$

6. $\int (3x + 2)^{-3}\,dx$

7. $\int \frac{1}{2x + 5}\,dx$

8. $\int e^{t+3}\,dt$

9. $\int \frac{\ln x}{x}\,dx$

10. $\int (x^3 - 3x + 1)^7(x^2 - 1)\,dx$

In Exercises 11–18, calculate the indefinite integral

11. $\int (x^2 + 1)^{22}x\,dx$

12. $\int x\sqrt{x^2 + 4}\,dx$

13. $\int x^2\,(1 - x^3)^{2/3}\,dx$

14. $\int \frac{1}{(2 - x)^3}\,dx$

15. $\int \frac{t^3}{1 + t^4}\,dt$

16. $\int \frac{x}{(3 - x^2)^7}\,dx$

17. $\int \frac{(\ln t)^2}{t}\,dt$

18. $\int e^{-2x+1}\,dx$

In Exercises 19–26, calculate the definite integral using the method of substitution.

19. $\int_1^5 \sqrt{5 - x}\,dx$

20. $\int_0^2 \frac{x}{\sqrt{x^2 + 1}}\,dx$

21. $\int_{-2}^{-1} \frac{1}{(x + 3)^5}\,dx$

22. $\int_0^1 \frac{x + 1}{3x^2 + 6x + 5}\,dx$

23. $\displaystyle\int_0^1 \frac{x}{(4-3x^2)^2}\,dx$

24. $\displaystyle\int_0^2 \frac{1}{2x+5}\,dx$

25. $\displaystyle\int_0^2 (2t+1)(t^2+t)^{3/2}\,dt$

26. $\displaystyle\int_{-a}^{a} xe^{-x^2/2}\,dx$

In Exercises 27—37, determine the indefinite integral.

27. $\displaystyle\int \frac{e^x}{2+e^x}\,dx$

28. $\displaystyle\int \frac{1}{x(1+\ln x)}\,dx$

29. $\displaystyle\int \frac{1-t^2}{(t^3-3t+1)^8}\,dt$

30. $\displaystyle\int \left(1+\frac{1}{x+2}\right)dx$

31. $\displaystyle\int \frac{1}{x^2}e^{1/x}\,dx$

32. $\displaystyle\int \frac{1}{2x+3}\,dx$

33. $\displaystyle\int \frac{1}{x\ln x}\,dx$

34. $\displaystyle\int xe^{1-x^2}\,dx$

35. $\displaystyle\int \frac{1}{e^{-x}-1}\,dx \quad \left(\textit{Hint}: \frac{1}{e^{-x}-1}=\frac{e^x}{1-e^x}\right)$

36. $\displaystyle\int \frac{x}{(1-x)^9}\,dx$ (*Hint*: Let $u=1-x$; then $x=1-u$.)

37. $\displaystyle\int (3x+2)^{16}x\,dx$ (*Hint*: If $u=3x+2$, then $x=(u-2)/3$.)

6.2 Advanced Integration Methods: Integration by Parts

In the previous section, we discussed the method of substitution, which turns around the chain rule for derivatives to get a rule for finding indefinite integrals. Here we will show how to invert the product formula for derivatives to obtain another integration technique, the method of **integration by parts**.

Recall the product formula for derivatives: if $u(x)$ and $v(x)$ are given, the derivative of their product $u(x)\cdot v(x)$ is

$$\frac{d}{dx}\big(u(x)v(x)\big) = u'(x)v(x) + u(x)v'(x)$$

Now take indefinite integrals of each term. Because

$$\int \frac{d}{dx}\big(u(x)v(x)\big)\,dx = u(x)v(x)$$

by the very definition of an indefinite integral, we obtain

$$u(x)v(x) = \int u'(x)v(x)\,dx + \int u(x)v'(x)\,dx$$

Subtracting $\int u'(x)v(x)\,dx$ from both sides, we obtain

$$\int u(x)v'(x)\,dx = u(x)v(x) - \int u'(x)v(x)\,dx \qquad [6]$$

the basic formula for integration by parts. Using the notation mentioned in Section 6.1, namely

$$dv = v'(x)\,dx \quad \text{and} \quad du = u'(x)\,dx$$

we may rewrite [6] in an equivalent but more commonly encountered form.

INTEGRATION BY PARTS FORMULA If $u(x)$ and $v(x)$ are differentiable functions, then

$$\int u\,dv = uv - \int v\,du \qquad [7]$$

In practice, if we want to find $\int f(x)\,dx$ for some function $f(x)$, we may be able to solve the problem if we can find two functions $u(x)$ and $v(x)$ such that $f(x)$ has the form

$$f(x) = u(x)v'(x) \quad \text{or equivalently} \quad f(x)\,dx = u\,dv$$

Here f is being broken up into two "parts": u and $v'(x)$, hence the name "integration by parts." If this can be done, we are assured by Formula [7] that

$$\int f(x)\,dx = \int u\,dv = u(x)v(x) - \int v\,du \qquad [8]$$

Finding $\int f(x)\,dx$ then reduces to the task of evaluating a *completely different* integral, namely $\int v\,du = \int v(x)u'(x)\,dx$, which may be much easier to handle if $u(x)$ and $v(x)$ are chosen carefully.

As a working example, consider the indefinite integral

$$\int x \ln x\,dx$$

It is not easy to guess an indefinite integral for $x \ln x$. And the substitution method is ineffective. But we may successfully apply integration by parts if we take

entries in the original integral: $u = \ln x \qquad dv = x\,dx$

entries in the new integral: $du = \dfrac{1}{x}\,dx \qquad v = \displaystyle\int x\,dx = \frac{1}{2}x^2$

The new integral is quite easy to evaluate,

$$\begin{aligned}\int x \ln x \, dx &= u(x)v(x) - \int v \, du \\ &= \frac{x^2}{2} \cdot \ln x - \int \frac{x^2}{2} \cdot \frac{1}{x} \, dx \\ &= \frac{1}{2} x^2 \cdot \ln x - \frac{1}{2}\left(\frac{1}{2} x^2\right) + \text{const} \\ &= \frac{1}{2} x^2 \cdot \ln x - \frac{1}{4} x^2 + \text{const}\end{aligned}$$

This answer may be checked by differentiation.

Example 4 Find $\int x^2 e^x \, dx$.

Solution We can decompose $f(x) \, dx$ in the form $f(x) \, dx = u \, dv$ by making various choices of u and v, such as

$$\begin{aligned} u &= e^x & dv &= x^2 \, dx \\ du &= e^x \, dx & v &= \int x^2 \, dx = \frac{1}{3} x^3 \end{aligned} \qquad [9]$$

or

$$\begin{aligned} u &= x^2 & dv &= e^x \, dx \\ du &= 2x \, dx & v &= \int e^x \, dx = e^x \end{aligned} \qquad [10]$$

Not all choices are equally effective. The first choice [9] gives a new integral more complicated than the one we started with:

$$\begin{aligned}\int x^2 e^x \, dx &= u(x)v(x) - \int v \, du \\ &= \frac{1}{3} x^3 e^x - \int \frac{1}{3} x^3 e^x \, dx\end{aligned}$$

This is not a happy choice of u and v. The second choice is much better:

$$\begin{aligned}\int x^2 e^x \, dx &= u(x)v(x) - \int v \, du \\ &= x^2 e^x - \int 2xe^x \, dx\end{aligned} \qquad [11]$$

This reduces evaluation of $\int x^2 e^x \, dx$ to evaluation of $\int 2xe^x \, dx$. Our situation has improved, but we are not yet finished. What works once often works twice; another application of integration by parts should reduce $\int 2xe^x \, dx$ to a manageable problem. To do this second step we try

$$u = 2x \qquad dv = e^x\,dx$$

$$du = 2\,dx \qquad v = \int e^x\,dx = e^x$$

and we obtain

$$\int 2xe^x\,dx = 2xe^x - \int 2e^x\,dx = 2xe^x - 2e^x + \text{const} \qquad [12]$$

Substituting [12] into [11] we get a complete evaluation of the original integral:

$$\int x^2e^x\,dx = x^2e^x - (2xe^x - 2e^x) + \text{const}$$
$$= x^2e^x - 2xe^x + 2e^x + \text{const}$$

Notice that it took two applications of the integration by parts formula to obtain the final result. ■

As this example shows, use of integration by parts requires some guesswork. There will be several ways to choose the "parts" $u(x)$ and $v(x)$ so that $f(x)\,dx = u\,dv$. The idea is to make this choice yield the simplest possible new integral $\int v\,du$.

Example 5 Compute $\int_1^2 (x^3 + 1)\ln x\,dx$.

Solution First calculate the indefinite integral. Because

$$\frac{d}{dx}(\ln x) = \frac{1}{x}$$

the new integral will be simpler than the original if we take

$$u(x) = \ln x \quad \left(\text{hence } du = \frac{1}{x}\,dx\right)$$

Accordingly, we must then take $dv = (x^3 + 1)\,dx$. Thus, let us try

$$u = \ln x \qquad dv = (x^3 + 1)\,dx$$

$$du = \frac{1}{x}\,dx \qquad v = \int (x^3 + 1)\,dx = \frac{1}{4}x^4 + x$$

Integration by parts then yields

$$\int (x^3 + 1)\ln x\,dx = u(x)v(x) - \int v\,du$$
$$= \left(\frac{1}{4}x^4 + x\right)\ln x - \int \left(\frac{1}{4}x^4 + x\right)\cdot\frac{1}{x}\,dx$$
$$= \left(\frac{1}{4}x^4 + x\right)\ln x - \frac{1}{16}x^4 - x + \text{const}$$

The definite integral is the difference of values at $x = 1$ and $x = 2$:

$$\int_1^2 (x^3 + 1)\ln x\, dx = \left(\frac{1}{4}x^4 + x\right)\ln x - \frac{1}{16}x^4 - x\Bigg|_1^2$$

$$= (6\ln(2) - 3) - \left(\frac{5}{4}\ln(1) - \frac{17}{16}\right)$$

$$= 2.2214$$

■

Exercises 6.2

In Exercises 1–3, evaluate the integral using integration by parts. Check your answer by differentiation.

1. $\int x^2 \ln x\, dx$ (*Hint*: Try $u = \ln x$, $dv = x^2\, dx$.)

2. $\int \frac{x}{e^{3x}}\, dx$ $\left(\textit{Hint}\text{: Try } u = x,\ dv = \frac{1}{e^{3x}}\, dx = e^{-3x}\, dx.\right)$

3. $\int \frac{\ln x}{x^2}\, dx$ $\left(\textit{Hint}\text{: Try } u = \ln x,\ dv = \frac{1}{x^2}\, dx.\right)$

In Exercises 4–23, evaluate each integral using integration by parts.

4. $\int xe^x\, dx$

5. $\int xe^{-4x}\, dx$

6. $\int (1 - x)e^x\, dx$

7. $\int_1^4 \sqrt{x} \ln x\, dx$

8. $\int (1 + x)e^{-2x}\, dx$

9. $\int_0^4 x^2 e^{-x}\, dx$

10. $\int_0^1 (4x - 1)e^{-x}\, dx$

11. $\int \frac{\ln x}{x^3}\, dx$

12. $\int_{-1}^1 (t + 1)e^t\, dt$

13. $\int x^3 \ln x\, dx$

14. $\int \ln x\, dx$ (*Hint*: Try $u = \ln x$, $dv = dx$.)

15. $\int x\sqrt{x + 5}\, dx$ (*Hint*: Try $u = x$, $dv = \sqrt{x + 5}\, dx$.)

16. $\int \frac{x}{\sqrt{1 - x}}\, dx$

17. $\int x^3 e^{-x^2/2}\, dx$

18. $\int x \ln(x^3)\, dx$

19. $\int_1^2 t \ln(3t)\, dt$

20. $\int (3x + 1)\ln x \, dx$

21. $\int (\ln x)^2 \, dx$ (*Hint*: You may have to apply integration by parts twice.)

22. $\int \frac{x}{e^x} \, dx$

23. $\int x^2 e^{-x} \, dx$ (*Hint*: Apply integration by parts twice.)

In Exercises 24–26, use integration by parts to find the indefinite integral. Integrals such as these, involving exponential functions, often turn up in probability density functions.

24. $\int x^3 e^{-x} \, dx$

25. $\int x^4 e^{-x} \, dx$

26. $\int x e^{kx} \, dx$ ($k \neq 0$ a fixed constant)

6.3 Improper Integrals

Definite integrals $\int_a^b f(x)\, dx$ allow us to calculate areas of regions of finite extent. In some applications, and particularly when dealing with probability distributions, we must extend the notion of integral to regions of infinite extent. This we do by introducing **improper integrals**.

Consider the shaded regions R shown in Figure 6.1a. These are bounded by the x axis, the vertical line $x = 1$, and the curves $y = 1/x^2$ or $y = 1/\sqrt{x} = x^{-1/2}$, respectively. The shaded regions extend "out to infinity," so we might be tempted to suppose that their areas are infinite. On the other hand, the widths taper off rapidly to the right; if they taper off rapidly enough, it is conceivable that their total areas are finite. We can decide the issue by introducing a "cutoff" at some point $x = r$. Wherever we place the cutoff, we get truncated regions R_r of finite extent, as in Figure 6.1b. The areas of the truncated regions are calculated in the usual way. In particular:

$$\begin{aligned} \text{Area}(R_r) &= \int_1^r f(x)\, dx \\ &= \int_1^r \frac{1}{x^2}\, dx \\ &= -\frac{1}{x}\bigg|_1^r = 1 - \frac{1}{r} \end{aligned} \qquad [13]$$

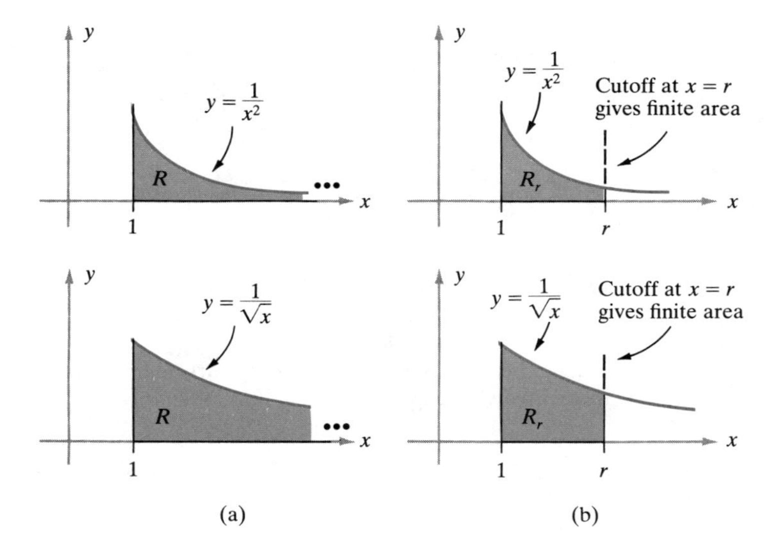

Figure 6.1
Figure 6.1a shows regions of infinite extent. Figure 6.1b shows truncated regions of finite extent.

for the region bounded by the curve $y = 1/x^2$, and

$$\begin{aligned} \text{Area}(R_r) &= \int_1^r f(x)\,dx \\ &= \int_1^r x^{-1/2}\,dx \\ &= 2\sqrt{x}\Big|_1^r = 2\sqrt{r} - 2 \end{aligned} \tag{14}$$

if $y = x^{-1/2}$. Now let the cutoff point $x = r$ move farther and farther to the right (indicated symbolically by writing "$r \to +\infty$"). There are two possibilities: Either the area Area(R_r) of the truncated region tends toward a finite limiting value L, in which case we indicate this state of affairs by the notation

$$\lim_{r \to +\infty} \int_1^r f(x)\,dx = L$$

or else the area gets larger and larger without any bound, which we indicate by writing

$$\lim_{r \to +\infty} \int_1^r f(x)\,dx = +\infty$$

In either case the limit value is called the improper integral of $f(x)$ from $x = 1$ to $x = +\infty$.

DEFINITION Let $f(x)$ be positive and continuous for all x to the right of some point $x = a$. The **improper integral** of $f(x)$ from $x = a$ to $x = +\infty$, denoted by the symbol

$$\int_a^{+\infty} f(x)\,dx$$

is defined to be the limit value of the cutoff integrals $\int_a^r f(x)\,dx$ as r increases:

$$\int_a^{+\infty} f(x)\,dx = \lim_{r \to +\infty} \int_a^r f(x)\,dx$$

If the limit does not exist, the improper integral is said to **diverge**, and we write

$$\int_a^{+\infty} f(x)\,dx = +\infty$$

In either case shown in Figure 6.1 we shall interpret the area of the full unbounded region R to be the limit value of the areas $\int_1^r f(x)\,dx$ of the truncated regions R_r

$$\text{Area}(R) = \int_1^{+\infty} f(x)\,dx$$

To calculate the values of improper integrals, and the areas associated with them, we must examine the behavior of the cutoff integrals $\int_a^r f(x)\,dx$ as r becomes larger and larger. This is not difficult if we appeal to a few simple principles:

(i) As r gets larger, its reciprocal $1/r$ gets smaller.
(ii) As r gets larger, any power r^a $(a > 0)$ gets larger. [15]

The second rule ensures that expressions such as $\sqrt{r}$, r^2, $r^{3/5}$, and so forth become larger and larger as r increases. Applying these ideas to the cutoff integrals in [13] and [14], we see that

$$\int_1^r \frac{1}{x^2}\,dx = \left(1 - \frac{1}{r}\right) \quad \text{approaches 1 as } r \text{ increases}$$

The constant 1 remains fixed, but $1/r$ becomes very small as r increases. Thus, for $y = 1/x^2$

$$\text{Area}(R) = \int_1^{+\infty} \frac{1}{x^2}\,dx = \lim_{r \to +\infty} \int_1^r \frac{1}{x^2}\,dx = 1 \quad \text{(finite!)}$$

On the other hand, for $y = x^{-1/2}$ the expression $\sqrt{r}$ gets very large as r increases, by [15ii], and so do $2\sqrt{r}$ and $2\sqrt{r} - 2$. Thus the integral diverges:

$$\text{Area}(R) = \int_1^{+\infty} \frac{1}{\sqrt{x}}\,dx = \lim_{r\to+\infty} (2\sqrt{r} - 2) = +\infty$$

In this case R contains subsets R_r with arbitrarily large area, so it is natural to say that Area(R) is infinite, written Area(R) $= +\infty$. Here are some additional examples.

Example 6 Find the value of the improper integrals

$$\int_1^{+\infty} 1\,dx \quad \text{and} \quad \int_2^{+\infty} \frac{1}{x^3}\,dx$$

Interpret the answers as areas.

Solution In the first integral, the cutoff integrals have values

$$\int_1^r 1\,dx = x\Big|_1^r = r - 1$$

Obviously, as r increases we have

$$\int_1^{+\infty} 1\,dx = \lim_{r\to+\infty} \int_1^r 1\,dx = \lim_{r\to+\infty} (r - 1) = +\infty$$

and the integral diverges. This is to be expected, because the area of the strip in Figure 6.2a should be infinite. In the second integral

$$\int_2^r \frac{1}{x^3}\,dx = -\frac{1}{2}x^{-2}\Big|_2^r = \frac{1}{2}\left(\frac{1}{2^2} - \frac{1}{r^2}\right) = \frac{1}{8} - \frac{1}{2r^2}$$

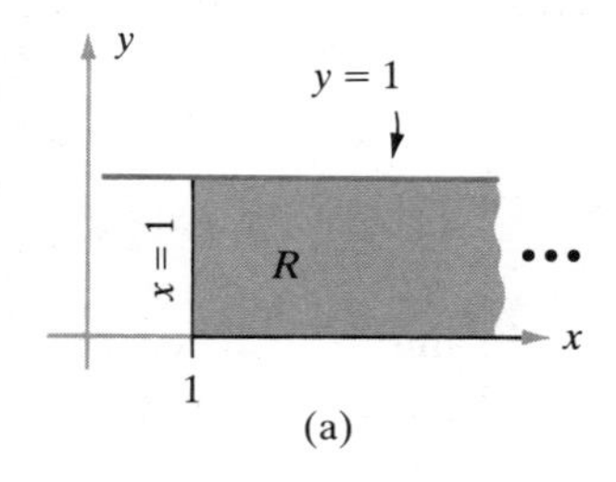

As r increases, r^2 takes on large positive values, and $1/r^2$ takes on very small positive values. Consequently, $-1/2r^2$ takes on small negative values tending toward zero, so that

$$\int_2^{+\infty} \frac{1}{x^3}\,dx = \lim_{r\to+\infty} \left(\frac{1}{8} - \frac{1}{2r^2}\right) = \frac{1}{8}$$

This is the area of the region shown in Figure 6.2b. ■

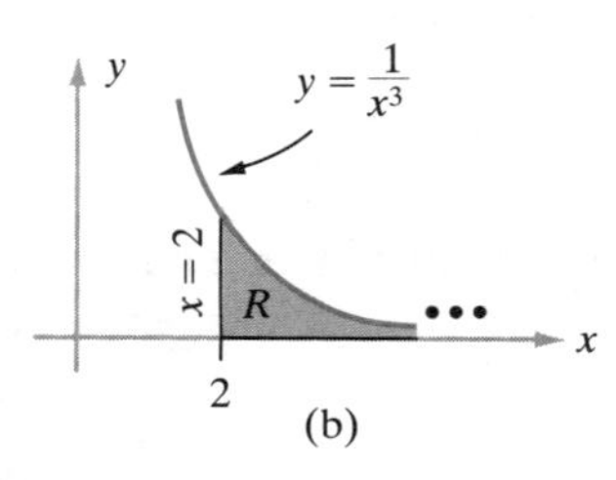

Figure 6.2
The regions whose areas correspond to the improper integrals $\int_1^{+\infty} 1\,dx$ and $\int_2^{+\infty} 1/x^3\,dx$.

If $f(x)$ is defined to the left of some point $x = a$, or on the entire number line, we can define corresponding improper integrals

$$\int_{-\infty}^{a} f(x)\,dx \quad \text{and} \quad \int_{-\infty}^{+\infty} f(x)\,dx$$

just as we defined improper integrals of the form $\int_a^{+\infty} f(x)\,dx$. Thus, if $f(x)$ is positive and continuous for $x \le a$, we take cutoffs to the left of $x = a$ and then move the cutoff point $x = r$ "toward $-\infty$," through larger and larger negative values, to define

$$\int_{-\infty}^{a} f(x)\,dx = \lim_{r\to-\infty} \int_{r}^{a} f(x)\,dx \qquad [16]$$

If $f(x)$ is positive and continuous for all x we take any convenient dividing point $x = a$, calculate the "one-sided" improper integrals $\int_{-\infty}^{a} f(x)\,dx$ and $\int_{a}^{+\infty} f(x)\,dx$ and then define

$$\int_{-\infty}^{+\infty} f(x)\,dx = \int_{-\infty}^{a} f(x)\,dx + \int_{a}^{+\infty} f(x)\,dx \qquad [17]$$

This integral is considered to diverge if either of the one-sided improper integrals diverges, in which case we write $\int_{-\infty}^{+\infty} f(x)\,dx = +\infty$.

In Section 5.6 we indicated that statistical problems may be handled by integrals and probability densities. Without doubt the most important density function in statistics is the **normal density function**

$$f(x) = \frac{1}{\sqrt{2\pi}} e^{-x^2/2} \quad \text{defined for all } x$$

whose graph is shown in Figure 6.3. One can show by means of advanced integration methods that

$$\int_{0}^{+\infty} \frac{1}{\sqrt{2\pi}} e^{-x^2/2}\,dx = \frac{1}{2} = \int_{-\infty}^{0} \frac{1}{\sqrt{2\pi}} e^{-x^2/2}\,dx$$

so that the total area under the graph equals 1:

$$\int_{-\infty}^{+\infty} \frac{1}{\sqrt{2\pi}} e^{-x^2/2}\,dx = 1$$

even though the region is unbounded. Because the exponential function is positive we see immediately that $f(x) \geq 0$, and therefore it satisfies all the conditions required of a probability density function. The variable x ranges over the entire number line, so improper integrals must be used to carry out a number of statistical calculations involving the normal density function. Recall that $\int_{a}^{b} f(x)\,dx$ gives the fraction of the population for which the variable x lies in the range $a \leq x \leq b$. Suppose we want to know the fraction of the population for

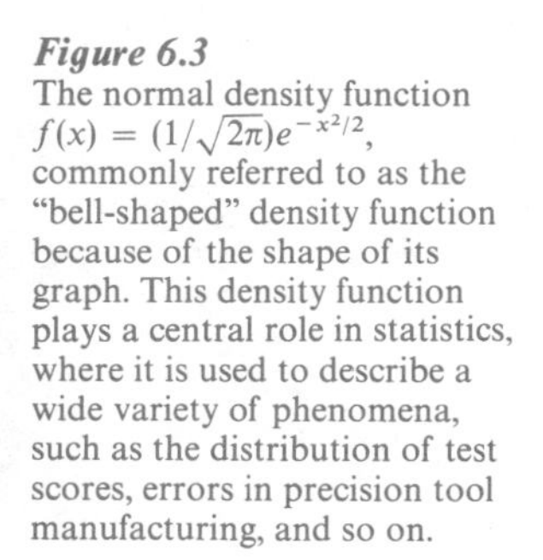
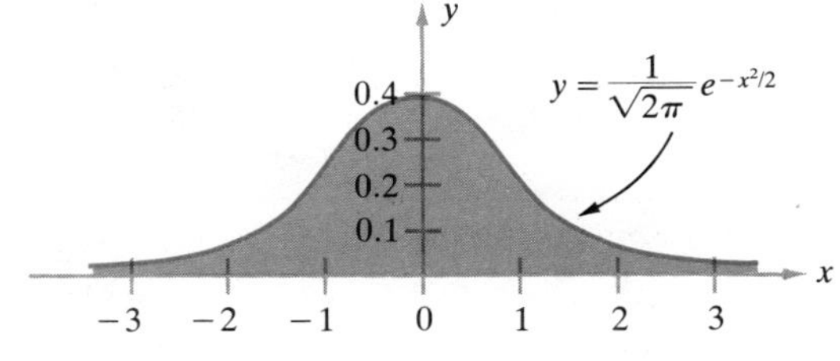

Figure 6.3
The normal density function $f(x) = (1/\sqrt{2\pi})e^{-x^2/2}$, commonly referred to as the "bell-shaped" density function because of the shape of its graph. This density function plays a central role in statistics, where it is used to describe a wide variety of phenomena, such as the distribution of test scores, errors in precision tool manufacturing, and so on.

which $x \geq a$. Intuitively, we should compute

$$\int_a^r f(x)\,dx = \text{fraction of population such that } a \leq x \leq r$$

and let the upper limit r get larger and larger. Thus the fraction of the population for which $x \geq a$ is given by an improper integral

$$\lim_{r \to +\infty} \int_a^r f(x)\,dx = \int_a^{+\infty} f(x)\,dx$$

Here is an example involving a slightly different density function, in which the desired improper integral may be computed by means at our disposal. (Integrals involving the normal density function are very difficult to compute, and are usually evaluated by numerical methods.)

Example 7 In a study of the lifetime t (in hours of use) of a type of 60-watt light bulb, the probability density function for lifetimes is found to be

$$f(t) = \frac{1}{1000} e^{-t/1000} \quad \text{defined for } t \geq 0$$

(In probability theory this type of density is called *exponential*.) What fraction of the bulbs will have lifetimes exceeding 500 hours?

Solution The density curve $y = f(t)$ is shown in Figure 6.4. The fraction of the population with lifetimes exceeding 500 hours is given by the improper integral

$$\int_{500}^{+\infty} f(t)\,dt$$

The cutoff integrals $\int_{500}^{r} f(t)\,dt$ are readily evaluated by using the standard integration formula

$$\int e^{kt}\,dt = \frac{1}{k} e^{kt} + \text{const}$$

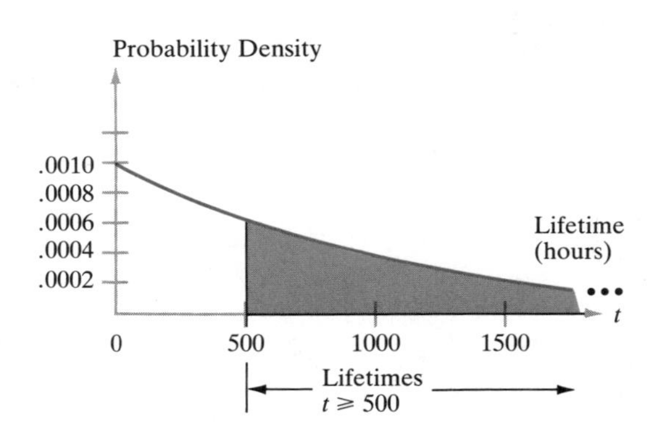

Figure 6.4
Graph of the probability density function $f(t) = 1/1000\, e^{-t/1000}$. Here the variable t stands for lifetime of a bulb (in hours). The fraction of the population with lifetimes exceeding 500 hours corresponds to the shaded area under the graph, and is given by an improper integral because the lifetimes could in principle be very large. Total area under the graph for $t \geq 0$ is one, as with any probability density function.

Taking $k = -1/1000$ here, we obtain

$$\int \frac{1}{1000} e^{-t/1000}\, dt = \frac{1}{1000}(-1000e^{-t/1000} + \text{const})$$
$$= -e^{-t/1000} + \text{const}$$

By the fundamental theorem of calculus, we then obtain the definite integral

$$\int_{500}^{r} f(t)\, dt = -e^{-t/1000}\Big|_{500}^{r} = (-e^{-r/1000}) - (-e^{-1/2})$$

for any $r > 500$. Letting $r \to +\infty$, we find that

$$\int_{500}^{+\infty} f(t)\, dt = \lim_{r \to +\infty} (e^{-1/2} - e^{-r/1000})$$
$$= e^{-1/2} \approx 0.6065$$

Here we have used a simple fact about the exponential function e^{-x}: As x gets larger and larger, e^{-x} tends very rapidly toward zero. Thus the term $e^{-r/1000}$ becomes very small as r increases, but the other term $e^{-1/2}$ remains constant. The limit is $e^{-1/2}$. In all, about 61% of the bulbs will burn for 500 hours or more. ■

Exercises 6.3

In Exercises 1–12, find the limit of the expression as $r \to +\infty$; that is, as r increases without bound.

1. r **2.** $\frac{1}{r}$ **3.** $\frac{1}{\sqrt{r}}$

4. e^r **5.** e^{-r} **6.** $\ln r$

7. $\frac{3}{2}r^{2/3} - 6$ **8.** $\frac{1}{2}\left(1 - \frac{1}{r^2}\right)$ **9.** $e^{-2} - e^{-r}$

10. $\sqrt[3]{r} - \frac{1}{r}$ **11.** $\frac{1}{\sqrt{r}} + \frac{1}{r}$ **12.** $1 - \frac{1}{r+1}$

In Exercises 13–18, evaluate the improper integral or show that it diverges.

13. $\int_{1}^{+\infty} \frac{1}{\sqrt[3]{x}}\, dx$ **14.** $\int_{5}^{+\infty} \left(1 + \frac{1}{x^2}\right) dx$

15. $\int_{1}^{+\infty} \frac{5}{x^4}\, dx$ **16.** $\int_{4}^{+\infty} x^{-3/2}\, dx$

17. $\int_{0}^{+\infty} e^{-x}\, dx$ **18.** $\int_{1}^{+\infty} \frac{1}{x}\, dx$

In Exercises 19–24, find the area above the x axis and under the graph of the function $y = f(x)$ for $x \geq 4$.

19. $f(x) = x^{-5/2}$

20. $f(x) = x$

21. $f(x) = \dfrac{4}{x^5}$

22. $f(x) = e^{3x}$

23. $f(x) = e^{-3x}$

24. $f(x) = \dfrac{1}{x \cdot \sqrt[3]{x}}$

25. Suppose the unit lengths along the x axis and y axis are measured in feet. Find the volume of paint required to cover the following regions under the curve $y = 1/x^3$ and above the x axis with a coat of paint 0.001 foot thick.
(i) the region above the interval $1 \leq x \leq 3$
(ii) the region above the interval $1 \leq x \leq 100$
(iii) the region above the interval $1 \leq x \leq 1{,}000{,}000$
(iv) the full region bounded by $y = 1/x^3$, the line $x = 1$, and the x axis.
(*Hint*: First find the area of each region, an integral involving $1/x^3$ in each case. Because of the choice of units, these areas will be measured in square feet. The desired volume is $0.001 \times$ (area).)

26. Repeat Exercise 25 for the curve $y = 1/\sqrt{x}$.

In Exercises 27–36, evaluate the improper integral or show that it diverges. (*Hint*: Use the method of substitution to find the appropriate indefinite integral in Exercises 31–36.)

27. $\displaystyle\int_1^{+\infty} \left(\frac{1}{x^2} + \frac{1}{\sqrt{x}}\right) dx$

28. $\displaystyle\int_1^{+\infty} \left(\frac{2}{x^2} - \frac{1}{x^3}\right) dx$

29. $\displaystyle\int_{-\infty}^{+\infty} e^{-x}\, dx$

30. $\displaystyle\int_{-\infty}^{0} e^{x}\, dx$

31. $\displaystyle\int_{-\infty}^{0} \frac{-x}{\sqrt{1 + x^2}}\, dx$

32. $\displaystyle\int_1^{+\infty} \frac{1}{(2x + 5)^2}\, dx$

33. $\displaystyle\int_{-\infty}^{0} \frac{1}{\sqrt{2 - 3x}}\, dx$

34. $\displaystyle\int_{-\infty}^{0} \frac{1}{(1 - 2x)^3}\, dx$

35. $\displaystyle\int_0^{+\infty} \frac{x}{(4 + x^2)^2}\, dx$

36. $\displaystyle\int_0^{+\infty} \frac{x}{4 + x^2}\, dx$

37. Show that the areas under the graphs of the functions

(i) $f(x) = \dfrac{1}{3} e^{-x/3}$ (ii) $f(x) = \dfrac{1}{2000} e^{-x/2000}$

defined for $x \geq 0$, satisfy the condition

$$\int_0^{+\infty} f(x)\, dx = 1$$

(*Note*: This shows that the area under each graph is 1, and that these functions are acceptable candidates for probability density functions defined on the unbounded interval $x \geq 0$.)

38. Functions of the form

$$f(x) = Ae^{-kx} \quad (A > 0 \text{ and } k > 0 \text{ constants})$$

defined for $x \geq 0$, occur frequently as probability density functions. If $f(x)$ is to be a density function, the total area under its graph for $0 \leq x < +\infty$ must equal *one*,

$$\int_0^{+\infty} f(x)\,dx = 1 \qquad [18]$$

If the constant k in the exponent is specified, show that the "normalizing constant" A must be taken to be

$$A = k$$

to ensure that [18] holds.

39. In Example 7, determine the fraction of light bulbs whose lifetimes exceed 2000 hours. What fraction has lifetimes in the range $0 \leq t \leq 2000$ hours? If we started with 10,000 bulbs, how many would we expect to last 2000 hours or more?

40. Insurance companies are interested in how long an automobile will be on the road once it is purchased: its "lifetime." Suppose the lifetime of a certain model is governed by the probability density function

$$f(x) = \frac{1}{7.5}e^{-x/7.5}$$

where $x =$ (lifetime in years from purchase). What fraction of these 1975 models are still operational in 1980 and in 1984?

41. A company monitors its outgoing phone calls, and has found that the duration t (in minutes) of a call is governed by the probability density function

$$f(t) = \frac{1}{4}e^{-t/4} \quad \text{for} \quad t \geq 0$$

What fraction of calls lasts for more than 6 minutes?

42. The time t in minutes between successive cars pulling up to a toll booth has the probability density function

$$f(t) = \frac{1}{3.5}e^{-t/3.5} \quad \text{for} \quad t \geq 0$$

What fraction of successive cars will be separated by more than 5 minutes?

6.4 Introduction to Differential Equations

Indefinite integration reconstructs a function from its derivative. For example, if $dy/dx = x^2$, then $y = \int x^2\,dx = \frac{1}{3}x^3 + \text{const}$. In this section we will see how to find a function, given information about it and its derivatives. As an example, suppose we want to find all functions $y = f(x)$ that satisfy the equation

$$\frac{dy}{dx} = \frac{x^2}{y^2} \qquad [19]$$

Now we cannot simply integrate, because the right side of [19] involves both the variable x and the unknown function y, whereas before it involved only x. Equation [19] is an example of a **differential equation**, an equation involving an unknown function *and its derivatives.*

Given a differential equation, we want to find its **solutions**: all functions $y = f(x)$ that reduce the equation to an identity if we set $y = f(x)$. We encountered simple differential equations such as

$$\frac{dy}{dx} = \frac{1}{2}y \qquad [20]$$

in studying growth problems (Section 4.4). There we were able to guess the solutions, based on our experience with exponential functions. The solutions have the form

$$y = Ce^{x/2} \quad \text{where } C \text{ is any constant}$$

To check, notice that

$$\frac{dy}{dx} = \frac{d}{dx}(Ce^{x/2}) = C\frac{d}{dx}(e^{x/2}) = C \cdot \frac{1}{2}e^{x/2} = \frac{1}{2}y$$

Notice that we obtain a whole family of solutions, one for each choice of C, as shown in Figure 6.5. This is typical: The solutions of a differential equation are

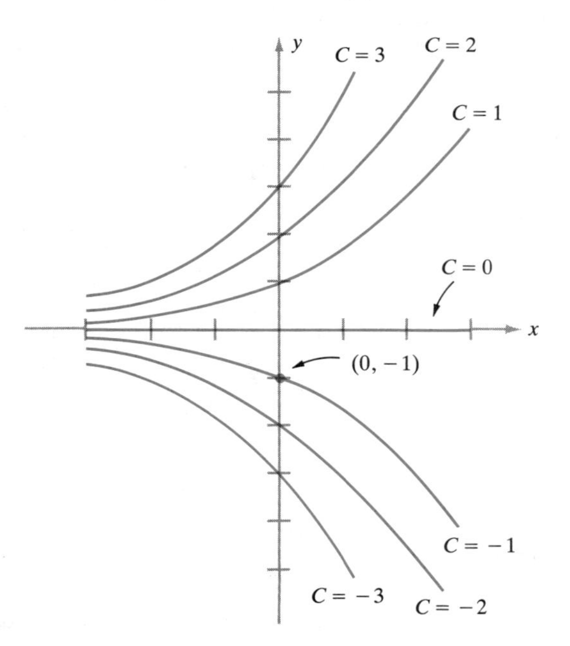

Figure 6.5
Graphs of $y = Ce^{x/2}$ for $C = -3, -2, \ldots, 2, 3$. Each is a solution of $dy/dx = \frac{1}{2}y$. Only $y = -e^{x/2}$ ($C = -1$) passes through the particular point $(0, -1)$, giving $y = -1$ when $x = 0$.

not unique. But in practice we are not interested in any old solution. Usually we want a particular solution that satisfies certain additional conditions; these conditions determine the exact solution we want.

Example 8 Find all solutions of $dy/dx = \frac{1}{2}y$ that have the value -1 when $x = 0$.

Solution The general solution of this equation has the form $y = Ce^{x/2}$. If we want $y = -1$ when $x = 0$, we need

$$-1 = Ce^{0/2} = C \cdot 1 = C$$

Only the choice $C = -1$ does the job; there is a unique solution of $dy/dx = \frac{1}{2}y$ satisfying the extra condition $y(0) = -1$. It is $y = -e^{x/2}$, as indicated in Figure 6.5. ■

Guesswork may have done the job in solving [20]. It will even produce some solutions of the more complicated equation [19]. But to find all solutions we will need a new idea.

Example 9 Show that

$$y = x$$

$$y = (x^3 + 5)^{1/3}$$

are both solutions of the equation $dy/dx = x^2/y^2$.

Solution If $y = x$, then

$$\frac{dy}{dx} = \frac{d}{dx}(x) = 1 \qquad \frac{x^2}{y^2} = \frac{x^2}{x^2} = 1$$

so $dy/dx = x^2/y^2$. If $y = (x^3 + 5)^{1/3}$, then by the chain rule

$$\frac{dy}{dx} = \frac{1}{3}(x^3 + 5)^{-2/3}\frac{d}{dx}(x^3 + 5) = x^2(x^3 + 5)^{-2/3}$$

$$\frac{x^2}{y^2} = \frac{x^2}{((x^3 + 5)^{1/3})^2} = x^2(x^3 + 5)^{-2/3}$$

These are equal, so Equation [19] is satisfied. ■

The solution $y = x$ could be found by trial and error. The other solution is found by a systematic procedure called *separation of variables*, which we now illustrate for [19] before we state it in general.

According to [19], a small change dx in the independent variable is related to the corresponding change dy in the dependent variable by

$$y^2\,dy = x^2\,dx \qquad [21]$$

obtained by multiplying each side of [19] by $y^2\,dx$ and treating dy/dx as a quotient. Equation [21] cries out for integral signs, which we now provide.

$$\int y^2\,dy = \int x^2\,dx \qquad [22]$$

Performing the integrations in [22] we obtain

$$\frac{1}{3}y^3 + c_1 = \frac{1}{3}x^3 + c_2 \quad (c_1, c_2 \text{ arbitrary constants})$$

or

$$y^3 = x^3 + c \quad (\text{where } c = 3(c_2 - c_1) \text{ is still an arbitrary constant})$$

Finally, we solve for y explicitly:

$$y = (x^3 + c)^{1/3} \qquad [23]$$

This is our general solution of [19]. We can verify that, for any c, this function is a solution of our equation just as in Example 9, which treats the cases when $c = 0$ or $c = 5$. The family of all solutions, one for each choice of c, is shown in Figure 6.6. Once again, if we specify a desired value for the solution at some base point, say $y(0) = 2$, we single out a unique particular solution.

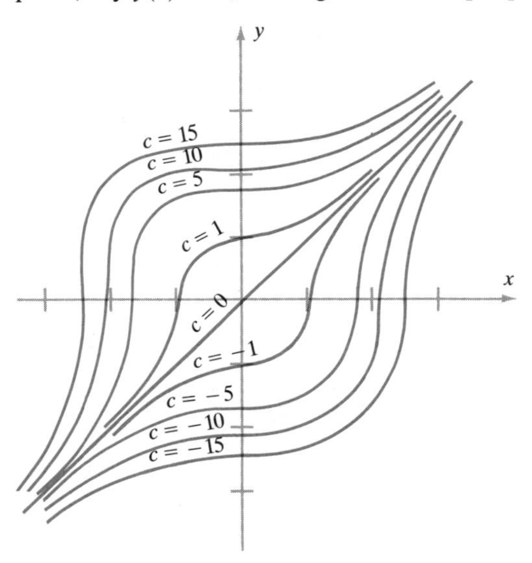

Figure 6.6
The family of solutions $y = (x^3 + c)^{1/3}$ of the differential equation $dy/dx = x^2/y^2$.

Example 10 Find all solutions of $dy/dx = x^2/y^2$ such that $y(0) = 2$.

Solution Any solution has the form $y = (x^3 + c)^{1/3}$. If we want $y(0) = 2$, this means

$$2 = y(0) = (0 + c)^{1/3} = c^{1/3}$$

so we must take $c = 2^3 = 8$. The particular solution is $y = (x^3 + 8)^{1/3}$. ■

We can employ the same procedure to solve a larger class of differential equations. A **differential equation with separable variables** is one that can be written in the special form

$$\frac{dy}{dx} = \frac{g(x)}{h(y)} \tag{24}$$

where $g(x)$ and $h(y)$ are functions of x and y, respectively. If we take $g(x) = \frac{1}{2}$ and $h(y) = 1/y$, we obtain the differential equation [20]: $dy/dx = \frac{1}{2}y$. If $g(x) = x^2$ and $h(y) = y^2$, we get Equation [19]: $dy/dx = x^2/y^2$. Thus both of these equations have separable variables.

Example 11 Which of these differential equations has separable variables?

(i) $\frac{dy}{dx} = x^2$ (iv) $x\frac{dy}{dx} = xy + y$

(ii) $\frac{dy}{dx} - 3y = 0$ (v) $\frac{dy}{dx} = \frac{y}{x^2}$

(iii) $\frac{dy}{dx} = y - x$

Solution All but (iii) are separable. To see this, take $g(x) = x^2$, $h(y) = 1$ for (i). Rewrite (ii) as $dy/dx = 3y = 3/(1/y)$ to see that $g(x) = 3$, $h(y) = 1/y$ do the job. Rewrite (iv) as

$$\frac{dy}{dx} = \frac{xy + y}{x} = \frac{y(x + 1)}{x} = \frac{\left(\frac{x+1}{x}\right)}{\left(\frac{1}{y}\right)}$$

and rewrite (v) as

$$\frac{dy}{dx} = \frac{\left(\frac{1}{x^2}\right)}{\left(\frac{1}{y}\right)}$$

to see that these are separable. There is no way to express $y - x$ in the form $g(x)/h(y)$; consequently, (iii) does not have separable variables. ■

To solve an equation [24] with separable variables, we multiply each side by $h(y)\,dx$ to obtain

$$h(y)\,dy = g(x)\,dx \tag{25}$$

In this expression the variables have been separated with all expressions involving y on the left and those involving x on the right. This is why equations of the form [24] are called "separable." Formula [25] strongly suggests that we take indefinite integrals,

$$\int h(y)\,dy = \int g(x)\,dx \tag{26}$$

This equality gives the desired solutions of [24] if we remember to introduce an arbitrary added constant (one will do) when we integrate.

Example 12 Find all solutions of the differential equation $x(dy/dx) - 3y = 0$.

Solution The variables are separable because the equation may be rewritten as

$$\frac{dy}{dx} = \frac{3y}{x} = \frac{\left(\frac{3}{x}\right)}{\left(\frac{1}{y}\right)} \qquad \left(\text{with } g(x) = \frac{3}{x} \text{ and } h(y) = \frac{1}{y}\right)$$

Separating the variables we find that

$$\frac{1}{y}\,dy = \frac{3}{x}\,dx \quad \text{or} \quad \int \frac{1}{y}\,dy = \int \frac{3}{x}\,dx$$

Therefore,

$$\ln y = 3 \ln x + c = \ln(x^3) + c$$

where c is an arbitrary constant. Taking the exponential of each side, we obtain the general solution

$$y = e^{\ln(x^3)+c} = e^c e^{\ln(x^3)} = Cx^3$$

Here $C\ (= e^c)$ is an arbitrary constant. ■

The next example illustrates how a physical phenomenon can be modeled as a differential equation.

Example 13 The speed of a skydiver before the parachute opens is given by the differential equation

$$\frac{dv}{dt} = 32 - 0.25v \qquad [27]$$

where t = (seconds from start of the jump) and the velocity v is measured in feet per second. Find the general solution of this equation. Then find the particular solution satisfying the obvious initial condition $v = 0$ when $t = 0$. How fast is the diver moving after 10 seconds of free fall? After 30 seconds?

Solution We separate variables by writing [27] as $dv/dt = 0.25(128 - v)$, so that

$$\frac{dv}{128 - v} = 0.25\,dt \quad \text{or} \quad \int \frac{1}{128 - v}\,dv = \int 0.25\,dt$$

We use the substitution method (Section 6.1) to find the indefinite integral on the left. If we set $u = 128 - v$, then

$$du = -dv \quad \text{and} \quad \frac{1}{128 - v} = \frac{1}{u}$$

so

$$\int \frac{1}{128 - v}\,dv = \int -\frac{1}{u}\,du \Bigg|_{u=128-v}$$

$$= -\ln u \Bigg|_{u=128-v}$$

$$= -\ln(128 - v)$$

Thus we obtain

$$-\ln(128 - v) + c = \int 0.25\,dt = 0.25t$$

or

$$\ln(128 - v) = -0.25t + c$$

where c is an arbitrary constant. Taking the exponential of each side, we obtain

$$128 - v = e^{-0.25t} \cdot e^{c} = Ce^{-0.25t} \quad (C \text{ an arbitrary constant})$$

so the general solution is

$$v = 128 - Ce^{-0.25t}$$

At the start of the jump, when $t = 0$, we have $v = 0$. This means $0 = 128 - C(e^{-0}) = 128 - C$, or $C = 128$. Thus, the particular solution describing the fall is

$$v = 128 - 128e^{-0.25t} = 128(1 - e^{-0.25t})$$

When $t = 10$, we have $v = 128(1 - e^{-2.5}) = 128(0.9179) = 117.5$ feet per second. Twenty seconds later, when $t = 30$, the speed will have increased only slightly: $v = 128(1 - e^{-7.5}) = 128(0.99945) = 127.9$ feet per second. ■

Our two final examples illustrate the use of differential equations in biology, ecology, and sociology.

Example 14 Growth and decay phenomena are modeled by the differential equation

$$\frac{dy}{dt} = ky \quad (k \text{ a fixed constant})$$

(Recall the discussion of Section 4.4). Use the method of separation of variables to solve this equation.

Solution We separate variables to obtain

$$\frac{1}{y}\,dy = k\,dt \quad \text{or} \quad \int \frac{1}{y}\,dy = \int k\,dt$$

Performing the integrations, we obtain

$$\ln y = kt + c \quad (c \text{ an arbitrary constant})$$

and taking exponentials of each side we obtain

$$y = e^{\ln y} = e^{kt+c} = e^c e^{kt} = Ce^{kt}$$

where C is an arbitrary constant. Thus $y = Ce^{kt}$ is the general solution. ■

Simple models of population growth are based on the differential equation $dy/dt = ky$ (k a constant) already discussed. But if deaths or external factors limiting population growth are to be taken into account, a more complicated equation must be used. One such modified growth equation has the form

$$\frac{dy}{dt} = ky(A - y) \qquad [28]$$

and is called the **logistic equation**. Here, k and A are constants that must be determined from observations. Equation [28] has the following features built into it:

(i) When the population y is relatively small, so that $A - y \approx A$, the rate of population increase is approximately proportional to y, and the population grows more or less exponentially.

(ii) As y increases toward the "asymptotic value" A, both $A - y$ and $ky(A - y)$ are very small, which forces the growth rate dy/dt to slow down and approach zero.

In the next example we show how the logistic equation can be solved, using separation of variables and a single algebraic trick.

Example 15 Find all solutions of the logistic equation $\dfrac{dy}{dt} = ky(A - y)$.

Solution By separating the variables y and t we obtain

$$\int \frac{1}{y(A - y)}\,dy = k\int dt$$

The antiderivative on the left can be evaluated by using the algebraic identity

$$\frac{1}{y(A - y)} = \frac{1}{A}\left[\frac{1}{y} + \frac{1}{A - y}\right] \qquad [29]$$

Thus

$$\begin{aligned}\int \frac{1}{y(A - y)}\,dy &= \frac{1}{A}\int\left(\frac{1}{y} + \frac{1}{A - y}\right)dy \\ &= \frac{1}{A}\int \frac{1}{y}\,dy + \frac{1}{A}\int \frac{1}{A - y}\,dy = \frac{1}{A}\ln y - \frac{1}{A}\ln(A - y) \\ &= \frac{1}{A}\ln\left(\frac{y}{A - y}\right)\end{aligned}$$

so the general solution is

$$\frac{1}{A}\ln\left(\frac{y}{A-y}\right) = kt + c \quad (c \text{ an arbitrary constant})$$

Multiplying by A and taking exponentials on each side, we get

$$\frac{y}{A-y} = e^{Ac}e^{Akt} = Ce^{Akt}$$

where C is an arbitrary constant. Solving explicitly for y, we have

$$y = (A - y)Ce^{Akt} = ACe^{Akt} - yCe^{Akt}$$

so that

$$y(1 + Ce^{Akt}) = ACe^{Akt}$$

Thus the general solution of the logistic equation is

$$y = \frac{ACe^{Akt}}{1 + Ce^{Akt}} \quad (C \text{ an arbitrary constant}) \qquad [30]$$

■

The graph of such a function, shown in Figure 6.7, is called *sigmoid* (literally, "S-shaped"), and the function arises quite often in realistic models of growth.

Here are some situations involving the logistic equation. Others are discussed in the exercises. In a population of size A, the rate at which a rumor spreads should be proportional to both

$y =$ (the number who have already heard it (the more suppliers, the faster the spread))

and

$A - y =$ (number who have yet to hear (the larger this is, the more likely the listener has not yet heard the news))

Thus the spread of the rumor is governed by a differential equation of logistic type

$$\frac{dy}{dt} = ky(A - y) \qquad [31]$$

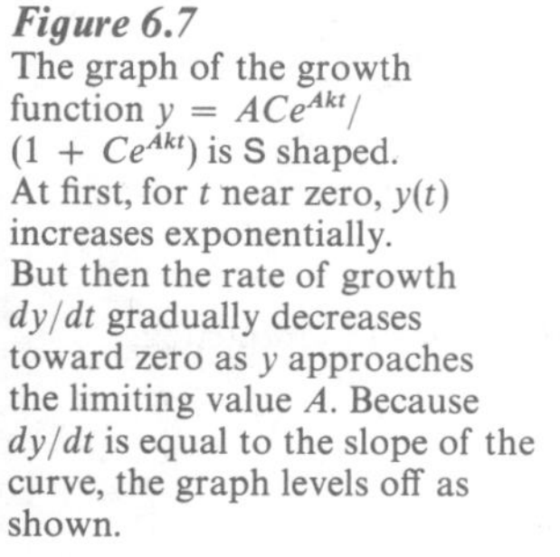

Figure 6.7
The graph of the growth function $y = ACe^{Akt}/(1 + Ce^{Akt})$ is S shaped. At first, for t near zero, $y(t)$ increases exponentially. But then the rate of growth dy/dt gradually decreases toward zero as y approaches the limiting value A. Because dy/dt is equal to the slope of the curve, the graph levels off as shown.

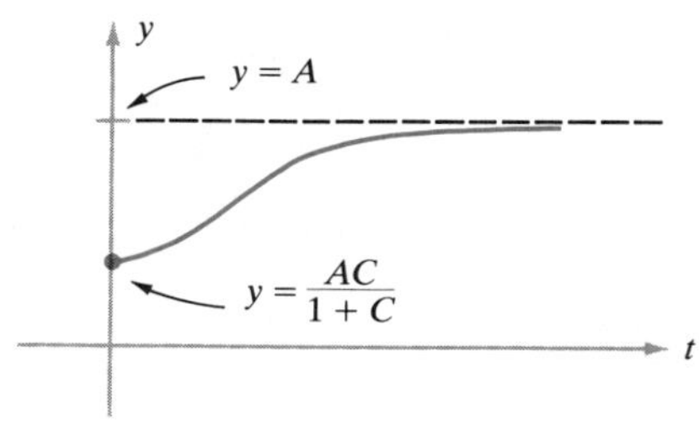

where k is some constant that must be determined by observation. Similarly, the spread of an epidemic should also be governed by the same sort of equation, where $y =$ (number who have contracted the disease) and A is the size of the total population.

Exercises 6.4

In Exercises 1–6, show that the function is a solution of the given differential equation.

1. $y = x$ is a solution of $y\dfrac{dy}{dx} = x$.

2. $y = 1 + e^{-x^2/2}$ is a solution of $\dfrac{dy}{dx} + xy = x$

3. $y = e^{-t} + 2e^{-2t}$ is a solution of $\dfrac{dy}{dt} + 2y = e^{-t}$

4. $y = 2e^{t} - e^{-t}$ is a solution of $\dfrac{d^2y}{dt^2} = y$

5. $y = (1 - x)^{-1}$ is a solution of $\dfrac{dy}{dx} = y^2$

6. $y = \dfrac{x^2}{1 - x}$ is a solution of $x^2\dfrac{dy}{dx} = y^2 + 2xy$

7. The differential equation $dy/dx = y - x$ is not separable. Advanced courses cover more sophisticated methods that yield the solutions

$$y = 1 + x + Ce^x \quad (C \text{ an arbitrary constant})$$

Show that this is a solution for each choice of C. Then find the particular solution such that $y = 2$ when $x = 1$.

In Exercises 8–15, find the general solution of the differential equation.

8. $\dfrac{dy}{dx} = x$

9. $\dfrac{dy}{dx} = x^3 + 1$

10. $\dfrac{dy}{dx} = x - 1$

11. $\dfrac{dy}{dt} = 50 - y$

12. $\dfrac{dy}{dt} = ty$

13. $\dfrac{dy}{dx} = xe^y$

14. $\dfrac{dy}{dx} - \dfrac{3x^2}{2y} = 0$

15. $\dfrac{dy}{dx} + x^2y = x^2$

In Exercises 16–21, find the particular solution satisfying the condition indicated.

16. $\dfrac{dy}{dx} = y$ with $y(0) = 2$

17. $\frac{dy}{dx} = y^2$ with $y = 1$ when $x = 2$

18. $\frac{dy}{dx} = -xy$ with $y = 5$ when $x = 1$

19. $\frac{dy}{dx} = \frac{0.5}{x + 10}$ passing through the point (1, 5)

20. $\frac{dy}{dx} = 0.3(y - 10)$ such that $y = 0$ when $x = 0$

21. $\frac{dy}{dx} = \frac{1}{xy}$ such that $y = 5$ when $x = 1$

22. The population N of a certain bacteria culture satisfies the differential equation

$$\frac{dN}{dt} = 0.1\sqrt{N}$$

If $N = 10{,}000$ when $t = 0$, find the population when $t = 100$.

23. If an object is kept in surroundings of constant temperature, say 50°F, its temperature gradually tends toward that of its surroundings. **Newton's law of cooling** states that the rate of change dU/dt in the temperature U of the object is proportional to the difference in temperatures $U - 50$ between the object and the medium:

$$\frac{dU}{dt} = k(U - 50) \quad \text{for some negative constant } k$$

Assume $k = -0.2$ and $U = 70$°F when $t = 0$. Find U when $t = 3$. Find the formula for $U(t)$.

24. In Exercise 23, how long does it take to cool an object halfway between its initial temperature 70°F and its ultimate temperature 50°F?

25. In a psychological test, subjects learn 50 nonsense words. The experimenter has a theory that each day the average person will forget 4% of the words remembered the previous day (assuming no further review of the word list). Thus if N = (number of words remembered after t days), behavior of N should be governed by the differential equation

$$\frac{dN}{dt} = -0.04N \quad \text{(negative coefficient because } N \text{ decreases with time)}$$

The solution is subject to the initial condition $N = 50$ when $t = 0$. Find the function $N(t)$ predicted by this model. How many words would the average person remember after 10 days?

26. A town has 10,000 street lights. New high-intensity lights are to be installed as the old-style bulbs burn out, starting April 1. Experience shows that about 2% of the surviving old-style bulbs will burn out each week. Thus, if we let t = (weeks elapsed from

April 1) and $y =$ (number of bulbs surviving), we must have

$$\frac{dy}{dt} = -0.02y$$

Find the general solution of this differential equation. Then find the particular solution such that $y = 10{,}000$ when $t = 0$. How many new lamps should be ordered to cover the first 8 weeks of the replacement program?

27. Rework Exercise 26 for a city with 1,000,000 street lamps, assuming that the old-style bulbs fail at the rate of 1% per week. How long will it take until half of the bulbs have been replaced?

28. Suppose the spread of a rumor among a population of 250 people is governed by the logistic type of equation

$$\frac{dy}{dt} = 0.002y(250 - y)$$

where $y =$ (number who have heard the rumor) and t is measured in days. Use the solution of the logistic equation worked out in Example 15 to find the function $y(t)$, assuming that 15 people have heard the rumor when $t = 0$: $y = 15$ when $t = 0$. How many have heard it after 10 days? After how many days will half the population hear the rumor?

29. In psychological testing, the response y varies with the amount of stimulus s. Furthermore, we often find that the *percentage change in response* $\Delta y/y$ is proportional to the *percentage change in stimulus* $\Delta s/s$, so that

$$\frac{\Delta y}{y} \approx k\frac{\Delta s}{s} \quad \text{or} \quad \frac{\Delta y}{\Delta s} \approx k\frac{y}{s} \quad (k \text{ is some constant})$$

for small increments Δs. That is, the stimulus–response relation $y = y(s)$ is approximately governed by the differential equation

$$\frac{dy}{ds} = k\frac{y}{s}$$

If the constant of proportionality has the value $k = 0.05$ and if $y = 1$ when $s = 1$, find y as a function of s.

30. Here is an equation similar to the logistic equation.

$$\frac{dy}{dx} = y^2 - 1 = (y - 1)(y + 1) \qquad [32]$$

By simple algebra we can verify the algebraic identity

$$\frac{1}{y^2 - 1} = \frac{1}{2(y - 1)} - \frac{1}{2(y + 1)}$$

Use the ideas discussed in Example 15 to find the general solution of equation [32].

Checklist of Key Topics

Substitution method
Integration by parts
Improper integrals
Divergent improper integrals
Applications of improper integrals to probability
Differential equations
Solution of a differential equation
Particular solutions
Differential equations with separable variables
Logistic equation
Applications of differential equations

Substitution formula:

$$\int f(x)\,dx = \int g(u)\,du\Big|_{u=u(x)}$$

Integration by parts formula:

$$\int u\,dv = uv - \int v\,du$$

Chapter 6 Review Exercises

In Exercises 1–12, find the indefinite integral.

1. $\int (3x+5)^7\,dx$
2. $\int \left(4-\frac{x}{3}\right)^{-7} dx$
3. $\int x\sqrt{1-x^2}\,dx$
4. $\int \left(2-\frac{x}{3}\right)^{2} dx$
5. $\int t^5 e^{-t^6}\,dt$
6. $\int \frac{\ln(x^2)}{x}\,dx$
7. $\int (2-x)\ln x\,dx$
8. $\int x^2\ln(x^3)\,dx$
9. $\int (5-3x)e^{-x}\,dx$
10. $\int \frac{t^2}{e^t}\,dt$
11. $\int \frac{1}{5x+1}\,dx$
12. $\int \frac{1}{(5x+1)^2}\,dx$

In Exercises 13–18, evaluate the definite integral.

13. $\int_0^\infty e^{-4t}\,dt$
14. $\int_0^2 xe^{-x}\,dx$
15. $\int_0^1 xe^{-x^2}\,dx$
16. $\int_2^4 x\ln x\,dx$
17. $\int_0^3 x\sqrt{x+1}\,dx$
18. $\int_0^\infty \frac{x}{(3x^2+2)^2}\,dx$

In Exercises 19–24, evaluate the improper integral or show that it diverges.

19. $\int_2^\infty x^{-1/5}\,dx$
20. $\int_2^\infty x^{-5}\,dx$

21. $\int_0^\infty \frac{x}{1+x^2}\,dx$ **22.** $\int_0^\infty \frac{x}{(1+x^2)^3}\,dx$

23. $\int_1^\infty te^t\,dt$ **24.** $\int_7^\infty e^{-5x}\,dx$

25. The time t (in minutes) that players spend on a certain video game is described by the probability density function $f(t) = \frac{1}{12}e^{-t/12}$ for $t \geq 0$.
(i) Show that $\int_0^\infty f(t)\,dt = 1$. (This shows that $f(t)$ is a bona fide probability density function.)
(ii) Find the fraction of players that play the game for more than 6 minutes.

In Exercises 26–31, find the general solution of the differential equation.

26. $\frac{dy}{dt} = 7t$ **27.** $\frac{dy}{dt} = 7y$

28. $y\frac{dy}{dx} = 2x - 3$ **29.** $\frac{1}{x^2}\frac{dy}{dx} - y = xy$

30. $\frac{dy}{dx} = \frac{3}{2}y^{1/3}$ **31.** $y^3\frac{dy}{dx} = x^5$

In Exercises 32–35, find the particular solution of the differential equation.

32. $\frac{dy}{dx} = y^3$ with $y(0) = 1$ **33.** $\frac{dy}{dx} = xy$ with $y(2) = 1$

34. $\frac{dy}{dx} = e^y$ passing through the point (1, 0)

35. $\frac{dy}{dx} = e^x$ such that $y = 2$ when $x = 0$

36. The amount A of money in a savings account paying 6% annual interest compounded continuously is governed by the equation $dA/dt = 0.06A$ (if t is measured in years). Find all solutions of this differential equation. Find the particular solution if the account initially contained $800.

37. A tank initially contains 5 pounds of salt dissolved in 100 gallons of water. Brine containing 1 pound of salt per gallon flows into the tank at the rate of 3 gallons per minute, while the mixed solution is drained from the tank at the same rate. If $y(t)$ is the amount of salt remaining in the tank at time t, then $y(t)$ satisfies the differential equation

$$\frac{dy}{dt} = 3 - 0.03y$$

Solve this differential equation. Then find the particular solution satisfying $y(0) = 5$. After 30 minutes, how much salt is in the tank?

CHAPTER 7

Calculus of Several Variables

7.1 Functions of Several Variables

If we have a carpenter build a rectangular box, we must independently specify several variables (namely, length x, width y, and height z) to describe what we want. Then the other features may be described in terms of these variables. For example, the volume of the box is (see Figure 7.1)

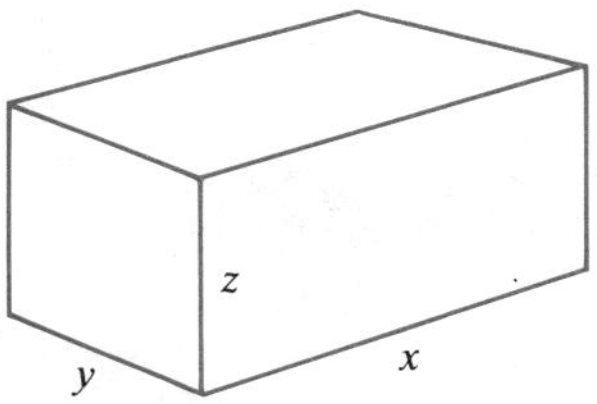

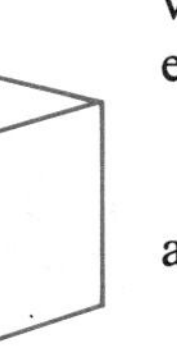

Figure 7.1
Rectangular box with dimensions x, y, z.

$$V = V(x, y, z) = xyz$$

and the surface area (sum of 6 rectangular sides) is

$$A = A(x, y, z) = 2xy + 2xz + 2yz$$

If the actual dimensions are $x = 5$, $y = 3$, and $z = 2$ feet, we find the corresponding volume and area by entering these values into the formulas for V and A,

$$V(5, 3, 2) = 5 \cdot 3 \cdot 2 = 30 \quad \text{cubic feet}$$

and

$$A(5, 3, 2) = 2(5)(3) + 2(5)(2) + 2(3)(2) = 62 \quad \text{square feet}$$

So far in our study of calculus we have confined our attention to functions having a single independent variable. We did this to keep things simple while we introduced such notions as graph of a function, derivative, and so on. But many problems in economics and biology involve more than one independent variable; indeed, in a situation as complex as an "input–output" model of the U.S. economy, we might have to distinguish hundreds or thousands of independent variables to obtain a realistic model. Thus we must come to grips with the way functions of several variables are handled in calculus.

Here are some examples of functions of two variables x and y. For each choice of x, y, there is a rule for computing the value of the function $f(x, y)$.

$$f(x, y) = xy - x^2$$
$$f(x, y) = \sqrt{x^2 + y^2}$$
$$f(x, y) = 24xy + 13x - 7y + 100$$
$$f(x, y) = 4 \quad \text{(constant function; same value for all } x, y\text{)}$$

Example 1 Consider the function $f(x, y) = 4 + xy - x^2$. Find its value $f(1, 3)$, when $x = 1$ and $y = 3$. Then find the values $f(1, 0)$ and $f(0, 1)$.

Solution If we enter $x = 1$ and $y = 3$ into the formula for f we obtain the value

$$f(1, 3) = 4 + (1)(3) - (1)^2 = 6$$

Similarly,

$$f(1, 0) = 4 + (1)(0) - (1)^2 = 3$$
$$f(0, 1) = 4 + (0)(1) - (0)^2 = 4$$

■

Here are examples of functions $f(x, y, z)$ of three independent variables x, y, and z.

$$f(x, y, z) = x^2 + y^2 - 4xz^2$$
$$f(x, y, z) = 14x - 21y + 6z^2x^2 - 2xyz^2$$
$$f(x, y, z) = 4 \quad \text{(a constant function)}$$

Values for particular choices of x, y, z are computed as with two variables. Although we can consider functions having any number of variables, we will restrict attention mostly to functions of two variables to keep things simple.

In applications, the functions of interest are often described verbally, and it is up to us to write down the appropriate formula. We illustrate this process with the following two-variable economic problem that shows we must take into consideration the **domain of definition**, or **feasible set**, as well as the formula for the function.

ILLUSTRATION Suppose a local bottler of soft drinks produces just two beverages: Brand X Cola and Total Tonic. He must meet monthly fixed costs of \$9500. Further, his plant produces bottles of Brand X at a cost of \$159 per thousand bottles, and Total Tonic at \$176 per thousand. How may we describe his monthly costs C and the feasible production levels for his operation?

Discussion Natural variables in this situation are the production levels

$$x = \text{(thousands of bottles of Brand X per month)}$$
$$y = \text{(thousands of bottles of Tonic per month)}$$

The monthly operating cost is a function $C = C(x, y)$ of these variables. It is the sum of the fixed costs $FC = 9500$ plus the variable costs

$$\begin{aligned} VC &= \text{(cost per thousand X)(thousands of X)} \\ &\quad + \text{(cost per thousand Tonic)(thousands of Tonic)} \\ &= 159x + 176y \end{aligned}$$

so that

$$\begin{aligned} C(x, y) &= \text{(fixed costs)} + \text{(variable costs)} \\ &= 9500 + 159x + 176y \end{aligned} \qquad [1]$$

whatever production levels x and y are chosen. Thus, if the plant is scheduled to produce $x = 70$ (thousand bottles) and $y = 35$ (thousand bottles), the costs for the month will be

$$C(70, 35) = 9500 + 159(70) + 176(35) = 26{,}790$$

If the plant is left idle (production levels $x = y = 0$), the costs reduce to the fixed cost

$$C(0, 0) = 9500 + 159(0) + 176(0) = 9500$$

The operating states of the plant may be represented graphically by thinking of the pairs of numbers (x, y) as coordinates of points in the plane whose coordinate axes are labeled x and y, as in Figure 7.2.

A manufacturer cannot set the production levels (x, y) arbitrarily. Certain **constraints** are imposed by plant capacity, for example. The production levels satisfying all production constraints form the feasible set for the cost function $C(x, y)$. If (x, y) lies outside of the feasible set, the plant cannot operate at these production levels, and it is not meaningful to speak of the operating costs C. As examples of constraints on the variables x and y, in the present problem, we must have $x \geq 0$ and $y \geq 0$; but there may be additional constraints.

Suppose the production lines for Brand X and Total Tonic can put out at most $x = 100$ and $y = 120$ (thousand bottles) per month, respectively. Then what is the domain of definition for the cost, revenue, and profit functions? We must have $x \geq 0$ and $y \geq 0$ as before, but we must also have $x \leq 100$, $y \leq 120$ because of production-line limitations. The feasible set now consists of all (x, y) satisfying

$$0 \leq x \leq 100 \quad \text{and} \quad 0 \leq y \leq 120$$

These constraints determine the shaded rectangular set shown in Figure 7.2b. ■

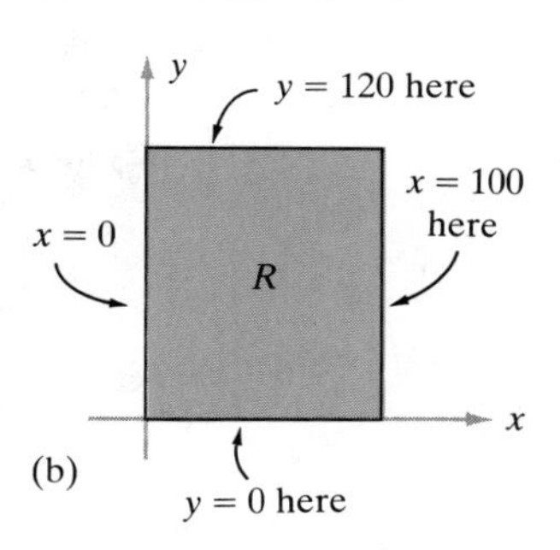

Figure 7.2
In (a) the possible operating states of a bottling plant are represented as points on a coordinate plane. The states in which $x = 70$, $y = 35$, and $x = 0$, $y = 0$ are shown as solid dots. The corresponding operating costs are calculated in the text. In (b) the shaded region R shows the feasible production levels determined by the constraints $0 \leq x \leq 100$ and $0 \leq y \leq 120$. The production variables x and y are measured in thousands of bottles per month.

Let us look at a few simple examples.

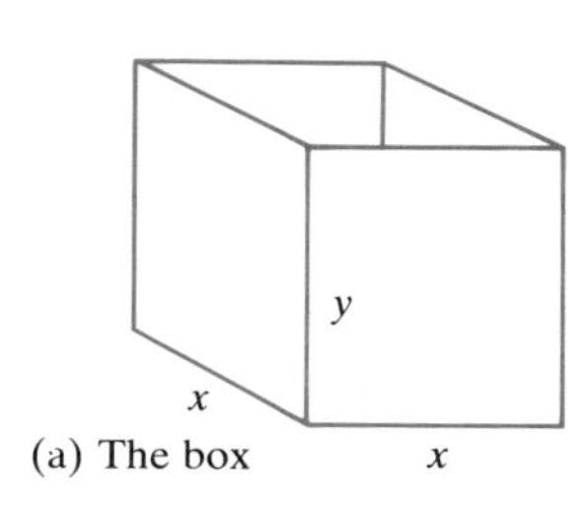

(a) The box

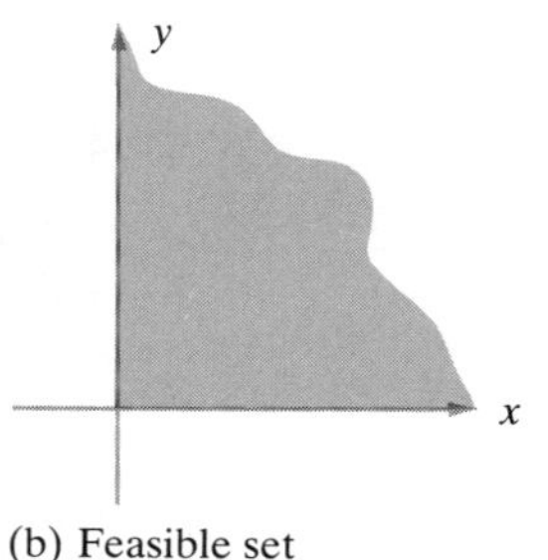

(b) Feasible set

Figure 7.3
The rectangular box of Example 2.

Example 2 The rectangular box shown in Figure 7.3 has a square bottom and no top. Its dimensions are x and y as shown.

(i) Find the total area of cardboard used to make the box.
(ii) If material for the sides costs 15¢ per square foot, and reinforced cardboard for the bottom costs 25¢ per square foot, find the total cost of the materials in the box.

What are the feasible values of x and y?

Solution The box has four sides, each with area xy, so there are $4xy$ square feet of material in the sides; clearly the cost is $15(4xy) = 60xy$ cents. The bottom has area x^2, and it costs $25x^2$ cents. The total area is

$$A(x, y) = 4xy + x^2 \quad \text{square feet}$$

and the cost of materials is

$$C(x, y) = 60xy + 25x^2 \quad \text{cents}$$

Only values $x > 0$ and $y > 0$ make sense in this setting. The feasible set is the shaded quadrant in the x, y plane shown in Figure 7.3b. ■

Example 3 The Whiztek Corporation is going to produce two versions of its new videogame "Spaceants." Producing each cartridge incurs costs for materials and labor. Let x and y be the number of units produced per week for Models I and II, respectively. If weekly fixed costs for this department are \$70,000 and if labor costs average out to \$12.50 per hour, find the total weekly costs $C(x, y)$. If every unit produced is sold at a price

\$100 for Model I   \$137 for Model II

use the following data to find the weekly revenue $R(x, y)$ and profit $P(x, y)$.

Model	Cost of material for one unit (chips, housing, wiring)	Labor required (hours)
Spaceant I	\$42.00	1.8
Spaceant II	\$53.00	1.9

Solution Fixed costs are $FC = 70{,}000$. Variable costs are $VC =$ (cost of materials) + (cost of labor). At production levels x, y we have

$$\begin{aligned}
(\text{cost of materials}) &= 42x + 53y \\
(\text{cost of labor}) &= 12.5 \times (\text{hours required}) \\
&= 12.5 \times (1.8x + 1.9y) = 22.5x + 23.75y
\end{aligned}$$

so the total variable cost is the sum of these expressions,

$$VC = 64.5x + 76.75y$$

and therefore

$$C(x, y) = FC + VC = 70{,}000 + 64.5x + 76.75y \quad \text{dollars per week.}$$

Revenue is

$$\begin{aligned} R(x, y) &= (\text{price per unit}) \times (\text{number of units}) \\ &= 100x + 137y \quad \text{dollars} \end{aligned}$$

and the net profit is

$$P(x, y) = R - C = -70{,}000 + 35.5x + 60.25y$$

Because no special restrictions were placed on the production levels, we consider the feasible set for each of these functions to be the quadrant consisting of points (x, y) satisfying the obvious requirements $x \geq 0$ and $y \geq 0$. ■

Exercises 7.1

Let $f(x, y) = 1 + 2x - y$ be defined for all x and y. Find the values of f at the following points in the coordinate plane.

1. $x = 0, \quad y = 0$

2. $x = 1, \quad y = 2$

3. $x = 2, \quad y = 1$

4. $(x, y) = (-1, 2)$

5. $(x, y) = (-2, -1)$

6. $(x, y) = (4.3, -5.7)$

Consider the function $P(x, y) = -1000 + 4.5x + 7.0y - 0.1xy$ defined for all $x \geq 0$ and $y \geq 0$. Evaluate P at each point listed below:

7. $x = 0, \quad y = 0$

8. $x = 15, \quad y = 0$

9. $x = 10, \quad y = 12$

10. $x = 3.8, \quad y = 7.2$

11. $x = 7.6, \quad y = 7.6$

12. $x = 3.2, \quad y = 11.4$

The next two problems refer to the economic illustration discussed in this section.

13. Evaluate the monthly cost function $C(x, y) = 9500 + 159x + 176y$ at the following production levels x, y.
(i) $x = 0, \quad y = 50$
(ii) $x = 50, \quad y = 0$
(iii) $x = 100, \quad y = 50$
(iv) $x = 75, \quad y = 75$

14. On graph paper make an accurate copy of the feasible set shown in Figure 7.2b. Plot the points (x, y) corresponding to the following production levels. Which ones are feasible?
(i) $x = 0, \quad y = 50$
(ii) $x = 50, \quad y = 0$
(iii) $x = -10, \quad y = 40$
(iv) $x = 100, \quad y = 50$
(v) $x = 100, \quad y = 100$
(vi) $x = 10, \quad y = 130$

In Exercises 15–22, evaluate each function at the points (2, 3) and (−1, 1).

15. $1 + x^2 + y^2$

16. $1 - x + y$

17. $\dfrac{2xy}{x^2 + y^2}$

18. $\dfrac{x + y}{xy}$

19. $5000 + 20x^2 + 100xy - 5y^2$

20. $30 - 2x^2 - 4y^3$

21. $\sqrt{5x^2 - 8xy + 5y^2}$

22. $e^{-(x^2+y^2)/2}$

23. An automobile fabricating plant producing two types of automobiles has found that its monthly profit P depends on the production levels

$$x = (\text{number of compacts}) \qquad y = (\text{number of standard models})$$

according to the formula

$$P(x, y) = -900{,}000 + (500x - 0.1x^2) + (700y - 0.2y^2) - 0.01xy$$

Find the profit if the following production levels are chosen.
(i) $x = 1500, \quad y = 2000$
(ii) $x = 2000, \quad y = 500$
(iii) $x = 2000, \quad y = 1000$

24. A pottery shop decides to produce pitchers and bowls for a week. Pitchers are sold for \$19.50 each and bowls for \$32.50. Revenue from sales is offset by production costs (materials, fuel, labor) of \$12.50 for each pitcher and \$18.50 for each bowl. In addition, the shop's fixed costs amount to \$1000 per week. Describe the cost function, revenue function, and profit function for this shop in terms of the production levels

$$x = (\text{number of pitchers}) \qquad y = (\text{number of bowls})$$

25. A factory produces two lines of shoes, styles A and B. Overhead costs amount to \$2000 per day, and production costs (labor and materials) amount to \$13 and \$20 per pair of shoes. All shoes produced are sold to retail outlets at prices of \$28 and \$35, respectively. If x and y are the numbers of shoes produced per day in each style, find
(i) the daily cost function $C(x, y)$
(ii) the daily revenue $R(x, y)$, assuming all shoes are sold
(iii) the daily profit $P(x, y)$

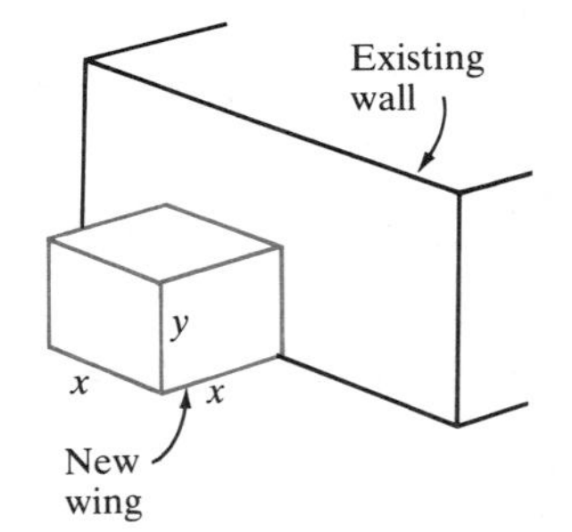

Figure 7.4
The structure in Exercises 26–29.

A new wing is to be added to an existing building, as shown in Figure 7.4. It will have a square base and dimensions x and y (in feet) as shown. Answer the following questions.

26. What is the volume enclosed by this structure?

27. What is the exterior surface area (roof plus three outside walls)?

28. Suppose daily heat loss through the roof is 8 thermal units per square foot, and through the side walls is 4 thermal units per square foot. Find a formula for the total daily heat loss (in thermal units) through the sides and roof of this structure.

29. Roof construction costs \$35 per square foot, side walls cost \$28 per square foot, and the concrete floor costs \$42 per square foot. What is the total cost of construction (roof, sides, and floor)?

For the following functions of three variables, find the values

$$\text{(i)}\ \ f(0, 0, 0) \qquad \text{(ii)}\ \ f(1, 3, 2) \qquad \text{(iii)}\ \ f(-1, 1, 0)$$

30. $f(x, y, z) = 3 - x - y + 2z$

31. $f(x, y, z) = 10 + x^2 - 4y^2z + z^2$

32. $f(x, y, z) = xy + yz + 3xz$

33. $f(x, y, z) = xyz - 6$

34. $f(x, y, z) = 4x^{0.2}y^{1.7}z^{1.2}$

35. A rectangular trough has dimensions x, y, z shown in Figure 7.5. Give formulas for:
(i) the volume of water it will hold
(ii) the total area of the sides and bottom

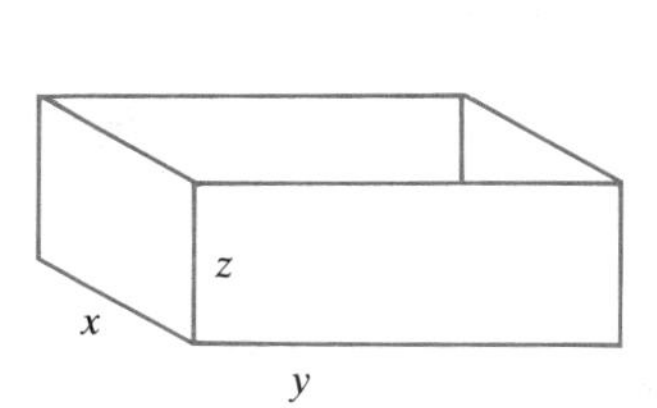

Figure 7.5
Exercise 35.

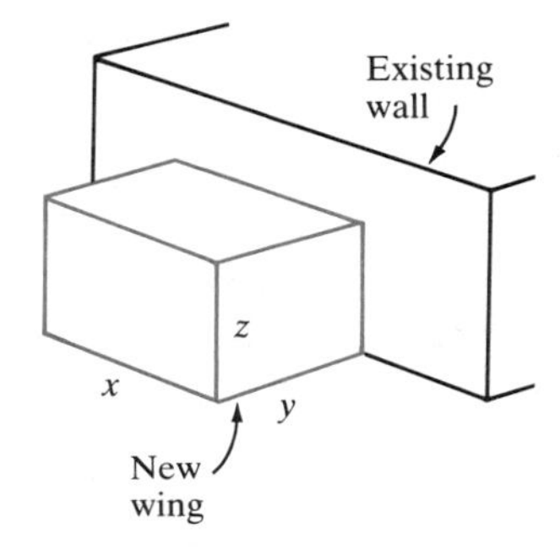

Figure 7.6
Exercises 36–38.

An extension is to be added to an existing building, as shown in Figure 7.6. Each dimension x, y, z can be varied independently. Find:

36. the volume $V(x, y, z)$ enclosed by this structure

37. the exterior surface area $A(x, y, z)$ (roof and three outside walls)

38. total construction costs $C(x, y, z)$ if the roof costs \$28 per square foot, side walls cost \$35 per square foot, and the concrete floor costs \$42 per square foot

7.2 Visualizing Functions of Several Variables: Graphs and Level Curves

One way to visualize a function $f(x, y)$ of two variables is to sketch the **graph** of the function, which is a surface in three-dimensional space, instead of a curve in the plane. We shall define this concept below. We shall also introduce the notion of **level curve**, which reduces many questions about the function to exercises in map reading. There is, unfortunately, no easy way to visualize functions of more than two variables; they must be handled algebraically, without the aid of accurate pictures. But our discussion of what happens with two variables does suggest what happens when f has many variables.

Before sketching the graph of a function $z = f(x, y)$ of two variables we must set up coordinates in three-dimensional space. We do this by marking an origin O and drawing three perpendicular coordinate axes through it. It is customary to identify the two horizontal axes with the independent variables x and y as shown in Figure 7.7a, and the vertical axis with the dependent variable z.

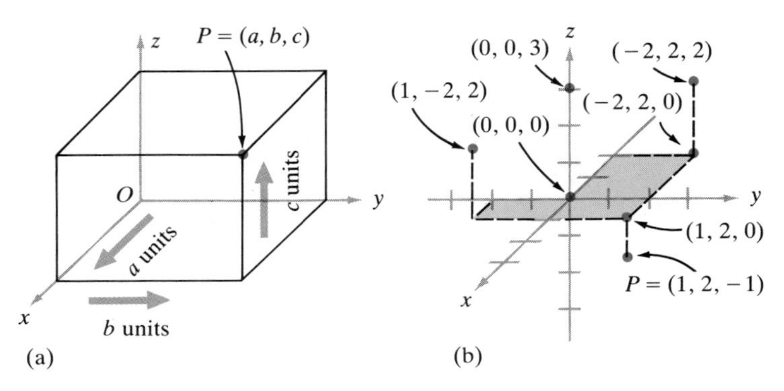

Figure 7.7
Cartesian coordinates in three-dimensional space. Unit lengths are marked off on each coordinate axis in (b). In (a) we show how to locate a typical point $P = (a, b, c)$ starting from the origin O. We also show the locations of a number of points. Dashed lines are visual guidelines showing how far to move parallel to each axis.

Each point P in space is then described by three coordinates (a, b, c). Starting from the origin, we reach $P = (a, b, c)$ by moving a units in the x direction, b units parallel to the y direction, and then c units parallel to the z axis, as in Figure 7.7a. To put it another way, we may locate $P = (a, b, c)$ by setting up a rectangular box with side lengths a, b, and c in the x, y, and z directions. Then $P = (a, b, c)$ is the corner point opposite the origin O (see Figure 7.7a).

For example, to reach the point $P = (1, 2, -1)$ shown in Figure 7.7b we move $+1$ unit in the x direction, then $+2$ units parallel to the y axis, and finally -1 unit (move downward) parallel to the z axis. The figure shows the locations of several other points, such as the ones whose coordinates are $(1, -2, 2)$, $(0, 0, 3)$, and $(-2, 2, 2)$. The origin O has coordinates $(0, 0, 0)$. The x axis and y axis together determine a horizontal plane, which we call the **xy plane.** It consists of all points $P = (x, y, z)$ such that $z = 0$; that is, it is the solution set of the simple equation $z = 0$. We do not move at all in the z direction to reach points in this plane. Similarly, we define the **xz plane** and the **yz plane**, which are determined by the equations $y = 0$ and $x = 0$, respectively.

If $f(x, y)$ is a function of two variables, its **graph** is a surface in three-dimensional space, determined as follows. First we sketch a copy of the feasible set for f in the xy plane by identifying pairs (x, y) for which f is defined with points $(x, y, 0)$ in the xy plane. In Figure 7.8 the feasible set for the function f is indicated by the shaded region in the xy plane. Over each point $Q = (x, y, 0)$ in this region, we then find the point $P = (x, y, z)$ lying a distance $z = f(x, y)$ above Q (or below Q if $z = f(x, y)$ is negative). As Q varies within the feasible set, the points $P = (x, y, z)$ with $z = f(x, y)$ trace out a surface. This surface is the graph of f.

Conversely, if we are given the graph of a function f, we may determine the value $z = f(x, y)$ geometrically. We simply locate the point $Q = (x, y, 0)$ in the xy plane and measure the distance z we must move up or down to reach the corresponding point $P = (x, y, z)$ on the surface. Then $z = f(x, y)$. (Remem-

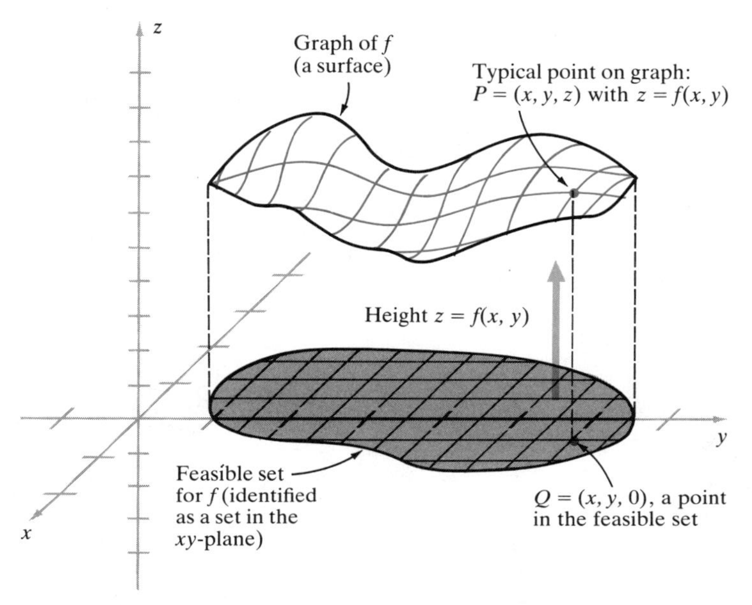

Figure 7.8
The graph of a typical function of two variables $f(x, y)$ is a surface. Points on the graph lie above the feasible set for f, the shaded region in the xy plane. Within the feasible set we show a grid of straight lines drawn parallel to the x axis and y axis. The shape of the surface is suggested by projecting this grid up onto the surface, as if a beam of light were shown up through the grid onto the surface. Not all of the surface can be seen in this perspective view; some parts lie out of sight behind the two "hilltops."

ber: We move *down* (move in the *negative* z direction) to reach points for which z is negative.)

Surfaces are hard to draw, requiring skills learned in more advanced courses. Consequently, we will not ask the reader to make sketches of the graphs of functions $f(x, y)$. But it is useful to be familiar with this concept. A sketch of the graph instantly provides a great deal of information about the behavior of the function. Furthermore, powerful computer programs can now plot perspective views of the graph of a given function. As a result, such sketches more and more frequently find their way into the decision-making process.

Another way to portray the behavior of f is to sketch its **level curves**. These are the curves in the xy plane consisting of all points (x, y) where $f(x, y)$ takes on a particular constant value, say $f(x, y) = c$. For each choice of the constant c there is a different level curve, the solution set of the equation $f(x, y) = c$. Information can be read out of the pattern of level curves for a function much as we would read a contour map. Figure 7.9 shows the level curves $H(x, y) = c$ for a certain function $H(x, y)$, choosing values for c in steps of 500. The curves for $c = 9000$, $c = 10{,}000$, $c = 11{,}000$, $c = 12{,}000$, and $c = 13{,}000$ are labeled. Given any point in the feasible set, such as the one marked A in Figure 7.9, we can estimate the value of H there by examining the nearest level curves. For this particular point, the nearest level curves have values $H = 11{,}000$ and $H = 11{,}500$, so the value

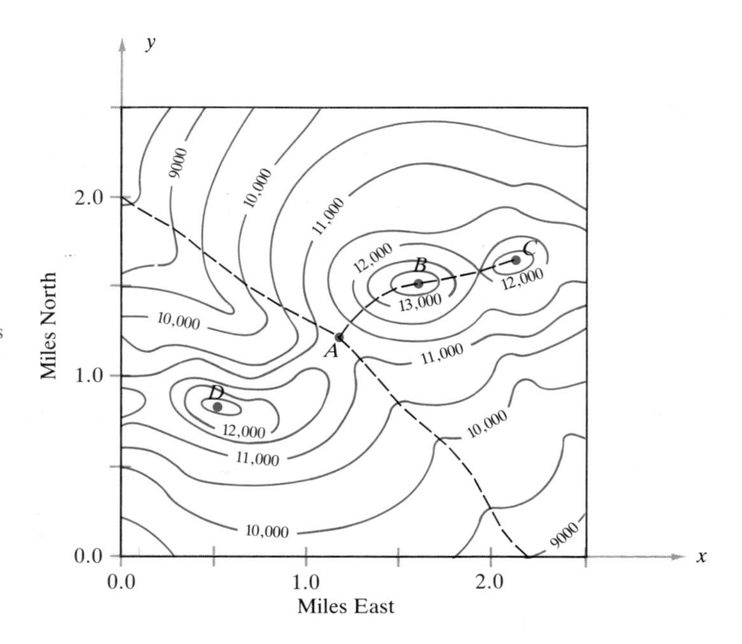

Figure 7.9
The function $H(x, y)$ gives the altitude above sea level as a function of position coordinates. In this contour map, solid curves are the level curves $H(x, y) = \text{const}$; on each level curve, the terrain has constant altitude. A level curve may consist of several pieces: Consider the contours where $H = 11{,}000$ or $H = 12{,}000$. Contours are given every 500 feet, but only the 1000-foot contours are labeled to keep the diagram from becoming too cluttered. In a very accurate map the contours might be drawn for every 100-foot change in altitude.

$H(A)$ is somewhere in between. A good estimate would be $H(A) \approx 11{,}250$. A finer pattern of level curves would allow us to make better estimates.

This diagram is, in fact, a contour map; $H(x, y)$ gives the altitude (in feet) for a certain part of the California Sierras. The curve $H(x, y) = c$ locates all points whose altitude is c. The coordinates x and y are measured in miles east and north of a geological reference marker located at the origin. Given only the pattern of level curves, it is not hard to read out a great deal of information about the terrain. It is certainly not necessary to sketch the graph of $z = H(x, y)$ (that is, to make a perspective drawing of the terrain) to see that the map depicts a mountain pass whose highest point A is about 11,250 feet above sea level. Points B, C, and D are mountain peaks. The dashed curve from bottom to top represents a trail from one side of the mountain range, across the pass, and down the other side. You might enjoy thinking over the following questions: (i) What happens if we follow the side trail ABC? (ii) Which of the peaks B, C, and D is the highest? (iii) What are the highest and lowest points on the map? (iv) In which part of the map is the terrain *steepest*?[†]

The graph of a function $f(x, y)$ and the pattern of level curves $f(x, y) = \text{const}$ are different ways of presenting the same information. Surfaces are hard to draw or visualize from perspective sketches, so in this book we will rely on patterns of level curves. One remark on the connection between surfaces and level curves may help you read level curve diagrams. We illustrate it with the

[†] Answers: (i) We climb up to B, then down and across a "saddle" (high pass), then up to peak C; (ii) B; (iii) Peak B is the highest point. Lowest points are in the upper left corner, elevation about 8250 feet; (iv) just north of D, where the contour lines bunch very closely together

surface shown in Figure 7.10. Suppose a model of this surface is immersed in water. The "shoreline" that appears on the surface, called a *contour line*, obviously consists of points with the same heights above the xy plane. Different water levels yield different contours, some of which are shown in Figure 7.10a. If these contours are projected down from the surface onto the xy plane, we obtain level curves for the function $f(x, y)$, as shown in Figure 7.10b. You should compare the

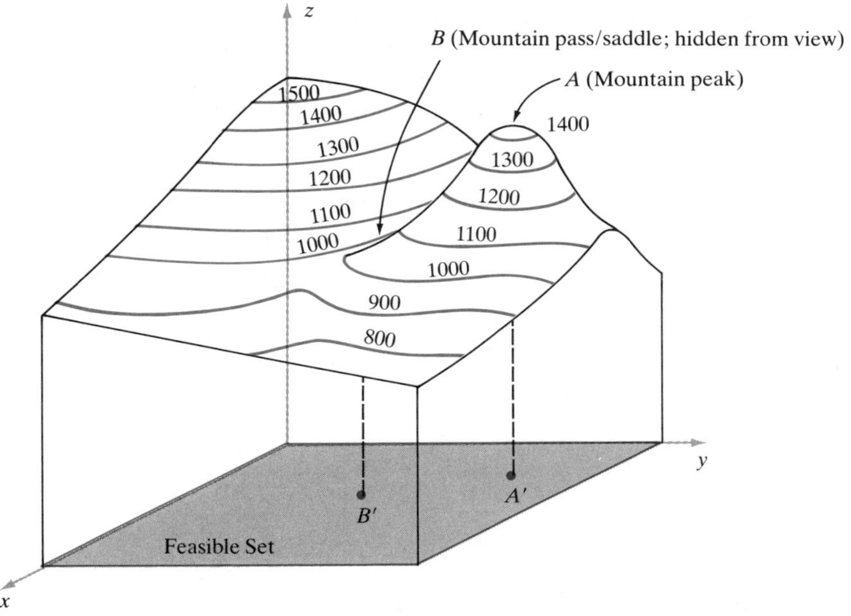

(a) Graph of $z = f(x, y)$ with various contour lines $z = 800$, $z = 900, \ldots, z = 15{,}000$ drawn in.

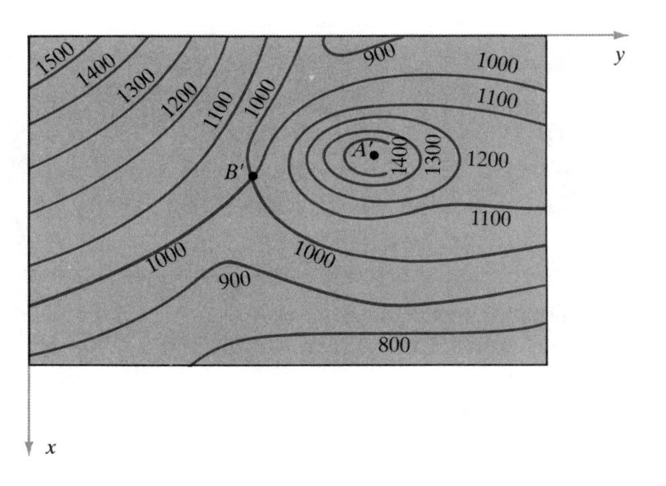

(b) The feasible set for $f(x, y)$ showing the level curves corresponding to the contours in Figure 7.10a.

Figure 7.10
In (a) we show the graph of a function $z = f(x, y)$. On the graph we have drawn in the contours where the surface has constant height above the xy plane. Contours for heights $z = 800$, $z = 900, \ldots, z = 1500$ are shown. When these contours are projected onto the feasible set in the xy plane, we obtain the pattern of level curves $f(x, y) = 800, \ldots, f(x, y) = 1500$ shown in (b). This is what we would see if we looked down at the surface along the z axis. The level curves reveal the behavior of the whole surface, even the parts obscured from view in (a).

behavior of the shoreline contours and the corresponding level curves near the hilltop at A and near the pass between adjacent hills at B. The contour where $z = 1000$ is particularly interesting. It is given by a water level that just reaches the pass at B. The corresponding level curve, made up of points in the feasible set for which $f(x, y) = 1000$, has a characteristic "X" shape right at the pass B'. Moving away from B', the altitude increases in two directions (we are moving toward one of the nearby hilltops), and the altitude decreases in two other directions as we move further down the valley between the hills.

The following examples illustrate an important use of level curve diagrams. By examining level curves we can determine (approximately) where a function attains its largest or smallest values.

Example 4 Level curves $f(x, y) = \text{const}$ are shown for the profit function of a certain automobile assembly plant (see Figure 7.11). The variables are

$$x = (\text{number of compacts per month})$$
$$y = (\text{number of standard models per month})$$

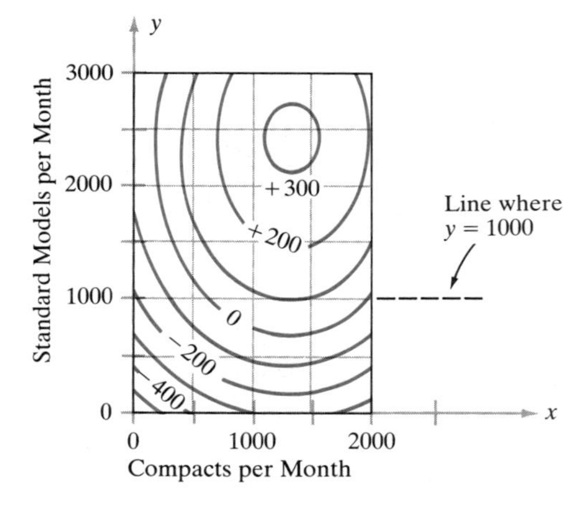

Figure 7.11 Profit function for an automobile assembly plant. Level curves $f(x, y) = \text{const}$ are given in steps of \$100,000 per month. Note the location of the "break-even" curve $f = 0$. Negative values of f stand for losses.

The feasible set is the rectangle $0 \leq x \leq 2000, 0 \leq y \leq 3000$. The profit $f(x, y)$ is measured in thousands of dollars per month.

Which production levels yield

(i) the maximum feasible profit?
(ii) the greatest possible loss (minimum profit)?

Describe what happens to $f(x, y)$ if y is held fixed at $y = 1000$ standard models per month and x is increased from 0 to 2000 per month.

Solution Profit is negative throughout much of the feasible set. Minimum profit $f = -600$ thousand dollars occurs at the lower left corner at production levels $x = 0, y = 0$ (plant idle; f = fixed costs). Profit is also low in the lower right corner where $x = 2000$, $y = 0$, and $f \approx -400$ thousand dollars; but the situation is not as bad as when $x = 0, y = 0$. Positive values of f occur toward the upper right, and maximum f occurs near $x = 1300$, $y = 2400$, where f is about 325 thousand dollars per month.

If production of standard models is fixed at $y = 1000$ and x is increased, the points $(x, y) = (x, 1000)$ representing the production levels move horizontally from left to right. The value of $f(x, y)$ starts at $f(0, 1000) \approx -200$ thousand, gradually increases until it reaches a maximum at $x = 1300$, where we have $f(1300, 1000) \approx +100$ thousand dollars; thereafter, $f(x, y)$ decreases steadily to its value at the right-hand edge, $f(2000, 1000) \approx 0$. ■

Example 5 In Figure 7.12 we show the level curves of a function $z = f(x, y)$ defined on a rectangular feasible set. Answer the following questions by examining the level curve pattern.

(i) Estimate the value of f at the points marked A, B, C, and D.
(ii) Estimate the maximum value achieved by $f(x, y)$ on the feasible set.
(iii) What are the x and y coordinates of the point where f achieves its maximum value?
(iv) What would happen to the value of f if we started from C and moved left? What if we moved vertically up from C?

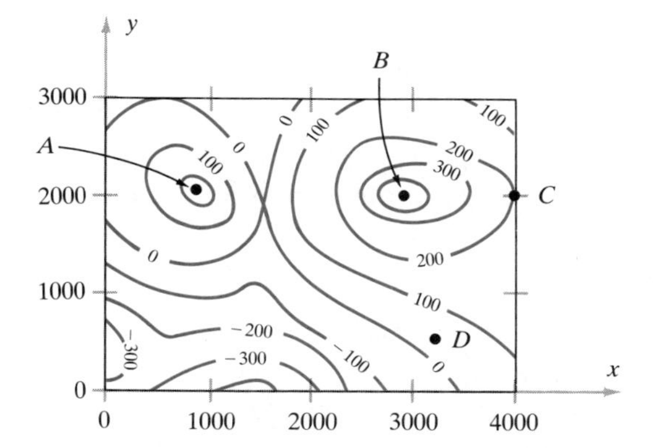

Figure 7.12 Level curves for the function $z = f(x, y)$ in Example 5. We are interested in the behavior of f at the points marked A, B, C, and D.

Solution By examining the nearest contours, we find that

At A the value of f is slightly larger than 200 (but smaller than 300).
At B the value of f is slightly larger than 400 (but smaller than 500).
At C the value of f is 200 because C lies on the level curve $f = 200$.
At D the value of f is about 50.

The largest value for f is attained at B, where f is slightly larger than 400. The coordinates of B are $x = 3000$, $y = 2000$. If we start at C and move left, the values of f increase from the value $f(C) = 200$; if we move upward, along the right-hand margin of the feasible set, the values of f decrease. ■

Example 6 A manufacturer producing two products can set his production levels x and y at any values in the shaded feasible set shown in Figure 7.13. His

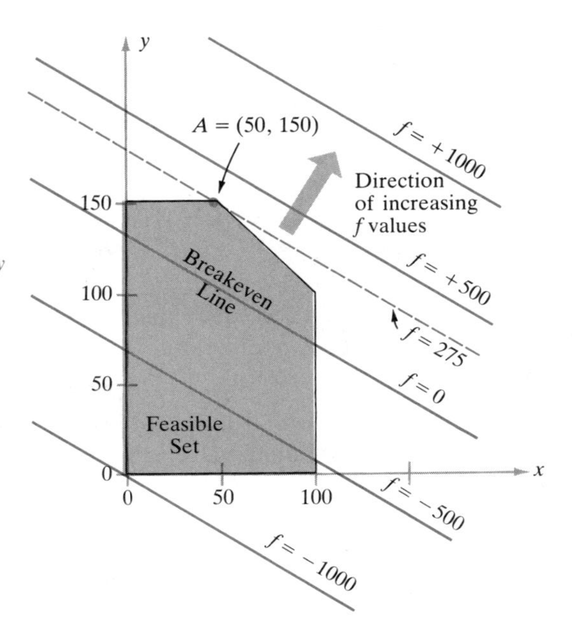

Figure 7.13
Level curves for the function $f(x, y) = -1000 + 4.5x + 7.0y$ are parallel straight lines. We may plot a typical line $-1000 + 4.5x + 7.0y = c$ by finding any two points on the line and drawing a straight line through them. We show the lines taking steps $\Delta c = 500$, starting with $c = -1000$. It should be clear that $f(x, y)$ increases as (x, y) moves upward and to the right, as shown by the arrow. Maximum profit occurs at $A = (50, 150)$ where the profit is $f(A) = f(50, 150) = 275$.

profit at production level (x, y) is

$$f(x, y) = -1000 + 4.5x + 7.0y \quad \text{dollars per week}$$

Sketch a few of the level curves $f(x, y) = \text{const}$ and use them to decide where in the feasible set the profit is maximized. In particular, show the level curve $f(x, y) = 0$ (the break-even curve for this operation).

Solution The level curve consisting of all points (x, y) for which

$$-1000 + 4.5x + 7.0y = c \quad (c \text{ a constant}) \qquad [2]$$

is a straight line in the xy plane. For each of these lines, slope $= -4.5/7.0 = -0.64$; the slope does not depend at all on the choice of c. Thus the level curves are parallel lines. The figure shows several of them superimposed on the feasible set, including the break-even line where $f = 0$. At points above this line, $f(x, y)$ is positive, and the operation is making money.

If we let (x, y) move about within the feasible set, $f(x, y)$ stays constant if (x, y) moves along one of the level curves and increases if (x, y) moves perpendicular to them in the direction of increasing f (indicated by the shaded arrow). Even from our crude pattern of level curves, it appears that the maximum profit is achieved at the point $A = (50, 150)$. This is made very clear if we draw in (dashed curve) the level curve passing through A. At A the function has the value $f(A) = f(50, 150) = 275$, so the dashed level curve is the one such that

$$f(x, y) = 275 \quad \text{or} \quad -1000 + 4.5x + 7.0y = 275$$ ■

We will seldom have to determine the level curves of a function more complicated than a linear one such as [2]. This would require skills usually discussed only in advanced courses. You are much more likely to encounter the (easier) converse problem: Given a detailed set of level curves, analyze the behavior of the function, as we did in the last examples. Economic data are often presented in this way.

Exercises 7.2

Using graph paper, set up a three-dimensional system of coordinate axes as in Figure 7.7 and plot the locations of the following points.

1. (0, 0, 1) **2.** (0, 0, −1) **3.** (1, −1, 1)

4. (0, 0, 0) **5.** (3, 4, 5) **6.** (−2, 2, 5)

7. (4, 0, −3) **8.** (−3, −4, 3) **9.** (0, 2, 2)

10. (0, 2, 0)

11. In Figure 7.14a we show level curves $f(x, y) = \text{const}$ for a certain function superimposed on a shaded triangular region R. By examining this pattern of curves, answer the following questions.
(i) Where on R is the maximum value of f achieved?
(ii) Where is the minimum value achieved?
(iii) What is the value of f at C?
(iv) How do the values vary if we move from B to A along the edge of R?

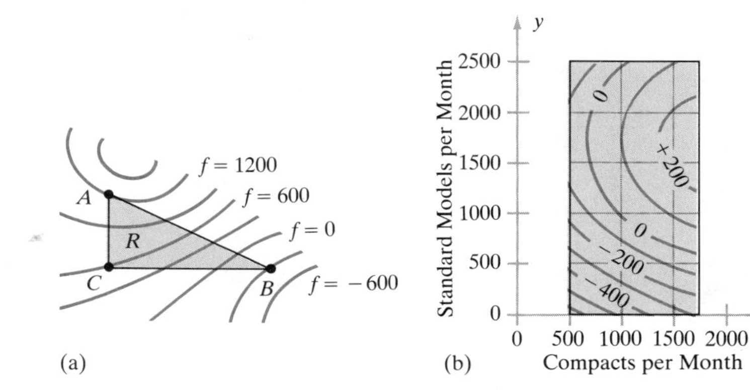

Figure 7.14

12. In Example 4, estimate the value of $f(x, y)$ at the following production levels (x, y), using the pattern of level curves shown in Figure 7.11.
(i) $x = 1000$, $y = 1000$
(ii) $x = 2000$, $y = 500$
(iii) $x = 500$, $y = 1500$

13. Suppose that production of compacts in Example 4 is held fixed at $x = 1000$ per month, and y is increased from 0 to 3000. How does the corresponding point $(x, y) = (1000, y)$ move within the feasible set? How does $f(x, y)$ behave as y increases in this situation? Which y value yields the greatest profit in this situation? What is this value of $f(x, y)$ (approximately)?

14. Consider an automobile manufacturer's profit function under the following circumstances. Production of compacts cannot exceed 1750 per month, and a minimum of 500 compacts per month must be produced to meet standing orders; thus $500 \leq x \leq 1750$. Production of standard models can be set anywhere in the range $0 \leq y \leq 2500$. These requirements yield the shaded feasible set in Figure 7.14b. Level curves for the profit function $f(x, y)$ are given in the figure. What is the maximum feasible profit, and at what production levels is it achieved?

15. Figure 7.15 shows level curves for altitudes within a portion of a regional park system. Altitudes $A(x, y)$ are measured in meters; x and y coordinates are measured in kilometers from a reference marker. The dashed curves are foot trails through the area.
(i) How does the altitude vary if you follow the trail AA' through the region?
(ii) How does the altitude vary if you follow the trail BB'?
In all of the rectangular region mapped out here, where does the lowest altitude occur? The highest? (Estimate the x and y coordinates.) If you follow trail BB', what are the lowest and highest altitudes encountered?

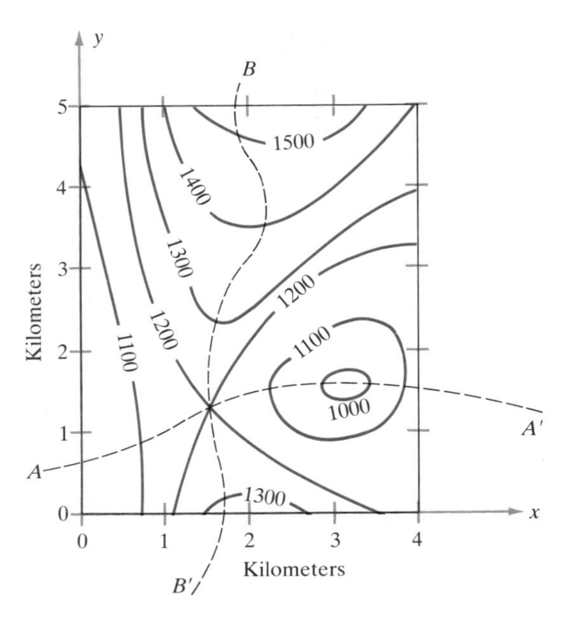

Figure 7.15
Level curves for altitude in Exercise 15.

For each function sketch the level curves where $f = 0, f = 1, f = 2$.

16. $f(x, y) = 2x + 3y - 1$

17. $f(x, y) = x - y + 2$

18. $f(x, y) = 2.5x + 1.8y$

19. $f(x, y) = x^2 - 2y + 1$

20. $f(x, y) = y^2 - 2x$

21. Draw the level curves where $P = 0, P = 2000$, and $P = 6000$ for the profit function $P(x, y) = 400x + 300y - 10{,}000$.

22. Answer the questions posed in Example 6, using the same feasible set, but a different profit function

$$f(x, y) = -1000 + 6.5x + 4.0y$$

(*Hint*: Make an accurate copy of the feasible set on graph paper. This will not be difficult, because the vertices have coordinates that are whole numbers. Then draw in the level curves $f = c$ taking steps of $\Delta c = 500$, starting with $c = -1000$.)

23. Find the maximum value of $f(x, y) = 10 + x - y$ on the feasible set determined by $0 \le x \le 8$ and $0 \le y \le 2$ (a rectangle). (*Hint*: First sketch the level lines where $f = c$ superimposed on a picture of the feasible set. Use graph paper.)

7.3 Partial Derivatives

If $f(x, y)$ is a function of two variables and $P = (x, y)$ is a fixed base point in its feasible set, we may move in many directions to reach nearby points $Q = (x + \Delta x, y + \Delta y)$. For example, if Δx is arbitrary and $\Delta y = 0$, we move horizontally, parallel to the x axis, to reach the adjacent point $Q' = (x + \Delta x, y)$. If Δy is arbitrary and $\Delta x = 0$, then we move vertically from P to reach $Q'' = (x, y + \Delta y)$, as shown in Figure 7.16a. However, the increments Δx and Δy are independent of one another, and we may reach any point by choosing Δx and Δy appropriately. For example, if $P = (1, 1)$ as in Figure 7.16b, and we wish to reach $Q = (4, -1)$, we should take $\Delta x = 3$ and $\Delta y = -2$ to get $Q = (1 + \Delta x, 1 + \Delta y) = (4, -1)$. As we can see, a "displacement" from one point P to another point Q is a quantity that has both a magnitude and a direction.

Now let $f(x, y)$ be any function defined at and near a point $P = (x, y)$. The value of the function there is just $f(P) = f(x, y)$. At a nearby point $Q = (x + \Delta x, y + \Delta y)$ its value is $f(Q) = f(x + \Delta x, y + \Delta y)$. Thus the change in the value of f is a real number

$$\Delta f = f(Q) - f(P) = f(x + \Delta x, y + \Delta y) - f(x, y)$$

but the change in the variable (the displacement from P to Q) is not a number. We cannot form averages

$$\frac{\Delta f}{\Delta P} = \frac{\text{change in value of } f}{\text{change in variable}}$$

as we did in Chapter 2, where we considered functions of one variable. We need a new idea to define rates of change and derivatives for functions of several variables.

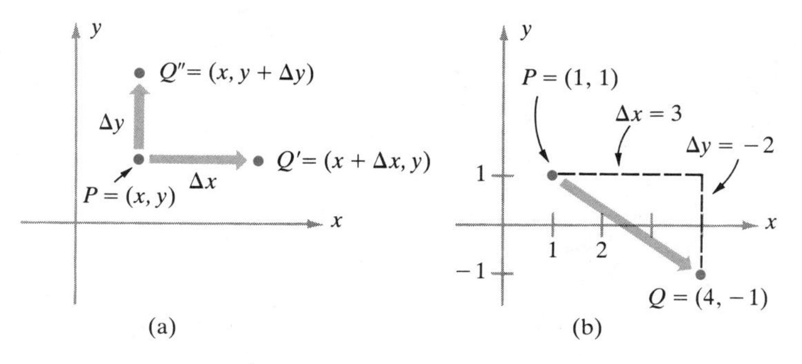

Figure 7.16
In (a) we show typical displacements from a base point P to nearby points Q' or Q''. In (b) we move from $P = (1, 1)$ to $Q = (4, -1) = (1 + \Delta x, 1 + \Delta y)$ by taking increments $\Delta x = 3$ and $\Delta y = -2$. The magnitude of the displacement (the total distance moved going from P to Q) is given by the Pythagorean theorem: Distance from P to $Q = \sqrt{(\Delta x)^2 + (\Delta y)^2} = \sqrt{13} = 3.605\ldots$

The idea is this: If a base point $P = (x, y)$ is given and we restrict attention to displacements from P in the x direction only (Δx arbitrary, $\Delta y = 0$), then the average rate of change in the x direction

$$\frac{\Delta f}{\Delta x} = \frac{f(x + \Delta x, y) - f(x, y)}{\Delta x} = \frac{f(Q) - f(P)}{\Delta x}$$

is defined for all small displacements $\Delta x \neq 0$. If these averages have a limit value as Δx approaches zero, we obtain the **rate of change of f with respect to x** (holding y fixed) at the point P. This rate of change, if it exists, is denoted by

$$\frac{\partial f}{\partial x}(x, y) = \frac{\partial f}{\partial x}(P) = \lim_{\Delta x \to 0} \left(\frac{f(x + \Delta x, y) - f(x, y)}{\Delta x} \right) \qquad [3]$$

It is called the **partial derivative of f with respect to x**. Similarly, we may define the **partial derivative of f with respect to y** as

$$\frac{\partial f}{\partial y}(x, y) = \frac{\partial f}{\partial y}(P) = \lim_{\Delta y \to 0} \left(\frac{f(x, y + \Delta y) - f(x, y)}{\Delta y} \right) \qquad [4]$$

by considering displacements from P parallel to the y axis.

To see what this means, consider Figure 7.17 which shows the level curve pattern for some function $f(x, y)$ in the vicinity of a typical base point P. Suppose we move to the right along a line through P parallel to the x axis at a speed of one unit per second, reading off the values of $f(x, y)$ at each position on the line. How fast would the values be changing as we move past P? According to [3] the instantaneous rate of change would just be the partial derivative $\partial f/\partial x(P)$. Similarly, if we move upward, parallel to the y axis, we would find that the values of $f(x, y)$ change at a rate $\partial f/\partial y(P)$ units per second as we move past P. The

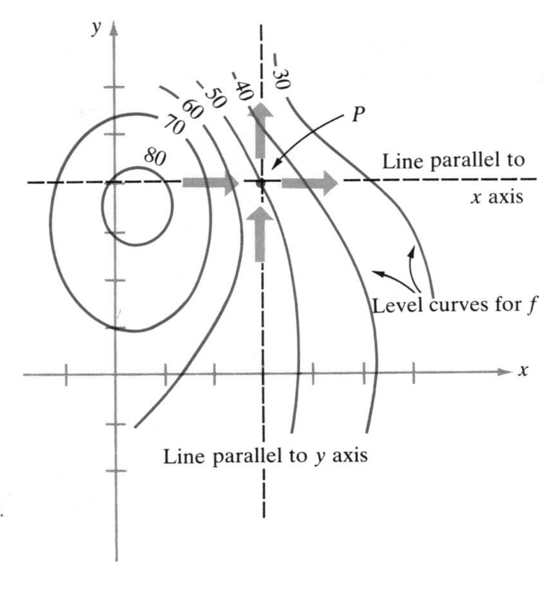

Figure 7.17
Behavior of a function $f(x, y)$ near a point P is described by the pattern of level curves. The partial derivatives $\partial f/\partial x(P)$ and $\partial f/\partial y(P)$ tell how fast values of f are changing as we move past P along the dashed lines parallel to the x axis and y axis, respectively. At P, both partial derivatives are *negative* numbers. Can you see why by examining the level curves near P?

partial derivatives $\partial f/\partial x$ and $\partial f/\partial y$ can have different values at different base points P. They are usually defined wherever f itself is defined, so these partial derivatives are also functions of two variables.

How do we actually calculate partial derivatives? Fortunately, if we think of the limits [3] and [4] in the right way, the problem of calculating $\partial f/\partial x$ and $\partial f/\partial y$ reduces to calculating ordinary derivatives of functions of one variable. Thus we can bring to bear all the techniques of Chapters 2 and 3.

> **RULE FOR CALCULATING PARTIAL DERIVATIVES** To calculate $\partial f/\partial x$, simply regard y as a constant wherever it appears in $f(x, y)$, and differentiate the resulting function of x in the ordinary way. The same rule, of course, applies to $\partial f/\partial y$: Regard x as a constant wherever it appears, and differentiate with respect to y.

Example 7 For $f(x, y) = 400 + x^2 + y^2 + 3xy$, calculate $\partial f/\partial x$ and $\partial f/\partial y$. What are the numerical values

$$\frac{\partial f}{\partial x}(1, 3) \qquad \frac{\partial f}{\partial y}(1, 3)$$

at the particular point $P = (1, 3)$? Where is $\partial f/\partial x$ equal to zero?

Solution We find $\partial f/\partial x$ as a function of x and y by applying the preceding rule: Carry out the differentiation

$$\frac{\partial f}{\partial x} = \frac{\partial}{\partial x}(400 + x^2 + y^2 + 3xy)$$

regarding y as a constant wherever it appears in the expression $400 + x^2 + y^2 + 3xy$. Using standard rules for derivatives in one variable, we obtain

$$\begin{aligned}\frac{\partial f}{\partial x} &= \frac{\partial}{\partial x}(400 + x^2 + y^2 + 3xy)\\ &= 400\cancel{\frac{\partial}{\partial x}(1)} + \frac{\partial}{\partial x}(x^2) + y^2\cancel{\frac{\partial}{\partial x}(1)} + 3y\frac{\partial}{\partial x}(x)\\ &= 2x + 3y\end{aligned}$$

Similarly, regarding x as a constant, we compute

$$\begin{aligned}\frac{\partial f}{\partial y} &= \frac{\partial}{\partial y}(400 + x^2 + y^2 + 3xy)\\ &= 400\cancel{\frac{\partial}{\partial y}(1)} + x^2\cancel{\frac{\partial}{\partial y}(1)} + \frac{\partial}{\partial y}(y^2) + 3x\frac{\partial}{\partial y}(y)\\ &= 3x + 2y\end{aligned}$$

Setting $x = 1$ and $y = 3$ we obtain the values of the partial derivatives at the base point $P = (1, 3)$,

$$\frac{\partial f}{\partial x}(1, 3) = 2(1) + 3(3) = 11$$

$$\frac{\partial f}{\partial y}(1, 3) = 3(1) + 2(3) = 9$$

Naturally the partial derivatives have other values at other base points. For example, at the origin $O = (0, 0)$ we obtain

$$\frac{\partial f}{\partial x}(0, 0) = 2(0) + 3(0) = 0$$

$$\frac{\partial f}{\partial y}(0, 0) = 3(0) + 2(0) = 0$$

If $\partial f/\partial x = 0$ at a point (x, y), this means precisely that $2x + 3y = 0$, or $y = -\frac{2}{3}x$. The solution set for the equation $\partial f/\partial x = 0$ consists of all points on the line determined by $2x + 3y = 0$. Notice that $\partial f/\partial x = 0$ at infinitely many points in the plane. ■

Example 8 Calculate $\partial f/\partial x$ and $\partial f/\partial y$ for $f(x, y) = 1000 + x^2 + 3xy + 5y + 7y^2$. Find all points $P = (x, y)$ where both partial derivatives are simultaneously equal to zero.

Solution Taking y as a constant everywhere it appears, we obtain

$$\begin{aligned}\frac{\partial f}{\partial x} &= \frac{\partial}{\partial x}(1000 + x^2 + 3xy + 5y + 7y^2)\\ &= 1000\frac{\partial}{\partial x}(1) + \frac{\partial}{\partial x}(x^2) + 3y\frac{\partial}{\partial x}(x) + 5y\frac{\partial}{\partial x}(1) + 7y^2\frac{\partial}{\partial x}(1)\\ &= 2x + 3y\end{aligned}$$

Similarly, treating x as a constant, we obtain

$$\begin{aligned}\frac{\partial f}{\partial y} &= \frac{\partial}{\partial y}(1000 + x^2 + 3xy + 5y + 7y^2)\\ &= 1000\frac{\partial}{\partial y}(1) + x^2\frac{\partial}{\partial y}(1) + 3x\frac{\partial}{\partial y}(y) + \frac{\partial}{\partial y}(5y + 7y^2)\\ &= 3x + 5 + 14y\end{aligned}$$

The solution set for $\partial f/\partial x = 0$ is the line given by $2x + 3y = 0$. The equation $\partial f/\partial y = 0$ gives a different line, determined by $3x + 14y = -5$. The partial derivatives are *both* zero when $P = (x, y)$ is a solution of the simultaneous equations

$$2x + 3y = 0$$
$$3x + 14y = -5$$

These are easy to solve. The first equation gives $y = -\frac{2}{3}x$; substituting this into the second equation, we obtain

$$-5 = 3x + 14\left(-\frac{2}{3}x\right) = -\frac{19}{3}x \quad \text{or} \quad x = \frac{15}{19}$$

Inserting $x = \frac{15}{19}$ into either equation, we find the corresponding value for y:

$$y = -\frac{2}{3}x = -\frac{2}{3}\left(\frac{15}{19}\right) = -\frac{10}{19}$$

Thus $P = (\frac{15}{19}, -\frac{10}{19})$ is the only point where $\partial f/\partial x = 0$ and $\partial f/\partial y = 0$ simultaneously. ■

More complicated functions $f(x, y)$ are handled using the differentiation rules for one variable, as discussed in Chapter 3. The next example uses the chain rule.

Example 9 Calculate $\partial f/\partial x$ and $\partial f/\partial y$ for $f(x, y) = \sqrt{x^2 - xy}$. How fast is f changing as we move right, parallel to the x axis, past the base point $P = (2, 1)$?

Solution It helps to write f in terms of exponents, so we can recognize it as a composite of $u^{1/2}$ and $u = x^2 - xy$,

$$f(x, y) = (x^2 - xy)^{1/2}$$

Having recognized that $f(x, y)$ is a composite function, we may use the chain rule on it. Treating y as a constant everywhere it appears, we obtain

$$\begin{aligned}\frac{\partial f}{\partial x} &= \frac{\partial}{\partial x}((x^2 - xy)^{1/2}) \\ &= \frac{1}{2}(x^2 - xy)^{-1/2}\frac{\partial}{\partial x}(x^2 - xy) \\ &= \frac{(2x - y)}{2\sqrt{x^2 - xy}}\end{aligned}$$

Similarly, regarding x as a constant, we obtain

$$\begin{aligned}\frac{\partial f}{\partial y} &= \frac{\partial}{\partial y}((x^2 - xy)^{1/2}) \\ &= \frac{1}{2}(x^2 - xy)^{-1/2}\frac{\partial}{\partial y}(x^2 - xy) \\ &= \frac{-x}{2\sqrt{x^2 - xy}}\end{aligned}$$

At the particular base point $P = (2, 1)$ the value of $\partial f/\partial x$ is

$$\frac{\partial f}{\partial x}(2, 1) = \frac{2(2) - 1}{2\sqrt{(2)^2 - (2)(1)}} = \frac{3}{2\sqrt{2}} \approx 1.061$$

so f increases about 1.061 units for every unit we move away from P parallel to the positive x axis. ■

Partial derivatives $\partial f/\partial x$, $\partial f/\partial y$, $\partial f/\partial z$ for a function $f(x, y, z)$ of three variables are handled in almost the same way. There is one partial derivative for each coordinate direction. To compute $\partial f/\partial x$, we treat all other variables y and z as constants and take the ordinary derivative with respect to x. The other partial derivatives $\partial f/\partial y$ and $\partial f/\partial z$ are computed the same way.

Example 10 Compute the partial derivatives $\partial f/\partial x$, $\partial f/\partial y$, $\partial f/\partial z$ for

$$f(x, y) = 2xy + 4z - x^2z^2$$

Find their numerical values when $x = 1$, $y = 0$, and $z = -2$.

Solution Treating y and z as constants and differentiating with respect to x, we obtain

$$\begin{aligned}\frac{\partial f}{\partial x} &= \frac{\partial}{\partial x}(2xy + 4z - x^2z^2) \\ &= 2y\frac{\partial}{\partial x}(x) + 4z\cancel{\frac{\partial}{\partial x}(1)} - z^2\frac{\partial}{\partial x}(x^2) \\ &= 2y - 2xz^2\end{aligned}$$

Similarly,

$$\frac{\partial f}{\partial y} = \frac{\partial}{\partial y}(2xy + 4z - x^2z^2) = 2x$$

$$\frac{\partial f}{\partial z} = \frac{\partial}{\partial z}(2xy + 4z - x^2z^2) = 4 - 2x^2z$$

Entering the values $x = 1$, $y = 0$, $z = -2$ into these formulas we obtain

$$\frac{\partial f}{\partial x}(1, 0, -2) = 2(0) - 2(1)(-2)^2 = -8$$

$$\frac{\partial f}{\partial y}(1, 0, -2) = 2(1) = 2$$

$$\frac{\partial f}{\partial z}(1, 0, -2) = 4 - 2(1)^2(-2) = 8$$

■

Some books use other notations for partial derivatives. Typical alternatives for the symbols $\partial f/\partial x$ and $\partial f/\partial y$ are

$$f_x \text{ and } f_y \qquad D_x f \text{ and } D_y f \qquad D_1 f \text{ and } D_2 f$$

We shall use notations $\partial f/\partial x$ and $\partial f/\partial y$ throughout this book. The functions $\partial f/\partial x$ and $\partial f/\partial y$ together play the role played by the ordinary derivative df/dx when we studied functions of one variable. They will be very useful when we search for maximum and minimum values of f.

Exercises 7.3

1. If $f(x, y) = x + x^3y^2$, calculate $\partial f/\partial x$ and $\partial f/\partial y$. What are their values at the following base points?

(i) $P = (1, 2)$ (iii) $P = (0, 1)$
(ii) $P = (1, 0)$ (iv) $P = (-3, 1)$

2. Given $f(x, y) = 2 + x + 3y - x^3y + 3xy^3$, find the partial derivatives $\partial f/\partial x$ and $\partial f/\partial y$ as functions of x and y. Then find their values at particular base points.

(i) $\dfrac{\partial f}{\partial x}(0, 0)$ (iii) $\dfrac{\partial f}{\partial x}(-1, 1)$

(ii) $\dfrac{\partial f}{\partial y}(0, 0)$ (iv) $\dfrac{\partial f}{\partial y}(-1, 1)$

3. For the function $f(x, y) = 1 + 2x + 3y + x^2y - 3xy$, find the partial derivatives $\partial f/\partial x$ and $\partial f/\partial y$ as functions of x and y. Then locate all points (x, y) at which *both* partial derivatives are equal to zero.

For each of the following elementary functions, find $\partial f/\partial x$ and $\partial f/\partial y$ as functions of x and y.

4. $f(x, y) = 3$ (constant function)

5. $f(x, y) = x$

6. $f(x, y) = -y + 1$

7. $f(x, y) = 2x + 3y$

8. $f(x, y) = xy + x^2 - y^2$

9. $f(x, y) = \dfrac{x}{y}$

10. $f(x, y) = 10xy - \dfrac{3}{y^2}$

In Exercises 11–28, find $\partial f/\partial x$ and $\partial f/\partial y$ for each function.

11. $f(x, y) = (x + y)^5$

12. $f(x, y) = \sqrt{x^2 + 2x - y^2}$

13. $f(x, y) = x + \sqrt{x^2 + 2y^2}$

14. $f(x, y) = \dfrac{1}{\sqrt{x^2 + 2x + y^2 + 1}}$

15. $f(x, y) = 15x^{1/3}y^{4/5}$

16. $f(x, y) = \dfrac{1 + x}{1 - y}$

17. $f(x, y) = x - \dfrac{1}{y^2}$

18. $f(x, y) = \dfrac{1 - x^2}{4 + y^2}$

19. $f(x, y) = \dfrac{x^2y + 1}{x + y}$

20. $f(x, y) = 40 - 16x^2 - 4xy - 25y^2$

21. $f(x, y) = -900{,}000 + (500x - 0.1x^2) + (700y - 0.2y^2) - 0.01xy$

22. $f(x, y) = e^{x^2 + xy}$

23. $f(x, y) = e^{-(x^2 + y^2)/2}$

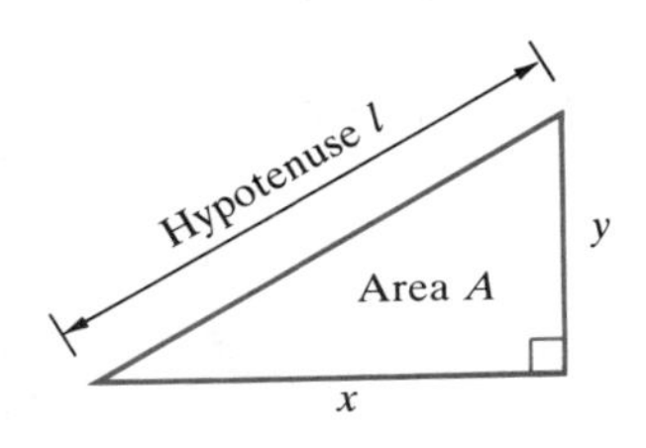

Figure 7.18

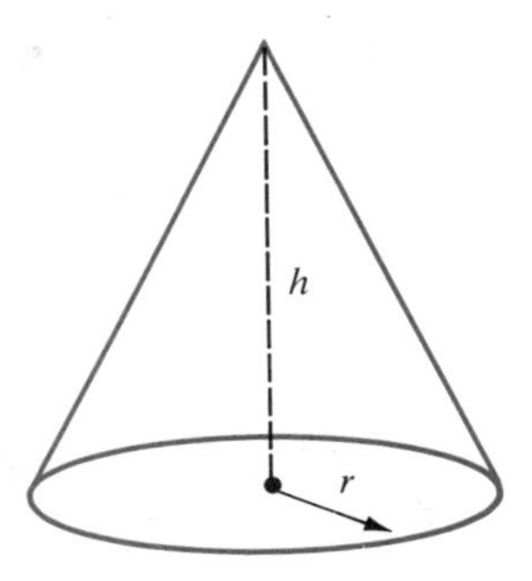

Figure 7.19

24. $f(x, y) = x \ln y$

25. $f(x, y) = \ln(4x + 3y + 1)$

26. $f(x, y) = 5 \ln(x - 2y)$

27. $f(x, y) = -1 + e^{4x-3y}$

28. $f(x, y) = \dfrac{x + y}{xy} - x^3 + y$

29. Given a right triangle with side lengths x and y as in Figure 7.18, the area A of the triangle and the length l of its hypotenuse (slanted side) are given by the formulas

$$A(x, y) = \frac{1}{2}xy \qquad l(x, y) = \sqrt{x^2 + y^2}$$

From this information:
(i) Find the partial derivatives of $A(x, y)$ and $l(x, y)$ as functions of x and y.
(ii) What are the values of the partial derivatives if the dimensions of the triangle are $x = 120$ and $y = 160$?

30. The volume of a cone with radius r and altitude h (see Figure 7.19) is

$$V(r, h) = \frac{\pi}{3} \cdot r^2 h$$

Calculate $\partial V/\partial r$ and $\partial V/\partial h$ as functions of r and h. Find their numerical values when $r = 2$ and $h = 10$.

31. A manufacturer's cost function is $C(x, y) = 900 + 100x^2 + 200y^2 + 10xy$, where x and y are the production levels for his two main products. Find the partial derivatives $\partial C/\partial x$ and $\partial C/\partial y$. What are the values of these partial derivatives at the production levels $x = 50$, $y = 75$? What are the values at the production levels $x = 60$, $y = 40$?

The concentration of penicillin in the blood t hours after a dose of x units is administered depends on the elapsed time t and the initial dose x:

$$(\text{concentration}) = A(x, t) = \frac{1}{4}xe^{-t/3} \quad \text{units per liter of blood}$$

Answer the questions in Exercises 32–35.

32. Find $\partial A/\partial x$ and $\partial A/\partial t$ as functions of x and t.

33. What are the numerical values of these partial derivatives if $x = 100$ and $t = 6$ hours?

34. If the initial dose is 100 units, what is the *concentration* A after 3 hours?

35. If the initial dose is 100 units, *how fast* is the concentration changing after 3 hours? Which partial derivative provides this information, and which base point (x, t) must you examine?

36. The average yield (in pounds) of a new variety of tomato plant is determined by the combined effect of the variables

$$t = (\text{growing temperature (in degrees Celsius)})$$
$$q = (\text{quantity of fertilizer applied weekly (in grams)})$$

according to the formula

$$Y(t, q) = (0.6 + 0.386q - 0.093q^2) \cdot (2t - 0.05t^2)$$

Find the partial derivatives $\partial Y/\partial t$ and $\partial Y/\partial q$. (*Hint*: Do *not* multiply out the expression before applying $\partial/\partial t$ or $\partial/\partial q$. When you take partial derivatives, one block of terms $(\ldots)$ will be regarded as a constant.)

An ideal gas in a container of volume V satisfies Boyle's Law:

$$PV = kT$$

where P = pressure, V = volume, and T = absolute temperature. Here $k > 0$ is a physical constant, measured experimentally. Answer the following questions:

37. Express P as a function of the variables V and T.

38. Calculate $\partial P/\partial V$ and $\partial P/\partial T$ as functions of V and T.

39. It turns out that $\partial P/\partial T$ is always *positive*: $\partial P/\partial T > 0$. What does this say about the behavior of the pressure if the volume of the container is held fixed while we increase the temperature? Does this agree with your intuition?

40. If the temperature is held fixed and we increase the volume of the container, the pressure of the confined gas decreases. What does this tell us about the sign of the partial derivative $\partial P/\partial V$?

In Exercises 41–49, functions of three variables are given. Compute $\partial f/\partial x$, $\partial f/\partial y$, $\partial f/\partial z$ for each function. Then find the values of each partial derivative when $x = 1$, $y = -2$, $z = 4$.

41. $x^2 + y^2 + z^2$

42. $12xy + 28xz - 7y^2z^2$

43. $4x^2 - y^2 + 10xz^2$

44. xyz

45. $\dfrac{xy}{z}$

46. $\dfrac{xy^2}{1 + z^2}$

47. $\sqrt{1 - xy + 4z^2}$

48. $e^{(x^2 + y^2 - z^2)}$

49. $\dfrac{1}{\sqrt{x^2 + y^2 + z^2}}$

50. The most general linear function of x and y has the form $f(x, y) = Ax + By + C$, where A, B, C are fixed constants. Show that the partial derivatives are the constant functions

$$\frac{\partial f}{\partial x} = A \qquad \frac{\partial f}{\partial y} = B$$

7.4 The Approximation Principle for Several Variables

If we examine the behavior of a function $f(x, y)$ near a base point $P = (x_0, y_0)$, the partial derivatives at the base point $\partial f/\partial x(P)$ and $\partial f/\partial y(P)$ tell us how fast f is changing as we move away from P in the x direction or the y direction. This follows from the very definition of partial derivatives. Even more is true: A simple formula allows us to estimate the value of f at any nearby point $Q = (x_0 + \Delta x, y_0 + \Delta y)$ no matter how the independent increments Δx and Δy are chosen. As we will see, this approximation formula has many uses.

THE APPROXIMATION PRINCIPLE FOR TWO VARIABLES If $P = (x_0, y_0)$ is a base point and $Q = (x_0 + \Delta x, y_0 + \Delta y)$ any nearby point in the plane, the resulting change in f, namely

$$\Delta f = f(Q) - f(P) = f(x_0 + \Delta x, y_0 + \Delta y) - f(x_0, y_0),$$

is closely approximated by the following expression

$$\Delta f \approx \frac{\partial f}{\partial x}(P) \cdot \Delta x + \frac{\partial f}{\partial y}(P) \cdot \Delta y \tag{5}$$

for all small increments Δx and Δy. Adding $f(P)$ to both sides, we get an approximation formula for the value of f at the nearby point,

$$\begin{aligned} f(Q) &= f(x_0 + \Delta x, y_0 + \Delta y) \\ &\approx f(P) + \frac{\partial f}{\partial x}(P) \cdot \Delta x + \frac{\partial f}{\partial y}(P) \cdot \Delta y \end{aligned} \tag{6}$$

for all small increments.

Once the base point P is specified $f(P)$, $\partial f/\partial x(P)$, and $\partial f/\partial y(P)$ are constants in Formulas [5] and [6]. Only the increments Δx and Δy vary, and Δf is just a simple linear combination of these increments. Because Δx and Δy may be chosen independently, the formulas tell us (approximately) how f behaves as we move away from the base point in *any* direction.

Two special cases of Formula [5] arise frequently, and are worth noting: If we move away from P in the x direction only (Δx arbitrary, $\Delta y = 0$) we obtain

$$\Delta f \approx \frac{\partial f}{\partial x}(P) \cdot \Delta x \quad \text{(for small increments in the } x \text{ direction)} \tag{7}$$

Similarly if we move only in the y direction ($\Delta x = 0$, Δy arbitrary),

$$\Delta f \approx \frac{\partial f}{\partial y}(P) \cdot \Delta y \quad \text{(for small increments in the } y \text{ direction)} \tag{8}$$

These simplified versions of the approximation formula are closely connected with the definitions of partial derivatives. By definition, $\partial f/\partial x(P) = \lim_{\Delta x \to 0} (\Delta f/\Delta x)$ for increments Δx in the x direction. So for small increments we have

$$\frac{\partial f}{\partial x}(P) \approx \frac{\Delta f}{\Delta x} \quad \left(\text{that is, } \Delta f \approx \frac{\partial f}{\partial x}(P) \cdot \Delta x\right)$$

and similarly for small increments in the y direction,

$$\frac{\partial f}{\partial y}(P) \approx \frac{\Delta f}{\Delta y} \quad \left(\text{that is, } \Delta f \approx \frac{\partial f}{\partial y}(P) \cdot \Delta y\right)$$

Thus we obtain [7] and [8] right from the definitions. The general formula [5] for arbitrary directions requires a little more discussion, which we leave to more advanced courses.

Example 11 Consider $f(x, y) = x^2 + 5xy - 2y^2$ near the base point $P = (2, 3)$. Write out the approximation formulas [5] and [6] for Δf and the nearby value $f(2 + \Delta x, 3 + \Delta y)$. What are the values of f and its partial derivatives at P? Use the approximation principle to estimate the following nearby values:

$$f(2.01, 3) \qquad f(2, 3.02) \qquad f(2.01, 3.02)$$

Solution First, write down the partial derivatives as functions of x and y.

$$\frac{\partial f}{\partial x} = \frac{\partial}{\partial x}(x^2 + 5xy - 2y^2) = 2x + 5y$$

$$\frac{\partial f}{\partial y} = \frac{\partial}{\partial y}(x^2 + 5xy - 2y^2) = 5x - 4y$$

At the base point $P = (2, 3)$, we have

$$f(P) = f(2, 3) = (2)^2 + 5(2)(3) - 2(3)^2 = 16$$

$$\frac{\partial f}{\partial x}(P) = 2(2) + 5(3) = 19$$

$$\frac{\partial f}{\partial y}(P) = 5(2) - 4(3) = -2$$

Formulas [5] and [6] then become

$$\Delta f \approx \frac{\partial f}{\partial x}(P) \cdot \Delta x + \frac{\partial f}{\partial y}(P) \cdot \Delta y = 19(\Delta x) - 2(\Delta y)$$

and

$$\begin{aligned} f(2 + \Delta x, 3 + \Delta y) &\approx f(P) + \frac{\partial f}{\partial x}(P) \cdot \Delta x + \frac{\partial f}{\partial y}(P) \cdot \Delta y \\ &= 16 + 19(\Delta x) - 2(\Delta y) \end{aligned}$$

To estimate $f(2.01, 3)$, we set $\Delta x = 0.01$, $\Delta y = 0$ to obtain

$$f(2.01, 3) \approx 16 + 19(0.01) - 2(0) = 16.19 \quad (\text{thus } \Delta f \approx 0.19)$$

Similarly,

$$f(2, 3.02) \approx 16 + 19(0) - 2(0.02) = 15.96 \quad (\text{thus } \Delta f \approx -0.04)$$

$$f(2.01, 3.02) \approx 16 + 19(0.01) - 2(0.02) = 16.15 \quad (\text{thus } \Delta f \approx 0.15)$$

■

Example 12 The dimensions of a rectangular plot of land have been measured at 1235 feet by 619 feet, yielding a calculated area $A = 1235 \times 619 = 764{,}465$ square feet. The error in the longer dimension is at most 3.5 feet, and in the shorter dimension at most 2 feet. Estimate the maximum error that might occur in our value for the area of this plot.

Solution If we let x and y denote the actual dimensions of a rectangular plot, the area is $A = xy$. Our measured base value for x is 1235 feet, with an error $\Delta x = x - 1235$ of at most ± 3.5 feet. Similarly, our measured value of 619 for y has a possible error $\Delta y = y - 619$ of at most ± 2 feet. The corresponding error in our figure for A is

$$\Delta A = A(x, y) - A(1235, 619) = A(1235 + \Delta x, 619 + \Delta y) - A(1235, 619)$$

Taking base point $P = (1235, 619)$ in the approximation formula, note that

$$\frac{\partial A}{\partial x} = \frac{\partial}{\partial x}(xy) = y \qquad \frac{\partial A}{\partial y} = \frac{\partial}{\partial y}(xy) = x$$

so that

$$\frac{\partial A}{\partial x}(P) = 619 \qquad \frac{\partial A}{\partial y}(P) = 1235$$

Using these values in Formula [5], we obtain

$$\begin{aligned} \Delta A &\approx \frac{\partial A}{\partial x}(P) \cdot \Delta x + \frac{\partial A}{\partial y}(P) \cdot \Delta y \\ &= 619(\Delta x) + 1235(\Delta y) \end{aligned}$$

The worst possible error occurs when $\Delta x = 3.5$, $\Delta y = 2$, in which case

$$\Delta A \approx 619(3.5) + 1235(2) = 4636.5 \quad \text{square feet}$$

■

Exercises 7.4

1. For the function $f(x, y) = 2 + x^2 - y^3$ and base point $P = (1, 1)$, compute $\partial f/\partial x(P)$ and $\partial f/\partial y(P)$ and show that in this situation

$$f(1 + \Delta x, 1 + \Delta y) \approx 2 + 2(\Delta x) - 3(\Delta y)$$

for small increments Δx, Δy away from this base point. Then calculate $f(0.9, 1.15)$ exactly and compare it with the approximation given by this formula. Do the same for the value $f(1.2, 1.1)$.

2. Consider $f(x, y) = x^2 + 5xy - 2y^2$ near the base point $P = (1, -2)$. Write out the approximation formulas [5] and [6] for Δf and for $f(1 + \Delta x, -2 + \Delta y)$. Then estimate the value of $f(0.9, -2.05)$.

For each function $f(x, y)$ and base point $P = (x_0, y_0)$ in Exercises 3–6, write out the approximation formulas [5] and [6]. Then estimate the change Δf corresponding to increments $\Delta x = 0.2$, $\Delta y = -0.1$ away from the given base point.

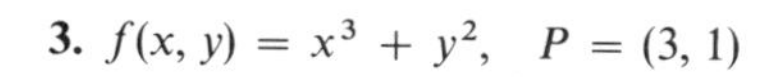

3. $f(x, y) = x^3 + y^2, \quad P = (3, 1)$

4. $f(x, y) = x^2y + y^3, \quad P = (2, 2)$

5. $f(x, y) = \dfrac{x}{x^2 + y^2}, \quad P = (-3, 4)$

6. $f(x, y) = e^{-x-2y}, \quad P = (1, 1)$

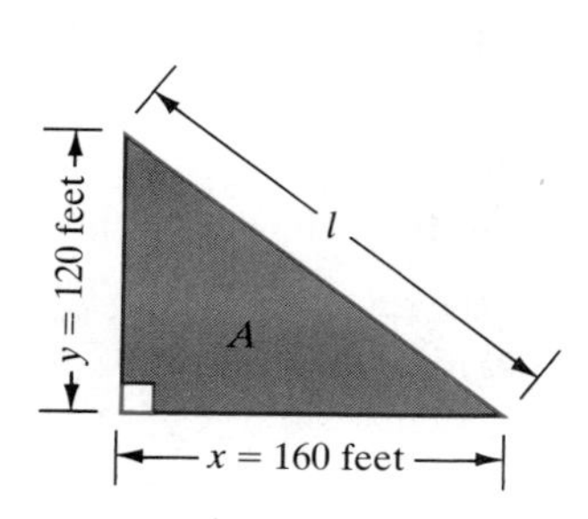

Figure 7.20
The triangular plot of ground in Exercise 7.

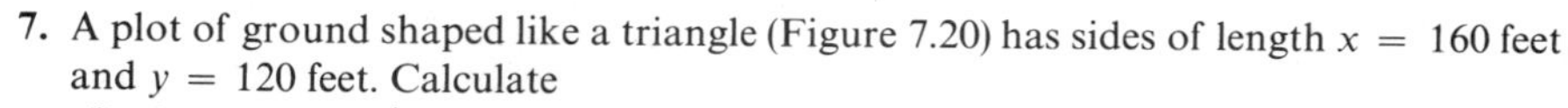

7. A plot of ground shaped like a triangle (Figure 7.20) has sides of length $x = 160$ feet and $y = 120$ feet. Calculate
   (i) the area $A = \frac{1}{2}xy$ of this plot
   (ii) the length $l = \sqrt{x^2 + y^2}$ of the slanted edge (hypotenuse)
   Then use the approximation principle to estimate the errors ΔA and Δl corresponding to errors $\Delta x = 0.5$ and $\Delta y = 0.4$ in measuring x and y.

8. The surface area of a solid cylinder with height h and diameter d (inches) is

$$A(h, d) = \pi hd + \frac{\pi}{2}d^2 \quad \text{square inches}$$

   Find the area if $d = 10$ inches and $h = 12$ inches. Approximately how much does the area change if
   (i) The diameter is increased by 1 inch ($\Delta d = 1$), keeping h fixed?
   (ii) The height is increased by 1 inch, ($\Delta h = 1$), keeping d fixed?
   (iii) The height is increased by 2 inches, but the diameter is decreased by 1 inch?

9. A manufacturer of chain-link fences produces two grades of fencing. Suppose his profit function is

$$P(x, y) = -700 + 0.5x + 0.3y - 0.00004xy$$

   where x and y are the number of yards of heavy-duty and standard fencing produced (and sold) per week. Suppose his plant is operating at production levels $x = 400$ and $y = 4000$.
   (i) Find the rates of change $\partial P/\partial x$ and $\partial P/\partial y$ with respect to x and y at these production levels.
   (ii) Is it profitable to produce more standard fencing, maintaining the present level for heavy-duty fencing?
   (iii) Is it profitable to produce more heavy-duty fencing, maintaining the present level for standard fencing?

10. Repeat Exercise 9, assuming a new profit function

$$P(x, y) = -700 + 0.5x + 0.3y - 0.001xy$$

   and that the current production levels are $x = 2000$ and $y = 400$.

The BBD & D Ad Agency has come up with the following model predicting the weekly sales of one of their fashion magazines as a function of the variables

$$p = (\text{selling price per copy (in dollars)})$$
$$a = (\text{total weekly advertising budget (in dollars)})$$

Their model predicts weekly sales of

$$S(p, a) = 117{,}000\left(1 + \frac{1}{100}\sqrt{a}\right) \cdot e^{2-p}$$

If the current price per copy is $p = \$1.50$ and the advertising budget is $a = \$20{,}000$, answer the following questions.

11. How many copies a week will they sell under these conditions?

12. What are the values of the partial derivatives $\partial S/\partial p$ and $\partial S/\partial a$ under these conditions?

13. Roughly how much will sales change if the price is increased by 10¢, keeping the same advertising budget?

14. Roughly how much will sales change if the advertising budget is raised by \$1000 and the price is held at \$1.50?

15. Roughly how much will sales change if the price is raised by 10¢ and the advertising budget is raised by \$1000?

Ten grams of helium gas are enclosed in a container (see Figure 7.21). The volume V of the space confining the helium is related to the pressure and temperature of the gas by a formula called Boyle's law:

$$P = 2.034\frac{T}{V}$$

where P = (pressure, in atmospheres), V = (volume, in liters), and T = (temperature, in °K = degrees Kelvin).† Answer the following questions, using the approximation principle where appropriate.

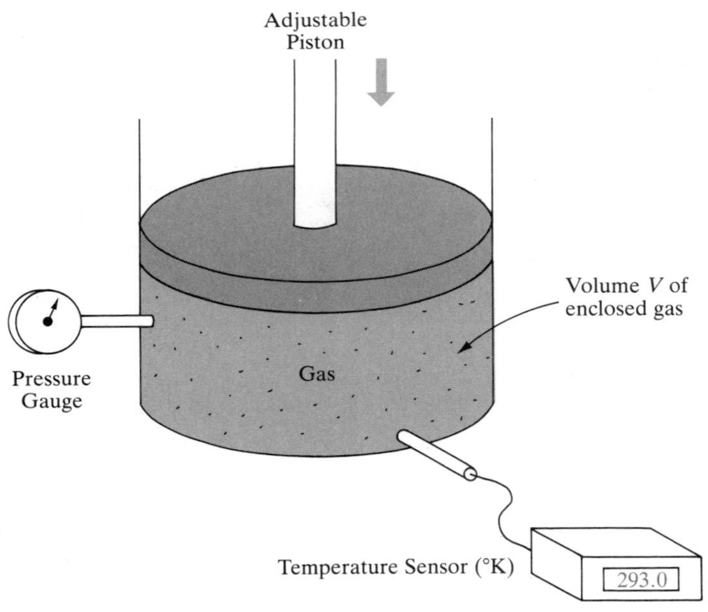

Figure 7.21
The apparatus in Exercises 16–21. Temperature is measured in (°K) = 273 + (°C). The gas consists of 10 grams of helium.

† Absolute temperatures are measured in °K = (degrees Kelvin). The absolute temperature is related to the familiar Centigrade, or Celsius, temperature by the formula °K = °C + 273.

16. Find the partial derivatives $\partial P/\partial V$ and $\partial P/\partial T$ as functions of T and V.

17. If $V = 10$ liters and $T = 293°K$ ($=20°$ Celsius, room temperature), what is the pressure in the container?

18. What are the numerical values of $\partial P/\partial V$ and $\partial P/\partial T$ when $V = 10$ and $T = 293°K$?

19. About how much will P change if the volume is kept fixed, $V = 10$, but the temperature is raised to $303°K$ ($= 30°C$)? Does P increase or decrease? Does your result agree with your intuition?

20. By how much will the pressure change if we start with $V = 10$, $T = 293°K$ and change the experimental conditions to $V = 9.5$, $T = 273°K$?

21. Suppose the temperature is $T = 293°K$, but there is a possible error of ± 0.05 liters in the estimated volume $V = 10$ liters. What is the corresponding uncertainty in the pressure P predicted by Boyle's law?

7.5 General Optimization Problems in Two Variables

Finding the maxima and minima of a given function $f(x, y)$ is one of the most important applications of calculus. This task is referred to as an *optimization problem.* The basic definitions used in discussing optimization are

> **Maxima and Minima of a Function $f(x, y)$** Given a function $f(x, y)$ and a base point P, we say that
>
> (i) The function has a **local maximum** at P if $f(P) \geq f(Q)$ for all points Q in the feasible set lying near P. That is, the value of f is as large at P as it is at any *nearby* point where f is defined.
> (ii) The function has an **absolute maximum** at P if $f(P) \geq f(Q)$ for *all* points Q in the feasible set, whether they lie close to P or not.
>
> **Local (absolute) minima** are defined similarly: $f(P) \leq f(Q)$ for all nearby points (all points) Q in the feasible set. If either a maximum or minimum occur at P, we sometimes combine these possibilities by saying that f has an **extremum** at P.

Near a local maximum or minimum, the graph of $z = f(x, y)$ has the shape shown in Figure 7.22. Clearly an absolute maximum is also a local maximum, though the reverse need not be true; likewise for minima. A function may achieve its extreme values on the boundary of the feasible set (insofar as f is actually defined on this boundary) or at points interior to the feasible set. An **interior point** P is one such that all points near P lie within the feasible set; the other feasible points are called **boundary** points. Some possible locations of extrema are shown in Figure 7.23. There is an absolute minimum over point A on the boundary of the feasible set. The absolute maximum is achieved at an interior point B. In

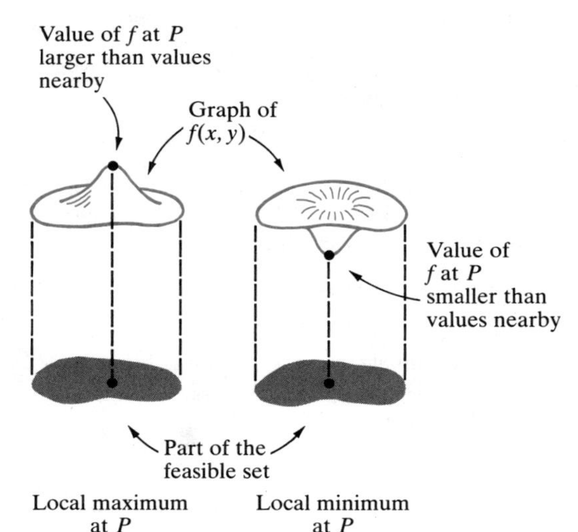

Figure 7.22
Shape of the graph near a local maximum or minimum for $f(x, y)$.

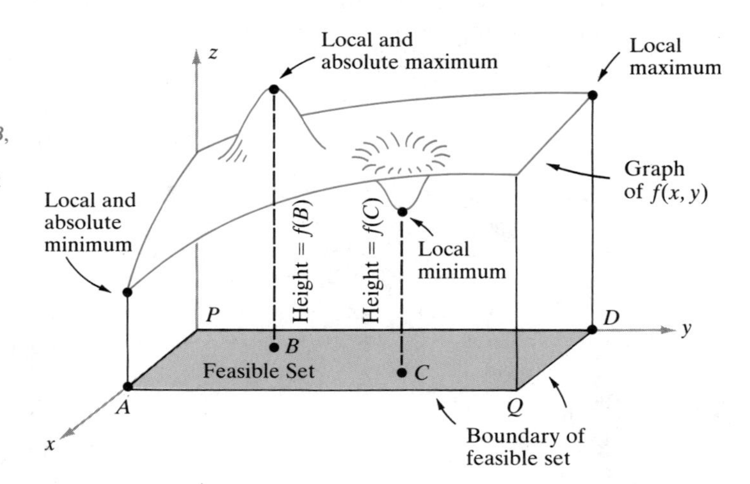

Figure 7.23
A surface exhibiting various kinds of maxima and minima. The shaded rectangle (feasible set) lies in the x, y plane. An absolute maximum occurs at B, interior to the feasible set; a local minimum occurs at C. At boundary point D there is a local maximum that is not an absolute maximum, because $f(B) > f(D)$. The absolute minimum occurs at A. Note that $f(A) < f(C)$. At P and Q there is *no* local extremum. (Consider what happens to $f(x, y)$ as we move along one edge of the feasible set to P and then depart along the other edge; likewise for Q.)

addition, there are local extrema that are not absolute extrema. For example, the value of f at C is lower than all nearby values, but is not lower than the value at the corner point A; thus, a local minimum occurs at C.

The **critical points** of a function $f(x, y)$ are those points P for which

$$\frac{\partial f}{\partial x}(P) = 0 \quad \text{and} \quad \frac{\partial f}{\partial y}(P) = 0 \quad \text{simultaneously}$$

The extrema occurring at points interior to the feasible set must appear among the critical points, as we shall explain. Since there are not many critical points in most cases, this observation greatly simplifies our task of locating the extrema of f.

> EXTREMA AND CRITICAL POINTS (OPTIMIZATION PRINCIPLE) An extremum for a function $f(x, y)$ of two variables can only occur
>
> (i) at a critical point interior to the feasible set
>
> or
>
> (ii) at a boundary point of the feasible set
>
> In particular, every extremum interior to the feasible set reveals its presence by being a critical point.

Sketch Proof: Consider a point $P = (x_0, y_0)$ not on the boundary and not a critical point. Then P is an interior point, and all small displacements keep us within the feasible set. Because P is not a critical point, at least one of the partial derivatives $\partial f/\partial x(P)$ or $\partial f/\partial y(P)$ is nonzero. By the approximation principle, the value of f at a nearby point $Q = (x_0 + \Delta x, y_0 + \Delta y)$ is

$$f(Q) \approx f(P) + \frac{\partial f}{\partial x}(P) \cdot \Delta x + \frac{\partial f}{\partial y}(P) \cdot \Delta y$$

So there must be points Q_1 near P for which $f(Q_1) > f(P)$ and other nearby points Q_2 where $f(Q_2) < f(P)$. (For example, if $\partial f/\partial x(P) \neq 0$, take $\Delta y = 0$ and Δx arbitrary and see what happens when $\Delta x > 0$ or $\Delta x < 0$.) Therefore, f cannot have a local maximum or minimum at P. By default, the only points P where f can have an extremum are those satisfying either (i) or (ii). ■

In an **optimization problem** our goal is to find the maximum or minimum of a function $f(x, y)$ of two variables. By the optimization principle, maxima or minima within the feasible set show up at critical points, so the first thing to do is locate all the feasible critical points. As the previous discussion shows, the extrema we seek sometimes do lie on the boundary of the feasible set. Problems like this are complicated and must be left for more advanced courses in optimization.† We will confine our attention to problems in which the extrema occur within the feasible set and not on the boundary. For example, if f is defined everywhere, there are no boundary points at all. More to the point, in many applications it is intuitively clear that conditions on the boundary of the feasible set are so far from being optimal that we can safely ignore the question of boundary extrema.

When boundary extrema do not occur, we use the following procedure to solve optimization problems.

† But the method of Lagrange multipliers, discussed in Section 7.8, can be applied to find boundary extrema.

OPTIMIZATION PROCEDURE Suppose $f(x, y)$ has an absolute maximum that does not occur on the boundary of the feasible set. It may be found as follows.

Step 1 Find the critical points $P_1, \ldots, P_k$ within the feasible set by solving the system of equations

$$\frac{\partial f}{\partial x} = 0 \qquad \frac{\partial f}{\partial y} = 0$$

These are the only candidates for solutions.

Step 2 If there is just one candidate, the absolute maximum must occur there. If there are several candidates $P_1, \ldots, P_k$ then tabulate the values $f(P_1), \ldots, f(P_k)$ of f at each of these points and compare. The largest value in this list is the absolute maximum for f.

A similar procedure works for finding absolute minima.

Example 13 **We can show that $f(x, y) = 10 + x^2 + \frac{1}{2}y^2 - x + xy$ has an absolute minimum. Where is it located, and what is the absolute minimum value for f?**

Solution The feasible set consists of the whole coordinate plane. There are no boundary points (or boundary extrema) to worry about. Because

$$\frac{\partial f}{\partial x} = 2x - 1 + y \qquad \frac{\partial f}{\partial y} = y + x$$

the simultaneous equations locating the critical points are

$$0 = \frac{\partial f}{\partial x} = 2x + y - 1$$

$$0 = \frac{\partial f}{\partial y} = x + y$$

This system has just one solution, the point $P = (1, -1)$ where $x = 1$ and $y = -1$. We have nothing to do in Step 2. The minimum is $f(1, -1) = 9.5$. ■

Example 14 **The function $f(x, y) = 2x^2 - x^4 - y^2$, defined for all x and y, has an absolute maximum. Find it.**

Solution Again, there are no boundary points and no boundary extrema to worry about.

First we locate the critical points, where $\partial f/\partial x = 0$ and $\partial f/\partial y = 0$ simultaneously. Because

$$\frac{\partial f}{\partial x} = 4x - 4x^3 \quad \text{and} \quad \frac{\partial f}{\partial y} = -2y$$

a point (x, y) is critical if

$$4x - 4x^3 = 0 \quad \text{and} \quad -2y = 0$$

From the first equation we see that

$$4x - 4x^3 = 4x(1 - x^2) = 4x(1 - x)(1 + x) = 0$$

and the solutions are $x = 0$, $x = 1$, $x = -1$. The second equation has only one solution, $y = 0$. Thus, both partial derivatives are zero at the following points

$$P_1 = (0, 0) \qquad P_2 = (1, 0) \qquad P_3 = (-1, 0)$$

These are the critical points.

Now compute the values of f at these points:

$$f(P_1) = 0$$
$$f(P_2) = 2(1)^2 - (1)^4 - (0)^2 = 1$$
$$f(P_3) = 2(-1)^2 - (-1)^4 - (0)^2 = 1$$

There is a tie. The maximum value of f is 1, and it occurs at each of the points $P_2 = (1, 0)$ and $P_3 = (-1, 0)$. ■

Example 15 A manufacturer sells two products, Brand X and Brand Y, at prices of \$23 and \$22 per unit, respectively. His costs are

$$C(x, y) = 400 + 10x + 12y + 0.01(x^2 + xy + 2y^2)$$

where x and y are the production levels. Find the most profitable production levels.

Solution The profit is

$$\begin{aligned}(\text{profit}) &= f(x, y) = (\text{revenue}) - (\text{cost}) \\ &= (23x + 22y) - C(x, y) \\ &= -400 + 13x + 10y - 0.01(x^2 + xy + 2y^2)\end{aligned}$$

so the critical points are the solutions of the system of equations

$$0 = \frac{\partial f}{\partial x} = 13 - 0.02x - 0.01y$$

$$0 = \frac{\partial f}{\partial y} = 10 - 0.01x - 0.04y$$

The only solution is $x = 600$, $y = 100$, so the maximum must occur at these production levels. ■

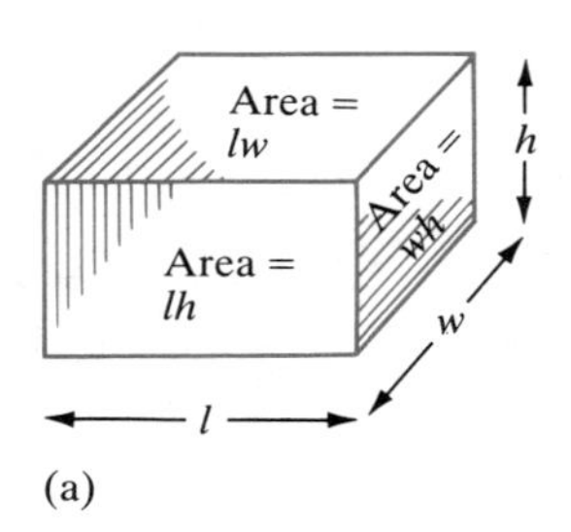

(a)

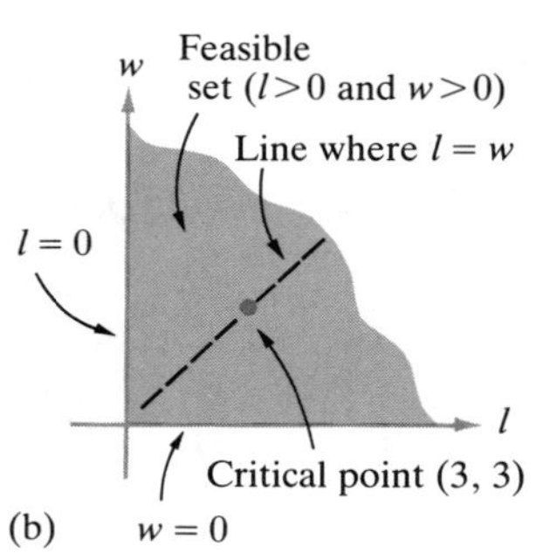

(b)

Figure 7.24
The box in Example 16 is shown in (a). The areas of the visible sides are marked; there are six sides in all. We take l and w as independent variables. The feasible set for the area function $A(l, w)$ consists of all (l, w) with $l > 0$ and $w > 0$. The only critical point is (3, 3), so minimum area occurs when $l = 3$, $w = 3$.

Example 16 A closed rectangular box (Figure 7.24) is to contain 27 cubic feet. Find the dimensions that yield the least surface area for the box. (If the cost of making the box is proportional to the amount of material used—the surface area—this gives the most economical shape.)

Solution Let l = length, w = width, and h = height, measured in feet. Because the volume must be 27 cubic feet, we have

$$V = lwh = 27 \qquad [9]$$

so the variables l, w, and h are not independent. We may use [9] to eliminate one of them, say h. If l and w are chosen independently, we get a box of the correct volume if we take $h = 27/lw$. For any choice of l and w the surface area of the corresponding box (six sides in all) is a function of l and w:

$$\begin{aligned} A(l, w) &= 2lw + 2lh + 2hw = 2lw + 2l\left(\frac{27}{lw}\right) + 2\left(\frac{27}{lw}\right)w \\ &= 2lw + \frac{54}{w} + \frac{54}{l} \end{aligned} \qquad [10]$$

The feasible choices for l and w are $l > 0$ and $w > 0$, the shaded quadrant shown in Figure 7.24b.

Next we calculate the partial derivatives and find the critical points:

$$\frac{\partial A}{\partial l} = 2w - \frac{54}{l^2} \quad \text{and} \quad \frac{\partial A}{\partial w} = 2l - \frac{54}{w^2}$$

These are simultaneously zero when

$$2w - \frac{54}{l^2} = 0 \quad \text{and} \quad 2l - \frac{54}{w^2} = 0 \qquad [11]$$

From the second equation we obtain $l = 27/w^2$. Substituting this into the first equation, we obtain

$$0 = 2w - \frac{54}{\left(\dfrac{27}{w^2}\right)^2} = 2w - \frac{2w^4}{27} \quad \text{or} \quad 2w\left(1 - \frac{w^3}{27}\right) = 0$$

This product can be zero only if one of its factors is zero. One possibility is $w = 0$, which does not correspond to a feasible point. The other is

$$\left(1 - \frac{w^3}{27}\right) = 0 \quad \text{or} \quad w^3 = 27$$

which yields $w = 27^{1/3} = 3$. To obtain the corresponding value of l, substitute this into either of the equations [11]; we obtain $l = 27/w^2 = 3$. Thus, we have located the one and only critical point inside the feasible set: $l = 3$, $w = 3$.

This critical point, the only possible candidate for a solution, must be the location of the minimum area. For the critical values $l = 3$, $w = 3$ the corresponding value of h is $h = 27/lw = 27/(3)(3) = 3$, so the optimal dimensions are

$$l = 3 \qquad w = 3 \qquad h = 3$$

These are equal, so the optimal box is a *cube*. In fact this is true regardless of the volume V of the box, as you might expect. ■

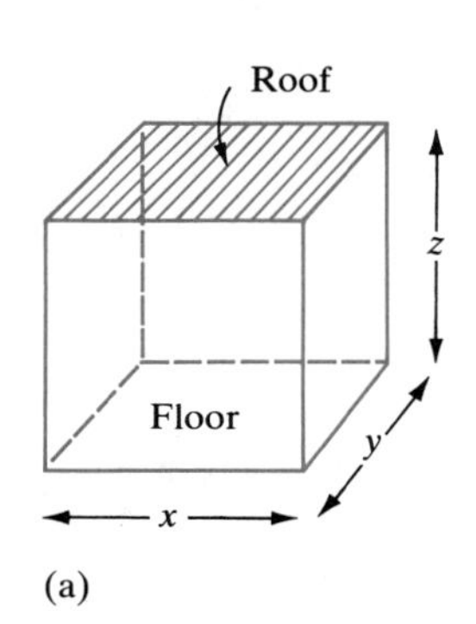

(a)

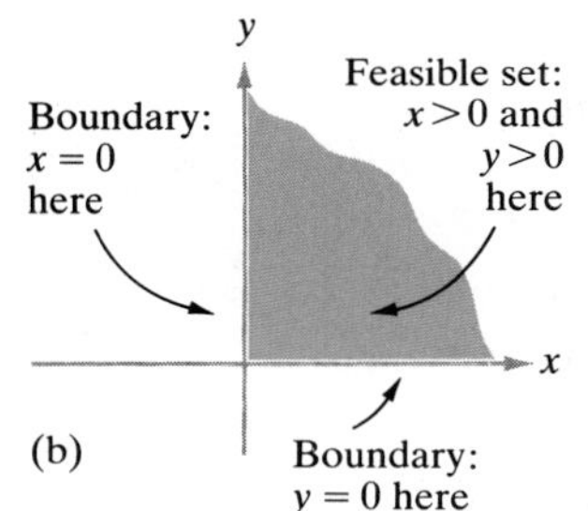

(b)

Figure 7.25
In (a) the warehouse in Example 17. In (b) we show the feasible set of values for x and y: those with $x > 0$ and $y > 0$.

Example 17 A rectangular warehouse (Figure 7.25) with dimensions

$$x = (\text{length}) \qquad y = (\text{width}) \qquad z = (\text{height})$$

is to enclose 16,000 cubic feet of space. Construction costs are different for roof, sides, and floor:

$$\begin{aligned}(\text{cost for roof}) &= \$7.50 \text{ per square foot}\\ (\text{cost for sides}) &= \$5.00 \text{ per square foot}\\ (\text{cost for concrete floor}) &= \$12.50 \text{ per square foot.}\end{aligned}$$

How should the dimensions be chosen to minimize construction costs?

Solution Again, it is obvious that optimal dimensions exist. Because

$$(\text{volume}) = xyz = 16{,}000$$

we can always write one variable, say z, in terms of the others

$$z = \frac{16{,}000}{xy} \qquad [12]$$

The feasible values of x and y are $x > 0$, $y > 0$; the corresponding z value is then given by [12]. We cannot set $x = 0$ or $y = 0$; if one of the dimensions is zero, the volume cannot be 16,000.

The construction cost is given in terms of x and y by the formula

$$\begin{aligned}C(x, y) &= 7.5(\text{roof area}) + 5.0(\text{area of sides}) + 12.5(\text{floor area})\\ &= 7.5xy + 5.0(2xz + 2yz) + 12.5(xy)\\ &= 20xy + 10xz + 10yz\end{aligned}$$

which takes the form below when we use [12] to eliminate z,

$$\begin{aligned}C(x, y) &= 20xy + 10x\left(\frac{16{,}000}{xy}\right) + 10y\left(\frac{16{,}000}{xy}\right)\\ &= 20xy + \frac{160{,}000}{y} + \frac{160{,}000}{x}\end{aligned} \qquad [13]$$

The critical points are the solutions of the system of equations

$$\begin{aligned}0 &= \frac{\partial C}{\partial x} = 20y - \frac{160{,}000}{x^2}\\ 0 &= \frac{\partial C}{\partial y} = 20x - \frac{160{,}000}{y^2}\end{aligned} \qquad [14]$$

From the second equation we get $x = 8000/y^2$. Entering this into the first equation, we obtain

$$0 = 20y - \frac{160,000}{\left(\frac{8000}{y^2}\right)^2} = 20y - \frac{20y^4}{8000} = 20y - \frac{y^4}{400} = y\left(20 - \frac{y^3}{400}\right)$$

Hence, either $y = 0$ (not a feasible value for y), or else

$$y^3 = 8000$$
$$y = \sqrt[3]{8000} = 20$$

Then we obtain

$$x = \frac{8000}{y^2} = 20$$

from Equation [14], so the only feasible critical point occurs where $x = 20$, $y = 20$. This must be the solution to our problem: The optimal dimensions are

$$x = 20 \qquad y = 20 \qquad z = \frac{16,000}{xy} = 40$$

The (minimized) construction costs are obtained from [13]:

$$C(20, 20) = \$24,000$$

■

Exercises 7.5

In Exercises 1–10 find all critical points (if any) for the function $f(x, y)$.

1. $f(x, y) = x + y$

2. $f(x, y) = 3x^2 + y$

3. $f(x, y) = xy$

4. $f(x, y) = 4x - xy$

5. $f(x, y) = x^2 - 2xy + 2y^2$

6. $f(x, y) = x^2 + xy + y^2$

7. $f(x, y) = x^2 - 2xy + 2y^2 - 4x + 6y + 5$

8. $f(x, y) = x^2 + 4xy + y^2 - x - 2$

9. $f(x, y) = \dfrac{1}{x} + \dfrac{1}{y} + xy$

10. $f(x, y) = \dfrac{1}{\sqrt{1 + x^2 + y^2}}$

In Exercises 11–27 find all points where it is possible for the function $g(x, y)$ to have an extremum.

11. $g(x, y) = 3xy - 2y^2 + y$

12. $g(x, y) = 2x^2 - 6x + xy$

13. $g(x, y) = 3x^2 + y^2 - x + 4y + 3$

14. $g(x, y) = 7 - 2x + 6y - \frac{1}{2}x^2 - 3y^2$

15. $g(x, y) = x^3 + y^3 - 3x^2 - 9x - 3y$

16. $g(x, y) = x^3 + y^3 + 3y^2 - 9y - 1$

17. $g(x, y) = x^3 - 2xy + y^2 - 4$

18. $g(x, y) = y^4 - 4xy + \frac{1}{2}x^2$

19. $g(x, y) = 4x + 3y + y^2$

20. $g(x, y) = x^4 - x^2 + 3x - y$

21. $g(x, y) = x^2 + 2xy + y^2$

22. $g(x, y) = xe^{-(x^2+y^2)/2}$ (*Hint*: $e^a \neq 0$ for every a.)

23. $\dfrac{1 + x}{1 - y}$

24. $2x^3 + 3x^2 - 12x + y^2 - y + 2$

25. $\dfrac{1}{1 + x^2 + y^2}$ (*Hint*: $1 + x^2 + y^2 > 0$ for all x, y.)

26. $e^{-(x^2+y^2)/2}$ (*Hint*: $e^a > 0$ for any a.)

27. $xe^{(x^2-y^2)/2}$ (*Hint*: $e^a > 0$ for any a.)

28. The function $f(x, y) = 10 + 4x - x^4 + 4y^3 - y^4$, defined for all x and y, has an absolute maximum. Find it.

29. The function $f(x, y) = x^2 + x - xy + y^2$ has an absolute minimum. Find it.

30. Find all critical points for $f(x, y) = x^2 - 12y^2 - 4y^3 + 3y^4$. We can show that f actually has an absolute minimum. Where is it, and what is the minimum value for f?

31. The function

$$f(x, y) = \frac{x}{1 + x^2 + y^2}$$

has an absolute maximum and an absolute minimum. Find these extrema.

32. Lawn Master, Inc. produces two types of lawn mowers, one a standard model and the other motorized. Their market analysts have set up a model predicting the net profit achieved when these models are offered at various prices:

$$P(x, y) = -55{,}700 + 250y + 200x + xy - 0.5y^2 - 2x^2$$

where x = (price of the standard model) and y = (price of the motorized model) in dollars. Find the prices that maximize profit.

33. A record store has found that its earnings (in thousands of dollars) can be predicted by the formula

$$E(x, y) = -86 + 30x + 24y - 3x^2 + 2xy - 6y^2$$

where x = (investment in inventory) and y = (investment in advertising) in thousands of dollars. Find the maximum earnings and the amounts that should be spent on inventory and advertising to achieve it.

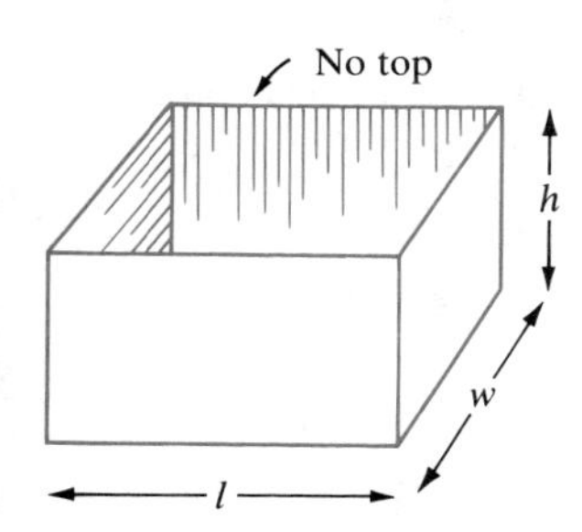

Figure 7.26
The box in Exercises 34 and 35. Because the volume is 108 cubic feet, the variable h may be written in terms of l and w.

34. A rectangular box without a top is to contain 108 cubic feet. Find the dimensions that yield the lowest possible surface area for the box (*Hint*: This amounts to minimizing $A = 2lh + 2wh + lw$ for $l > 0$, $w > 0$, $h > 0$).

35. A rectangular box without a top is to contain 108 cubic feet. The material for the base costs \$8 per square foot, but material for the sides costs \$1 per square foot. Find the dimensions yielding the minimum *cost of materials*. (*Hint*: The dimensions of the box are shown in Figure 7.26. In terms of these dimensions l,w,h find the total cost of materials $C(l, w)$ = (cost of bottom) + (cost of 4 sides).)

36. A rectangular warehouse is to enclose 500,000 cubic feet of space. Its four brick walls cost \$5 per square foot to erect. Every square foot of roof costs \$1.50 per square foot, and the poured concrete floor costs \$3.50 per square foot. Find the building dimensions (height, width, and length) yielding the lowest construction costs.

37. A laboratory sink is to be constructed containing 12,000 cubic inches. It has the form of a rectangular box without a top. Reinforced material for the bottom costs 4¢ per square inch. Material for the side against the wall costs 1¢ per square inch, and material for the other three sides costs 3¢ per square inch. Find the dimensions yielding the minimum cost of materials.

38. Three towns in Siberia have position coordinates on a reference map as shown in Figure 7.27. They are to be connected by rail lines joining at some common point $P = (x, y)$ as shown. No natural obstacles occur in this flat terrain. Construction materials must be shipped greater and greater distances along the lines from A, B, C as each rail line is extended; therefore, construction costs for each line are proportional to the *square* of its length

$$(\text{cost of each line}) = 100{,}000 \times (\text{length in miles})^2$$

Where should the junction P be placed to minimize construction costs? That is, which choice of x and y minimizes the total cost? (*Hint*: Use the Pythagorean theorem to obtain the distance from each town to P; then set up the cost function $C(x, y)$.)

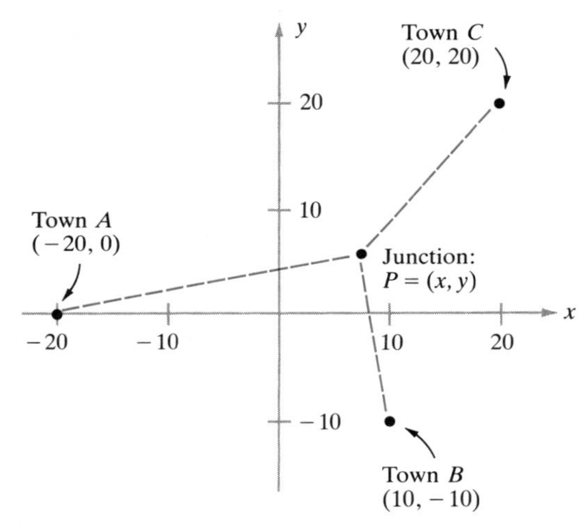

Figure 7.27
Location of the towns in Exercise 38. Units are in miles from the reference point located at the origin. Proposed rail lines are dashed.

39. Find the minimum value of the function

$$f(x, y) = (x - 1)^2 + (y - 1)^2 + x^2 + y^2$$

defined for all x and y. (*Note*: This function arises in a geometric problem: $f(x, y) = s^2$, where s is the distance in three-dimensional space from the point $Q = (1, 1, 0)$ to an arbitrary point (x, y, z) on the surface $z = \sqrt{x^2 + y^2}$, a vertical cone with vertex at the origin. The minimum of f corresponds to the minimum distance from Q to the surface. It should be intuitively obvious that there is such a minimum distance.)

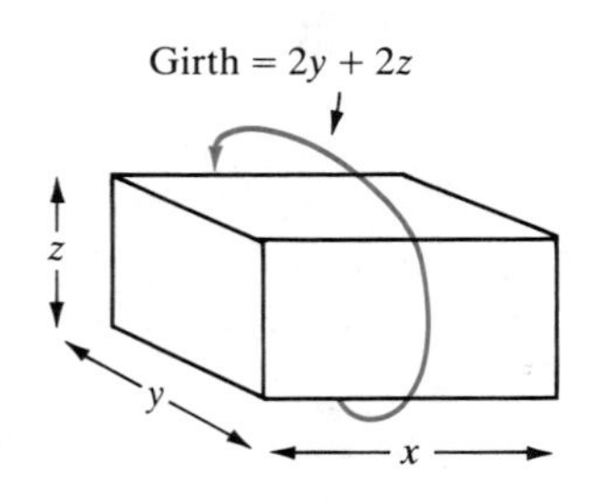

Figure 7.28
The package in Exercise 41. Girth is the total distance around the middle of the package, as shown. The length x of the package is its longest dimension.

40. Suppose the profit function for a facility that manufactures two products is given in terms of the production levels x and y by

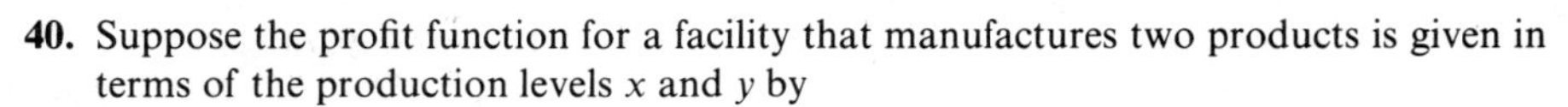

$$P(x, y) = -180{,}000 + 600x - 0.1x^2 + 800y - 0.2y^2 - 0.015xy$$

Find the production levels that yield maximum profit. What is the maximum value of P?

41. Postal regulations require that (length) + (girth) of a package be no more than 84 inches. Girth is the distance around the middle (Figure 7.28) for a rectangular package. Find the dimensions x, y, z yielding the largest possible volume for the package, subject to this regulation. What is the volume of the optimal package?

7.6 Application: Least Squares

Here is a typical problem in setting up a mathematical model from real-life data. We shall solve it using the optimization methods of the last section.

A marketing research group has been hired to help a manufacturer of video discs formulate his pricing policy. Table 7.1 shows the result when the discs were offered at different prices in selected test cities of the same size. These data points are plotted in Figure 7.29. The data points are somewhat scattered; this is to be expected in real-life observations. Although price is a major influence on sales, other factors—such as cultural differences or differences in the residents' disposable income—cannot be controlled even if the test cities are reasonably well matched.

Data Point P_i	Price (dollars) x_i	Number Sold (thousands) y_i
P_1	32	2.7
P_2	38	1.9
P_3	45	1.9
P_4	50	1.3
P_5	55	0.9

Table 7.1 Selling price and number of video discs in selected trial markets. These data are plotted in Figure 7.29. The problem is to find a linear function $y = ax + b$ that best predicts the level of sales y in terms of the price x.

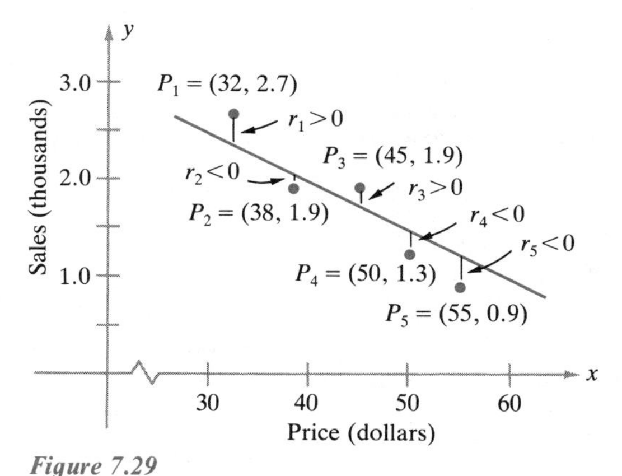

Figure 7.29
A plot of the data points listed in Table 7.1. We have drawn in a straight line that seems to fit the points reasonably well. For the ith data point $P_i = (x_i, y_i)$, the residual r_i is the vertical distance from the data point to the line, taken positive if P_i lies above the line and negative otherwise. Various lines that seem to fit the data will give different residuals $r_1, \ldots, r_5$. The least-squares line is the one that makes the sum of squared residuals $r_1^2 + \cdots + r_5^2$ as small as possible, and is uniquely determined.

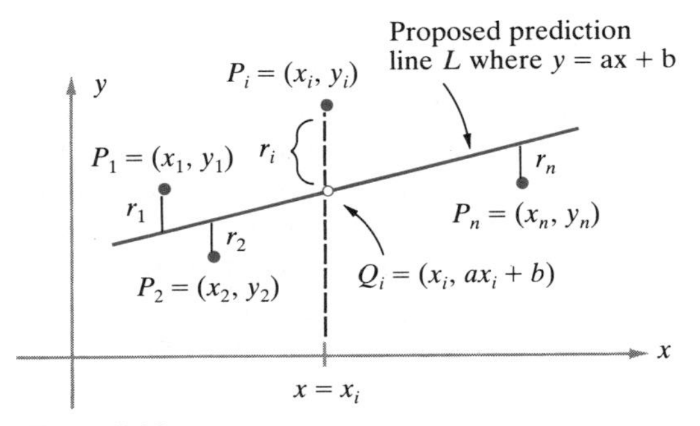

Figure 7.30
Some data points and a proposed prediction line L. For each data point $P_i = (x_i, y_i)$, the residual r_i is the vertical distance from the point to the line. If the equation of the line is $y = ax + b$, the corresponding point on L is $Q_i = (x_i, ax_i + b)$; so the residual is $r_i = y_i - (ax_i + b)$. The residual is positive if P_i lies above L; it is negative otherwise.

The data points seem to fall more or less along a straight line, so it seems reasonable to try to interpolate a straight line that describes the relationship between price and sales. This line, or its equation, could then be used to estimate sales at various other prices. But the plotted points do not line up exactly, so it is not clear which of several possible lines does the best job. The same problem of fitting a "prediction line" to a given set of data points arises in many other situations. Here is what statisticians do to find the best possible line fitted to the data. It is called the **method of least squares.**

We start with data points $P_1 = (x_1, y_1), \ldots, P_n = (x_n, y_n)$, much as in Figure 7.29. Given a possible line $y = ax + b$, each data point $P_k = (x_k, y_k)$ lies directly above or below a corresponding point on the line $Q_k = (x_k, y'_k)$, both having the same x coordinate $x = x_k$, as in Figure 7.30. Obviously, $y'_k = ax_k + b$. The vertical distance from Q_k to P_k,

$$r_k = y_k - y'_k$$

is called the *residual* for the kth data point. The sum of the squares of the residuals

$$S = r_1^2 + \cdots + r_n^2 = (y_1 - y'_1)^2 + \cdots + (y_n - y'_n)^2$$

is a convenient way of measuring how well the given line $y = ax + b$ fits the data. The larger the residuals, the larger S will be. We would like to choose the line so S is as small as possible. On what does S depend? Once the line is specified, by its coefficients a and b, the residuals and S are determined. Thus S is a function of the variables a and b, and we want to find the choice of a and b that minimizes $S(a, b)$. Remarkably, we always have an optimal choice, giving the **least squares line** best fitted to the data points. Let us see how we can use calculus to find it.

Example 18 In a drug test the number of mutations y in a sample of 100 mice depends on the dosage x of drug. Testing yields the data values shown in Table 7.2. Find the line $y = ax + b$ giving the best least-squares fit. Use this line to predict how many mutations might be expected if the dosage were $x = 10$ units.

Observation Number i	Drug Dosage x_i	Number of Mutations y_i	$y_i' = ax_i + b$	Residual r_i	(Residual)^2 r_i^2
1	0	2	b	$(2 - b)$	$(2 - b)^2$
2	1	4	$a + b$	$(4 - a - b)$	$(4 - a - b)^2$
3	2	5	$2a + b$	$(5 - 2a - b)$	$(5 - 2a - b)^2$
4	3	7	$3a + b$	$(7 - 3a - b)$	$(7 - 3a - b)^2$
5	4	9	$4a + b$	$(9 - 4a - b)$	$(9 - 4a - b)^2$

Table 7.2 The first three columns show the number of mutations observed in a set of 100 mice in response to a drug dosage of x units. Some mutations occur spontaneously in inbred mice used for these experiments, which accounts for the two mutations observed when $x = 0$ (no drug administered). Given a trial prediction line $y = ax + b$, we then compute the entries in the last three columns, which involve the undetermined coefficients a and b.

Solution We must choose a and b to minimize the sum of squared residuals

$$S(a, b) = r_1^2 + \cdots + r_5^2$$

For each data point $P_k = (x_k, y_k)$, the corresponding point on the line $y = ax + b$ has x value $x = x_k$ and y value $y = y_k' = ax_k + b$. These values y_k' are given in the fourth column of the table, along with the residuals $r_k = y_k - y_k'$ and their squares $r_k^2 = (y_k - y_k')^2$; of course they involve a and b. Thus we want to minimize

$$\begin{aligned} S(a, b) &= r_1^2 + \cdots + r_5^2 \\ &= (2 - b)^2 + (4 - (a + b))^2 + (5 - (2a + b))^2 \\ &\quad + (7 - (3a + b))^2 + (9 - (4a + b))^2 \end{aligned}$$

a polynomial in the variables a and b. The minimum for S must occur at a critical point, where $\partial S/\partial a = 0$ and $\partial S/\partial b = 0$. The partial derivatives are

$$\begin{aligned} \frac{\partial S}{\partial a} &= 0 - 2(4 - (a + b)) - 4(5 - (2a + b)) \\ &\quad - 6(7 - (3a + b)) - 8(9 - (4a + b)) \\ &= -142 + 60a + 20b \end{aligned}$$

$$\begin{aligned} \frac{\partial S}{\partial b} &= -2(2 - b) - 2(4 - (a + b)) - 2(5 - (2a + b)) \\ &\quad - 2(7 - (3a + b)) - 2(9 - (4a + b)) \\ &= -54 + 20a + 10b \end{aligned}$$

This calculation can be done using the chain rule on each term in s; for example

$$\begin{aligned} \frac{\partial}{\partial a}((5 - (2a + b))^2) &= 2(5 - (2a + b))\frac{\partial}{\partial a}(5 - (2a + b)) \\ &= 2(5 - (2a + b))(-2) \\ &= -20 + 8a + 4b \end{aligned}$$

The resulting system of equations

$$0 = \frac{\partial S}{\partial a} = -142 + 60a + 20b$$

$$0 = \frac{\partial S}{\partial b} = -54 + 20a + 10b$$

has $a = 1.7$ and $b = 2$ as its only solution. This is the only critical point. Thus the line giving the best "least-squares" fit to the data is $y = 1.7x + 2$. Setting $x = 10$ in this linear equation, we get the expected number of mutations $y = 1.7(10) + 2 = 19$ per hundred mice when the dosage is 10 units. ■

Finding the least-squares line as the solution of an optimization problem is simple if there are a few data points. But this construction is done so often in statistics that we would like to know if calculus can be applied once and for all to find an explicit formula for the coefficients a, b of the least-squares line $y = ax + b$ in terms of the original data points. We are in luck. If we patiently carry out arguments like those in Example 18 (we omit the details), we arrive at a formula that works for any set of data. Once we have the formula, the calculus has done its job, and the rest is algebra.

LEAST-SQUARES FORMULA Given a set of observed data points $P_1 = (x_1, y_1), \ldots, P_n = (x_n, y_n)$, we find the coefficients of the line $y = ax + b$ giving the best least squares fit to these data as follows.

Step 1 Make a table showing the values of

$$x_i, \quad y_i, \quad x_i y_i, \quad x_i^2$$

for each data point.

Step 2 From the table, compute the averages

$$\bar{x} = \frac{1}{n}(x_1 + \cdots + x_n) \qquad \bar{y} = \frac{1}{n}(y_1 + \cdots + y_n)$$

and the quantities

$$p = x_1 y_1 + \cdots + x_n y_n \quad \text{(sum of products of } x \text{ and } y\text{)}$$
$$s = x_1^2 + \cdots + x_n^2 \quad \text{(sum of squares of } x \text{ values)}$$

Step 3 The coefficients of the least-squares line are then

$$a = \frac{p - n \cdot \bar{x} \cdot \bar{y}}{s - n(\bar{x})^2} \qquad b = \frac{\bar{y} \cdot s - \bar{x} \cdot p}{s - n(\bar{x})^2} \qquad [15]$$

Here, n is the number of data points.

Notice that the denominators are the same in each case and require only a single computation.

Example 19 At the beginning of this section, we discussed the situation of a manufacturer of video discs. His test-market data are duplicated in Table 7.3. Calculate the least-squares line fitted to these data using Formula [15]. What level of sales would you predict if the price is set at $40 per disc?

i	x_i	y_i	$x_i y_i$	x_i^2
1	32	2.7	86.4	1024
2	38	1.9	72.2	1444
3	45	1.9	85.5	2025
4	50	1.3	65.0	2500
5	55	0.9	49.5	3025
Sums:	220	8.7	358.6	10,018

Table 7.3 The first three columns list the x and y values of the $n = 5$ data points $P_i = (x_i, y_i)$. From these we calculate the products $x_i y_i$ (fourth column) and squares x_i^2 (fifth column). Sums of the column entries, at bottom, are used to find the least-squares line fitted to these data points by formula [15].

Solution The steps are easily executed if we enlarge the table, adding columns that list the values $x_i y_i$ and x_i^2, as in Table 7.3. There are $n = 5$ data points.

Step 2: Adding up the second column entries x_i and dividing by $n = 5$, we get

$$\bar{x} = \frac{1}{5}(32 + 38 + 45 + 50 + 55) = \frac{220}{5} = 44.0$$

Similarly, the entries in the third column give

$$\bar{y} = \frac{1}{5}(2.7 + 1.9 + 1.9 + 1.3 + 0.9) = \frac{8.7}{5} = 1.74$$

The fourth and fifth columns add up to p and s, respectively:

$$p = (x_1 y_1 + \cdots + x_5 y_5) = 86.4 + 72.2 + 85.5 + 65.0 + 49.5$$
$$= 358.6$$

and

$$s = x_1^2 + \cdots + x_5^2 = 1024 + 1444 + 2025 + 2500 + 3025$$
$$= 10{,}018$$

Step 3: From [15] we obtain the coefficients

$$a = \frac{p - n \cdot \bar{x} \cdot \bar{y}}{s - n(\bar{x})^2} = \frac{358.6 - 5(44)(1.74)}{10{,}018 - 5(44)^2} = \frac{-24.2}{338} = -0.07160$$

$$b = \frac{\bar{y} \cdot s - \bar{x} \cdot p}{s - n(\bar{x})^2} = \frac{1.74(10{,}018) - 44(358.6)}{10{,}018 - 5(44)^2}$$

$$= \frac{1652.92}{338} = 4.890$$

so the equation of the least-squares line is $y = -0.0716x + 4.89$. If $x = 40$, we would expect sales of approximately

$$y = -0.0716(40) + 4.89 = 2.026 \quad \text{thousand discs}$$

In the first few of the following exercises we ask you to find the least-squares line directly, using calculus. The rest are exercises in using the general formula [15].

Exercises 7.6

1. Consider the data points

$$P_1 = (-5, 105) \qquad P_2 = (0, 47) \qquad P_3 = (8, -80) \qquad P_4 = (15, -200)$$

and the straight line $y = -15x + 40$. For each P_i compute the coordinates of the corresponding point Q_i directly above or below it on the line. Then compute the residuals r_1, r_2, r_3, r_4 and the sum of their squares.

2. Repeat Exercise 1 using the same data points, but a slightly different straight line $y = -15.5x + 40$.

In Exercises 3–8 we give several sets of data points in the plane. Use the technique of Example 18 to find the equation $y = ax + b$ of the straight line giving the best least-squares fit to the data points.

3. $P_1 = (-1, 2)$ $\quad P_2 = (4, 7)$

4. $P_1 = (1, 2.5)$ $\quad P_2 = (5, 1.8)$

5. $P_1 = (1, 4)$ $\quad P_2 = (3, 7)$ $\quad P_3 = (5, 9)$

6. $P_1 = (1, 4.2)$ $\quad P_2 = (3, 6.8)$ $\quad P_3 = (5, 9.1)$

7. $P_1 = (-1, -2)$ $\quad P_2 = (0, 1)$ $\quad P_3 = (2, 6)$ $\quad P_4 = (4, 14)$

8. $P_1 = (-5, 105)$ $\quad P_2 = (0, 47)$ $\quad P_3 = (8, -80)$ $\quad P_4 = (15, -200)$

9. Here are some data on cost vs. production level for a manufacturer. Use the method of Example 18 to find the least-squares line $C(x) = ax + b$ best fitted to these four data points. What do you expect his costs to be when $x = 2$?

Production level x (hundreds)	0	1	4	5
Costs C (thousands of dollars)	2.0	4.0	7.0	9.0

10. Administration of a daily dose of x milligrams of the drug diazide to a hypertensive patient resulted in the diastolic blood pressures y listed in the table. Use the method of

x	0	15	50	100
y	109	105	97	83

Example 18 to find the least-squares line best fitted to these four data points. (The resulting formula $y = ax + b$ gives the linear relationship between dosage x and blood pressure y most consistent with the given data.)

11. Find the least-squares line requested in Exercise 9 using the least-squares formula. What do you expect the costs to be if $x = 2.5$?

12. Find the least-squares line requested in Exercise 10 using the least-squares formula. What do you expect the blood pressure to be if the dosage is $x = 30$ milligrams per day?

In Exercises 13–17 use the least-squares formula to find the straight line best fitted to each set of data points.

13.

x	0	1	2	3	4
y	4	2	0	-1	-3

14.

x	0	1	2	3	4
y	400	200	0	-100	-300

15.

x	0	1000	2000	3000	4000
y	4.0	2.0	0.0	-1.0	-3.0

16.

x	1.10	1.95	3.15	4.00	4.80
y	2.75	3.00	3.75	3.90	4.25

17.

x	2.75	3.00	3.75	3.90	4.25
y	1.10	1.95	3.15	4.00	4.80

18. Silver futures contracts issued on April 30 have prices (cents per troy ounce) that depend on the delivery date (end of the month indicated). These prices reflect traders' estimates of the market price of silver on the delivery date. Letting $x =$ (months elapsed after April 30), $y =$ (price for delivery date x), find the least-squares line fitted to these data points.

Delivery Date	May	June	July	Sept	Dec	Jan
Price	470.5	473.0	473.5	479.0	486.5	489.0

19. The table gives the death rates from lung cancer per 1000 U.S. females in a certain age range; x is the number of years elapsed since 1950. Compute the least-squares line giving y in terms of x.

x	6	8	10	12	14
Rate y	0.24	0.25	0.24	0.26	0.29

20. The rate at which a certain species of cricket chirps varies with the temperature in a roughly linear way. A biologist has tabulated observations on this species. Find the least-squares line fitted to these data. If you heard one of these crickets chirping 18 times per minute, what would be your guess about the temperature?

Temperature (°F)	88.6	93.3	80.6	75.2	69.7	83.3
Chirps per minute	20.0	19.8	17.1	15.5	14.7	16.2

21. In a drug-testing experiment, pure-bred mice were given measured daily doses of a new drug. Dosage was measured as x = (milligrams of drug per kilogram of body weight). Investigators want to know if the drug has possible side effects, and have noted the number of mice exhibiting chromosomal abnormalities after exposure for 6 months (see the table, where y is the number of mice with abnormalities per thousand mice). Compute the least-squares line $y = ax + b$ best fitted to these 8 data points.

x	0.3	1.0	1.5	2.0	3.0	3.7	4.0	4.5
y	0.8	1.6	2.4	3.3	4.8	5.9	6.0	7.0

7.7 Higher-Order Partial Derivatives and the Search for Extrema

If a function $f(x, y)$ has partial derivatives $\partial f/\partial x$ and $\partial f/\partial y$, we may apply the operations $\partial/\partial x$ and $\partial/\partial y$ to each of these functions of two variables, thereby obtaining the **second partial derivatives** of f:

$$\frac{\partial^2 f}{\partial x^2} = \frac{\partial}{\partial x}\left(\frac{\partial f}{\partial x}\right) \qquad \frac{\partial^2 f}{\partial y^2} = \frac{\partial}{\partial y}\left(\frac{\partial f}{\partial y}\right)$$

$$\frac{\partial^2 f}{\partial x\,\partial y} = \frac{\partial}{\partial x}\left(\frac{\partial f}{\partial y}\right) \qquad \frac{\partial^2 f}{\partial y\,\partial x} = \frac{\partial}{\partial y}\left(\frac{\partial f}{\partial x}\right)$$

For all functions we deal with, the "mixed" second partial derivatives agree:

$$\frac{\partial^2 f}{\partial x\,\partial y} = \frac{\partial^2 f}{\partial y\,\partial x} \qquad [16]$$

This shortens the list of partial derivatives to keep track of.

We could go on to apply the operations $\partial/\partial x$ and $\partial/\partial y$ to these second partial derivatives, obtaining higher-order partial derivatives of f such as

$$\frac{\partial^3 f}{\partial x^3} = \frac{\partial}{\partial x}\left(\frac{\partial^2 f}{\partial x^2}\right), \qquad \frac{\partial^3 f}{\partial y\,\partial x\,\partial y} = \frac{\partial}{\partial y}\left(\frac{\partial^2 f}{\partial x\,\partial y}\right)$$

and so on. We will not make use of higher-order partial derivatives. Second partial derivatives are important by themselves in the physical sciences and in certain areas of economics. However, our main application will be to test for maxima and minima.

Here are some examples of calculations involving second partial derivatives.

Example 20 Calculate the first and second partial derivatives of

$$f(x, y) = 4x^2 - 2xy + y^3 + 10$$

as functions of x and y. Find the values of the second partial derivatives at the particular points $P_1 = (1, 3)$ and $P_2 = (-1, 0)$.

Solution Regarding x or y as a constant, we obtain the first order partial derivatives

$$\frac{\partial f}{\partial x} = \frac{\partial}{\partial x}(4x^2 - 2xy + y^3 + 10) = 4(2x) - 2y(1) + 0 + 0 = 8x - 2y$$

$$\frac{\partial f}{\partial y} = \frac{\partial}{\partial y}(4x^2 - 2xy + y^3 + 10) = 0 - 2x(1) + 3y^2 + 0 = -2x + 3y^2$$

Now apply $\partial/\partial x$ and $\partial/\partial y$ to each of these functions to obtain the second partial derivatives:

$$\frac{\partial^2 f}{\partial x^2} = \frac{\partial}{\partial x}\left(\frac{\partial f}{\partial x}\right) = \frac{\partial}{\partial x}(8x - 2y) = 8 \quad \text{(constant function)}$$

$$\frac{\partial^2 f}{\partial x\,\partial y} = \frac{\partial}{\partial x}\left(\frac{\partial f}{\partial y}\right) = \frac{\partial}{\partial x}(-2x + 3y^2) = -2 \quad \text{(constant function)}$$

and

$$\frac{\partial^2 f}{\partial y\,\partial x} = \frac{\partial}{\partial y}\left(\frac{\partial f}{\partial x}\right) = \frac{\partial}{\partial y}(8x - 2y) = -2 \quad \text{(constant function)}$$

$$\frac{\partial^2 f}{\partial y^2} = \frac{\partial}{\partial y}\left(\frac{\partial f}{\partial y}\right) = \frac{\partial}{\partial y}(-2x + 3y^2) = 6y$$

Notice that the mixed partials are equal, as stated in [16], so we could have skipped one of these computations.

At $P_1 = (1, 3)$ the values of the second partial derivatives are

$$\frac{\partial^2 f}{\partial x^2}(1, 3) = 8 \qquad \frac{\partial^2 f}{\partial x\,\partial y}(1, 3) = \frac{\partial^2 f}{\partial y\,\partial x}(1, 3) = -2$$

$$\frac{\partial^2 f}{\partial y^2}(1, 3) = 6(3) = 18$$

At $P_2 = (-1, 0)$ we have

$$\frac{\partial^2 f}{\partial x^2}(-1, 0) = 8 \qquad \frac{\partial^2 f}{\partial x\,\partial y}(-1, 0) = \frac{\partial^2 f}{\partial y\,\partial x}(-1, 0) = -2$$

$$\frac{\partial^2 f}{\partial y^2}(-1, 0) = 6(0) = 0$$

■

Example 21 For the function $f(x, y) = x^3 + x^2y - 3y^2$, the point $P = (-9, 27/2)$ is a critical point, where both $\partial f/\partial x$ and $\partial f/\partial y$ are equal to zero. Verify this property. Then calculate the second partial derivatives of f, and find their values at P.

Solution Regarding x or y as constant we obtain the respective first partial derivatives

$$\frac{\partial f}{\partial x} = \frac{\partial}{\partial x}(x^3 + x^2y - 3y^2) = 3x^2 + 2xy$$

$$\frac{\partial f}{\partial y} = \frac{\partial}{\partial y}(x^3 + x^2y - 3y^2) = x^2 - 6y$$

Substituting $x = -9$, $y = 27/2$ into these formulas, we see that these partial derivatives are both zero; so P is a critical point. Next, treat y as a constant, and compute the second partial derivatives

$$\frac{\partial^2 f}{\partial x^2} = \frac{\partial}{\partial x}\left(\frac{\partial f}{\partial x}\right) = \frac{\partial}{\partial x}(3x^2 + 2xy) = 6x + 2y$$

$$\frac{\partial^2 f}{\partial x\,\partial y} = \frac{\partial}{\partial x}\left(\frac{\partial f}{\partial y}\right) = \frac{\partial}{\partial x}(x^2 - 6y) = 2x$$

Taking x as a constant, we obtain the remaining second partial derivative

$$\frac{\partial^2 f}{\partial y^2} = \frac{\partial}{\partial y}\left(\frac{\partial f}{\partial y}\right) = \frac{\partial}{\partial y}(x^2 - 6y) = -6 \quad \text{(constant function)}$$

Values at the critical point $P = (-9, 27/2)$ are obtained by setting $x = -9$ and $y = 27/2$ in these formulas:

$$\frac{\partial^2 f}{\partial x^2}(P) = -27 \qquad \frac{\partial^2 f}{\partial x\,\partial y}(P) = -18 \qquad \frac{\partial^2 f}{\partial y^2}(P) = -6$$

■

For a function $f(x, y)$ of two variables, the first partial derivatives $\partial f/\partial x$ and $\partial f/\partial y$ taken together play the same role as the ordinary derivative df/dx when f is a function of one variable. Similarly, the whole set of second partial derivatives plays the role of d^2f/dx^2. The second partial derivatives of $f(x, y)$ are often presented in a little two-by-two array called the **Hessian matrix** for f:

$$\begin{bmatrix} \dfrac{\partial^2 f}{\partial x^2} & \dfrac{\partial^2 f}{\partial x\,\partial y} \\ \dfrac{\partial^2 f}{\partial y\,\partial x} & \dfrac{\partial^2 f}{\partial y^2} \end{bmatrix}$$

Its determinant, called the **discriminant** of f, is given by the following combination of second partial derivatives

$$\text{discriminant } D(x, y) = \left(\frac{\partial^2 f}{\partial x^2}\right)\left(\frac{\partial^2 f}{\partial y^2}\right) - \left(\frac{\partial^2 f}{\partial x\,\partial y}\right)^2 \qquad [17]$$

and is another function of x and y. The discriminant will be used to test for maxima and minima when we come to optimization problems.

Example 22 Find the discriminant of $f(x, y) = x^3 - x + y^2 + xy$ as a function of x and y. What is its value at the particular point $P = (-\frac{1}{2}, \frac{1}{4})$?

Solution Calculating partial derivatives in the usual way, we get

$$\frac{\partial f}{\partial x} = 3x^2 - 1 + y$$

$$\frac{\partial f}{\partial y} = 2y + x$$

and

$$\frac{\partial^2 f}{\partial x^2} = \frac{\partial}{\partial x}(3x^2 - 1 + y) = 6x$$

$$\frac{\partial^2 f}{\partial x\,\partial y} = \frac{\partial^2 f}{\partial y\,\partial x} = \frac{\partial}{\partial y}(3x^2 - 1 + y) = 1 \quad \text{(constant function)}$$

$$\frac{\partial^2 f}{\partial y^2} = \frac{\partial}{\partial y}(2y + x) = 2 \quad \text{(constant function)}$$

By definition of the discriminant,

$$\begin{aligned} D(x, y) &= \left(\frac{\partial^2 f}{\partial x^2}\right)\left(\frac{\partial^2 f}{\partial y^2}\right) - \left(\frac{\partial^2 f}{\partial x\,\partial y}\right)^2 \\ &= (6x)(2) - (1)^2 \\ &= 12x - 1 \end{aligned}$$

At the particular point $P = (-\frac{1}{2}, \frac{1}{4})$, its value is

$$D\left(-\frac{1}{2}, \frac{1}{4}\right) = 12\left(-\frac{1}{2}\right) - 1 = -6 - 1 = -7$$

■

Suppose we want information about the extrema of a function $f(x, y)$. In certain applications the *local* extrema are precisely what we want to find. For example, if x and y describe the state (position) of a mechanical system and $f(x, y)$ is the energy of the system, then the local minima are of great interest because they are the states of the system that are **stable**. The system will return to such a state if perturbed slightly. This notion of stability also applies to chemical, ecologic, and economic systems.

In an optimization problem our task is to find the *absolute* maximum or minimum of f, as when we want to find the most profitable production levels for a manufacturing plant or optimal growing conditions in an agricultural experiment. The simple methods of Section 7.5 apply only if we have some preliminary information or intuition about the existence of extrema and the possibility of boundary extrema. What can we do in more complicated problems? Whatever the circumstances, the optimization principle tells us that maxima and minima within the feasible set show up at critical points. So it is clear that first we must locate all the feasible critical points. Then the following test allows us to sort out the local maxima and minima among the critical points; these are candidates for solutions. The ultimate resolution of the optimization problem might require a study of possible boundary extrema, but this can be very difficult and must be left to advanced courses.

SECOND DERIVATIVE TEST Suppose $P = (x_0, y_0)$ is a critical point for $f(x, y)$, so that

$$\frac{\partial f}{\partial x}(P) = 0 \quad \text{and} \quad \frac{\partial f}{\partial y}(P) = 0$$

To test P we form the **discriminant**

$$D(P) = \frac{\partial^2 f}{\partial x^2}(P) \cdot \frac{\partial^2 f}{\partial y^2}(P) - \left(\frac{\partial^2 f}{\partial x\, \partial y}(P)\right)^2 \quad [18]$$

out of the second partial derivatives. Then

Step 1 If $D(P) < 0$, the function does not have an extremum at P.
Step 2 If $D(P) > 0$ and $\partial^2 f/\partial x^2(P) < 0$, a local maximum occurs at P.
Step 3 If $D(P) > 0$ and $\partial^2 f/\partial x^2(P) > 0$, a local minimum occurs at P.

If $D(P) = 0$, the test gives no information about P.

We shall not prove the validity of the test here; the following examples illustrate its use.

Example 23 Find the critical points for the profit function

$$f(x, y) = -900{,}000 + (500x - 0.1x^2) + (700y - 0.2y^2) - 0.01xy$$

and test them.

Solution At a critical point we have

$$\frac{\partial f}{\partial x} = 500 - 0.2x - 0.01y = 0$$

$$\frac{\partial f}{\partial y} = 700 - 0.4y - 0.01x = 0 \quad [19]$$

Solve this system of equations to find the critical points. From the first equation we obtain

$$0.01y = 500 - 0.2x \quad \text{or} \quad y = 50{,}000 - 20x$$

Substituting this into the second equation, we can solve for x

$$0 = 700 - 0.4y - 0.01x = 700 - 0.4(50{,}000 - 20x) - 0.01x$$

$$0 = -19{,}300 + 7.99x$$

$$x = 2415.5$$

Entering this x value into either of the equations [19], we obtain the corresponding y value at the critical point. Using the first equation, we obtain

$$0 = 500 - 0.2(2415.5) - 0.01y$$
$$y = 1690.0$$

Thus, the only critical point is at $(x_0, y_0) = (2415.5, 1690.0)$.

To test it we need the second partial derivatives. These are

$$\frac{\partial^2 f}{\partial x^2} = \frac{\partial}{\partial x}(500 - 0.2x - 0.01y) = -0.2 \quad \text{(constant function)}$$

$$\frac{\partial^2 f}{\partial x\, \partial y} = \frac{\partial}{\partial x}(700 - 0.4y - 0.01x) = -0.01 \quad \text{(constant function)}$$

$$\frac{\partial^2 f}{\partial y^2} = \frac{\partial}{\partial y}(700 - 0.4y - 0.01x) = -0.4 \quad \text{(constant function)}$$

The discriminant has constant value everywhere

$$\begin{aligned} D &= \frac{\partial^2 f}{\partial x^2}\frac{\partial^2 f}{\partial y^2} - \left(\frac{\partial^2 f}{\partial x\, \partial y}\right)^2 = (-0.2)(-0.4) - (-0.01)^2 \\ &= 0.08 - 0.0001 \\ &= 0.07999 \end{aligned}$$

At the critical point $P = (x_0, y_0)$, we have $D(P) > 0$ and $\partial^2 f/\partial x^2(P) = -0.2 < 0$; so a local maximum occurs at P, as expected. ■

Up to now, we would have found such maxima using a computer-generated pattern of level curves. By comparison, we need only do a little algebra to come to the same conclusion using the second derivative test.

Example 24 Test all critical points of $f(x, y) = 1 - \frac{1}{2}x^2 + \frac{1}{2}y^2$. Can this function have a local maximum? Can it have an absolute maximum?

Solution We have

$$\frac{\partial f}{\partial x} = -x \qquad \frac{\partial f}{\partial y} = y$$

so that $Q = (0, 0)$ is the only critical point. Because

$$\frac{\partial^2 f}{\partial x^2} = -1 \qquad \frac{\partial^2 f}{\partial y^2} = 1 \qquad \frac{\partial^2 f}{\partial x^2\, \partial y^2} = 0 \quad \text{(constant functions)}$$

the discriminant is always negative, $D = (-1)(1) - (0)^2 = -1$. Thus f has neither a local maximum nor a local minimum at Q. Because there are no other critical points where a local maximum could occur, there are no local maxima and hence no absolute maxima. ■

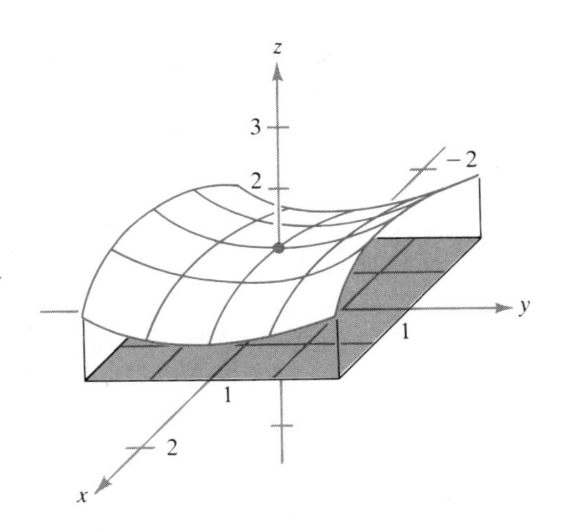

Figure 7.31
At $x = 0$, $y = 0$ there is a critical point for $f(x, y) = 1 - \frac{1}{2}x^2 + \frac{1}{2}y^2$ that is neither a local maximum nor a local minimum for f. We show part of the surface $z = f(x, y)$, which is defined for all x and y. The solid dot on the surface lies above the critical point (0, 0). The curves on the surface, which reveal its shape, lie above the lines in the xy plane where $x =$ const ($x = -1.0, -0.5, 0, 0.5, 1.0$) and $y =$ const ($y = -1.0, -0.5, 0, 0.5, 1.0$).

It is worth looking at the graph of $f(x, y)$ above the critical point (0, 0) to see what is going on. As in Figure 7.31, the surface is "saddle shaped." Values of f increase as we move away from Q in either direction along the y axis and decrease if we move off along the x axis. Thus neither a maximum nor a minimum occurs at $Q = (0, 0)$ for this function.

This example shows that critical points can occur where f has neither maxima nor minima. Step 1 of the second derivative test sorts out these extraneous critical points; Steps 2 and 3 examine the honest extrema that remain and classify them as local maxima or minima. Here is one more example of this sorting process. It also illustrates some algebraic techniques for locating critical points; testing them once they are in hand is routine.

Example 25 Find all local minima for $f(x, y) = x^3 - x + y^2 + xy$.

Solution First we find the critical points, the choices of x and y that simultaneously satisfy the equations

$$0 = \frac{\partial f}{\partial x} = 3x^2 + y - 1$$
$$0 = \frac{\partial f}{\partial y} = x + 2y \qquad [20]$$

Taking the simpler equation, we see that $x = -2y$ for any critical point. Substituting the condition $x = -2y$ (hence $x^2 = 4y^2$) into the first equation, we obtain

$$12y^2 + y - 1 = 0$$

Now solve for y by using the quadratic formula:

$$y = \frac{-1 \pm \sqrt{1 + 48}}{24}$$

so that

$$y = -\frac{1}{3} \quad \text{or} \quad y = \frac{1}{4}$$

Thus, if (x, y) is a critical point we must have $y = -\frac{1}{3}$ or $y = \frac{1}{4}$. Substitute these y values into the second equation [20] to obtain the corresponding x values $x = \frac{2}{3}$ or $x = -\frac{1}{2}$. This shows that there are just two critical points

$$P_1 = \left(\frac{2}{3}, -\frac{1}{3}\right) \quad \text{and} \quad P_2 = \left(-\frac{1}{2}, \frac{1}{4}\right)$$

Next, calculate the second partial derivatives

$$\frac{\partial^2 f}{\partial x^2} = 6x \qquad \frac{\partial^2 f}{\partial x\, \partial y} = 1 \qquad \frac{\partial^2 f}{\partial y^2} = 2$$

and apply the second derivative test. At P_1 we get

$$D(P_1) = 4(2) - (1)^2 = 7 > 0 \quad \text{and} \quad \frac{\partial^2 f}{\partial x^2}(P_1) = 6\left(\frac{2}{3}\right) = 4 > 0$$

so there is a local minimum at P_1. At P_2 we obtain

$$D(P_2) = -3(2) - (1)^2 = -7 < 0$$

so no local extremum occurs at P_2. Because neither P_1 nor P_2 is a local maximum, there are no local maxima for f. ■

Exercises 7.7

1. Calculate all first and second order partial derivatives of

$$f(x, y) = 4x^3 + 7xy^2 - x^2y^2 + \frac{1}{y} - 7$$

Give their numerical values at the particular point $P = (-3, 1)$.

In Exercises 2–9 find all second partial derivatives for each function.

2. $f(x, y) = 4x + 3y$

3. $f(x, y) = 4x + 3y - 10xy^2$

4. $f(x, y) = 2xy + 4x^3 - 6y^2$

5. $f(x, y) = \dfrac{x}{y}$

6. $f(x, y) = (x^2 + 4)e^{-y}$

7. $f(x, y) = \ln(4x^2 - y^2)$

8. $f(x, y) = \sqrt{x^2 - 2y^2}$

9. $f(x, y) = \sqrt{\dfrac{x}{y}}$

In Exercises 10–14, compute the discriminant $D(x, y)$ for each of the functions. Then find its value $D(P)$ at the particular point indicated.

10. $x^2 - 2y^2 + x^2y \quad P = (0, 0)$

11. $x^2 + 4y^2 + 1 \quad P = (0, 0)$

12. $x^3 - xy - y^3 \quad P = \left(-\frac{1}{3}, \frac{1}{3}\right)$

13. $x^2y - xy^2 - x \quad P = \left(\frac{2}{3}, 1\right)$

14. $x^2 + 3xy + y^2 \quad P = (1, 2)$

In Exercises 15–18, find the value of the discriminant $D(P)$ at $P = (1, 2)$ for each of these functions.

15. $\dfrac{1 + x}{1 - y}$

16. $y^2 + 3x^4 - 4x^3 - 12x^2 + 14$

17. $5e^{y-2x}$

18. $\ln(y - x)$

In Exercises 19–30, find all critical points, and test for local maxima and minima using the second derivative test.

19. $x^2 - 2y^2 + 10$

20. $x^2 + 4y^2 + 2xy + x - y + 1$

21. $x^2 + 3xy + y^2$

22. $x^2 + xy + 4y^2 - 3x + 4y + 1$

23. $40x + 60y + xy - \frac{1}{2}x^2 - 2y^2$

24. $-2000 + 250x + 190y + \frac{1}{4}xy$

25. $x^3 + x^2y - y$

26. $x^2y - xy^2 - x$

27. $x^3 - xy - y^3$

28. $y^2 + 2xy + 3x^4 + 4x^3 - 11x^2 + 24$

29. $f(x, y) = 4x^2 - y^2 + 5$

30. $f(x, y) = -86 + 30x + 24y - 3x^2 + 2xy - 6y^2$

31. Show that the function $f(x, y) = x^2 - 3xy + y^2$ has neither an absolute maximum nor an absolute minimum in the xy plane. (*Hint*: Are there any local extrema?)

32. Repeat Exercise 31 for the function $f(x, y) = xy^2 - x^2y - x$.

33. The annual profit of a small manufacturer of personal computers (in millions of dollars) is given by

$$P(x, y) = -3x^2 - 2y^2 + 4x + 2y - 2xy + 20$$

where x and y are the annual amounts (hundreds of thousands of dollars) spent on television and magazine advertising, respectively. Find the combination x, y of advertising expenditures that maximizes the profit.

34. The total cost of performing the carpentry work in an apartment building renovation project depends on the number x of master carpenters and the number y of apprentices employed. The architect estimates that the total cost for the work will depend on x and y, according to the formula

$$C(x, y) = 12{,}000 + 5x^2 - 6xy + 8y^2 - 18x - 14y$$

How many carpenters of each type should be hired to minimize the cost?

7.8 Lagrange Multipliers

Finding extrema of a function $f(x, y)$ on the boundary of a feasible set usually reduces to finding extrema when the variables x and y are obliged to satisfy some constraint equation

$$g(x, y) = c \quad (c \text{ a fixed constant}) \qquad [21]$$

that describes part of the boundary. Thus, f is defined on the curve determined by the equation $g(x, y) = c$, and we seek the largest (or smallest) value of f on this curve. In Figure 7.32, we show a curve C determined by some constraint equation $g(x, y) = c$. Superimposed we show some of the level curves $f(x, y) = \text{const}$ of the function under investigation. We seek the maximum value of $f(x, y)$ *on the curve C*; this curve is the feasible set in this optimization problem. Note that if attention were not confined to points on the curve, the maximum for f would occur at the point labeled D, some distance away. This is not what we want, because D is not on the curve (the feasible set)! By examining the pattern of level curves for f, we can see that the constrained maximum occurs at the point P, where the value of f is $f(P) = +50$.

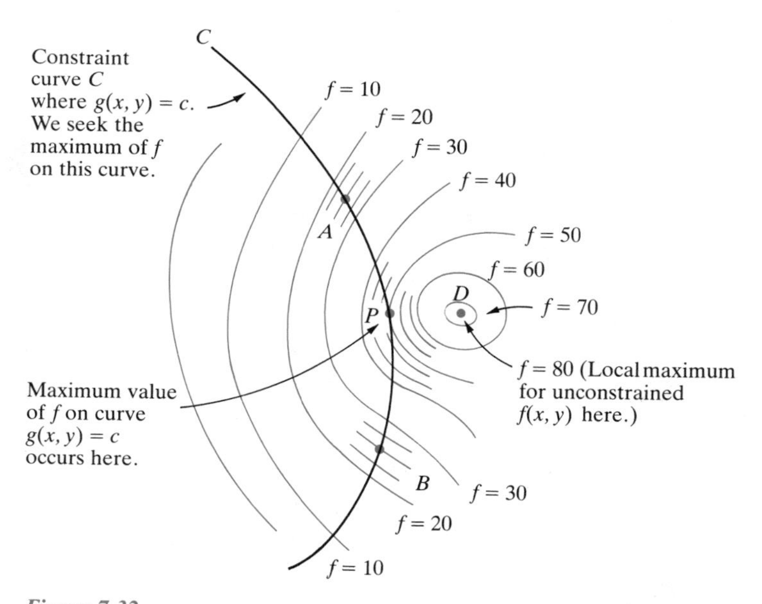

Figure 7.32
A constraint curve $g(x, y) = c$ with level curves of $f(x, y)$ superimposed. We seek the constrained maximum of f—the largest value of f on the curve. Level curves $f(x, y) = \text{const}$ are given in steps of $\Delta f = 10$, but near the particular points A, B, P we have drawn in additional level curves. The constrained maximum we seek occurs at P. If we were to drop the constraint condition, the maximum for f would occur at D some distance from the curve (unconstrained maximum for f).

If we consider the behavior of f at other points on the constraint curve, certain useful facts emerge. Near point A, the level curves for f are transverse to the constraint curve. If we move past A along the curve, the values of f increase or decrease steadily depending on which direction we move; a constrained extremum cannot occur at A. Similarly, at any other point—such as B—where the level curves for f are transverse to the constraint curve, there cannot be a constrained extremum of f. At point P things are different: The level curves for f become *tangent to* the constraint curve $g(x, y) = c$, and in fact a constrained maximum for f occurs at P. This observation contains a general truth:

> For a function $f(x, y)$ to have a constrained extremum at a point P on the constraint curve $g(x, y) = c$, the level curve for f passing through P must be tangent to the constraint curve. [22]

This leads to the method of Lagrange multipliers stated below. Given a point P on the constraint curve, we may use partial derivatives to decide whether the level curve through P is tangent to the constraint curve. For tangency, it can be shown that the partial derivatives of $f(x, y)$ must be *proportional* to those of $g(x, y)$ at P.

METHOD OF LAGRANGE MULTIPLIERS Suppose we want to find the absolute maximum of $f(x, y)$ on the set of points satisfying a constraint equation $g(x, y) = c$. Its location may be found as follows (similar methods apply for minima).

Step 1 Form a new function $F(x, y, \lambda)$ with three independent variables x, y, λ

$$F(x, y, \lambda) = f(x, y) - \lambda \cdot [g(x, y) - c]$$

This new function is defined for all values of x, y, λ; its variables are not subject to constraints.

Step 2 Find the critical points of $F(x, y, \lambda)$. These are the points $P = (x, y, \lambda)$ for which

$$\frac{\partial F}{\partial x}(x, y, \lambda) = 0 \qquad \frac{\partial F}{\partial y}(x, y, \lambda) = 0 \qquad \frac{\partial F}{\partial \lambda}(x, y, \lambda) = 0$$

List the coordinates $P_1 = (x_1, y_1, \lambda_1), \ldots, P_k = (x_k, y_k, \lambda_k)$ of these critical points.

Step 3 Tabulate the points in the xy-plane $Q_1 = (x_1, y_1), \ldots, Q_k = (x_k, y_k)$ given by the first two coordinates of the critical points $P_1, \ldots, P_k$. These automatically satisfy the constraint equation $g(x, y) = c$.

Step 4 Compare values $f(Q_1), \ldots, f(Q_k)$ of $f(x, y)$ at these points. The largest is the absolute maximum value of f on the curve $g(x, y) = c$.

That is, there must be a constant of proportionality λ such that†

$$\frac{\partial f}{\partial x}(P) = \lambda \frac{\partial g}{\partial x}(P) \quad \text{and} \quad \frac{\partial f}{\partial y}(P) = \lambda \frac{\partial g}{\partial y}(P) \qquad [23]$$

By various algebraic transformations, this criterion may be reformulated as the Lagrange multiplier procedure stated above. In effect, we take a constrained extremum problem with two variables x and y, and by introducing an extra independent variable λ convert it into an *unconstrained* extremum problem in the three variables x, y, λ. Standard methods can be used to attack the resulting unconstrained problem.

Example 26 The function $f(x, y) = 10xy$ has an absolute maximum on the line determined by the equation $x + y = 200$. Find it using Lagrange multipliers.

Solution The line is described by the constraint equation

$$g(x, y) = 200 \quad \text{where} \quad g(x, y) = x + y \quad (\text{and } c = 200)$$

Thus we should examine the auxiliary function

$$F(x, y, \lambda) = f(x, y) - \lambda[g(x, y) - c] = 10xy - \lambda(x + y - 200)$$

and solve the following system of equations to locate its critical points,

$$0 = \frac{\partial F}{\partial x} = 10y - \lambda$$

$$0 = \frac{\partial F}{\partial y} = 10x - \lambda$$

$$0 = \frac{\partial F}{\partial \lambda} = 200 - x - y$$

The first two equations give $y = \lambda/10$ and $x = \lambda/10$; inserting these into the last equation, we find that

$$0 = 200 - \left(\frac{\lambda}{10}\right) - \left(\frac{\lambda}{10}\right) = 200 - \frac{\lambda}{5}$$

so that $\lambda = 1000$. Hence,

$$x = \frac{\lambda}{10} = 100 \quad \text{and} \quad y = \frac{\lambda}{10} = 100$$

† The symbol λ is the Greek letter "lambda," the equivalent of the Roman letter L (as in "Lagrange"). The constant λ in [23] is called the *Lagrange multiplier* for the constrained extremum point P.

and $(x, y, \lambda) = (100, 100, 1000)$ is the only critical point for $F(x, y, \lambda)$. It must correspond to the maximum we seek. The first two coordinates of the critical point, namely $x = 100$ and $y = 100$, give the point on L where the maximum occurs. The maximum value of f is therefore $f(100, 100) = 10xy = 10(100)(100) = 100{,}000$. ■

Example 27 A contractor is going to build an enclosure in which the floor area must be 3250 square feet (see Figure 7.33). The wall facing the street will be finished in brickface and costs \$800 per linear foot. The other three walls will be composed of cinder block, costing \$500 per linear foot. What dimensions x and y minimize construction costs?

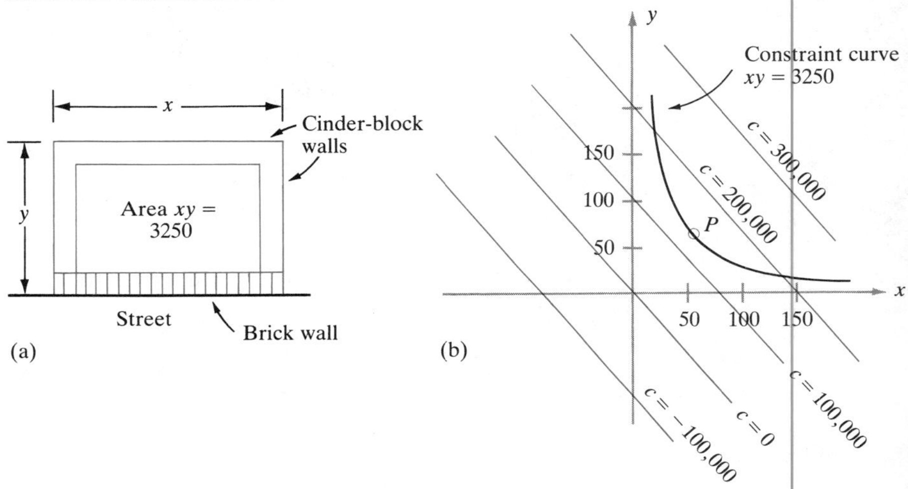

Figure 7.33
In (a) we show the dimensions of the enclosure discussed in Example 27. In (b) is the constraint curve $xy = 3250$. A few level curves for the cost function we are trying to minimize, $C(x, y) = 1300x + 1000y$, are superimposed; they are parallel straight lines. The constrained minimum occurs at $P = (50, 65)$. Only at P are the level curves for f tangent to the constraint curve.

Solution The area requirement imposes the constraint $xy = 3250$, or

$$g(x, y) = 3250 \quad \text{where} \quad g(x, y) = xy$$

Meanwhile, for any choice of x and y, the cost is

$$\begin{aligned} C(x, y) &= (\text{cost of brick wall}) + (\text{cost of 3 cinder-block walls}) \\ &= 800x + 500(x + 2y) \\ &= 1300x + 1000y \end{aligned}$$

To apply Lagrange's method we must find the critical points of the auxiliary function

$$\begin{aligned} F(x, y, \lambda) &= C(x, y) - \lambda[g(x, y) - c] \\ &= 1300x + 1000y - \lambda(xy - 3250) \end{aligned}$$

These critical points are the points (x, y, λ) satisfying the system of equations

$$0 = \frac{\partial F}{\partial x} = 1300 - \lambda y$$

$$0 = \frac{\partial F}{\partial y} = 1000 - \lambda x$$

$$0 = \frac{\partial F}{\partial \lambda} = -(xy - 3250)$$

From the first two equations we see that x and y can be written in terms of λ,

$$1300 = \lambda y \qquad y = \frac{1300}{\lambda}$$

$$1000 = \lambda x \qquad x = \frac{1000}{\lambda} \tag{24}$$

Entering these values into the third equation, we obtain

$$\frac{1300}{\lambda} \cdot \frac{1000}{\lambda} = 3250$$

so that

$$\lambda^2 = 400$$

This means that two possible choices exist for λ, namely $\lambda = +20$ and $\lambda = -20$. For each choice we find the corresponding x and y values by entering these values for λ into [24]:

If $\lambda = 20$, then:

$$x = \frac{1000}{20} = 50$$

$$y = \frac{1300}{20} = 65$$

If $\lambda = -20$, then:

$$x = \frac{1000}{-20} = -50$$

$$y = \frac{1300}{-20} = -65$$

The latter case does not yield feasible values of x and y, which must be positive numbers. The other critical point must therefore give the solution to our problem. It has coordinates $(x, y, \lambda) = (50, 65, 20)$, and the first two coordinates give the optimal building dimensions

$$x = 50 \text{ feet} \qquad y = 65 \text{ feet}$$

The (minimized) cost is $C(50, 65) = \$130{,}000$. ■

Example 28 A pet food manufacturer wants to minimize the material in a can containing 16 cubic inches. How should he choose the shape (radius r and height h as shown in Figure 7.34) to do this? What is the optimal ratio of h/r?

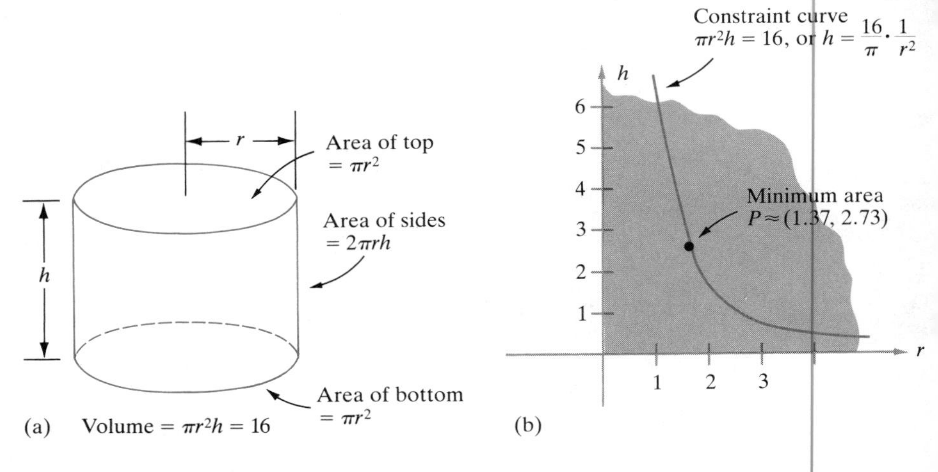

Figure 7.34
The dimensions of the can in Example 28 are shown in (a). Because the volume $V = \pi r^2 h$ must equal 16, the legitimate values of r and h are confined to the constraint curve shown in (b): $\pi r^2 h = 16$ (or $h = (16/\pi)\cdot(1/r^2)$). On this curve, the area $A = 2\pi r^2 + 2\pi rh$ achieves its minimum value at $(r, h) = (1.37, 2.73) = P$.

Solution The problem is to minimize the area of the can, which is

$$\begin{aligned} A(r, h) &= (\text{area of top}) + (\text{area of sides}) + (\text{area of bottom}) \\ &= \pi r^2 + 2\pi rh + \pi r^2 \\ &= 2\pi r^2 + 2\pi rh \end{aligned}$$

Legitimate choices of r and h are constrained by the volume requirement:

$$16 = (\text{volume}) = (\text{area of base}) \cdot (\text{height } h) = \pi r^2 h$$

which may be rewritten in the form

$$g(r, h) = \pi r^2 h = 16$$

To apply Lagrange multipliers, we examine the function

$$\begin{aligned} F(r, h, \lambda) &= A(r, h) - \lambda[g(r, h) - c] \\ &= 2\pi r^2 + 2\pi rh - \lambda(\pi r^2 h - 16) \end{aligned}$$

and look for critical choices of (r, h, λ), those satisfying the equations

$$\begin{aligned} 0 &= \frac{\partial F}{\partial r} = 4\pi r + 2\pi h - 2\pi\lambda rh \\ 0 &= \frac{\partial F}{\partial h} = 2\pi r - \pi\lambda r^2 \\ 0 &= \frac{\partial F}{\partial \lambda} = 16 - \pi r^2 h \end{aligned} \qquad [25]$$

To solve this system we start with the simplest equation in it and use it to eliminate one variable. The second equation says

$$0 = 2\pi r - \pi\lambda r^2 = \pi r(2 - \lambda r) \qquad [26]$$

One possibility is that $r = 0$, but this does not correspond to a feasible choice of r

and h in our problem, so we turn to the other possibility inherent in [26], namely

$$2 - \lambda r = 0 \quad \text{or} \quad r = \frac{2}{\lambda} \tag{27}$$

Entering this value for r into the third equation of [25], we find h in terms of λ.

$$0 = \pi\left(\frac{2}{\lambda}\right)^2 h - 16 \quad \text{or} \quad h = \frac{16\lambda^2}{4\pi} = \frac{4\lambda^2}{\pi} \tag{28}$$

Entering [27] and [28] into the first equation of [25], we find the actual value for λ,

$$\begin{aligned} 0 &= 4\pi\left(\frac{2}{\lambda}\right) + 2\pi\left(\frac{4\lambda^2}{\pi}\right) - 2\pi\lambda\left(\frac{2}{\lambda}\right)\left(\frac{4\lambda^2}{\pi}\right) \\ &= \frac{8\pi}{\lambda} - 8\lambda^2 \end{aligned}$$

which yields

$$\lambda^3 = \pi \quad \text{or} \quad \lambda = (\pi)^{1/3} \approx 1.46459$$

We now recover the optimal values of r and h by substituting this value for λ into [27] and [28], obtaining

$$r = \frac{2}{\lambda} = 2(\pi)^{-1/3} \approx 1.36557$$

$$h = \frac{4\lambda^2}{\pi} = \frac{4}{\pi}(\pi)^{2/3} = 4(\pi)^{-1/3} \approx 2.73114$$

The ratio h/r is more informative because it does not involve π,

$$\frac{h}{r} = \frac{4(\pi)^{-1/3}}{2(\pi)^{-1/3}} = 2$$

The optimal shape is obtained if (height) $= 2 \cdot$ (radius) or (height) $=$ (diameter). This outcome is actually valid for a can of *any* volume. ■

There is a useful interpretation of the Lagrange multiplier λ in economic problems. Suppose we are trying to maximize or minimize some quantity $f(x, y)$ determined by production levels x and y. A constraint of the form

$$g(x, y) = c \tag{29}$$

confines the production variables to some curve, as in Figure 7.35, usually because of shortages of money, supplies, manpower, and so on. If more supplies become available, the allowable production levels shift to a somewhat different constraint curve

$$g(x, y) = c + \Delta c \tag{30}$$

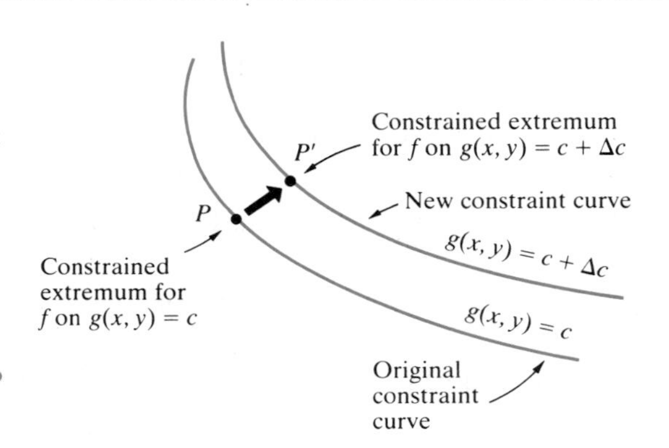

Figure 7.35
A function $f(x, y)$ has its optimal value on the constraint curve $g(x, y) = c$ at the point P. If we shift operations to a nearby constraint curve $g(x, y) = c + \Delta c$, the optimal value of f is achieved at a different point P'. The Lagrange multiplier at P tells us how the change $\Delta f = f(P') - f(P)$ is related to the change Δc in the constraint equation.

where the change Δc in the constant on the right represents the amount of additional supplies now available. On each curve there will be a constrained extremum for f, at P and P' in Figure 7.35.

How much does the constrained optimum value of f change because of the loosening of our constraint? This is obviously important in deciding whether it is worth trying to round up more supplies, so that operations may be shifted to the new curve. To a strategist, the most informative piece of information is the ratio

$$\frac{\Delta f}{\Delta c} = \frac{f(P') - f(P)}{\Delta c}$$

which tells him how much additional production will result per unit of additional supplies provided. The Lagrange multiplier provides this information.

Meaning of the Lagrange Multiplier For small changes Δc in the constraint, the change in the optimized value of f is approximately equal to $\lambda \cdot \Delta c$:

$$\frac{\Delta f}{\Delta c} \approx \lambda \qquad [31]$$

where λ is the Lagrange multiplier associated with the original constrained extremum point P.

Example 29 In Example 27, the floor area of an enclosure was specified as $xy = 3250$ square feet; costs were minimized by taking dimensions $x = 50$, $y = 65$ feet. Yesterday, the owner asked us to see how much more it would cost to make the floor area 10% larger: $xy = 3575$ square feet. Can we answer his question without reworking the whole problem?

Solution The Lagrange multiplier associated with the solution $P = (50, 65)$ of the original problem

$$\text{minimize } C(x, y) = 1300x + 1000y \quad \text{subject to } xy = 3250 \qquad [32]$$

was $\lambda = 20$. We made no use of the multiplier in that example, but now it tells us about the solution to today's problem

$$\text{minimize } C(x, y) = 1300x + 1000y \quad \text{subject to } xy = 3250 + 325 \quad [33]$$

(in which $\Delta c = 325$, $c = 3250$). Because $\Delta C/\Delta c \approx \lambda = 20$, the change in cost is approximately

$$\Delta C = C(P') - C(P) \approx \lambda \cdot \Delta c = \lambda \cdot (325) = 20(325) = \$6500,$$

and the net optimal cost for the revised project is $C(P') = C(P) + \Delta C \approx 130{,}000 + 6500 = \$136{,}500$. Because the owner did not ask about such things as the optimal dimensions $P' = (x, y)$ in the new problem, we are finished. ■

Exercises 7.8

Level curves for the profit function $P(x, y)$ of a certain manufacturer are shown in Figure 7.36. Because of supply difficulties, feasible production levels for the month are constrained to satisfy $x + y = 3000$; that is, they must lie on the slanted line L shown in the figure. Answer Exercises 1–3 by examining the level curve pattern.

1. Which production levels (x, y) maximize P subject to this constraint?
2. What is the maximum possible value of P in this situation?
3. Next month the supply situation will be worse: Feasible production levels will have to satisfy $x + y = 2000$. Make a sketch showing the new constraint curve. Estimate the optimal production levels (x, y) and the maximum profit in this situation.

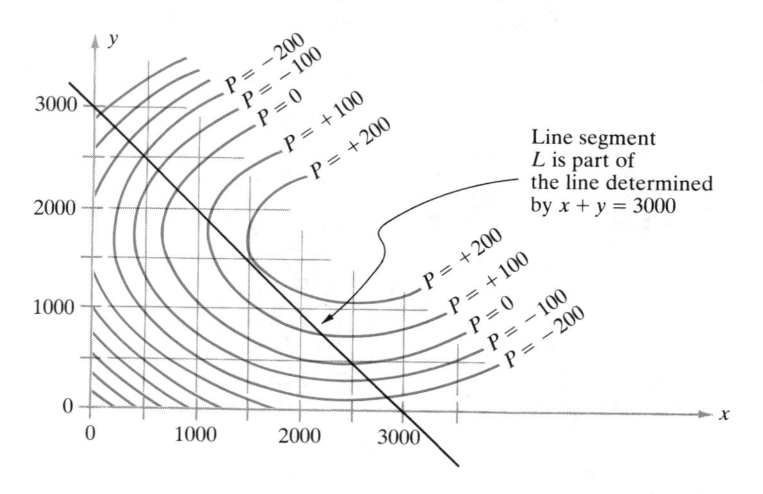

Figure 7.36 Level curves for the profit function in Exercises 1–3. Because of supply difficulties, feasible production levels are confined to the slanted line segment $x + y = 3000$ shown.

Solve Exercises 4–9 using Lagrange multipliers.

4. Maximize the function $f(x, y) = xy$ on the straight line $x + 2y - 2 = 0$.
5. Find the maxima and minima of $f(x, y) = 3x + 4y$, subject to the constraint $x^2 + y^2 = 100$.

6. Find the minimum of the function $f(x, y) = 10 + x^2 + y^2$, subject to the constraint $3x + 4y = 5$.

7. Find real numbers x and y such that their sum is 50 and their product xy is as large as possible.

8. Find the maximum value of $f(x, y) = xy$ on the circle $x^2 + y^2 = 1$. (There is also a minimum, but we don't ask for that.)

9. Find the maxima and minima of the function $f(x, y) = 3xy$ on the ellipse determined by $x^2 + 4y^2 = 1$.

10. If the profit function for a manufacturer is

$$P(x, y) = 15x - 0.1x^2 + 20y - 0.3y^2 + 0.1xy \quad [34]$$

and if supply difficulties confine the feasible production levels to those pairs (x, y) satisfying the condition $x + y = 1000$, find the production levels that maximize profit in this constrained situation.

11. In Exercise 10, the Lagrange multiplier turns out to be $\lambda = -93.5$. By approximately how much does the maximum profit change if we shift to a new constraint curve $x + y = 900$? What if the new constraint curve were $x + y = 1050$?

12. Suppose the production levels in Exercise 10 were *not* constrained. Where would the profit function [34] achieve its maximum? Do these unconstrained optimal production levels lie on the constraint curve mentioned in Exercise 10?

13. The surface $z = \frac{1}{4}x^2 + \frac{1}{9}y^2 + 10$ lies above the x, y plane as shown in Figure 7.37. In the x, y plane, consider points (x, y) lying on the line $x + y = 1$. Find the point on the line over which the height of the surface is a minimum. What is the minimum height? (This amounts to minimizing $f(x, y) = \frac{1}{4}x^2 + \frac{1}{9}y^2 + 10$ subject to the constraint $x + y = 1$.)

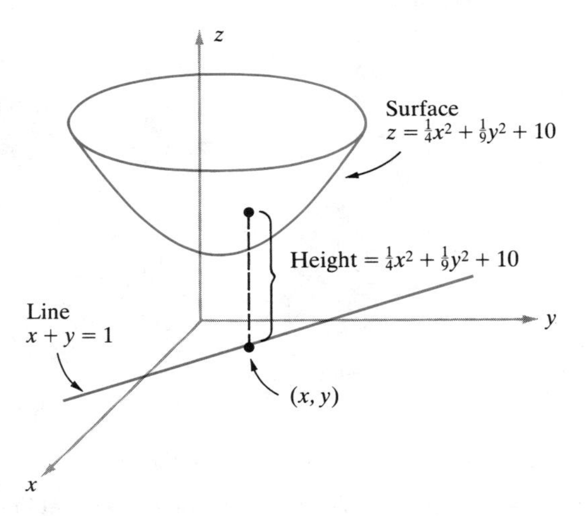

Figure 7.37
Exercise 13.

14. A building contractor budgets \$16,000 per week for labor. The wage rate is \$400 per week for skilled labor and \$300 per week for semiskilled. If x and y are the numbers of

skilled and semiskilled employed, the contractor's profit for the week is $P(x, y) = 1000x + 750y + 12xy - 30{,}000$ dollars. Use Lagrange multipliers to find the values of x and y that maximize P, subject to the weekly budget constraint.

15. Repeat Exercise 14, assuming that wages have been increased to \$425 and \$320, everything else being the same.

16. In Exercise 14, the Lagrange multiplier turns out to be $\lambda = 3.30$. By approximately how much does the maximum profit change if the contractor budgets an additional \$2000 for labor?

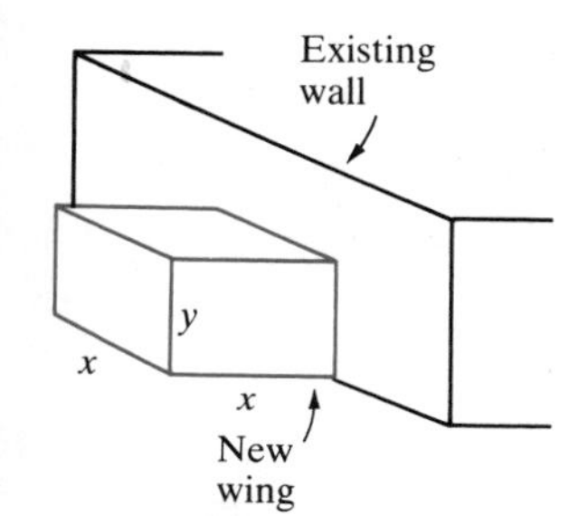

Figure 7.38
Exercises 17–20.

A new wing is to be added to an existing building, as shown in Figure 7.38. It has a square base and is to enclose 36,000 cubic feet of space. Answer the questions in Exercises 17–20.

17. Construction costs for the wing are \$28 per square foot of roof, \$35 per square foot for the side walls, and \$42 per square foot for the concrete floor. Find the dimensions x, y that minimize the total construction cost (roof, floor, and three sides).

18. Daily heat losses through the roof and sides are 8 thermal units per square foot for the roof and 4 thermal units per square foot for the sides. Losses through the existing wall and floor are negligible. Find the dimensions x, y that minimize the *total* daily heat loss (roof and three exterior walls).

19. In Exercise 17, the Lagrange multiplier turns out to be $\lambda = 3.5$. By approximately how much will costs increase if the volume enclosed is increased by 5000 cubic feet?

20. In Exercise 18, the Lagrange multiplier turns out to be $\lambda = 0.4$. By approximately how much will heat loss increase if the volume enclosed is increased by 5000 cubic feet?

The distance from the origin to a point (x, y) in two-dimensional space is given by the Pythagorean formula $d(x, y) = \sqrt{x^2 + y^2}$. In many geometric problems, we wish to minimize this distance—which is the same as minimizing the squared distance $x^2 + y^2$. This leads to geometric interpretations of the following problems.

21. Minimize $x^2 + y^2$ on the line $3x + 4y = 25$. (This amounts to minimizing the distance from a point (x, y) on the line to the origin.)

22. Minimize $(x - 1)^2 + (y - 2)^2$ on the parabola $y = x^2$. (This amounts to finding the point on the parabola closest to the point $(1, 2)$.) (*Hint*: Guess a root of the resulting cubic equation in λ.)

7.9 Multiple Integrals

In Section 5.3 we discussed the definite integral $\int_a^b f(x)dx$ of a function of one variable over an interval $a \le x \le b$, which may be interpreted geometrically as the area under the graph of the function $f(x)$. There is a similar notion of definite integral

$$\iint_R f(x, y)\,dx\,dy \qquad R \text{ some rectangular region } a \le x \le b, c \le y \le d. \qquad [35]$$

for functions of two (or more) variables. These are referred to as **multiple integrals** because they involve more than one variable. The integral [35] has a natural interpretation as the volume underneath the surface $z = f(x, y)$ determined by the function $f(x, y)$. By way of this interpretation, multiple integrals find many applications, especially in statistics. We will first discuss the definition and calculation of such integrals, and will then indicate a few simple applications.

Suppose $f(x, y)$ is defined on a rectangular region R $(a \leq x \leq b, c \leq y \leq d)$ in the x, y plane. We define the **iterated integral** of f

$$\int_a^b \left[\int_c^d f(x, y)\, dy \right] dx$$

as follows. If we hold x fixed, $f(x, y)$ becomes a function of y only, defined for $c \leq y \leq d$. We may form the usual one-variable definite integral, thinking of x as a fixed constant and integrating with respect to y:

$$\int_c^d f(x, y)\, dy = F(x)$$

When we do this, the variable of integration y is integrated out of existence; the result is a function $F(x)$ depending only on x. Next, take the definite integral with respect to x

$$\int_a^b \left[\int_c^d f(x, y)\, dy \right] dx = \int_a^b F(x)\, dx \qquad [36]$$

Now both variables have been integrated, and we are left with a number [36], the iterated integral of $f(x, y)$. We could also integrate x first (regarding y as a constant everywhere it appears) and then integrate the resulting function of y. This yields a different iterated integral, denoted by

$$\int_c^d \left[\int_a^b f(x, y)\, dx \right] dy \qquad [37]$$

The notation tells you which integration to do first—the inner one, in square brackets.

Example 30 Compute both iterated integrals for $f(x, y) = 2x^2 + xy$ defined for $0 \leq x \leq 1, 0 \leq y \leq 4$.

Solution For $\int_0^1 [\int_0^4 (2x^2 + xy)\, dy]\, dx$ we first integrate with respect to y, regarding x as a constant throughout the integration:

$$\begin{aligned}\int_0^4 (2x^2 + xy)\, dy &= 2x^2 \int_0^4 dy + x \int_0^4 y\, dy \\ &= \left(2x^2 \cdot y + x \cdot \frac{y^2}{2} \Bigg|_{y=0}^{y=4} \right) \\ &= \left(2x^2 \cdot 4 + x \cdot \frac{16}{2} \right) - (0) = 8x^2 + 8x\end{aligned}$$

Then integrate with respect to x from $x = 0$ to $x = 1$ to obtain

$$\int_0^1 \left[\int_0^4 (2x^2 + xy)\,dy\right] dx = \int_0^1 (8x^2 + 8x)\,dx$$

$$= \left(\frac{8}{3}x^3 + \frac{8}{2}x^2 \Big|_{x=0}^{x=1}\right) = \frac{20}{3}$$

For the other iterated integral, regard y as a constant and integrate with respect to x:

$$\int_0^1 (2x^2 + xy)\,dx = \int_0^1 2x^2\,dx + y \cdot \int_0^1 x\,dx$$

$$= \left[2 \cdot \frac{x^3}{3} + y \cdot \frac{x^2}{2} \Big|_{x=0}^{x=1}\right]$$

$$= \left(2 \cdot \frac{1}{3} + y \cdot \frac{1}{2}\right) - (2 \cdot 0 + y \cdot 0)$$

$$= \frac{2}{3} + \frac{y}{2}$$

Then integrate the resulting function of y from $y = 0$ to $y = 4$.

$$\int_0^4 \left[\int_0^1 (2x^2 + xy)\,dx\right] dy = \int_0^4 \left(\frac{2}{3} + \frac{y}{2}\right) dy$$

$$= \left(\frac{2}{3}y + \frac{y^2}{4} \Big|_{y=0}^{y=4}\right) = \frac{20}{3}$$ ■

Notice that the result $\frac{20}{3}$ was the same regardless of the order in which the integrations were performed. It is a general fact that the iterated integrals must agree. Therefore we are justified in making the following definition.

THE DOUBLE INTEGRAL $\iint_R f(x, y)\,dx\,dy$. If a function $f(x, y)$ is defined on a rectangle R ($a \le x \le b, c \le y \le d$) the double integral of f over this rectangle is the common value of the two possible iterated integrals.

$$\iint_R f(x, y)\,dx\,dy = \int_a^b \left[\int_c^d f(x, y)\,dy\right] dx$$
$$= \int_c^d \left[\int_a^b f(x, y)\,dx\right] dy \qquad [38]$$

The iterated integrals yield the same value for $\iint_R f(x, y)\,dx\,dy$ but arrive at it in different ways. You are free to choose whichever one you wish in calculating double integrals.

Example 31 Calculate the double integral $\iint_R (1 + ye^{-x})\,dx\,dy$ over the rectangular region R, $0 \le x \le 5$ and $-1 \le y \le 2$.

Solution Let us perform the integration with respect to y first, regarding x (and e^{-x}) as a constant,

$$\begin{aligned}\int_{-1}^{2} (1 + ye^{-x})\,dy &= \int_{-1}^{2} dy + e^{-x}\cdot \int_{-1}^{2} y\,dy \\ &= \left(y + e^{-x}\cdot \frac{y^2}{2}\Bigg|_{y=-1}^{y=2}\right) \\ &= (2 + 2e^{-x}) - \left(-1 + \frac{1}{2}e^{-x}\right) \\ &= 3 + \frac{3}{2}e^{-x}\end{aligned}$$

Then perform the integration in the remaining variable x.

$$\begin{aligned}\int_0^5 \left[\int_{-1}^{2} (1 + ye^{-x})\,dy\right] dx &= \int_0^5 \left(3 + \frac{3}{2}e^{-x}\right) dx \\ &= 3\int_0^5 dx + \frac{3}{2}\int_0^5 e^{-x}\,dx \\ &= \left(3x - \frac{3}{2}e^{-x}\Bigg|_{x=0}^{x=5}\right) \\ &= \left(15 - \frac{3}{2}e^{-5}\right) - \left(0 - \frac{3}{2}e^{0}\right) \\ &= 15 + \frac{3}{2}(1 - 0.00674) = 16.4899\end{aligned}$$

■

Figure 7.39 shows the graph of a function of two variables $z = f(x, y)$. This, you will recall, is a surface lying above the xy plane (or below it if $f(x, y)$ is

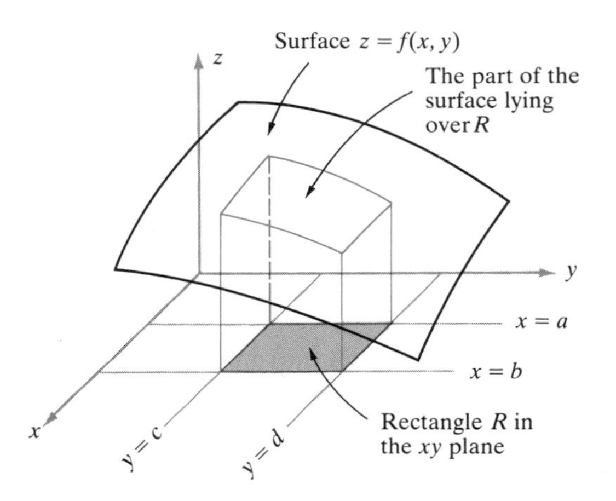

Figure 7.39
The graph of a typical function $z = f(x, y)$ is a curved surface. We show the part of the surface over a rectangle R ($a \le x \le b$, $c \le y \le d$) in the xy-plane. The solid region lies between the rectangle R and the surface.

negative); z gives the height of the surface above (or below) the point with coordinates (x, y) in the xy plane. The shaded rectangular region R in the xy plane is given by the inequalities $a \leq x \leq b$, $c \leq y \leq d$. In this situation the double integral $\iint_R f(x, y)\,dx\,dy$ gives the volume of the solid region lying between the rectangle R and the surface.

Example 32 The equation $z = f(x, y) = 10 - 5x^2 - 5y^2$ describes a parabolic surface, part of which is shown in Figure 7.40. Consider the solid region shown in the figure: it is bounded between the rectangle R $(-1 \leq x \leq 1, -1 \leq y \leq 1)$ and the surface. Find its volume.

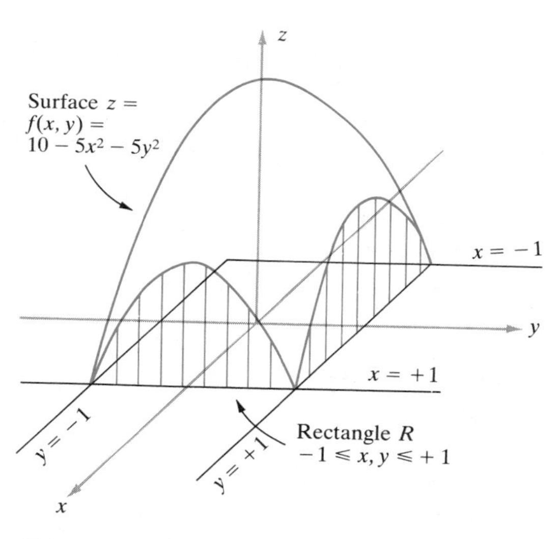

Figure 7.40

Solution We have

$$(\text{volume}) = \iint_R f(x, y)\,dx\,dy$$

$$= \int_{-1}^{1} \left[\int_{-1}^{1} (10 - 5x^2 - 5y^2)\,dy \right] dx$$

First integrate with respect to y, treating x as a constant.

$$\int_{-1}^{1} (10 - 5x^2 - 5y^2)\,dy = \left(10y - 5x^2y - \frac{5}{3}y^3 \Big|_{y=-1}^{y=1} \right)$$

$$= \left(10 - 5x^2 - \frac{5}{3} \right) - \left(10(-1) - 5x^2(-1) - \frac{5}{3}(-1) \right)$$

$$= \frac{50}{3} - 10x^2$$

The result is a function only of x, which we now integrate.

$$\int_{-1}^{1}\left(\frac{50}{3} - 10x^2\right)dx = \left(\frac{50}{3}x - \frac{10}{3}x^3\Big|_{x=-1}^{x=1}\right)$$
$$= \left(\frac{50}{3} - \frac{10}{3}\right) - \left(\frac{50}{3}(-1) - \frac{10}{3}(-1)^3\right) = \frac{80}{3}$$

The volume is $80/3 = 26.667$ cubic units. ■

Example 33 An exhibition hall has a rectangular base and a modernistic saddle-shaped roof (Figure 7.41). If x and y coordinates are laid out on the ground as shown, the height of the roof above the point $P = (x, y)$ is

$$f(x, y) = 30 - 5\left(\frac{x}{25}\right)^2 + 10\left(\frac{y}{50}\right)^2 = 30 - \frac{x^2}{125} + \frac{y^2}{250}$$

To estimate normal operating costs (heating, air conditioning) we need to know the volume of the hall. Compute it using double integrals.

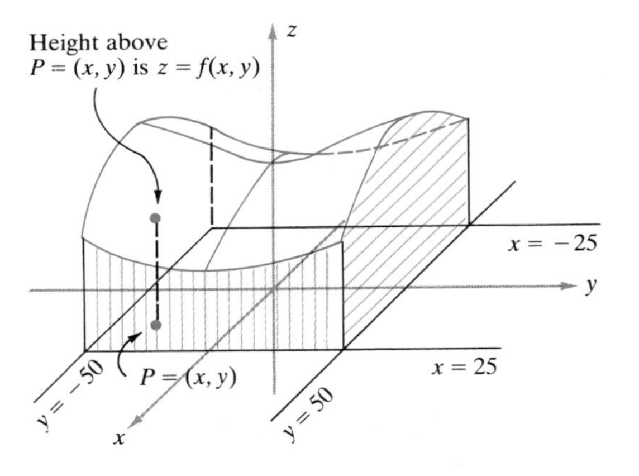

Figure 7.41

Solution In the surveyor's coordinates x, y, the base of the building is the rectangle R with $-25 \le x \le 25$ and $-50 \le y \le 50$. The volume is therefore

$$\iint_R f(x, y)\,dx\,dy = \int_{-25}^{25}\left[\int_{-50}^{50}\left(30 - \frac{x^2}{125} + \frac{y^2}{250}\right)dy\right]dx$$

We first compute the inner integral, with respect to y, treating x as a constant.

$$\int_{-50}^{50}\left(30 - \frac{x^2}{125} + \frac{y^2}{250}\right)dy = \left(30y - \frac{x^2}{125}\cdot y + \frac{y^3}{750}\Big|_{y=-50}^{y=50}\right)$$
$$= \left(30(50) - \frac{x^2\cdot 50}{125} + \frac{(50)^3}{750}\right) - \left(30(-50) - \frac{x^2(-50)}{125} + \frac{(-50)^3}{750}\right)$$
$$= \frac{10{,}000}{3} - \frac{100}{125}x^2$$

Then integrate this function of x from $x = -25$ to $x = 25$.

$$\int_{-25}^{25}\left(\frac{10{,}000}{3} - \frac{100}{125}x^2\right)dx = \left[\frac{10{,}000}{3}x - \frac{100}{125}\frac{x^3}{3}\Big|_{x=-25}^{x=25}\right]$$

$$= \frac{475{,}000}{3} \approx 158{,}333 \quad \text{cubic feet.}$$

■

Exercises 7.9

In Exercises 1–7, evaluate the two iterated integrals $\int_a^b [\int_c^d f(x, y)\, dy]\, dx$ and $\int_c^d [\int_a^b f(x, y)\, dx]\, dy$. Are they equal?

1. $f(x, y) = 1$ (constant for all x and y); $a = 0, b = 1, c = 0, d = 2$

2. $f(x, y) = 1$ (constant for all x and y); $a = -1, b = 1, c = 1, d = 3$

3. $f(x, y) = y$; $a = 0, b = 2, c = 1, d = 5$

4. $f(x, y) = x + 2y$; $a = -2, b = -1, c = 0, d = 1$

5. $f(x, y) = x^2y^2$; $a = 0, b = 2, c = -1, d = 1$

6. $f(x, y) = x^2 - 2xy + 3y^2$; $a = -2, b = 1, c = -2, d = 0$

7. $f(x, y) = e^{x+y} = e^x e^y$; $a = 0, b = 2, c = 0, d = 1$

8. If h is a positive constant, the graph of the constant function $f(x, y) = h$ is a horizontal plane lying h units above the xy plane. A rectangular block is bounded by this surface and the rectangle $0 \le x \le b, 0 \le y \le d$ in the xy plane (b and d positive constants). Set up and evaluate the double integral of $f(x, y)$ corresponding to this volume. Does your answer agree with your geometric intuition?

9. Evaluate the double integral $\iint_R (x + y)\, dx\, dy$ over the rectangle R $(0 \le x \le 1, 0 \le y \le 1)$. Interpret this double integral as a volume.

In Exercises 10–13, find the volume of the solid between the surface $z = f(x, y)$ and the rectangle R in the xy plane.

10. $f(x, y) = 2x + 3y - 1$; $1 \le x \le 3, 0 \le y \le 4$

11. $f(x, y) = x^2y$; $0 \le x \le 1, 0 \le y \le 1$

12. $f(x, y) = (x + y)^2 = x^2 + 2xy + y^2$; $-2 \le x \le 2, -3 \le y \le 3$

13. $f(x, y) = -\dfrac{x^2}{10} - \dfrac{y^2}{5} + 15$; $-4 \le x \le 4, -5 \le y \le 5$

14. A warehouse 40 feet wide and 100 feet long has a parabolic roof as shown in Figure 7.42. With respect to survey coordinates shown in the figure, the height of the roof (in feet) above a point with coordinates (x, y) is

$$z = f(x, y) = 40 - 15\left(\frac{x}{20}\right)^2; \qquad -20 \le x \le 20, 0 \le y \le 100$$

Find the volume of this warehouse.

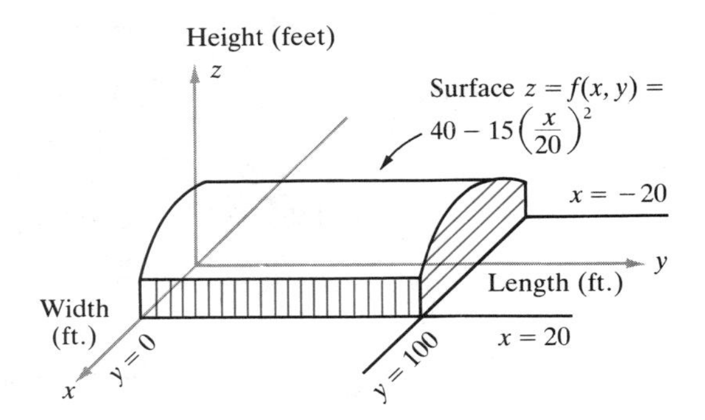

Figure 7.42

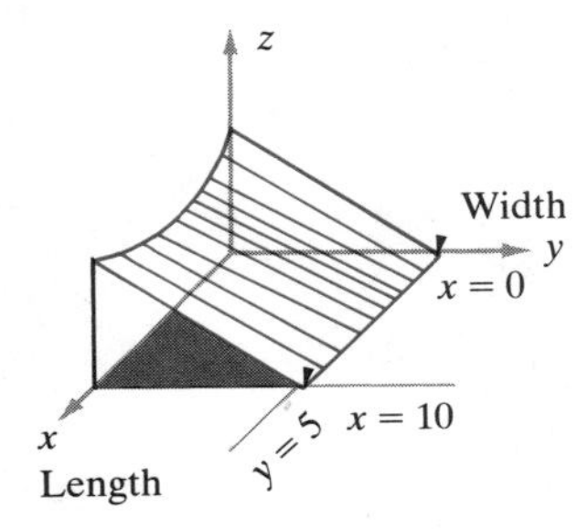

Figure 7.43

15. A portable shelter is shown in Figure 7.43. If it is set up over a 5 × 10 foot plot of ground as shown, the height of the canvas top above the ground is given by the equation

$$z = 5 - \frac{2}{5}x - y + \frac{x^2}{25} + \frac{2}{25}xy - \frac{x^2y}{125};$$

$$0 \leq x \leq 10, 0 \leq y \leq 5$$

Calculate the volume of the protected region under the tarp.

16. Consolidated coal company of Rhode Island owns the mineral rights on a 1-mile-square piece of flat land. In survey coordinates, this square is given by $0 \leq x \leq 1$, $0 \leq y \leq 1$ (in miles). It is estimated that recoverable coal deposits extend from the surface down to a depth $z = 0.03 - 0.004x - 0.007y$ miles. Find the volume of coal in this deposit.

Checklist of Key Topics

Function of several variables
Constraints and feasible sets
Graph of a function of several variables (a surface)
Level curves
Partial derivatives $\partial f/\partial x$ and $\partial f/\partial y$
How to calculate partial derivatives
The approximation principle for several variables
Optimization problems
Absolute maxima and minima
Local maxima and minima
Boundary extrema
Critical points and their importance in optimization problems
The basic optimization procedure
Word problems
Application to least squares; least-squares formula
Higher-order partial derivatives
The second derivative test for local maxima and minima
Constrained maxima and minima
The method of Lagrange multipliers
Significance of the Lagrange multiplier
Double integral
Calculating double integrals as iterated integrals
Volumes under a surface

Chapter 7 Review Exercises

1. Let $f(x, y) = 7 - 3x + 2y$. Find $f(0, 0)$, $f(1, 3)$, $f(3, 1)$, $f(-1, 3)$. In the x, y plane, sketch the level curves $f(x, y) = -2$, $f(x, y) = 0$, $f(x, y) = 2$, and $f(x, y) = 4$.

2. Find the value of the function $f(x, y) = \dfrac{x - 2y}{\sqrt{3x^2 + 2y^2 + 5}}$ at the following points:

(i) (0, 0) (ii) (3, 1) (iii) (1, 3)

3. Find the value of the function of three variables $f(x, y) = xe^{y-z}$ at the points:

(i) (0, 0, 0)
(ii) (3, 1, 0)
(iii) (1, 0, 3)
(iv) (1, −1, 1)
(v) (2, 1, −1)
(vi) (1, 0, 3)

4. A dress factory produces two lines, the moderately priced J. C. Smith and the expensive Fleur models. Each J. C. Smith dress requires 1.5 hours of labor and uses \$17 worth of materials; a Fleur dress requires 3 hours of labor and \$40 worth of materials. Labor costs are \$4.50 per hour. The finished dresses sell for \$60 and \$105, respectively. If x and y are the weekly production levels for each style,
(i) find the weekly cost function $C(x, y)$.
(ii) find the weekly revenue function $R(x, y)$.
(iii) find the weekly profit $P(x, y)$.

In Figure 7.44, we show the level curves for a function $f(x, y)$. Answer Exercises 5–14 by inspecting this curve pattern.

5. Estimate the value of f at the point marked S.

6. Does $f(x, y)$ increase or decrease as we move away from Q in the direction indicated by the arrow?

7. Is $\partial f/\partial x$ positive, negative, or zero at the point Q?

8. Is $\partial f/\partial y$ positive, negative, or zero at the point S?

9. What is the maximum value achieved by $f(x, y)$ on the rectangular feasible set shown?

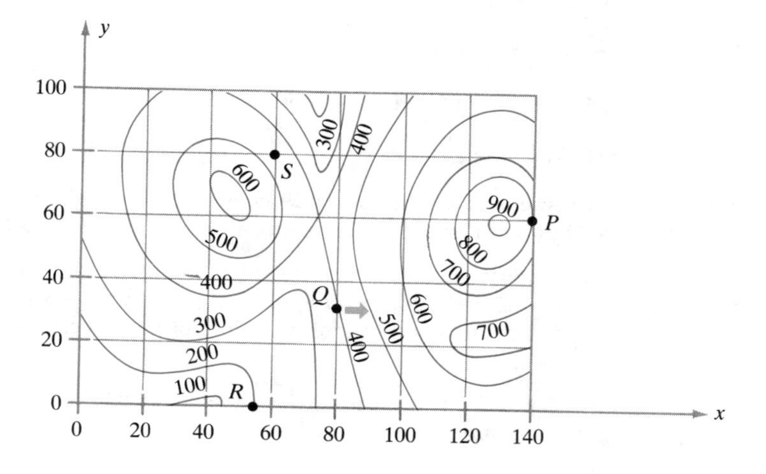

Figure 7.44

10. What are the coordinates of the point (x, y) where the absolute maximum occurs?

11. What is the absolute minimum value of $f(x, y)$, and what are the coordinates of the point where it is achieved?

12. Does a local maximum occur at the point marked R?

13. Does a local extremum occur at the point marked P?

14. Does a local extremum occur at the point with coordinates (0, 100)?

15. Consider the function $f(x, y) = x - 2y + 5$ on the feasible set determined by $0 \leq x \leq 5$ and $0 \leq y \leq 12.5$.
   (i) Sketch the feasible set on a piece of graph paper.
   (ii) On your sketch, superimpose the level lines $f(x, y) = c$ corresponding to $c = 0$, ± 10, ± 20.
   (iii) By examining this pattern of level curves on the feasible set, find the maximum and minimum values of f. Where do these extrema occur?

16. On graph paper sketch the level curves $f(x, y) = c$ for the function $f(x, y) = 3x - 2y + 4$; take $c = -3, 0, 3, 6, 9, 12, 15$. Superimpose these lines on the rectangular feasible set determined by $1 \leq x \leq 4$ and $1 \leq y \leq 5$. Where does the maximum value of f occur? Where does the minimum occur?

In Exercises 17–20, find the partial derivatives $\partial f/\partial x$ and $\partial f/\partial y$. Then compute the values of f, $\partial f/\partial x$, and $\partial f/\partial y$ at the point (2, 2).

17. $f(x, y) = xy - 2x^2 - 4y^2$

18. $f(x, y) = \dfrac{1 - x}{1 + y^2}$

19. $f(x, y) = \sqrt{x^2 + 3y^2}$

20. $f(x, y) = \ln(x^2 + y^2)$

In Exercises 21 and 22, find the partials $\partial f/\partial x$, $\partial f/\partial y$, and $\partial f/\partial z$.

21. $f(x, y, z) = x^2 + y^2 - z(x^3 + xy - y^2 - 1)$

22. $f(x, y, z) = \ln(x + y^2) - z(x + y - e^{x+y})$

In Exercises 23–27, find the partials $\partial f/\partial x$, $\partial f/\partial y$, $\partial^2 f/\partial x^2$, $\partial^2 f/\partial y^2$, and $\partial^2 f/\partial y \partial x$.

23. $f(x, y) = 3x - 7y + 12$

24. $f(x, y) = x^2 - 5xy + 2y^2 + 3x + 1$

25. $f(x, y) = x^4 - 4x^2y - x^2y^3$

26. $f(x, y) = \dfrac{x}{x + y}$

27. $f(x, y) = e^{2x+y}$

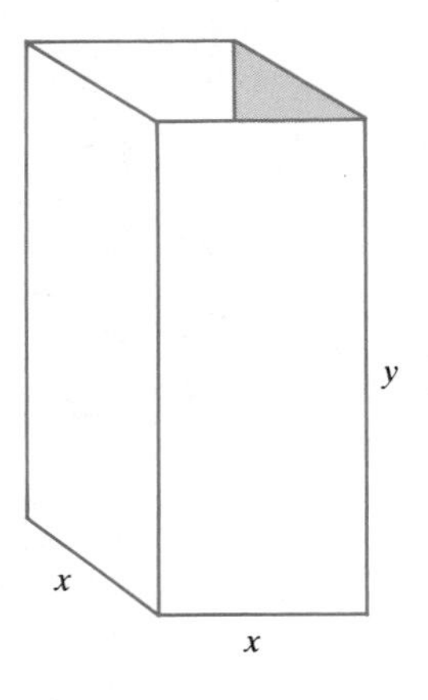

Figure 7.45

In answering Exercises 28 and 29, consider a rectangular box with square base and no top as shown in Figure 7.45. Its volume is $V = x^2y$, and its surface area is $A = x^2 + 4xy$.

28. Suppose the dimensions of the box in Figure 7.45 are specified as $x = 4$ inches and $y = 12$ inches. The error in the fabricating process is at most 0.1 inch in each dimension. Use the approximation formula to estimate the maximum error in the volume of the box.

29. Repeat Exercise 28, using the approximation formula to estimate the maximum error in the surface area of the box.

In Exercises 30 and 31, find all critical points of the function.

30. $f(x, y) = x^2 + y^2 + xy + 2y - 3x$

31. $f(x, y) = x^3 - xy + y^2$

In Exercises 32 and 33, find all points where it is possible for the function to have an extremum.

32. $f(x, y) = x^4 - 7x^3 - \frac{1}{2}x + 3 - y + xy + y^2$

33. $f(x, y) = x^2 + 2xy + 2y^2 - 4x$

34. A rectangular box with a top is to contain 3000 cubic inches. The material for the base and top costs 3¢ per square inch; the material for the sides costs 8¢ per square inch. Find the dimensions yielding the minimum *cost of materials.*

35. A manufacturer of computer chips estimates that his yearly earnings can be predicted by the formula

$$E(x, y) = 16{,}000 + 4000x + 3000y + 12xy - 6x^2 - 8y^2$$

where x = (monthly research expenditures) and y = (monthly salaries of sales staff). Find the values of x and y that maximize earnings.

In Exercises 36–40, find all critical points, and test for local maxima and minima using the second derivative test.

36. $f(x, y) = x^2 - y^2 + 4x + 6y + xy$

37. $f(x, y) = x^2 + y^2 - 4x - 6y + xy$

38. $f(x, y) = 2xy - 3x^2 - y^2 + 2x - y + 1$

39. $f(x, y) = 3xy - x^3 - y^3$

40. $f(x, y) = e^{x^2 - y^2}$

In Exercises 41–45, we give sets of data points. Find the equation $y = ax + b$ of the straight line giving the best least squares fit to the data points.

41. $P_1 = (0, 4)$ $P_2 = (3, 1)$

42. $P_1 = (0, 3)$ $P_2 = (1, 2)$ $P_3 = (2, 0)$

43. $P_1 = (1, 2)$ $P_2 = (2, 1)$ $P_3 = (3, 3)$ $P_4 = (4, 4)$

44.

x	1	5
y	2	4

45.

x	0	2	3
y	0	1	3

In Exercises 46–49, use the method of Lagrange multipliers.

46. Minimize $f(x, y) = 2x^2 + 3y^2 + 5$ on the line $x + 3y - 7 = 0$.

47. Find the maximum and minimum of $f(x, y) = xy$ on the circle $x^2 + y^2 = 4$.

48. Find the extrema of $f(x, y) = x - y$ subject to the constraint $x^2 - xy + y^2 = 1$.

49. Find the extrema of $f(x, y) = xy$ subject to the constraint $x^2 + y^2 - 2y = 1$.

50. A brokerage firm comprises a stock department and a bond department. If x and y are the weekly budgets for stocks and bonds, respectively, the weekly profit is $P(x, y) = 60x^{3/4}y^{1/4}$ dollars. Use Lagrange multipliers to find the values of x and y that maximize P, given that the firm's total budget for the week is $40,000.

51. Evaluate the iterated integral $\int_0^4 [\int_1^2 (3x - y)\, dy]\, dx$.

52. Evaluate the double integral $\iint_R (3x^2 + 4xy + y^2)\, dx\, dy$ over the rectangle R given by $-1 \leq x \leq 3$ and $0 \leq y \leq 1$.

APPENDIX 1

Basic Review Material

In this section we review a few very basic facts. These topics form the background for the discussion in Chapter 1, and will be familiar to most readers. The main point here is to review powers, exponent laws, and the handling of inequalities.

The Real Number System; Inequalities

In doing calculations with real numbers, two things should be kept in mind. First

Division by zero is never permitted!

If you try it on a calculator, you will receive an immediate error message. In doing algebra, you must watch out for the possibility that the denominator in an expression could be zero.

Second, you should recall the basic laws of arithmetic whenever you do hand calculations with fractions.

(i) $\dfrac{a}{b} + \dfrac{c}{d} = \dfrac{ad + bc}{bd}$

(ii) $\dfrac{a}{b} \cdot \dfrac{c}{d} = \dfrac{ac}{bd}$

(iii) $\dfrac{\left(\dfrac{a}{b}\right)}{\left(\dfrac{c}{d}\right)} = \dfrac{a}{b} \cdot \dfrac{d}{c} = \dfrac{ad}{bc}$

Example 1 Write as simple fractions:

(i) $\dfrac{1}{5} - \dfrac{1}{4}$ (ii) $\dfrac{1}{5} \cdot \dfrac{2}{3}$ (iii) $\dfrac{\left(\frac{1}{2}\right)}{\left(\frac{3}{5}\right)}$

Solution In (i)

$$\frac{1}{5} - \frac{1}{4} = \frac{1}{5} + \frac{-1}{4} = \frac{4(1) + (-1)(5)}{5(4)} = -\frac{1}{20}$$

Similarly

$$\frac{1}{5} \cdot \frac{2}{3} = \frac{2}{15}$$

$$\frac{\left(\frac{1}{2}\right)}{\left(\frac{3}{5}\right)} = \frac{1}{2} \cdot \frac{5}{3} = \frac{5}{6}$$

■

Real numbers can be represented as points on a straight line. Starting with an unmarked straight line, choose one point and label it with the number zero (this becomes the **origin**). Then choose a second point and label it with the number 1, as in Figure A.1. These two points determine a "unit length" in the line and also specify a positive direction along the line (to the right in Figure A.1). By stepping off successive intervals of unit length to the right or left of the origin we can identify all positive or negative integers with points on the line. By dividing the unit interval, which extends from 0 to 1, into subintervals of equal length we can locate the point on the line corresponding to any fraction. For example, to locate $\frac{3}{2}$ divide the unit interval in half and, moving to the right from the origin, step off three copies of this half interval. Arbitrary real numbers may be identified with points on the line using their decimal expansions. For example, you have probably seen the first few terms in the decimal expansion of $\sqrt{2}$,

$$\sqrt{2} = 1.4142135\ldots$$

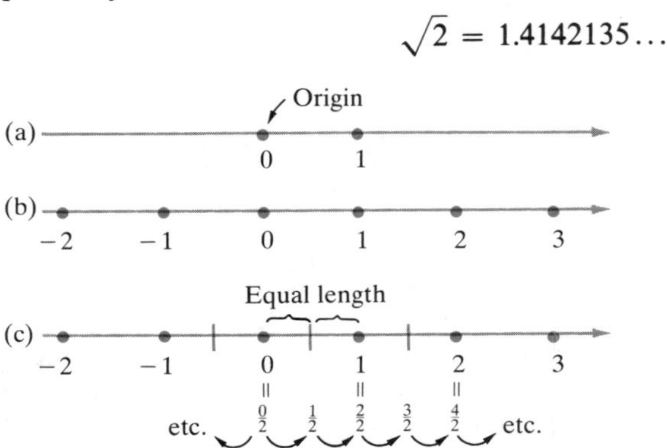

Figure A.1

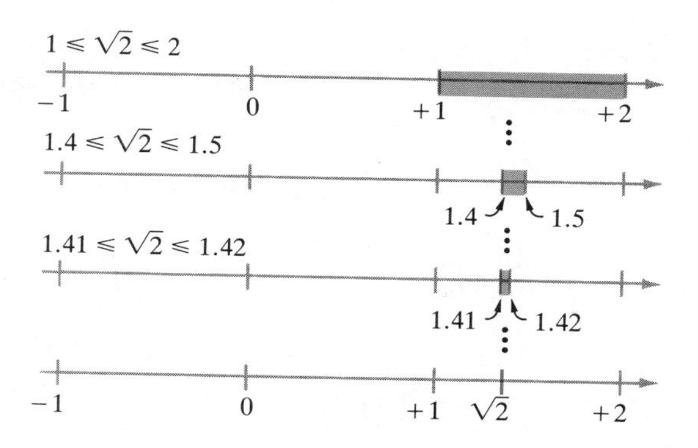

Figure A.2
Locating $\sqrt{2} = 1.414\ldots$ within smaller and smaller intervals (shaded) in the number line. In the next step we would locate $\sqrt{2}$ with an accuracy of 0.001, namely $1.414 \leq \sqrt{2} \leq 1.415$.

The position of $\sqrt{2}$ in the number line is located as follows. The number 1 to the left of the decimal point means that $\sqrt{2}$ lies between the integers 1 and 2. The digit 4 in the tenths' place gives the additional information that $\sqrt{2}$ lies between 1.4 and 1.5. The digit 1 in the hundredths' place locates $\sqrt{2}$ with an accuracy of 0.01, between 1.41 and 1.42. The actual location is pinned down by taking more and more decimal entries into account, as shown in Figure A.2. The exact location is determined by the full unending decimal 1.4142..., but the first few terms suffice to locate the number accurately enough for all practical purposes. The same method applies just as well to locate any number once it is given in decimal form. Since real numbers may be identified with points on a line, the system of real numbers is often referred to as the **number line.**

We use the following symbols to express the fact that a number x is larger than a number y:

$$x > y \quad \text{or} \quad y < x$$

Sometimes we know that x is either greater than y or equal to y, but we don't know which relation is true. In that case we use the symbols

$$x \geq y \quad \text{or} \quad y \leq x$$

Geometrically, **inequalities** describe the relative positions of two numbers x and y on the number line

$x < y$ means that x lies strictly to the left of y
(or equivalently, y lies strictly to the right of x) [1]

The symbol $x < y$ is read "x is less than y" or "y is greater than x." See Figure A.3 for some examples. Do not be confused by the minus signs in the inequality $-3 < -1$. Clearly -3 lies to the left of -1.

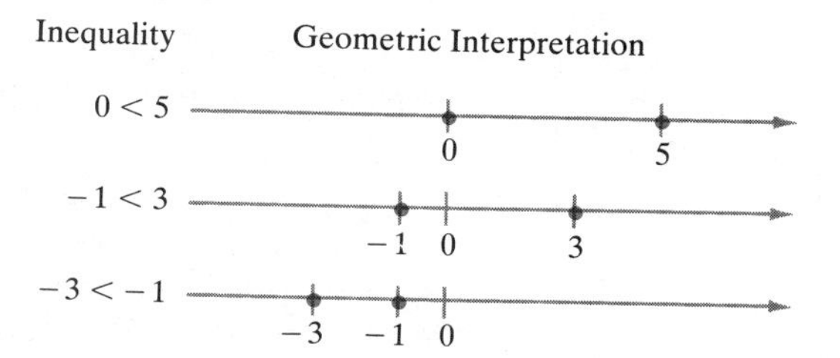

Figure A.3

The **positive** numbers are those to the right of the origin, the negative numbers are those to the left. In terms of inequalities

$$x \text{ is positive} \quad \text{if } x > 0$$
$$x \text{ is negative} \quad \text{if } x < 0$$

The sign (positive or negative) of the product or quotient of two numbers is determined by the familiar **rule of signs**

$$(+)\cdot(+) = (+) \quad (+)\cdot(-) = (-) \quad (-)\cdot(+) = (-) \quad (-)\cdot(-) = (+)$$
$$(+)/(+) = (+) \quad (+)/(-) = (-) \quad (-)/(+) = (-) \quad (-)/(-) = (+)$$

We indicate that a number x satisfies a *pair* of inequalities, such as $a \le x$ and $x < b$, by writing the inequalities together as a single symbol: $a \le x < b$. Note that the pair of inequalities must be in the same direction. This convention is handy for specifying sets in the number line such as intervals. A **bounded interval** is defined by taking all x lying between two points $a < b$. We get slightly different intervals depending on whether or not we include the **endpoints** a and b. Thus, bounded intervals are determined by pairs of inequalities such as $-\frac{1}{2} < x < \frac{3}{2}$ or $-2 \le x \le 1$, as shown in Figure A.4.

Intervals are indicated by symbols like $[a, b]$, which stands for the interval $a \le x \le b$ (endpoints included). When a square bracket is replaced with a rounded bracket, that means the endpoint is excluded. Thus, (a, b) stands for the interval $a < x < b$, $[a, b)$ stands for $a \le x < b$, and so forth.

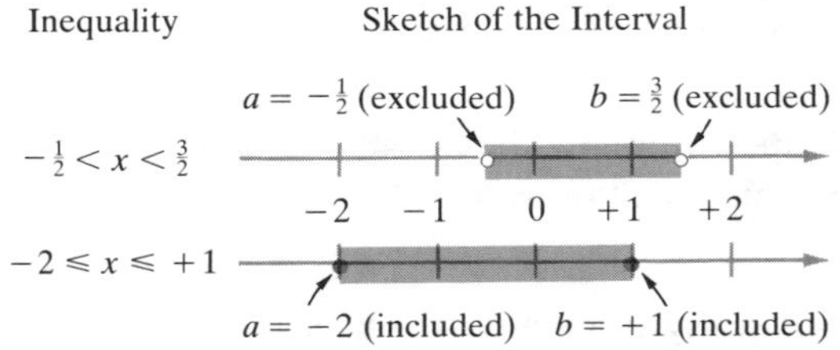

Figure A.4
Some bounded intervals. The points in the interval are shaded. An open dot indicates an excluded endpoint, and a solid dot indicates that the point is included as part of the interval.

Inequalities arise in many situations. Here we show how to find the solution set of an inequality involving one unknown. As an example, suppose a publisher is preparing a book for which initial typesetting and editorial costs of $40,000 must be offset by revenue of $12.80 for each copy sold. If he sells x copies, his profit will be

$$P(x) = -40{,}000 + 12.80x$$

Obviously, he wants to know the values of x for which he makes a profit instead of a loss. Mathematically, he wants to determine when the inequality $P \ge 0$ is satisfied, that is

$$-40{,}000 + 12.80x \ge 0 \qquad [2]$$

The set of x for which this is true is called the **solution set** of the inequality. Such inequalities are solved using the following simple rules.

RULES OF INEQUALITIES

(i) Inequalities are preserved when the same number is added to each side.

$$\text{If} \quad a < b \quad \text{and if} \quad c \quad \text{is any number, then} \quad a + c < b + c \qquad [3]$$

This is so whether the numbers are positive, negative, or of mixed type.

(ii) Inequalities are preserved when multiplied by a positive number. Let c be a positive number, $c > 0$. Then

$$\text{If} \quad a < b \quad \text{and if} \quad c \quad \text{is positive, then} \quad ac < bc \qquad [4]$$

As we shall see, positivity of c is essential here.

(iii) Inequalities are not preserved if we multiply both sides by -1 or by any other negative number. In fact, this *reverses* the original inequality.

$$\text{If} \quad a < b \quad \text{and if} \quad c \quad \text{is negative, then} \quad ac > bc \qquad [5]$$

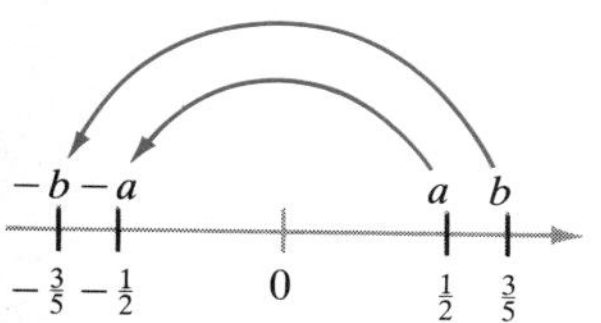

Figure A.5
Effect on an inequality of multiplication by -1.

As an example of [5], if $a = 1/2$ and $b = 3/5$ then $a < b$. But $(-1) \cdot a = -1/2$ is greater than $(-1) \cdot b = -3/5$, so that $-a > -b$ as shown in Figure A.5.

From algebra you are familiar with solutions of equations. Inequalities, though they may be less familiar, are solved in much the same way. By applying the rules given above we can show that a typical inequality such as

$$3x - 7 > x + 2$$

is equivalent to a much simpler inequality

$$x > \frac{9}{2}$$

in the sense that they have exactly the same solution set.

Example 2 Describe the solution set for the inequality $3x - 7 > x + 2$.

Solution The following inequalities are equivalent, for the reasons indicated:

$$3x - 7 > x + 2$$

Add $+7$ to both sides (Rule [3])

$$3x > x + 9$$

Add $-x$ to both sides (Rule [3])

$$2x > 9$$

Multiply both sides by $\frac{1}{2}$ (Rule [4])

$$x > \frac{9}{2}$$

In the last line the original inequality has been reduced, without altering its solution set, to an inequality so simple that the solution set can be sketched immediately. A number x satisfies $x > \frac{9}{2}$ if it lies strictly to the right of $\frac{9}{2}$ on the number line. The solution set is an "interval" extending to the right of $\frac{9}{2}$, the

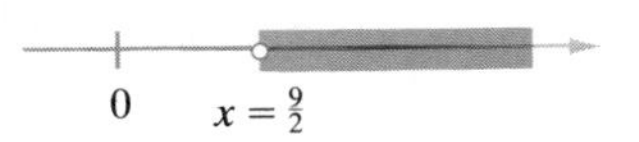

Figure A.6
The shaded portion of the line is the solution set of $x > \frac{9}{2}$.

shaded portion of the line shown in Figure A.6. The point $x = \frac{9}{2}$ itself is *not* a solution. Had we started with the inequality $3x - 7 \geq x + 2$, which by the same reasoning is equivalent to $x \geq \frac{9}{2}$, the solution set would have been the shaded portion of the line together with the endpoint $x = \frac{9}{2}$. ■

By similar methods we can find the solution of the publisher's problem [2]. He makes a profit whenever

$$0 \leq -40{,}000 + 12.80x$$

Add 40,000 to both sides

$$40{,}000 \leq 12.80x$$

Divide by 12.80

$$3125 \leq x$$

The break-even point occurs when 3125 copies are sold.

Example 3 Find the solution set of the inequality $-4x + 2 \geq \frac{1}{2}$.

Solution First simplify the inequality

$$-4x + 2 \geq \frac{1}{2}$$

Add -2 to both sides

$$-4x \geq -\frac{3}{2}$$

Multiply both sides by -1; this reverses the inequality!

$$4x \leq \frac{3}{2}$$

Multiply by $\frac{1}{4}$

$$x \leq \frac{3}{8}$$

The solution set consists of all x lying to the left of $\frac{3}{8}$. ■

The **absolute value** $|x|$ of a number x is, essentially, the number with its sign stripped away. Officially

$$|x| = \begin{cases} x & \text{if } x > 0 \\ 0 & \text{if } x = 0 \\ -x & \text{if } x < 0 \end{cases}$$

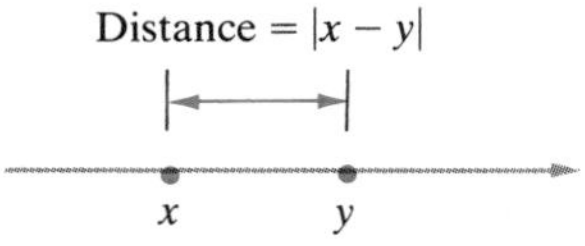

Distance = $|x - y|$

x y

Figure A.7
The geometric distance between two points x and y in the number line is given by $|x - y|$. Here we have labeled the points so x lies to the left of y; even if we reverse the labels, the distance is still given by $|x - y|$.

Thus $|3| = 3$, $|-4| = 4$, $|0| = 0$, and so on. Expressions such as $|x - y|$ turn up frequently because they have a natural geometric interpretation. The difference $x - y$ between two numbers x and y tells us how far we must move in the number line, starting from y, to get to x. But $x - y$ could be positive or negative, so it is the absolute value $|x - y|$ that gives the distance between x and y as a non-negative number (see Figure A.7). Using this geometric interpretation of $|x - y|$, we can solve inequalities involving absolute values.

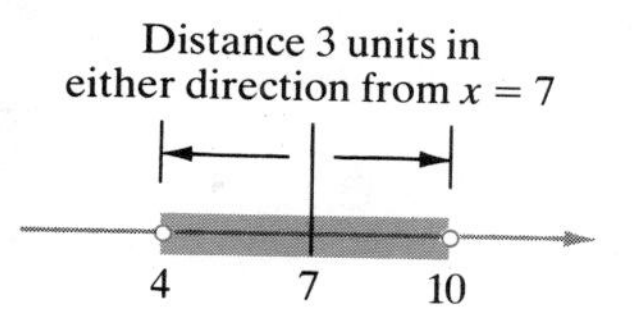

Figure A.8
The solution set for $|x - 7| < 3$ consists of all x whose distance from 7 is less than 3. The endpoints are excluded.

Example 4 Find the solution set of $|x - 7| < 3$.

Solution This consists of all points whose distance from 7 is less than 3. This is just an interval extending from $7 - 3 = 4$ to $7 + 3 = 10$, as in Figure A.8. It is all x such that $4 < x < 10$. ■

Example 5 Solve (i) $|x| \geq \frac{2}{3}$, and (ii) $|x + 6| \leq 1$.

Solution Write $x = x - 0$ and $|x| = |x - 0| =$ (distance from x to the origin). Then (i) holds when the distance from x to 0 is greater than or equal to $\frac{2}{3}$. The solution set consists of two intervals $(-\infty, -\frac{2}{3}]$ and $[\frac{2}{3}, +\infty)$ as shown in Figure A.9a. For (ii), write $x + 6 = x - (-6)$ and

$$|x + 6| = |x - (-6)| = (\text{distance from } x \text{ to } -6)$$

The solution set consists of the interval extending from $-6 - 1 = -7$ to $-6 + 1 = -5$, including the endpoints: all x such that $-7 \leq x \leq -5$, as in Figure A.9b. ■

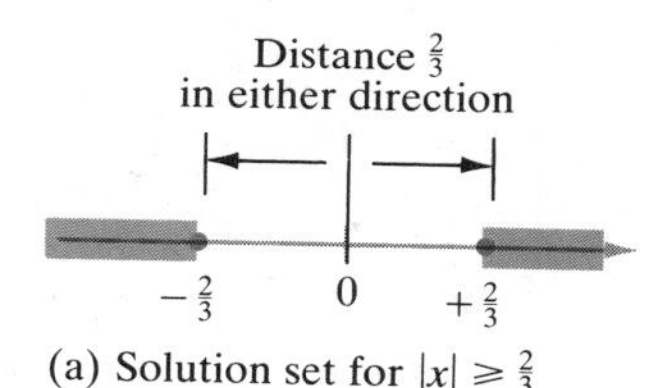

(a) Solution set for $|x| \geq \frac{2}{3}$

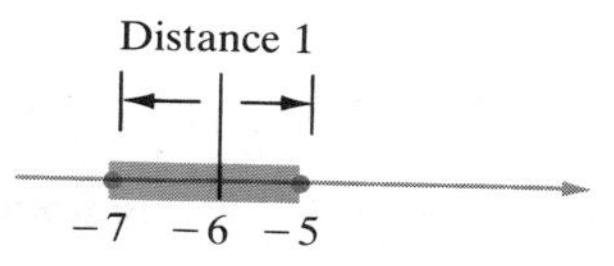

(b) Solution set for $|x - (-6)| \leq 1$

Figure A.9

Powers of a Real Number and Exponent Laws

If n is a positive integer, the ***n*th power** of a real number a is obtained by multiplying a by itself n times

$$a^n = \underbrace{a \cdots a}_{n \text{ times}} \qquad [6]$$

As examples, $1^5 = 1$, $2^4 = 16$, $(1/7)^2 = \frac{1}{49}$. Obviously, $a^1 = a$ for any a. In the expression a^n, the number a is called the **base** and n is called the **exponent**. Powers have two important algebraic properties, called **laws of exponents**.

$$a^m \cdot a^n = a^{m+n}$$
$$(a^m)^n = a^{mn} \qquad [7]$$

These are easily deduced from the definition [6]. For example

$$a^m \cdot a^n = \underbrace{(a \cdots a)}_{m \text{ times}} \cdot \underbrace{(a \cdots a)}_{n \text{ times}} = \underbrace{a \cdots a}_{m+n \text{ times}} = a^{m+n}$$

Similarly,

$$(a^m)^n = \underbrace{(a^m) \cdots (a^m)}_{n \text{ times}} = \underbrace{(a \cdots a)}_{m \text{ times}} \cdot \underbrace{(a \cdots a)}_{m \text{ times}} \cdots \underbrace{(a \cdots a)}_{m \text{ times}}$$

Here there are n blocks, each containing m copies of a, so in all there are mn copies of a multiplied together. Thus, $(a^m)^n = a^{mn}$.

If a is not zero, we define a^n for $n = 0, -1, -2, \ldots$ by taking

$$a^0 = 1$$
$$a^{-n} = \frac{1}{a^n} \quad \text{for} \quad n = 1, 2, 3, \ldots \qquad [8]$$

Thus, for example

$$5^0 = 1 \qquad 5^{-1} = \frac{1}{5} = 0.2 \qquad 5^{-2} = \frac{1}{25} = 0.04$$

and so on. If we define a^n in this way for all integers n, the laws of exponents remain valid.

We can define **fractional powers** a^r for any fraction $r = m/n$ (where m and n are integers with $n \neq 0$) if we restrict our attention to positive numbers $a > 0$. Fractional powers of the form $a^{1/n}$ (n a positive integer) are interpreted as follows:

$$a^{1/n} = \text{the positive } n\text{th root of } a \qquad [9]$$

This interpretation is not arbitrary. It is strongly suggested if we want to preserve the laws of exponents. Thus $a^{1/n}$ is the unique positive number whose nth power is a because the exponent law [7] requires that

$$(a^{1/n})^n = a^1 = a$$

For example, $a^{1/2} = \sqrt{a}$ (the square root), $a^{1/3} = \sqrt[3]{a}$ (the cube root), and so on. In particular

$$2^{1/2} = \sqrt{2} = 1.414\ldots \qquad 8^{1/3} = \sqrt[3]{8} = 2 \qquad 25^{1/4} = \sqrt[4]{25} = \sqrt{5} = 2.236\ldots$$

Once we know how to interpret powers of the form $a^{1/n}$ we can interpret $a^{m/n}$ as follows:

$$a^{m/n} = (a^{1/n})^m = (a^m)^{1/n} \qquad [10]$$

Thus there are two ways to calculate $a^{m/n}$

(i) Take a, form the positive nth root $a^{1/n}$, then take the mth power $(a^{1/n})^m = (a^{1/n}) \cdots (a^{1/n}) = a^{m/n}$.
(ii) Take a, form the mth power a^m (also a positive number), then take the positive nth root $(a^m)^{1/n} = a^{m/n}$. [11]

but both procedures yield the same value for $a^{m/n}$.

The exponent laws remain valid for fractional powers. As a result, it is no more difficult to manipulate fractional powers of a number than it is to manipulate integral powers a^n. Several laws of exponents that we have not yet mentioned are also valid. We give the full set of laws here.

THE LAWS OF EXPONENTS Here, $a, b, \ldots$ stand for positive real numbers and $r, s, \ldots$ for fractions.

$$\text{(i)}\ a^r \cdot a^s = a^{r+s} \qquad \text{(iv)}\ a^{-r} = \frac{1}{a^r}$$
$$\text{(ii)}\ (a^r)^s = a^{rs} \qquad \text{(v)}\ \left(\frac{a}{b}\right)^r = \frac{a^r}{b^r} \qquad [12]$$
$$\text{(iii)}\ (ab)^r = a^r \cdot b^r$$

Rather than prove these laws (time consuming, but not terribly difficult) we shall work out a few examples.

Example 6 Show that $4^{5/2} = 32$.

Solution Although the procedures [11i] and [11ii] for evaluating $a^{m/n}$ always lead to the same answer, in this case it is easier to take the square root first, and then the 5th power. Compare the calculations:

$$4^{5/2} = (4^{1/2})^5 = 2^5 = 2 \cdot 2 \cdot 2 \cdot 2 \cdot 2 = 32$$
$$4^{5/2} = (4^5)^{1/2} = (1024)^{1/2} = \sqrt{1024} = 32$$

If we take the power first, we get the large number 1024, whose square root is not as easy to recognize. ■

Example 7 Suppose that the population of a certain midwestern city t years after the 1950 census is given by the formula

$$N = 30{,}000{,}000 \cdot 2^{t/30}$$

In order to evaluate N we must calculate fractional powers of 2:

Initially, when $t = 0$,	$N = 30{,}000{,}000 \cdot (2^0)$
After $t = 10$ years,	$N = 30{,}000{,}000 \cdot (2^{1/3})$
After $t = 15$ years,	$N = 30{,}000{,}000 \cdot (2^{1/2})$
After $t = 45$ years,	$N = 30{,}000{,}000 \cdot (2^{3/2})$

Calculate the value of N after 15 years, after 45 years, and after 60 years.

Solution Here $2^{15/30} = 2^{1/2} = \sqrt{2} = 1.414\ldots$, so when $t = 15$ we obtain

$$N = 30{,}000{,}000 \cdot (1.414) = 42.42 \text{ million}$$

Similarly, $2^{45/30} = 2^{3/2} = (2^{1/2})^3 = \sqrt{2}\cdot\sqrt{2}\cdot\sqrt{2} = 2\sqrt{2} = 2.828\ldots$, so when $t = 45$

$$N = 30{,}000{,}000 \cdot (2.828) = 84.84 \text{ million}$$

Finally, $2^{60/30} = 2^2 = 4$, so $N = 30{,}000{,}000(4) = 120$ million when $t = 60$. ■

Example 8 Evaluate $(\frac{1}{7})^{3/2}$ and $(\frac{1}{7})^{-3/2}$.

Solution We are free to use any of the exponent laws [12]. First apply [12v] to obtain

$$\left(\frac{1}{7}\right)^{3/2} = \frac{1^{3/2}}{7^{3/2}}$$

From the definition [10] we get $1^{3/2} = (1^{1/2})^3 = 1^3 = 1$ and $7^{3/2} = (7^{1/2})^3 = (\sqrt{7})^3 = \sqrt{7}\cdot\sqrt{7}\cdot\sqrt{7} = 7\sqrt{7}$. Thus

$$\left(\frac{1}{7}\right)^{3/2} = \frac{1}{7^{3/2}} = \frac{1}{7\sqrt{7}} = 0.05399\ldots$$

We need not wrestle with similar calculations to determine $(\frac{1}{7})^{-3/2}$; just use [12iv]

$$\left(\frac{1}{7}\right)^{-3/2} = \frac{1}{\left(\frac{1}{7}\right)^{3/2}} = \frac{1}{\left(\frac{1}{7\sqrt{7}}\right)} = 7\sqrt{7} = 18.5202\ldots$$ ■

Scientific Notation

In practical calculations we often encounter very large or very small numbers, such as

$$x = 2{,}875{,}000{,}000 \quad \text{or} \quad y = 0.0000000437$$

These are not very convenient to write down or calculate with; it is very easy to make a mistake keeping track of the decimal place. For example, try calculating $x\cdot y$. Keeping track of the decimal place is made easier by **scientific notation**. Most numbers can be written as a number of modest size times a suitable power of 10. Thus, after counting decimal places, we may write x and y as

$$x = 2.875 \times 10^9 \quad \text{or} \quad 28.75 \times 10^8 \quad \text{or} \quad 0.2875 \times 10^{10}$$
$$y = 4.37 \times 10^{-8} \quad \text{or} \quad 43.7 \times 10^{-9} \quad \text{or} \quad 0.437 \times 10^{-7}$$

Now we can handle the powers of 10 separately in computations, which is a great convenience. For example

$$\begin{aligned} x \cdot y &= (2.875 \times 10^9) \cdot (4.37 \times 10^{-8}) \\ &= (2.875 \times 4.37) \cdot (10^9 \times 10^{-8}) \\ &= 12.564 \times 10^1 = 125.64 \end{aligned}$$

Example 9 Compute

$$\left.\begin{array}{l} \text{(i) } \dfrac{x}{y} \\ \text{(ii) } x^2 \\ \text{(iii) } \sqrt{y} \end{array}\right\} \quad \text{where} \quad \begin{array}{l} x = 2.875 \times 10^9 \\ y = 4.37 \times 10^{-8} \end{array}$$

Solution In (i)

$$\begin{aligned} \frac{x}{y} &= \frac{2.875 \times 10^9}{4.37 \times 10^{-8}} = \frac{2.875}{4.37} \times 10^9 \times \frac{1}{10^{-8}} \\ &= 0.6579 \times 10^9 \times 10^8 \\ &= 0.6579 \times 10^{17} \end{aligned}$$

Similarly,

$$\begin{aligned} x^2 &= (2.875 \times 10^9)^2 = (2.875)^2 \times (10^9)^2 \\ &= 8.266 \times 10^{18} \\ \sqrt{y} &= y^{1/2} = (4.37 \times 10^{-8})^{1/2} \\ &= (4.37)^{1/2} \times (10^{-8})^{1/2} \\ &= 2.090 \times 10^{-4} \end{aligned}$$

■

Exercises A.1

Write the following as fractions.

1. $\frac{1}{2} + \frac{1}{4}$ **2.** $\frac{2}{3} + \frac{1}{7}$ **3.** $\frac{4}{5} - \frac{5}{4}$

4. $1 + \frac{2}{3}$ **5.** $\frac{2}{5} \cdot \frac{1}{2}$ **6.** $-\frac{2}{5} \cdot \frac{4}{7}$

7. $\dfrac{1}{\left(\frac{2}{3}\right)}$ **8.** $\dfrac{\left(\frac{1}{3}\right)}{2}$ **9.** $\dfrac{\left(\frac{1}{3}\right)}{\left(\frac{2}{5}\right)}$

10. On graph paper make a sketch of the number line and plot the following points.

(i) -1 (ii) 2.25 (iii) -2.43 (iv) 4.1 (v) -3.66 (vi) -0.75

11. Indicate whether true or false

(i) $1 < 3$ (ii) $1 \leq 3$ (iii) $-1 < -3$ (iv) $-3 < -1$ (v) $2 > 2$ (vi) $-2 \geq \frac{1}{2}$

Sketch the intervals in the number line determined by the following inequalities

12. $0 < x < 1$ **13.** $-1 \leq x < 3$ **14.** $-3 \leq x < 3$

15. $-4 \leq x \leq -1$ **16.** $-3 < x \leq 0$ **17.** $-\sqrt{2} \leq x \leq \sqrt{2}$

A single inequality such as $a < x$ or $x < a$ determines an *unbounded interval*, one stretching out to infinity to one side or the other of a. Sketch the following unbounded intervals.

18. $x \geq 1$ **19.** $x > 0$ **20.** $x \leq -1$

21. $x \leq 4$ **22.** $x < 4$ **23.** $x \geq -2$

24. Suppose Company A charges \$10 per day and 10¢ per mile to rent a certain type of automobile, and Company H charges \$13 per day and 8¢ per mile. If you expect to drive x miles per day, for which values of x is it advantageous to rent from Company A? From Company H?

25. A car dealer employs five salesmen, each paid a weekly salary of \$250. The dealer's costs for rent, insurance, and so on are \$1000 per week. He makes \$300 on every car sold. How many cars must be sold per week to break even?

26. A manufacturing plant has weekly costs of $C(q) = 54{,}000 + 160q$ dollars, operating at a production level of q units per week. The weekly revenue is $R(q) = 218q$ (\$218 for each unit produced). Determine the values of q for which the profit $P = R - C$ is (i) positive, and (ii) greater than \$10,000 per week.

Sketch the solution sets in the number line for the following inequalities.

27. $x > -1$ **28.** $x < -1$

29. $x + 2 > 0$ **30.** $x + 2 < 0$

31. $x - 3 \geq 0$ **32.** $2 - x < 7$

33. $2 - x < 2x + 11$ **34.** $2 - x \geq 2x + 2$

35. $-\frac{7}{8}x + 1 \leq 2x + \frac{1}{2}$ **36.** $2.0x - 500 \leq 100 + 1.5x$

37. $3x + 3 \geq 2x + 2$ **38.** $5 - 3x < 9 - 8x$

39. Four salesmen are on the road for 5 days each week. As travel expenses they are allowed \$75 per salesman per diem plus \$0.18 per mile. Write a formula expressing the weekly combined expenses in terms of total mileage covered by all salesmen. If the weekly sales budget for these costs is \$3850, what is the maximum combined mileage in a week?

40. If we want to duplicate 200 copies of a one-page report, what should we do given the following alternatives?
(i) Make photo copies at 4¢ per copy.
(ii) Make a photo offset stencil, initial cost \$1.50, that produces copies at 1.5¢ per copy.
Would the conclusion change if the report were 12 pages long?

41. Ten thousand copies of a book have been printed. Copies sell at \$8.50 each, and after 2 years all unsold copies will be "remaindered" at \$1.00 per copy. This revenue must be balanced against publishing and editorial costs that amount to \$57,000. How many copies must be sold before remaindering to break even?

Find the absolute values of the following numbers.

42. 7 **43.** -7 **44.** -3.142

Which numbers x satisfy the following equalities (use the geometric interpretation of $|x - y|$).

45. $|x - 1| = 2$ **46.** $|x - 3| = 5$

47. $|x + 6| = 2$ **48.** $|x + 6| = 3$

49. $|x + 1| = 4$ **50.** $|x - 2.73| = 0.55$

Sketch the solution sets of the following inequalities.

51. $|x - 1| < 2$ **52.** $|x - 1| \leq 2$

53. $|x - 3| > 5$ **54.** $|x - 3| \leq 5$

55. $|x + 6| \leq 0.5$ **56.** $|x + 6| > 1$

Write the numbers in Exercises 57–62 as fractional powers.

57. $\sqrt{3}$ **58.** $\sqrt[3]{2}$ **59.** $\sqrt[5]{5}$

60. $\dfrac{1}{\sqrt{7}}$ **61.** $\sqrt[3]{(28)^2}$ **62.** $\dfrac{1}{\sqrt[3]{(28)^2}}$

In Exercises 63–72, evaluate the fractional powers.

63. 4^0 **64.** $1^{4/34}$ **65.** $0^{4/35}$

66. 1^{-100} **67.** 3^{-1} **68.** 5^{-2}

69. $16^{3/4}$ **70.** $4^{-2.50}$ **71.** $\left(\dfrac{8}{27}\right)^{5/3}$

72. $(0.01)^{3/2}$

73. Given $2^{1/3} = 1.2599\ldots$, what is the population

$$N = 30{,}000{,}000 \cdot 2^{t/30}$$

in Example 7 when $t = 10$ years? 20 years? 30 years? 40 years?

74. A small population of pheasants transported to an isolated island increases according to the formula

$$N = 10(2^{3t/2}) = 10(2^{1.50t}) \quad (t \text{ in years})$$

Find the population N in successive years, when $t = 0, 1, 2, 3$.

75. As sunlight penetrates a murky lake, the light intensity I falls off exponentially according to the formula

$$I = I_0 2^{-0.5x}$$

where x is the depth below the surface (in feet) and I_0 is the light intensity at the surface. What fraction of the light penetrates to a depth of 1 foot? 2 feet? 5 feet? At what depth is 75% of the light absorbed (i.e., only 25% of the light penetrates)?

In Exercises 76–83, write the numbers in the form shown, where (. . .) indicates a suitable power of 10. For example, $0.00011 = 11 \times 10^{-5}$ and $1100 = 1.1 \times 10^3$.

76. $0.000098 = 98 \times (\ldots)$

77. $9{,}842{,}000{,}000 = 9.842 \times (\ldots)$

78. $980{,}000 = 0.98 \times (\ldots)$

79. $10 = 1 \times (\ldots)$

80. $4421 = 4.421 \times (\ldots)$

81. $\dfrac{1}{100{,}000} = 1 \times (\ldots)$

82. $\dfrac{43}{10{,}000} = 4.3 \times (\ldots)$

83. $\dfrac{0.2}{100{,}000} = 2 \times (\ldots)$

In Exercises 84–89, evaluate these expressions involving powers of 10.

84. $\dfrac{3.172 \times 10^{13}}{6.231 \times 10^{23}}$

85. $(15.1 \times 10^4) \cdot (0.23 \times 10^{-7})$

86. $\dfrac{15 \times 10^{-4}}{3 \times 10^{-6}}$

87. $(0.22 \times 10^{-5}) \cdot (62.5 \times 10^{-3})$

88. $(3.1 \times 10^{13})^2$

89. $\sqrt{144 \times 10^6}$

If a, b, c are positive numbers in Exercises 90–95, simplify these expressions by using the exponent laws.

90. $\dfrac{a^{10}}{a^5}$

91. $a^{-3} \cdot a^7$

92. $(a^3)^2$

93. $\dfrac{(a^2 \cdot b^3 \cdot c)^4}{(a^3 \cdot b^{-1} \cdot c^2)^3}$

94. $(a^2 \cdot b^{-3} \cdot c^{-1}) \cdot (a^4 \cdot b^3 \cdot c^7)$

95. $\left(\dfrac{a^{1/3}b^{5/6}}{a^{4/3}b^{-1/6}}\right)^4$

96. Multiply out the following expressions:

(i) $(a + b)^2$
(ii) $(a + b)^3$
(iii) $(a + b)^4$
(iv) $(a + b)(a - b)$
(v) $(a - b)^2$

APPENDIX 2

Tables of Natural Logarithms ($\ln x$) and Exponentials (e^x)

Table 1 **Natural Logarithms ($\ln x$)**

x	0.00	0.01	0.02	0.03	0.04	0.05	0.06	0.07	0.08	0.09
1.0	0.00000	0.00995	0.01980	0.02956	0.03922	0.04879	0.05827	0.06766	0.07696	0.08618
1.1	0.09531	0.10436	0.11333	0.12222	0.13103	0.13976	0.14842	0.15700	0.16551	0.17395
1.2	0.18232	0.19062	0.19885	0.20701	0.21511	0.22314	0.23111	0.23902	0.24686	0.25464
1.3	0.26236	0.27003	0.27763	0.28518	0.29267	0.30010	0.30748	0.31481	0.32208	0.32930
1.4	0.33647	0.34359	0.35066	0.35767	0.36464	0.37156	0.37844	0.38526	0.39204	0.39878
1.5	0.40547	0.41211	0.41871	0.42527	0.43178	0.43825	0.44469	0.45108	0.45742	0.46373
1.6	0.47000	0.47623	0.48243	0.48858	0.49470	0.50078	0.50682	0.51282	0.51879	0.52473
1.7	0.53063	0.53649	0.54232	0.54812	0.55389	0.55962	0.56531	0.57098	0.57661	0.58222
1.8	0.58779	0.59333	0.59884	0.60432	0.60977	0.61519	0.62058	0.62594	0.63127	0.63658
1.9	0.64185	0.64710	0.65233	0.65752	0.66269	0.66783	0.67294	0.67803	0.68310	0.68813
2.0	0.69315	0.69813	0.70310	0.70804	0.71295	0.71784	0.72271	0.72755	0.73237	0.73716
2.1	0.74194	0.74669	0.75142	0.75612	0.76081	0.76547	0.77011	0.77473	0.77932	0.78390
2.2	0.78846	0.79299	0.79751	0.80200	0.80648	0.81093	0.81536	0.81978	0.82418	0.82855
2.3	0.83291	0.83725	0.84157	0.84587	0.85015	0.85442	0.85866	0.86289	0.86710	0.87129
2.4	0.87547	0.87963	0.88377	0.88789	0.89200	0.89609	0.90016	0.90422	0.90826	0.91228
2.5	0.91629	0.92028	0.92426	0.92822	0.93216	0.93609	0.94001	0.94391	0.94779	0.95166
2.6	0.95551	0.95935	0.96317	0.96698	0.97078	0.97456	0.97833	0.98208	0.98582	0.98954
2.7	0.99325	0.99695	1.0006	1.0043	1.0080	1.0116	1.0152	1.0189	1.0225	1.0260

Table 1 *(continued)*

x	**0.00**	**0.01**	**0.02**	**0.03**	**0.04**	**0.05**	**0.06**	**0.07**	**0.08**	**0.09**
2.8	1.0296	1.0332	1.0367	1.0403	1.0438	1.0473	1.0508	1.0543	1.0578	1.0613
2.9	1.0647	1.0682	1.0716	1.0750	1.0784	1.0818	1.0852	1.0886	1.0919	1.0953
3.0	1.0986	1.1019	1.1053	1.1086	1.1119	1.1151	1.1184	1.1217	1.1249	1.1282
3.1	1.1314	1.1346	1.1378	1.1410	1.1442	1.1474	1.1506	1.1537	1.1569	1.1600
3.2	1.1632	1.1663	1.1694	1.1725	1.1756	1.1787	1.1817	1.1848	1.1878	1.1909
3.3	1.1939	1.1970	1.2000	1.2030	1.2060	1.2090	1.2119	1.2149	1.2179	1.2208
3.4	1.2238	1.2267	1.2296	1.2326	1.2355	1.2384	1.2413	1.2442	1.2470	1.2499
3.5	1.2528	1.2556	1.2585	1.2613	1.2641	1.2670	1.2698	1.2726	1.2754	1.2782
3.6	1.2809	1.2837	1.2865	1.2892	1.2920	1.2947	1.2975	1.3002	1.3029	1.3056
3.7	1.3083	1.3110	1.3137	1.3164	1.3191	1.3218	1.3244	1.3271	1.3297	1.3324
3.8	1.3350	1.3376	1.3403	1.3429	1.3455	1.3481	1.3507	1.3533	1.3558	1.3584
3.9	1.3610	1.3635	1.3661	1.3686	1.3712	1.3737	1.3762	1.3788	1.3813	1.3838
4.0	1.3863	1.3888	1.3913	1.3938	1.3962	1.3987	1.4012	1.4036	1.4061	1.4085
4.1	1.4110	1.4134	1.4159	1.4183	1.4207	1.4231	1.4255	1.4279	1.4303	1.4327
4.2	1.4351	1.4375	1.4398	1.4422	1.4446	1.4469	1.4493	1.4516	1.4540	1.4563
4.3	1.4586	1.4609	1.4633	1.4656	1.4679	1.4702	1.4725	1.4748	1.4771	1.4793
4.4	1.4816	1.4839	1.4861	1.4884	1.4907	1.4929	1.4952	1.4974	1.4996	1.5019
4.5	1.5041	1.5063	1.5085	1.5107	1.5129	1.5151	1.5173	1.5195	1.5217	1.5239
4.6	1.5261	1.5282	1.5304	1.5326	1.5347	1.5369	1.5390	1.5412	1.5433	1.5454
4.7	1.5476	1.5497	1.5518	1.5539	1.5560	1.5581	1.5603	1.5624	1.5644	1.5665
4.8	1.5686	1.5707	1.5728	1.5749	1.5769	1.5790	1.5810	1.5831	1.5852	1.5872
4.9	1.5892	1.5913	1.5933	1.5953	1.5974	1.5994	1.6014	1.6034	1.6054	1.6074
5.0	1.6094	1.6114	1.6134	1.6154	1.6174	1.6194	1.6214	1.6233	1.6253	1.6273
5.1	1.6292	1.6312	1.6332	1.6351	1.6371	1.6390	1.6409	1.6429	1.6448	1.6467
5.2	1.6487	1.6506	1.6525	1.6544	1.6563	1.6582	1.6601	1.6620	1.6639	1.6658
5.3	1.6677	1.6696	1.6715	1.6734	1.6752	1.6771	1.6790	1.6808	1.6827	1.6846
5.4	1.6864	1.6883	1.6901	1.6919	1.6938	1.6956	1.6975	1.6993	1.7011	1.7029
5.5	1.7047	1.7065	1.7083	1.7101	1.7119	1.7138	1.7156	1.7174	1.7191	1.7209
5.6	1.7227	1.7245	1.7263	1.7281	1.7298	1.7316	1.7334	1.7351	1.7369	1.7387
5.7	1.7404	1.7422	1.7439	1.7457	1.7474	1.7492	1.7509	1.7526	1.7544	1.7561
5.8	1.7578	1.7595	1.7613	1.7630	1.7647	1.7664	1.7681	1.7698	1.7715	1.7732
5.9	1.7749	1.7766	1.7783	1.7800	1.7817	1.7833	1.7850	1.7867	1.7884	1.7900
6.0	1.7917	1.7934	1.7950	1.7967	1.7984	1.8000	1.8017	1.8033	1.8050	1.8066
6.1	1.8082	1.8099	1.8115	1.8131	1.8148	1.8164	1.8180	1.8197	1.8213	1.8229
6.2	1.8245	1.8261	1.8277	1.8293	1.8309	1.8325	1.8341	1.8357	1.8373	1.8389
6.3	1.8405	1.8421	1.8437	1.8453	1.8468	1.8484	1.8500	1.8516	1.8531	1.8547
6.4	1.8563	1.8578	1.8594	1.8609	1.8625	1.8640	1.8656	1.8671	1.8687	1.8702
6.5	1.8718	1.8733	1.8748	1.8764	1.8779	1.8794	1.8809	1.8825	1.8840	1.8855
6.6	1.8870	1.8885	1.8901	1.8916	1.8931	1.8946	1.8961	1.8976	1.8991	1.9006
6.7	1.9021	1.9036	1.9050	1.9065	1.9080	1.9095	1.9110	1.9125	1.9139	1.9154
6.8	1.9169	1.9183	1.9198	1.9213	1.9227	1.9242	1.9257	1.9271	1.9286	1.9300
6.9	1.9315	1.9329	1.9344	1.9358	1.9373	1.9387	1.9401	1.9416	1.9430	1.9444
7.0	1.9459	1.9473	1.9487	1.9501	1.9516	1.9530	1.9544	1.9558	1.9572	1.9586
7.1	1.9600	1.9615	1.9629	1.9643	1.9657	1.9671	1.9685	1.9699	1.9713	1.9726
7.2	1.9740	1.9754	1.9768	1.9782	1.9796	1.9810	1.9823	1.9837	1.9851	1.9865
7.3	1.9878	1.9892	1.9906	1.9919	1.9933	1.9947	1.9960	1.9974	1.9987	2.0001
7.4	2.0014	2.0028	2.0041	2.0055	2.0068	2.0082	2.0095	2.0108	2.0122	2.0135

Table 1 *(continued)*

x	**0.00**	**0.01**	**0.02**	**0.03**	**0.04**	**0.05**	**0.06**	**0.07**	**0.08**	**0.09**
7.5	2.0149	2.0162	2.0175	2.0189	2.0202	2.0215	2.0228	2.0241	2.0255	2.0268
7.6	2.0281	2.0294	2.0307	2.0320	2.0334	2.0347	2.0360	2.0373	2.0386	2.0399
7.7	2.0412	2.0425	2.0438	2.0451	2.0464	2.0476	2.0489	2.0502	2.0515	2.0528
7.8	2.0541	2.0554	2.0566	2.0579	2.0592	2.0605	2.0617	2.0630	2.0643	2.0656
7.9	2.0668	2.0681	2.0693	2.0706	2.0719	2.0731	2.0744	2.0756	2.0769	2.0781
8.0	2.0794	2.0806	2.0819	2.0831	2.0844	2.0856	2.0869	2.0881	2.0893	2.0906
8.1	2.0918	2.0931	2.0943	2.0955	2.0967	2.0980	2.0992	2.1004	2.1016	2.1029
8.2	2.1041	2.1053	2.1065	2.1077	2.1090	2.1102	2.1114	2.1126	2.1138	2.1150
8.3	2.1162	2.1174	2.1186	2.1198	2.1210	2.1222	2.1234	2.1246	2.1258	2.1270
8.4	2.1282	2.1294	2.1306	2.1318	2.1329	2.1341	2.1353	2.1365	2.1377	2.1388
8.5	2.1400	2.1412	2.1424	2.1435	2.1447	2.1459	2.1471	2.1482	2.1494	2.1506
8.6	2.1517	2.1529	2.1540	2.1552	2.1564	2.1575	2.1587	2.1598	2.1610	2.1621
8.7	2.1633	2.1644	2.1656	2.1667	2.1679	2.1690	2.1702	2.1713	2.1724	2.1736
8.8	2.1747	2.1758	2.1770	2.1781	2.1792	2.1804	2.1815	2.1826	2.1838	2.1849
8.9	2.1860	2.1871	2.1883	2.1894	2.1905	2.1916	2.1927	2.1938	2.1950	2.1961
9.0	2.1972	2.1983	2.1994	2.2005	2.2016	2.2027	2.2038	2.2049	2.2060	2.2071
9.1	2.2082	2.2093	2.2104	2.2115	2.2126	2.2137	2.2148	2.2159	2.2170	2.2181
9.2	2.2192	2.2202	2.2213	2.2224	2.2235	2.2246	2.2257	2.2267	2.2278	2.2289
9.3	2.2300	2.2310	2.2321	2.2332	2.2343	2.2353	2.2364	2.2375	2.2385	2.2396
9.4	2.2407	2.2417	2.2428	2.2439	2.2449	2.2460	2.2470	2.2481	2.2491	2.2502
9.5	2.2512	2.2523	2.2533	2.2544	2.2554	2.2565	2.2575	2.2586	2.2596	2.2607
9.6	2.2617	2.2628	2.2638	2.2648	2.2659	2.2669	2.2679	2.2690	2.2700	2.2710
9.7	2.2721	2.2731	2.2741	2.2752	2.2762	2.2772	2.2782	2.2793	2.2803	2.2813
9.8	2.2823	2.2834	2.2844	2.2854	2.2864	2.2874	2.2884	2.2895	2.2905	2.2915
9.9	2.2925	2.2935	2.2945	2.2955	2.2965	2.2975	2.2985	2.2995	2.3005	2.3015
10.0	2.3025	2.3035	2.3045	2.3055	2.3065	2.3075	2.3085	2.3095	2.3105	2.3115
10.1	2.3125	2.3135	2.3145	2.3155	2.3164	2.3174	2.3184	2.3194	2.3204	2.3214
10.2	2.3223	2.3233	2.3243	2.3253	2.3263	2.3272	2.3282	2.3292	2.3302	2.3311
10.3	2.3321	2.3331	2.3340	2.3350	2.3360	2.3369	2.3379	2.3389	2.3398	2.3408
10.4	2.3418	2.3427	2.3437	2.3446	2.3456	2.3466	2.3475	2.3485	2.3494	2.3504
10.5	2.3513	2.3523	2.3532	2.3542	2.3551	2.3561	2.3570	2.3580	2.3589	2.3599
10.6	2.3608	2.3618	2.3627	2.3636	2.3646	2.3655	2.3665	2.3674	2.3683	2.3693
10.7	2.3702	2.3711	2.3721	2.3730	2.3739	2.3749	2.3758	2.3767	2.3776	2.3786
10.8	2.3795	2.3804	2.3814	2.3823	2.3832	2.3841	2.3850	2.3860	2.3869	2.3878
10.9	2.3887	2.3896	2.3906	2.3915	2.3924	2.3933	2.3942	2.3951	2.3960	2.3969

Table 2 **Values of e^x**

x	0.00	0.01	0.02	0.03	0.04	0.05	0.06	0.07	0.08	0.09
0.0	1.0000	1.0100	1.0202	1.0304	1.0408	1.0512	1.0618	1.0725	1.0832	1.0941
0.1	1.1051	1.1162	1.1275	1.1388	1.1502	1.1618	1.1735	1.1853	1.1972	1.2092
0.2	1.2214	1.2336	1.2460	1.2586	1.2712	1.2840	1.2969	1.3099	1.3231	1.3364
0.3	1.3498	1.3634	1.3771	1.3909	1.4049	1.4190	1.4333	1.4477	1.4622	1.4769
0.4	1.4918	1.5068	1.5219	1.5372	1.5527	1.5683	1.5840	1.5999	1.6160	1.6323
0.5	1.6487	1.6652	1.6820	1.6989	1.7160	1.7332	1.7506	1.7682	1.7860	1.8039
0.6	1.8221	1.8404	1.8589	1.8776	1.8964	1.9155	1.9347	1.9542	1.9738	1.9937
0.7	2.0137	2.0339	2.0544	2.0750	2.0959	2.1170	2.1382	2.1597	2.1814	2.2034
0.8	2.2255	2.2479	2.2705	2.2933	2.3163	2.3396	2.3631	2.3869	2.4109	2.4351
0.9	2.4596	2.4843	2.5092	2.5345	2.5599	2.5857	2.6117	2.6379	2.6644	2.6912
1.0	2.7182	2.7456	2.7731	2.8010	2.8292	2.8576	2.8863	2.9153	2.9446	2.9742
1.1	3.0041	3.0343	3.0648	3.0956	3.1267	3.1581	3.1899	3.2219	3.2543	3.2870
1.2	3.3201	3.3534	3.3871	3.4212	3.4556	3.4903	3.5254	3.5608	3.5966	3.6327
1.3	3.6693	3.7061	3.7434	3.7810	3.8190	3.8574	3.8961	3.9353	3.9749	4.0148
1.4	4.0552	4.0959	4.1371	4.1787	4.2207	4.2631	4.3059	4.3492	4.3929	4.4371
1.5	4.4816	4.5267	4.5722	4.6181	4.6645	4.7114	4.7588	4.8066	4.8549	4.9037
1.6	4.9530	5.0028	5.0530	5.1038	5.1551	5.2069	5.2593	5.3121	5.3655	5.4194
1.7	5.4739	5.5289	5.5845	5.6406	5.6973	5.7546	5.8124	5.8708	5.9298	5.9894
1.8	6.0496	6.1104	6.1718	6.2338	6.2965	6.3598	6.4237	6.4883	6.5535	6.6193
1.9	6.6858	6.7530	6.8209	6.8895	6.9587	7.0286	7.0993	7.1706	7.2427	7.3155
2.0	7.3890	7.4633	7.5383	7.6140	7.6906	7.7679	7.8459	7.9248	8.0044	8.0849
2.1	8.1661	8.2482	8.3311	8.4148	8.4994	8.5848	8.6711	8.7582	8.8463	8.9352
2.2	9.0250	9.1157	9.2073	9.2998	9.3933	9.4877	9.5830	9.6794	9.7766	9.8749
2.3	9.9741	10.074	10.175	10.277	10.381	10.485	10.590	10.697	10.804	10.913
2.4	11.023	11.133	11.245	11.358	11.473	11.588	11.704	11.822	11.941	12.061
2.5	12.182	12.304	12.428	12.553	12.679	12.807	12.935	13.065	13.197	13.329
2.6	13.463	13.599	13.735	13.873	14.013	14.154	14.296	14.439	14.585	14.731
2.7	14.879	15.029	15.180	15.332	15.486	15.642	15.799	15.958	16.119	16.281
2.8	16.444	16.609	16.776	16.945	17.115	17.287	17.461	17.637	17.814	17.993
2.9	18.174	18.356	18.541	18.727	18.915	19.105	19.297	19.491	19.687	19.885
3.0	20.085	20.287	20.491	20.697	20.905	21.115	21.327	21.541	21.758	21.977
3.1	22.197	22.421	22.646	22.873	23.103	23.336	23.570	23.807	24.046	24.288
3.2	24.532	24.779	25.028	25.279	25.533	25.790	26.049	26.311	26.575	26.842
3.3	27.112	27.385	27.660	27.938	28.219	28.502	28.789	29.078	29.370	29.665
3.4	29.964	30.265	30.569	30.876	31.186	31.500	31.816	32.136	32.459	32.785
3.5	33.115	33.448	33.784	34.123	34.466	34.813	35.163	35.516	35.873	36.234
3.6	36.598	36.966	37.337	37.712	38.091	38.474	38.861	39.251	39.646	40.044
3.7	40.447	40.853	41.264	41.679	42.097	42.521	42.948	43.380	43.816	44.256
3.8	44.701	45.150	45.604	46.062	46.525	46.993	47.465	47.942	48.424	48.910
3.9	49.402	49.898	50.400	50.906	51.418	51.935	52.457	52.984	53.517	54.054
4.0	54.598	55.146	55.701	56.260	56.826	57.397	57.974	58.556	59.145	59.739
4.1	60.340	60.946	61.559	62.177	62.802	63.434	64.071	64.715	65.365	66.022
4.2	66.686	67.356	68.033	68.717	69.407	70.105	70.809	71.521	72.240	72.966
4.3	73.699	74.440	75.188	75.944	76.707	77.478	78.257	79.043	79.838	80.640
4.4	81.450	82.269	83.096	83.931	84.774	85.626	86.487	87.356	88.234	89.121

Table 2 (*continued*)

x	**0.00**	**0.01**	**0.02**	**0.03**	**0.04**	**0.05**	**0.06**	**0.07**	**0.08**	**0.09**
4.5	90.017	90.921	91.835	92.758	93.690	94.632	95.583	96.544	97.514	98.494
4.6	99.484	100.48	101.49	102.51	103.54	104.58	105.63	106.69	107.77	108.85
4.7	109.94	111.05	112.16	113.29	114.43	115.58	116.74	117.91	119.10	120.30
4.8	121.51	122.73	123.96	125.21	126.46	127.74	129.02	130.32	131.63	132.95
4.9	134.28	135.63	137.00	138.37	139.77	141.17	142.59	144.02	145.47	146.93
5.0	148.41	149.90	151.41	152.93	154.47	156.02	157.59	159.17	160.77	162.38
5.1	164.02	165.67	167.33	169.01	170.71	172.43	174.16	175.91	177.68	179.46
5.2	181.27	183.09	184.93	186.79	188.67	190.56	192.48	194.41	196.36	198.34
5.3	200.33	202.35	204.38	206.43	208.51	210.60	212.72	214.86	217.02	219.20
5.4	221.40	223.63	225.87	228.14	230.44	232.75	235.09	237.46	239.84	242.25
5.5	244.69	247.15	249.63	252.14	254.67	257.23	259.82	262.43	265.07	267.73
5.6	270.42	273.14	275.88	278.66	281.46	284.29	287.14	290.03	292.94	295.89
5.7	298.86	301.87	304.90	307.96	311.06	314.19	317.34	320.53	323.75	327.01
5.8	330.29	333.61	336.97	340.35	343.77	347.23	350.72	354.24	357.80	361.40
5.9	365.03	368.70	372.41	376.15	379.93	383.75	387.61	391.50	395.44	399.41

Answers to Even-Numbered Exercises

Chapter 1

Section 1.1

2. (i) -1; (ii) 2; (iii) -2; (iv) 0 **4.** 15, 5, 5, 6, 50 **6.** 0, 2, $\frac{5}{8}$, 0, 112 **8.** 3620, 4000, 4099.6875, 4197.5, 4687.5 **10.** $\sqrt{84} = 9.1652$, 10, $\sqrt{99} = 9.9499$, $\sqrt{96} = 9.7979$, 0 **12.** -1.5, 0, 0, 9.9 **14.** 0.20, 0.50, 0.50, $\frac{1}{101} = 0.9901$ **16.** -0.8, -1, 1, $\frac{20}{101} = 0.19802$ **18.** 0, 1, $\sqrt{3} = 1.73205$, $\sqrt{12} = 3.4641$ **20.** 0, 1, $3^{3/2} = 5.1962$, $12^{3/2} = 41.5692$ **22.** (i) $C(x) = 90 + 0.15x$; (ii) $C(550) = 172.50$, $C(800) = 210.00$ **24.** $V(s) = 3s^2$ cu. in. **26.** $C(s) = 373.33s$ **28.** (i) -600; (ii) -22.5; (iii) 160; (iv) 2400 **30.** See Figure E.1 **32.** 2.5 **34.** 0 **36.** positive **38.** $x = -1$ **40.** $-2 \leq x \leq 3$ **42.** 50 **44.** $A = 125$ **46.** See Figure E.2 **48.** See Figure E.3

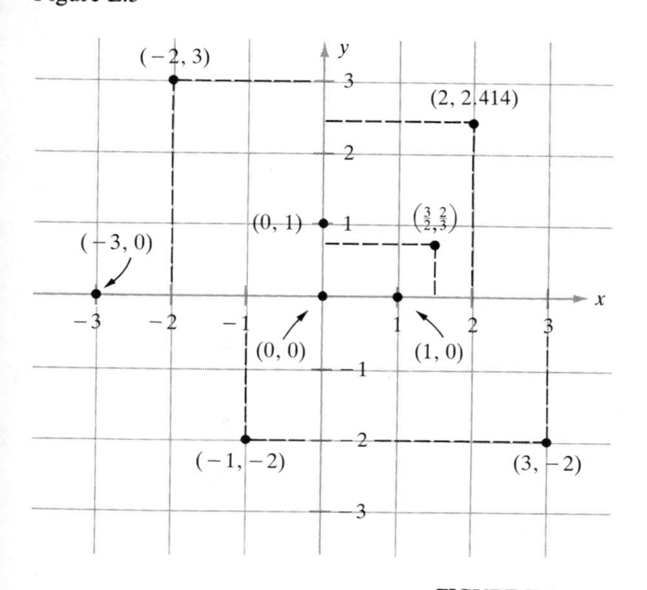

FIGURE E.1

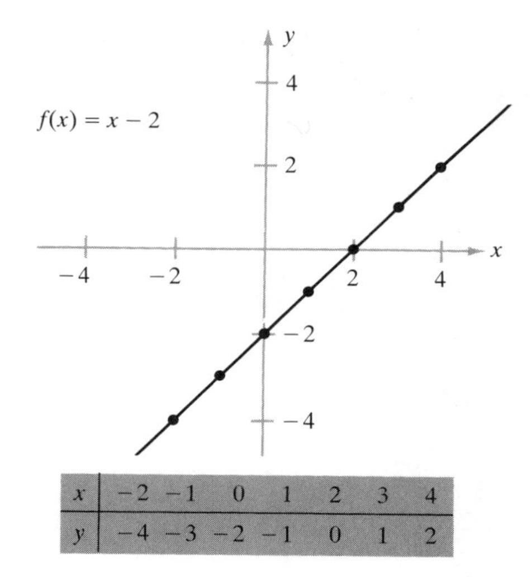

x	−2	−1	0	1	2	3	4
y	−4	−3	−2	−1	0	1	2

FIGURE E.2

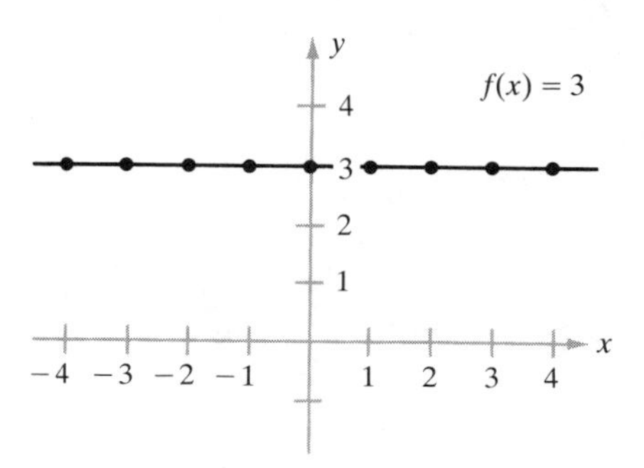

FIGURE E.3

50. See Figure E.4 **52.** See Figure E.5 **54.** See Figure E.6 **56.** See Figure E.7 **58.** (i) $x \neq 1$; (ii) $x \neq -3$; (iii) $x \neq 1$; (iv) $x > 0$; (v) $4 - x > 0$ or $x < 4$; (vi) $x \neq 0$ and $x \neq -1$ **60.** 1971; one

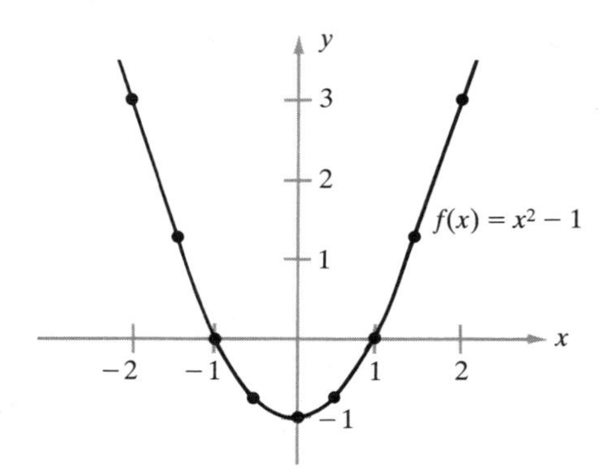

x	0	±0.5	±1.0	±1.5	±2.0
y	−1.0	−0.75	0	1.25	3.0

FIGURE E.4

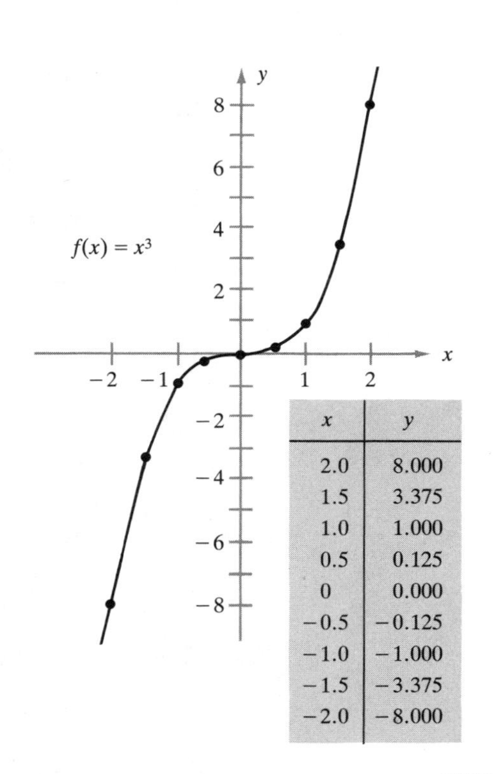

x	y
2.0	8.000
1.5	3.375
1.0	1.000
0.5	0.125
0	0.000
−0.5	−0.125
−1.0	−1.000
−1.5	−3.375
−2.0	−8.000

FIGURE E.5

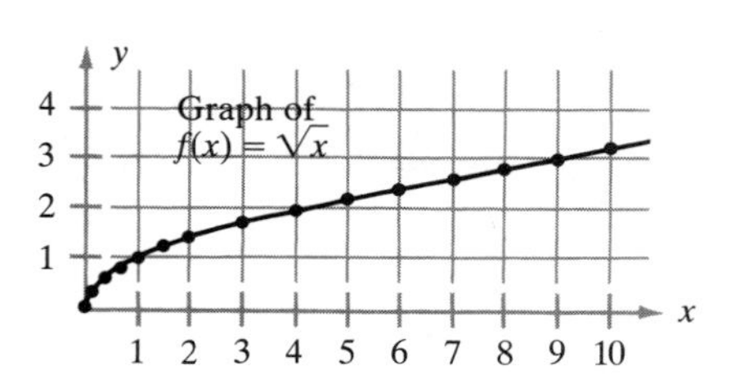

FIGURE E.6

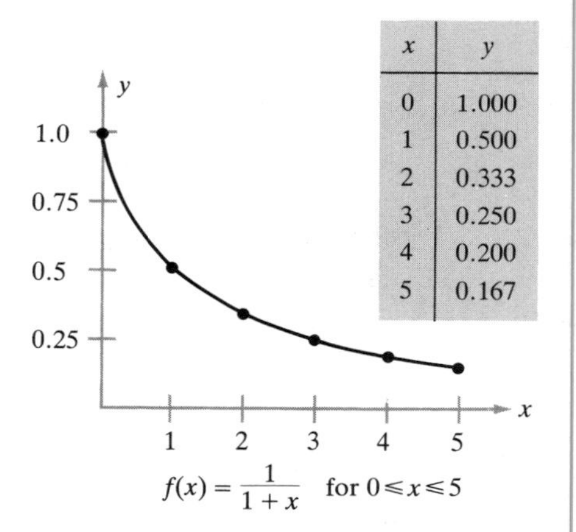

x	y
0	1.000
1	0.500
2	0.333
3	0.250
4	0.200
5	0.167

FIGURE E.7

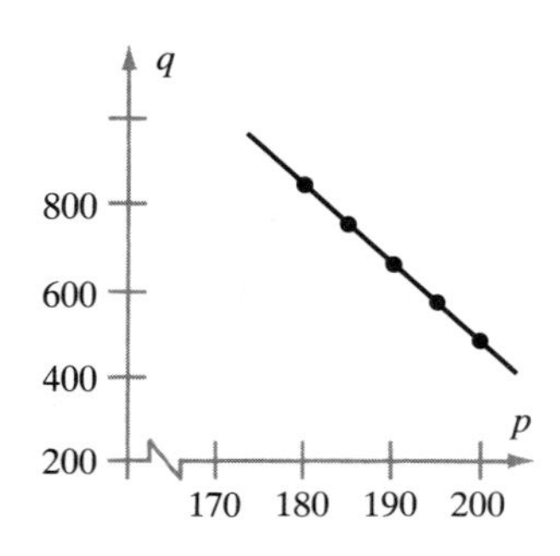

FIGURE E.8

1950 dollar bought as much as \$1.50 in 1971, or, one 1971 dollar $= \frac{1}{1.50} =$ \$0.667 in 1950 dollars.
62. When $I = 0.5$: $t = 1915, 1930, 1934, 1938–1940$.

Section 1.2

2. -3 **4.** 0 **6.** 3 **8.** $y = -\frac{3}{4}x + \frac{5}{4}$, slope $= -\frac{3}{4}$
10. $y = x + 1$ **12.** $y = -\frac{1}{2}x + \frac{9}{2}$
14. $y = -\frac{1}{3}x + 5$ **16.** $C(q) = 55q + 9500$
18. Value $= 1500t + 55{,}000$ after t years
20. Collinear along line $q = -18p + 4100$, $180 \leq p \leq 200$, see Figure E.8 **22.** (200, 0), (0, 100), slope $= -\frac{1}{2}$ **24.** Slope $= -\frac{1}{2}$, see Figure E.9
26. Slope $= 4$, see Figure E.10 **28.** Slope $= \frac{3}{2}$, see Figure E.11 **30.** Slope $= -\frac{5}{6} = -0.8333$, see Figure E.12 **32.** Points $(0, -C/B)$ and $(-C/A, 0)$ are on graph.

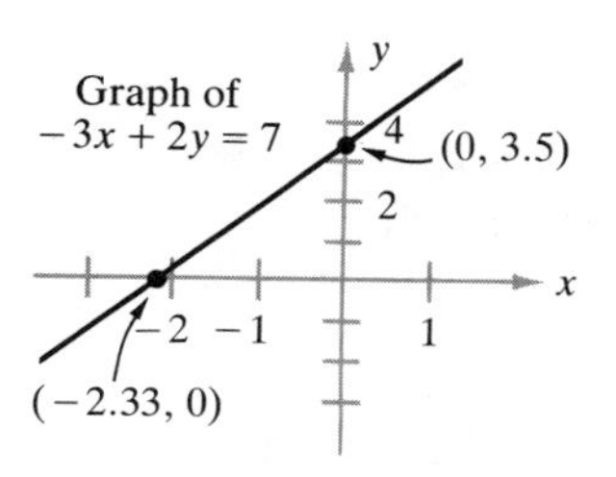

FIGURE E.11

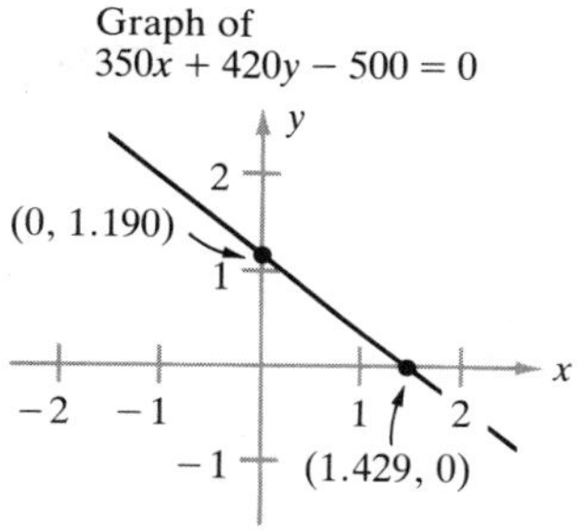

Graph of
$350x + 420y - 500 = 0$

(0, 1.190)

(1.429, 0)

FIGURE E.12

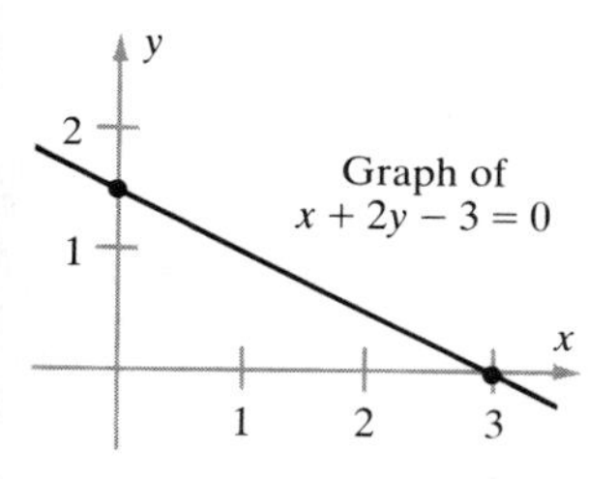

FIGURE E.9

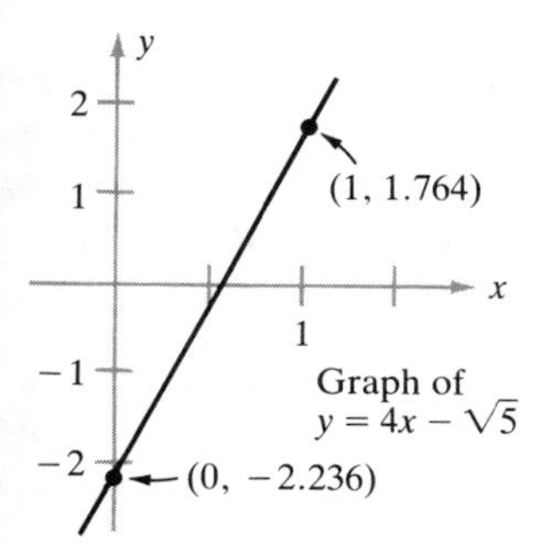

FIGURE E.10

Section 1.3

2. $C = 437 + 5.70x$ **4.** $C(147) = 1274.90$, $R(147) = 1614.06$, $P(147) = 339.16$ **6.** $C(q) = 750{,}000 + 29{,}000q$, $C(50) = 2{,}200{,}000$ **8.** $C(x) = 10{,}000 + 60x$, $C(0) = 10{,}000$; $C(10) = 10{,}600$, $C(500) = 40{,}000$, $C(1000) = 70{,}000$; $C = 19{,}000$ if $x = 150$ **10.** $C = 14$ if $0 \leq x \leq 200$, $C = 14 + 0.17(x - 200)$ if $x > 200$; $C(380) = \$44.60$
12. See Figure E.13 **14.** $C = 5000 + 27x$, $R = 37x$, $P = -5000 + 10x$ **16.** $A = 27 + (5000/x)$

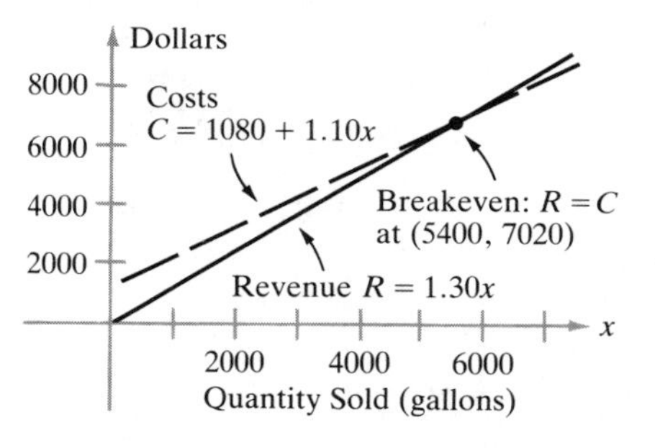

FIGURE E.13

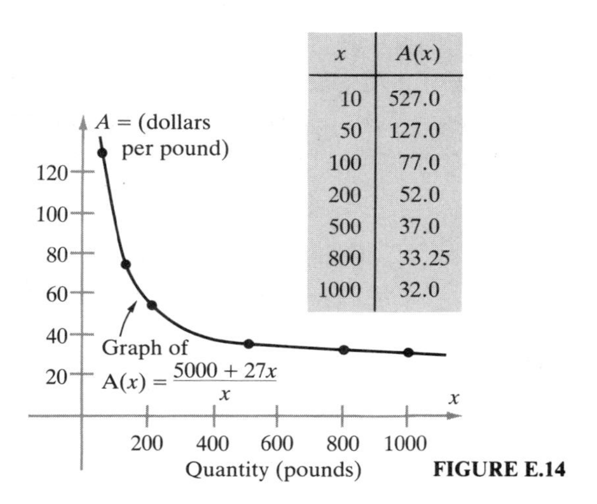

x	$A(x)$
10	527.0
50	127.0
100	77.0
200	52.0
500	37.0
800	33.25
1000	32.0

FIGURE E.14

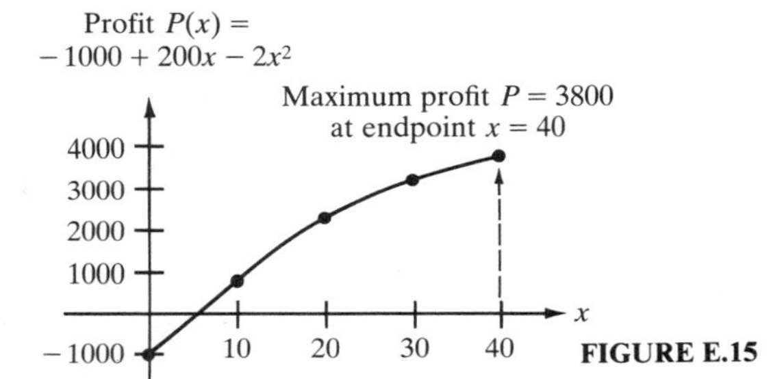

FIGURE E.15

18. A large if x small, A approaches 27 with $A > 27$ as x gets large, see Figure E.14 **20.** \$25 **22.** \$26 **24.** (i) \$1.3997; (ii) \$1.00; (iii) 2781.10 gal. **26.** (i) $C = 1500 + 35q$; (ii) $R = 48q - 0.01q^2$; (iii) $P = -1500 + 13q - 0.01q^2$; $R(100) = 4700$, $R(300) = 13{,}500$, $R(800) = 32{,}000$, $R(1000) = 38{,}000$; $P(100) = -300$, $P(300) = 1500$, $P(800) = 2500$, $P(1000) = 1500$ **28.** Max. $P = 3800$ at $x = 40$, see Figure E.15 **30.** $R = 1.50x - 0.00002x^2$, $P = -1080 + 0.40x - 0.00002x^2$; Max. P when $x = 10{,}000$; price = \$1.30 per gallon; see Figure E.16 **32.** See Figure E.17 **34.** \$142.50

Section 1.4

2. $P = 22{,}000{,}000(2^{t/50})$; in 1995, $P = 27.085$ million; in 2020, $P = 38.304$ million
4. $P = 45{,}000(2^{t/100})$; in 1975, $P = 53{,}514$; in 1990, $P = 59{,}378$ **6.** $N = 500(2^{t/5})$; $N(7.5) = 1414$, $N(10) = 2000$, $N(12) = 2639$, $N(24) = 13{,}929$
8. $N = 1000(2^t)$; $N(8) = 256{,}000$, $N(24) = 16.78 \times 10^9$, $N(48) = 2.815 \times 10^{17}$

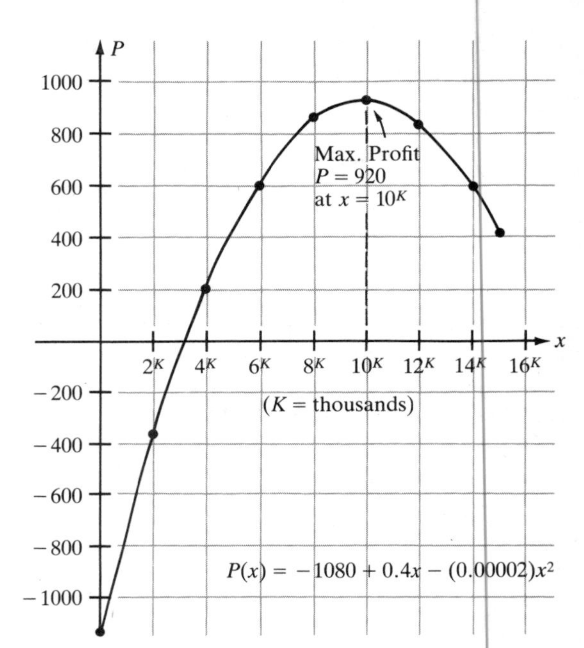

FIGURE E.16

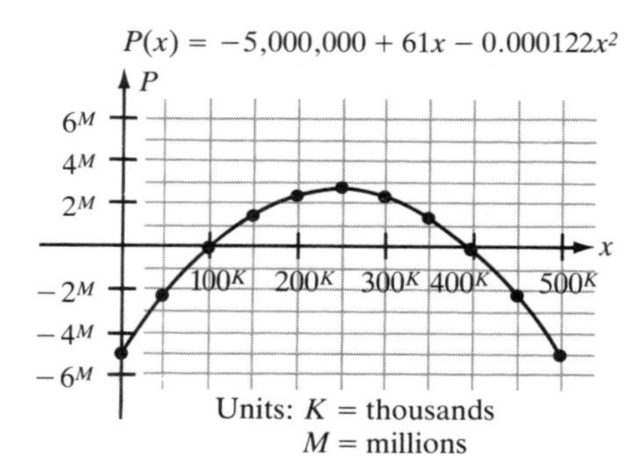

FIGURE E.17

10. $Q = 5(2^{-t/1600})$; $Q(800) = 3.5355$, $Q(2000) = 2.1022$ **12.** $Q(t) = 10(2^{-t/1.28})$; $Q(5) = 0.6670$, $Q(5)/Q(0) = 0.06670$ (or, 6.67% remains)
14. 2500, 7071 **16.** 100 **18.** $N(0) = 100$, $N(1) = 181.8$, $N(2) = 307.7$, $N(3) = 470.6$, $N(4) = 640$, $N(5) = 780.5$, $N(6) = 876.7$, $N(7) = 934.3$, $N(8) = 966.0$; see Figure E.18

Section 1.5

2. $(x, y) = (-\frac{15}{7}, \frac{10}{7}) = (-2.1429, 1.4286)$
4. $(x, y) = (1.7730, 9.7199)$ **6.** $(x, y) = (37.5, 212.5)$
8. $(q, p) = (6.4615, 1.3308)$ **10.** $(q, p) = (1.2, 1.15)$

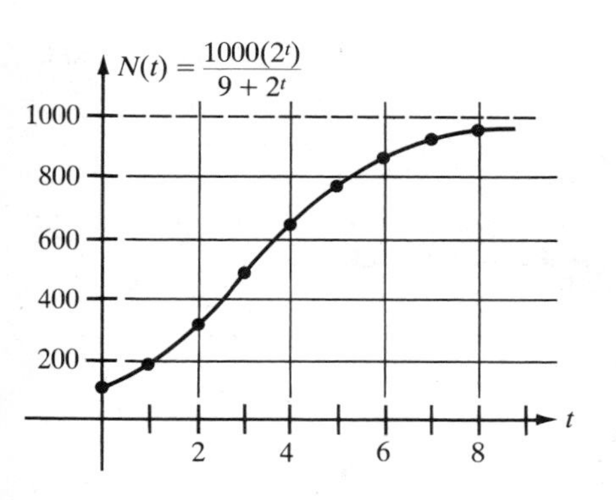

t	$N(t)$
0	100.0
1	181.8
2	307.7
3	470.6
4	640.0
5	780.0
6	876.7
7	934.3
8	966.0

FIGURE E.18

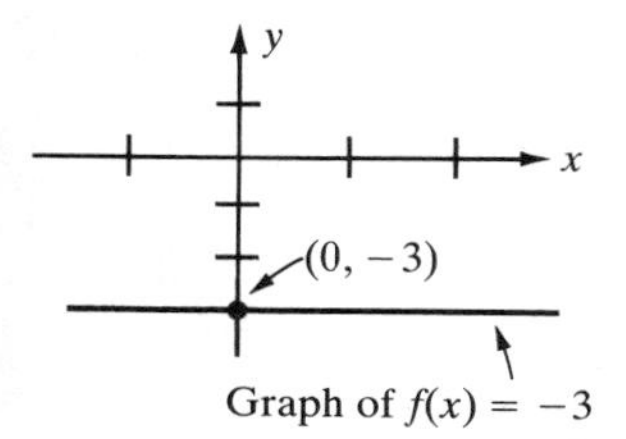

FIGURE E.19

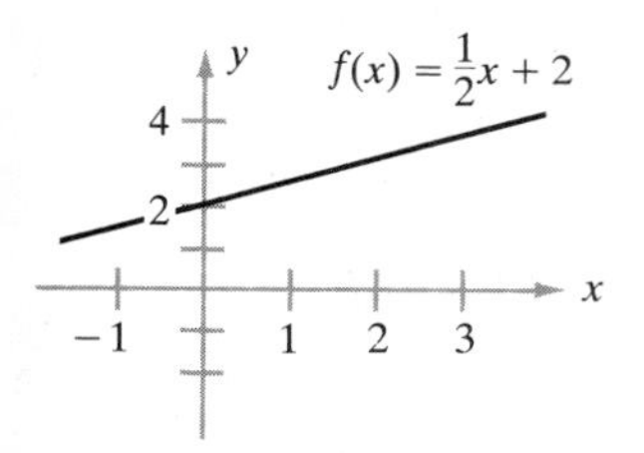

FIGURE E.20

12. No roots **14.** $x = \pm\frac{1}{20}$ **16.** No roots **18.** No roots **20.** No intersection: $B^2 - 4AC < 0$ **22.** (0, 1) **24.** $(1 - \sqrt{17}, 9 - 2\sqrt{17}) = (-3.1231, 0.7538)$ and $(1 + \sqrt{17}, 9 + 2\sqrt{17}) = (5.1231, 17.2462)$ **26.** No intersection **28.** 25.36 years (the other root 94.64 is out of range where the formula is valid) **30.** $R = 110x$, $P = -50{,}000 + 10x + 0.1x^2$, breakeven at $x = 658.87$ **32.** $(3 \pm \sqrt{5})/2$ **34.** 0, 2, −2 **36.** 1, −1 **38.** 0, 2 **40.** −1 **42.** $(x - (1 + \sqrt{3}))(x - (1 - \sqrt{3}))$ **44.** $(x - \sqrt{2})(x + \sqrt{2})$ **46.** $2(x - \frac{1}{2})(x + 3)$ **48.** No factorization

Section 1.6

2. (0, −3), see Figure E.19 **4.** (0, 2), see Figure E.20 **6.** See Figure E.21 **8.** See Figure E.22

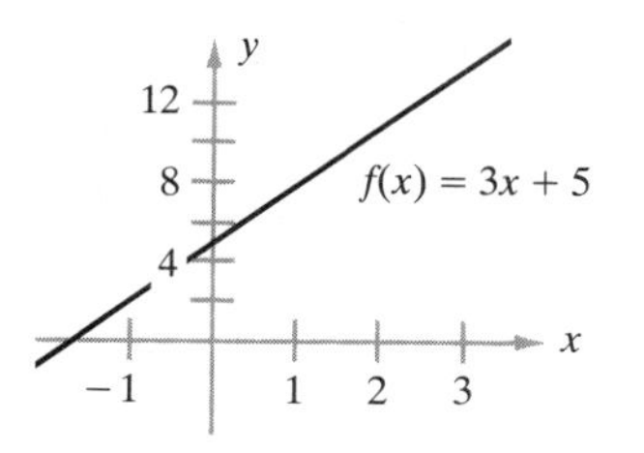

FIGURE E.21

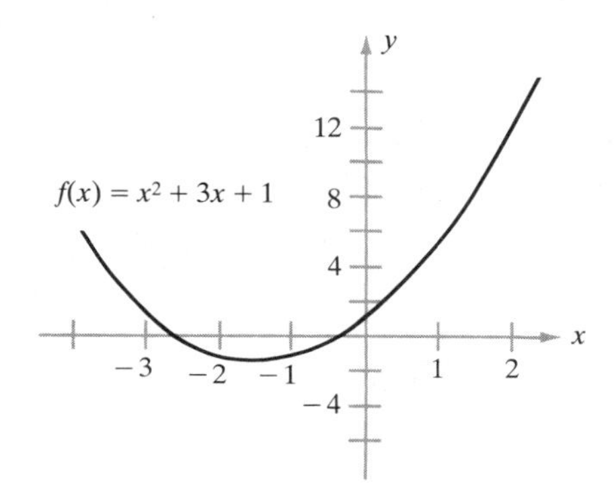

FIGURE E.22

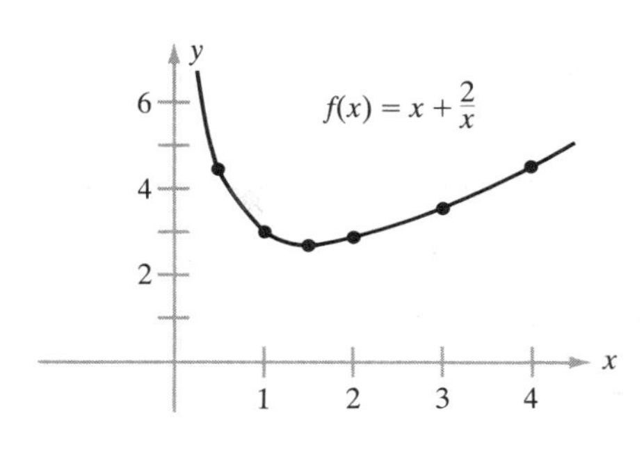

x	0.5	1.0	1.5	2.0	3.0	4.0
$f(x)$	4.50	3.00	2.83	3.00	3.67	4.50

FIGURE E.23

10. See Figure E.23 **12.** $a = \frac{10}{3}$ **14.** $c = 1$ **16.** $y = x + 1$ **18.** $y = \frac{1}{2}x + 2$ **20.** 1, −3 **22.** $0, (-1 + \sqrt{5})/2, (-1 - \sqrt{5})/2$ **24.** −1, 2 **26.** $C = 115{,}000 + 13{,}000x$; $C(40) = 635{,}000$ **28.** $P = -3500 + 24x - 0.02x^2$

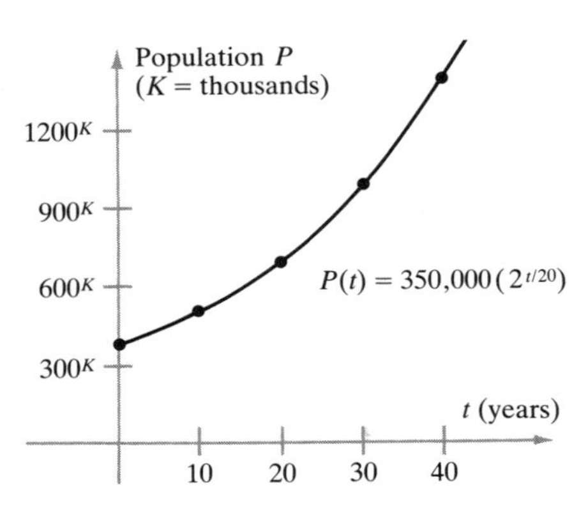

t	0	10	20	30	40
$P(t)$	350K	509K	700K	1018K	1400K

FIGURE E.24

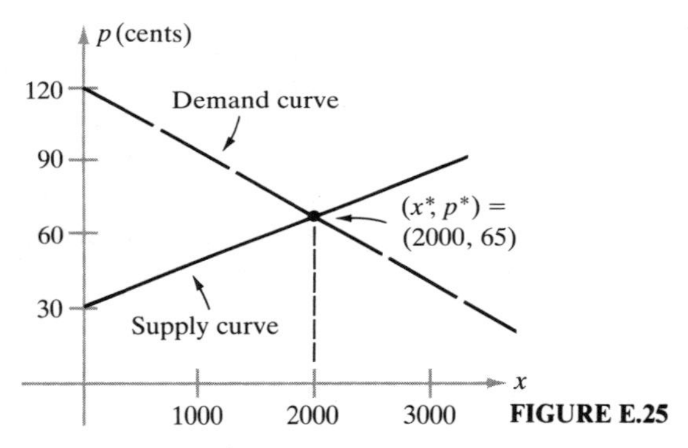

FIGURE E.25

30. $P(t) = 350{,}000(2^{t/20})$, see Figure E.24 **32.** $(\frac{5}{6} + \frac{1}{6}\sqrt{37}, \frac{7}{6} + \frac{5}{6}\sqrt{37}) = (1.8471, 6.2356)$ and $(\frac{5}{6} - \frac{1}{6}\sqrt{37}, \frac{7}{6} - \frac{5}{6}\sqrt{37}) = (-0.1805, -3.9023)$ **34.** $(x, p) = (2000, 65)$, see Figure E.25

Chapter 2

Section 2.1

2. 22.5 mi./gallon **4.** In dollars per year: (i) 18; (ii) 27; (iii) 22.5; (iv) 3.5 **6.** (i) 0; (ii) 2; (iii) 2; (iv) $3(\Delta x) + (\Delta x)^2$ **8.** (*i*) $-\frac{1}{6}$; (ii) $\frac{-1}{10}$; (iii) $\frac{1}{2}$; (iv) $\frac{1}{6}$ **10.** 4 **12.** 2 **14.** -9 **16.** $-\frac{1}{2}$ **18.** 0, 2 **20.** 64 ft./sec. **22.** (i) 10; (ii) $\sqrt{10} = 3.1623$; (iii) 1.310; (iv) 1.005; sensory overload perhaps **24.** $x_2 = 3$

Section 2.2

2. $\Delta y/\Delta x = 4 + \Delta x$; values are 4.01, 4.001, 4.0001 when $\Delta x = 0.01, 0.001, 0.0001$ **4.** Inst. rate = $2x - 2$ at base point x **6.** If $y = f(x)$ is constant, $\Delta y = 0$ for any base point x and any Δx, so $\Delta y/\Delta x = 0$ and so is its limit value. **8.** Inst. rate at x is $4x - 3$; value is 5 if $x = 2$, -15 if $x = -3$; value is 4 when $4x - 3 = 4$ (or $x = \frac{7}{4}$); value is zero if $4x - 3 = 0$ (or $x = \frac{3}{4}$) **10.** -3 (const.) **12.** $2x - 1$ **14.** $1 + 3x^2$ **16.** $3x^2$ **18.** $14x - \frac{1}{2}$ **20.** $-2 + 0.2x$ **22.** Inst. rate at x is $100 - 0.2x$; rate is \$50 per unit if $x = 250$

Section 2.3

2. $7x^6$ **4.** $12x^{11}$ **6.** $\frac{1}{3}x^{-2/3}$ **8.** $-\frac{1}{2}x^{-3/2}$ **10.** $118x^{117}$ **12.** $(3.2)x^{2.2}$ **14.** $\frac{11}{4}x^{7/4}$ **16.** $\frac{2}{27} = 0.07407$ **18.** $\frac{3}{4}$ **20.** $-x^{-1/2}$ **22.** $3x^8$ **24.** $12x^2 - \frac{1}{2}x^{-1/2}$ **26.** $-4x^3$ **28.** -26 **30.** 1 **32.** $1 + 2x + 3x^2$ **34.** $2x - e^x$ **36.** $5/x$ **38.** $3e^x - (4/x)$ **40.** $(3/x) - 5e^x + 7$ **42.** $50t^{-1/2} + 14t - 3t^2$ **44.** $0.3 - 0.002q$ **46.** $4\pi r^2$ **48.** $dP/dq = 61 - 0.000196q$; zero if $q = 311{,}244.5$ **50.** $N'(t) = 300 + 36t$; $N'(10) = 660$ **52.** $P = R - C$ so $dP/dx = dR/dx - dC/dx$, and $dP/dx = 0$ when $dR/dx = dC/dx$

Section 2.4

2. 3 **4.** 0 **6.** 3 **8.** -5 **10.** $y = 0$ **12.** $y = 2$ **14.** $y = 4x - 8$ **16.** $dy/dx = 2x - 12$; $dy/dx = 0$ if $x = 6$ **18.** $P_1 = (2, -3)$ and $P_2 = (-2, 13)$ **20.** $y = 2 - x$

Section 2.5

2. $N'(10) = 660$, $N'(120) = 4620$ **4.** $P'(t) = 475 - 20t$; (i) $P(10) = 7250$; (ii) $P'(10) = 275$; (iii) $t = 15$ **6.** $q(4) = 5.35$ mg., $q'(4) = -0.33125$ mg./hr. **8.** $N(10) = 20{,}000$; $N'(10) = 5000$ cases/day; $N'(25) = -6250$ cases/day **10.** $t = 1.5$ sec., $v = 48$ ft./sec. **12.** (i) $h'(1) = 32$, $h'(2) = 0$, $h'(3) = -32$ ft./sec. Falls if $h' < 0$; (ii) $t = 4$ sec.; (iii) impact: $h'(4) = -64$ ft./sec. **14.** $t = \sqrt{s/2.66}$ sec. **16.** $C'(1000) = 18$; $\Delta C = C(1001) - C(1000) = 17.993$ **18.** $dP/dx = 5 - 0.008x$; $P'(450) = \$1.40$ per unit – increase x; $P'(750) = \$-1.00$ per unit – decrease x **20.** (i) $dC/dx = 100 - 2x$; (ii) $P = -100 + 100x - x^2$; (iii) $dP/dx = 100 - 2x$ **22.** $C'(30) = \frac{20}{9} = 2.222$; $\Delta C \approx C'(30) \cdot (0.2) = 0.444$.

Section 2.6

2. $\Delta W \approx W'(30) \cdot \Delta t = 4(7) = 28$ gms.; $(\Delta W/W) \times 100\% \approx (28/1500) \times 100\% = 1.87\%$ **4.** $\Delta f \approx f'(3) \cdot \Delta x = 6(\Delta x)$; $\Delta f \approx 0.6$; $\Delta f \approx -1.2$
6. $\Delta f \approx f'(100) \cdot \Delta x = 70(\Delta x)$; $\Delta f \approx 7$; $\Delta f \approx -14$
8. $f(x) \approx f(1) + f'(1)(x-1) = 4x - 2$; $f(0.95) \approx 1.80$ (Δx is -0.05) **10.** $f(x) \approx f(-1) + f'(-1)(x+1) = -x$; $f(-1.1) \approx 1.1$ (Δx is -0.1)
12. $\Delta P \approx P'(500) \cdot \Delta x = 20(10) = \$200$
14. $\Delta V \approx V'(3950) \cdot \Delta x = (1.96 \times 10^8)(10) = 1.96 \times 10^9$ cu. mi. **16.** $\Delta C \approx C'(200) \cdot \Delta x = (57.8)(15) = \$867$ **18.** $\Delta A \approx A'(500) \cdot \Delta x = 1000(\pm 1.2) = \pm 1200$ sq. ft. **20.** $f(x) \approx f(1) + f'(1)(\Delta x) = 1 - \Delta x$; (i) 0.89205; (ii) 1.034; (iii) $1 - h$
22. $\Delta T \approx T(0) + T'(0) \cdot \Delta t = 680 - 25(2) = 630°$F **24.** $\Delta v \approx v'(0.01) \cdot \Delta r = 400(\Delta r) = \pm 0.4$ cm./sec.; $v \approx 2.0$ cm./sec.

Section 2.7

2. 0 **4.** 1 **6.** 5 **8.** 4 **10.** 0 **12.** $-\frac{1}{2}$ **14.** 0
16. $1/\sqrt{2}$ **18.** 0 **20.** -1 **22.** 0 **24.** 0
26. a, c, e, f **28.** Defined and cnts. except at $x = 1$, $x = -1$ **30.** Defined and cnts. except at $x = -1$
32. Defined and cnts. all x **34.** Discontinuous, see Figure E.26

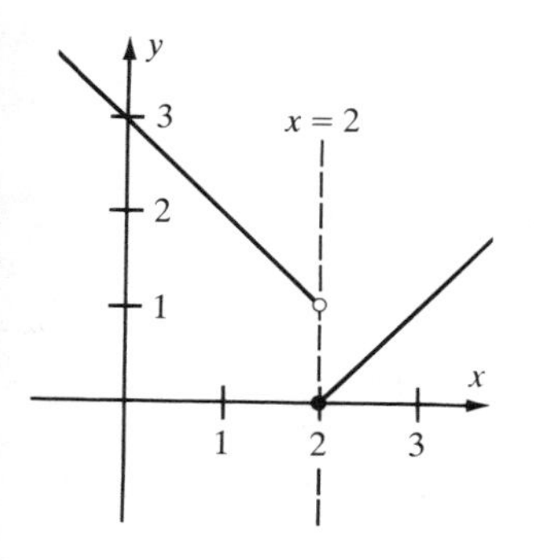

FIGURE E.26

Section 2.8

2. $\Delta y/\Delta x = -4$; $f'(1) = -4$ **4.** $\Delta y/\Delta x = -16$; $f'(1) = -14$ **6.** $\Delta y/\Delta x = \frac{25}{4}$; $f'(1) = 1$
8. $dC/dx = 72 - 0.02x$; $\Delta C/\Delta x = 51$
10. $-1 + 3x^2$ **12.** $24x^3 - 4x^2 + \frac{5}{2}x$
14. $f'(1) = 8$ **16.** $\frac{23}{8} = 2.875$ **18.** $-\frac{1}{2}x^{-3/2} + 10x^{1/4}$ **20.** $10x + 27$ **22.** $y = 1 + x$ **24.** $y = 7 - 6x$ **26.** $P = (3, 6)$ **28.** $P(4) = 65{,}518$, $P'(4) = 65{,}518$ **30.** $\Delta f \approx 0.06$ **32.** $\Delta f \approx 0.0222$
34. $f(1.03) \approx 6.94$ **36.** 5 **38.** 2

Chapter 3

Section 3.1

2. $4x^3$ **4.** $(4x^3 + 2x)(14x^3 - 3x + 2) + (x^4 + x^2 + 1)(42x^2 - 3) = 98x^6 + 55x^4 + 8x^3 + 33x^2 + 4x - 3$ **6.** $(1 - x^{-2})(x^{1/2} + 1) + (x + x^{-1})(\frac{1}{2}x^{-1/2}) = \frac{3}{2}x^{1/2} + 1 - \frac{1}{2}x^{-3/2} - x^{-2}$
8. $\frac{1}{3}x^{-2/3}(x^4 + 3x^2 + 1) + x^{1/3}(4x^3 + 6x) = \frac{13}{3}x^{10/3} + 7x^{4/3} + \frac{1}{3}x^{-2/3}$ **10.** $(2t + t^2)e^t$
12. $(2 + 3x + x^2)e^x$ **14.** $-1/x^2$ **16.** $2/(1-x)^2$
18. $45/(100 - 45x)^2$ **20.** $(-4x^2 - 2x + 5)/(x^2 - x + 1)^2$ **22.** $1/x^{1/2}(1 - x^{1/2})^2$
24. $(-3x^4 + 1)/(x^4 + 5x + 1)^2$ **26.** $f'(1) = -\frac{13}{36}$
28. $f'(1) = -\frac{1}{2}$ **30.** slope $y'(1) = \frac{1}{2}$, tangent $y = \frac{1}{2}x - \frac{1}{2}$ **32.** $dR/dx = p + x(dp/dx)$, so if $dp/dx = -p/x$, then $dR/dx = 0$ **34.** $g = xf(x)$, $g'(x) = f(x) + xf'(x)$, $g'(0) = f(0) + 0 \cdot f'(0) = f(0)$

Section 3.2

2. $1/(x^2 + 1)$ **4.** $(1 + x)^{1/2}(1 - x)^{-1/2}$ **6.** $x^2 + 2 + (x^2 + 1)^{1/2}$ **8.** $f(u) = u^5$, $u = 7 - x$
10. $f(u) = u^{1/3}$, $u = x^2 - x + 2$ **12.** $f(u) = u^{1/2}$, $u = 1 + \sqrt{x}$ **14.** $f(u) = u^{-3}$, $u = 5 - 4x$
16. (Exercise 7) $45(x^2 - x + 1)^{44}(2x - 1)$; **(Exercise 8)** $5(7 - x)^4(-1)$; **(Exercise 9)** $\frac{1}{2}(1 - x^2)^{-1/2}(-2x)$; **(Exercise 10)** $\frac{1}{3}(x^2 - x + 2)^{-2/3}(2x - 1)$; **(Exercise 11)** $-\frac{1}{2}(x^2 + x + 3)^{-3/2}(2x + 1)$; **(Exercise 12)** $\frac{1}{4}x^{-1/2}(1 + \sqrt{x})^{-1/2}$; **(Exercise 13)** $-e^{-x}$; **(Exercise 14)** $-2x/(1 - x^2)$; **(Exercise 15)** $12(5 - 4x)^{-4}$
18. $x/(x^2 + 1)^{1/2}$ **20.** $-10(1 + 2x)^{-6}$
22. $-x(x^2 + 1)^{-3/2}$ **24.** $(1 - 2x^2)/(1 - x^2)^{1/2}$
26. $(-x^3 - \frac{3}{2}x^2 - 3x - \frac{1}{2})(x^2 + x + 1)^{-3/2}$
28. $5e^{5x}$ **30.** $-(1 + x)^{-2}[(1 - x)/(1 + x)]^{-1/2}$
32. $(x^3 + 2x^2 + 1)/(1 + x^2)^{1/2}(x + 1)^2$
34. $dA/dr = 600(1 + r)^5$ **36.** $A'(t) = -0.0886e^{-0.0886t}$ mg./day; $A'(0) = -0.0886$, $A'(7) = -0.04765$ mg./day **38.** $R(3) = 84.27$, $R(5) = 32.16$ **40.** $R'(3) = -39.65$, decreasing

Section 3.3

2. $P'(40{,}000) = 0.04$ (increase x), $P'(70{,}000) = -0.02$ (decrease x) **4.** $x = \frac{1}{3}$, $x = 1$ **6.** $-\frac{1}{2}$

8. -1 **10.** 25,500 **12.** $f' = 3x^2 - 4x + 1$; f increasing in each case **14.** $dP/dq = 5 - 0.008q$; P increases if $q < 625$, decreases if $q > 625$, max. P is $P(625) = 562.50$ **16.** Increasing on $(-\frac{1}{2}, +\infty)$, decreasing on $(-\infty, -\frac{1}{2})$ **18.** Increasing on $(0, +\infty)$, decreasing on $(-\infty, 0)$ **20.** Increasing for all x **22.** Increasing for $q < 25{,}500$, decreasing for $q > 25{,}500$; **24.** $A' = 0.1 - (5500/x^2)$; $A' = 0$ at $x = \pm\sqrt{55{,}000} = \pm 234.52$; for $x > 0$, A increases if $x > 234.52$, decreases if $0 < x < 234.52$

Section 3.4

2. $\frac{7}{6}$ **4.** 1 **6.** $x = 0, 1, 3$ **8.** $x = -1, 1$ **10.** (i) $x = 0, 5, 10$; (ii) $x = 3, 7$; (iii) $x = 0$; (iv) $x = 3$; max. $= 3$, min. $= -2$ **12.** Critical: $x = -1, 0, 4$; max. 131 at $x = 4$, min. -114 at $x = -3$ **14.** Max. $\frac{3}{2}$ at $x = 1$, min. -3 at $x = -2$ **16.** Critical: $x = 1$; max. $\frac{10}{3} = 3.3333$ at $x = \frac{1}{3}$, min. 2 at $x = 1$ **18.** No critical x; max. 0 at $x = 0$, min. -14 at $x = -2$ **20.** Max. 7 at $x = 5$, min. 1 at $x = 2$ **22.** Critical: $x = 2$; max. 20 at $x = 4$, min. 12 at $x = 2$ **24.** Critical: $t = 1$; max. 2 at $t = 1$, min. 0 at $t = 0$ **26.** See Figure E.27; (i) no. abs. extrema; (ii) abs. min. 0 at $x = 0$, no abs. max.; (iii) no abs. extrema **28.** Critical $q = 311{,}224.5$; (i) max. $= 4{,}490{,}000$; (ii) max. $= 4{,}125{,}000$ (at right endpoint) **30.** Critical $x = 5625$ now feasible; max. $P = \$20{,}625$ at $x = \$5625$

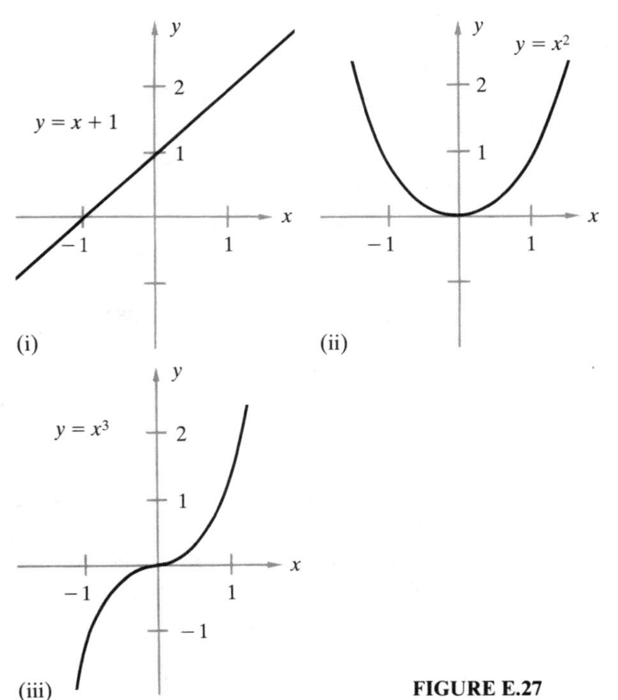

FIGURE E.27

Section 3.5

2. $y'' = 1$ **4.** $12x^2 - 30$ **6.** $-\frac{1}{4}x^{-3/2} - \frac{3}{4}x^{-5/2}$ **8.** $(x - 2)e^{-x}$ **10.** (i) $-\frac{1}{4}x^{-3/2}$; (ii) $-1/(1 + x^2)^{3/2}$; (iii) $(2x^3 + 3x)/(x^2 + 1)^{3/2}$ **12.** (i) 6, 1, 4; (ii) 1, $\frac{1}{2}$, $-\frac{1}{4}$; (iii) -2, 4, 14 **14.** Infl. at $x = 0$, $y'' = 6x$; see Figure E.28 **16.** Infl. at $x = 2$, $y'' = (x - 2)e^{-x}$; see Figure E.29 **18.** (i) $N' = 3000t/(1 + t^2)^2$; (ii) $N'' = (3000 - 9000t^2)/(1 + t^2)^3$; $N'' = 0$ if $t = 1/\sqrt{3} = 0.5774$ weeks **20.** (i) $720x^7 + 840x^3 - 6$; (ii) $6x^{-4}$; (iii) $(3 - x)e^{-x}$

y = x³ − 12x + 2
Signs of d²y/dx²
Inflection point
− − − − − − + + + +
x
Concave down
x = 0
Concave up

FIGURE E.28

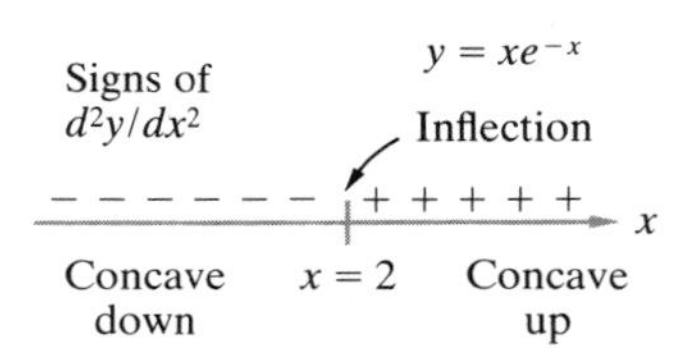

FIGURE E.29

Section 3.6

2. $y' < 0$, $y'' > 0$, y' never zero **4.** No critical x **6.** Critical: $x = 0, -4$; $y''(0) = 12$, loc. min.; $y''(-4) = -2$, loc. max. **8.** Critical: $x = -1, 0, 1$; $y''(-1) = 8$, loc. min.; $y''(0) = -4$, loc. max.; $y''(1) = 8$, loc. min. **10.** Loc. min. at $x = 7$ **12.** Loc. min. at $x = 0$, loc. max. at $x = 2$ **14.** Loc. max. at $x = -1/\sqrt{3}$, loc. min. at $x = 1/\sqrt{3}$ **16.** Loc. min. at $x = 1$ **18.** See Figure E.30 **20.** See Figure E.31 **22.** See Figure E.32 **24.** Abs. min. -36 at $x = 3$ **26.** Abs. min. 2 at $x = 1$ **28.** Abs. max. $1/e = 0.3679$ at $x = 1$

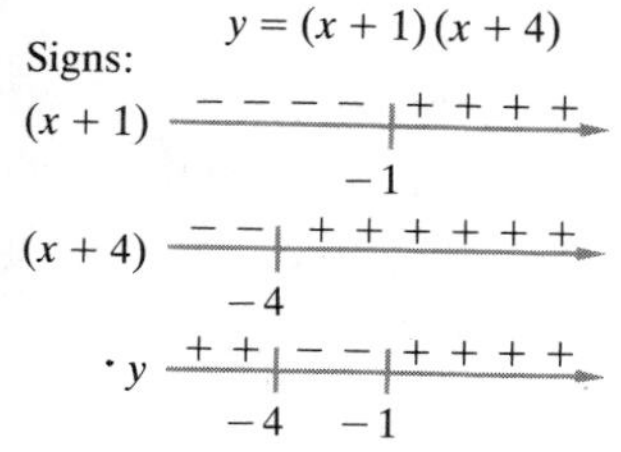

$y = (x + 1)(x + 4)$

Signs:

$(x + 1)$ − − − − | + + + + (at −1)

$(x + 4)$ − − | + + + + + + + (at −4)

y + + | − − | + + + + (at −4, −1)

FIGURE E.30

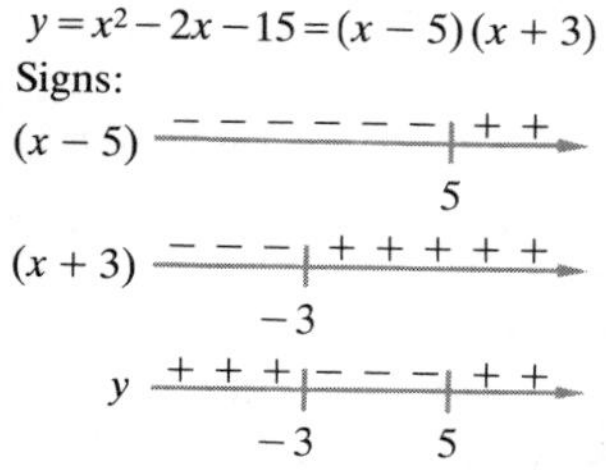

$y = x^2 - 2x - 15 = (x - 5)(x + 3)$

Signs:

$(x - 5)$ − − − − − − − | + + (at 5)

$(x + 3)$ − − − | + + + + + + (at −3)

y + + + | − − − | + + (at −3, 5)

FIGURE E.31

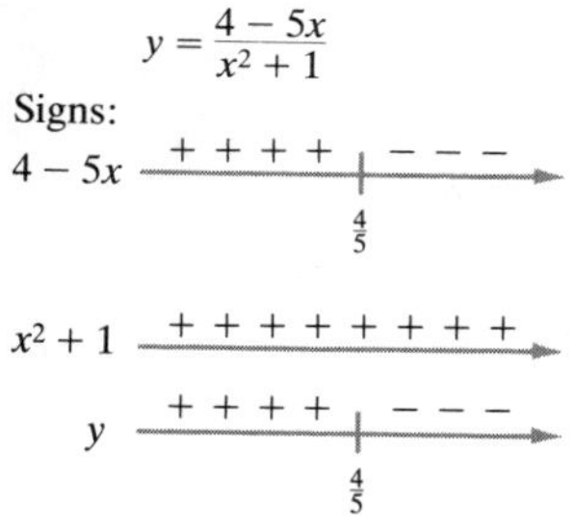

$y = \dfrac{4 - 5x}{x^2 + 1}$

Signs:

$4 - 5x$ + + + + | − − − (at $\frac{4}{5}$)

$x^2 + 1$ + + + + + + + + +

y + + + + | − − − (at $\frac{4}{5}$)

FIGURE E.32

30. $y'(0) = 0$, $y''(0) = 0$, second derivative test fails; but $y' = 4x^3$ is positive for $x > 0$, negative for $x < 0$, so there must be an abs. min. $y(0) = 0$ at $x = 0$ **32.** $R = 1$ **34.** Max. $P = 71.6$ at $x = 43¢$

Section 3.7

2. $x = 5, y = 5$; min. sum of squares $= 50$ **4.** $x = y = \sqrt{1000} = 31.62$ ft. **6.** $R = 50x - 10x^2$ for $1 \le x \le 5$; max. $R = \$62.50$ if $x = \$2.50$ **8.** $v = 400$ mph **10.** $x = y = \sqrt{5000} = 70.71$ ft. **12.** $L = 5x + 2y = 5x + (100{,}000/x)$ for $0 < x \le 125$; critical point $x = \sqrt{20{,}000} = 141.2$ is not feasible; L is decreasing on feasible set so min. L is 1425 ft. at $x = 125$, $y = 400$ **14.** $q = 400$ **16.** Width $= 5$ in., height $= 10$ in. **18.** At max. $dP/dx = dR/dx - dC/dx = 0$ **20.** $x = 200$, $y = 150$ ft. (In terms of x and y triangle area is $60{,}000 = 150x + 200y$) **22.** $V(x) = x(2 - 2x)^2$, max. $V = \frac{16}{27} = 0.5926$ at $x = \frac{1}{3}$ **24.** Area $36 = 4sh + s^2$; Volume $V = s^2h = (36s - s^3)/4$ is max. if $s = 2\sqrt{3} = 3.4641$, $h = (36 - s^2)/4s = \sqrt{3} = 1.7320$ **26.** Demand decreases by 357.14 tons/mo. **28.** $x = (a + b)/2$ **30.** $P = (10p - 130)e^{-p/15}$ million cents has max. at $p = 28¢$

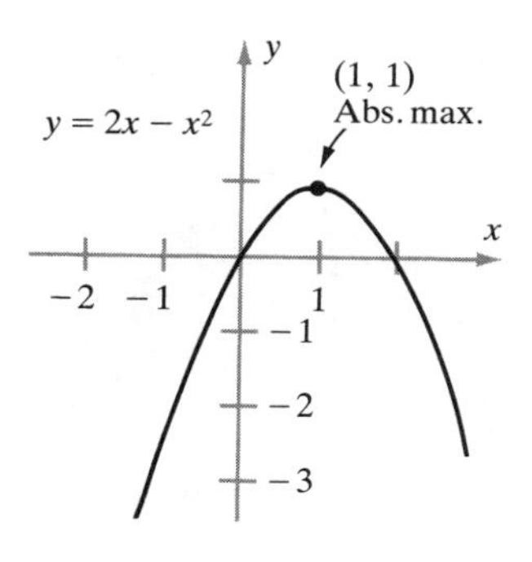

FIGURE E.33

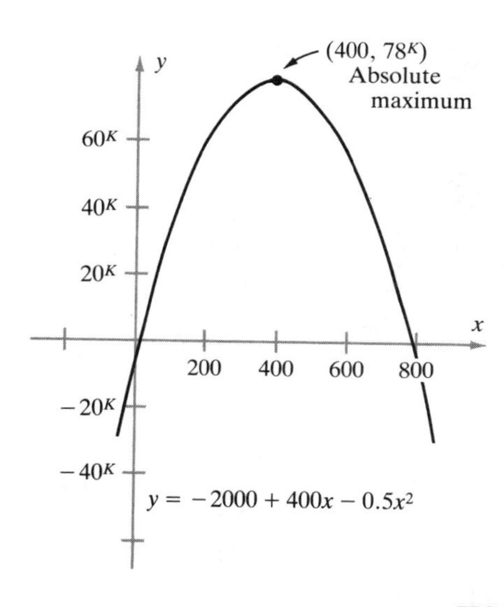

FIGURE E.34

Section 3.8

2. Critical: $x = 1$; no infl. pts.; $y' > 0$ if $x < 1$ and $y' < 0$ if $x > 1$; $y'' < 0$ all x; see Figure E.33 **4.** Critical: $x = 400$; no infl. pts.; $y' > 0$ on $(-\infty, 400)$, $y' < 0$ on $(400, +\infty)$; $y'' < 0$ all x; see Figure E.34

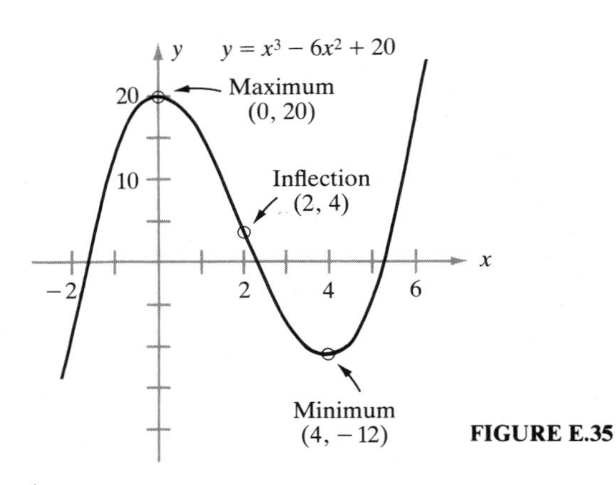

FIGURE E.35

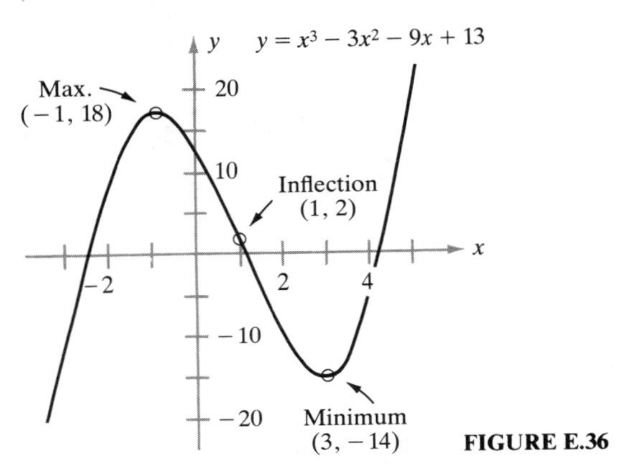

FIGURE E.36

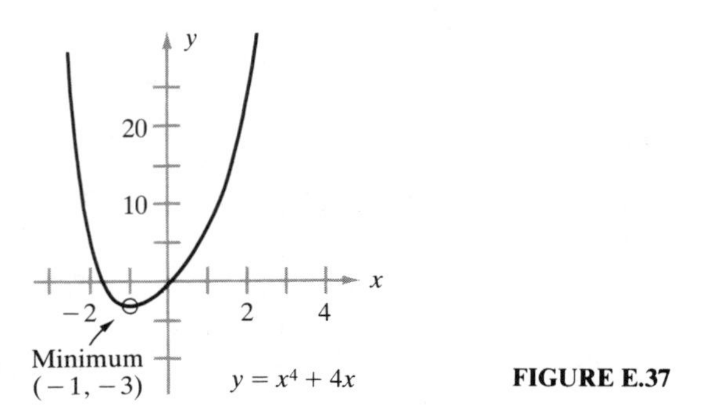

FIGURE E.37

6. Critical: $x = 0, 4$; infl. pt. at $x = 2$; $y' > 0$ on $(-\infty, 0)$ and $(4, +\infty)$, $y' < 0$ on $(0, 4)$; $y'' > 0$ on $(2, +\infty)$, $y'' < 0$ on $(-\infty, 2)$; see Figure E.35 **8.** Critical: $x = -1, 3$; infl. pt. at $x = 1$; $y' > 0$ on $(-\infty, -1)$ and $(3, +\infty)$, $y' < 0$ on $(-1, 3)$; $y'' > 0$ on $(1, +\infty)$, $y'' < 0$ on $(-\infty, 1)$; see Figure E.36 **10.** Critical $x = -1$; no infl. pts.; $y' > 0$ on $(-1, +\infty)$, $y' < 0$ on $(-\infty, -1)$; $y'' > 0$ all x except $x = -1$, where $y'' = 0$; see Figure E.37 **12.** Critical $x = 0, 1$; infl. pt. at $x = 0, \frac{2}{3}$; $y' > 0$ on $(1, +\infty)$, $y' < 0$ on $(-\infty, 0)$ and on $(0, 1)$; $y'' > 0$ on $(-\infty, 0)$ and $(\frac{2}{3}, +\infty)$, $y'' < 0$ on $(0, \frac{2}{3})$; see Figure E.38 **14.** Critical $x = -1, 0, 4$; infl. pts. at $x = 1 + \frac{1}{6}\sqrt{84} = 2.5257$ ($y = 77.8815$) and at $x = 1 - \frac{1}{6}\sqrt{84} = -0.5275$ ($y = 4.5615$); $y' > 0$ on $(-\infty, -1)$ and $(0, 4)$, $y' < 0$ on $(-1, 0)$ and $(4, +\infty)$; $y'' > 0$ on $(1 - \frac{1}{6}\sqrt{84}, 1 + \frac{1}{6}\sqrt{84})$, $y'' < 0$ elsewhere; see Figure E.39 **16.** Abs. min. at $(1, 3)$, see Figure E.40 **18.** Abs. max. at $(8, \frac{1}{4})$, see Figure E.41 **20.** No abs. extrema, see Figure E.42

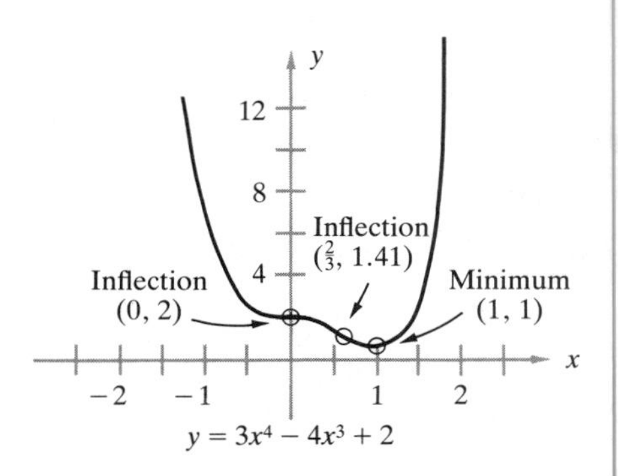

FIGURE E.38

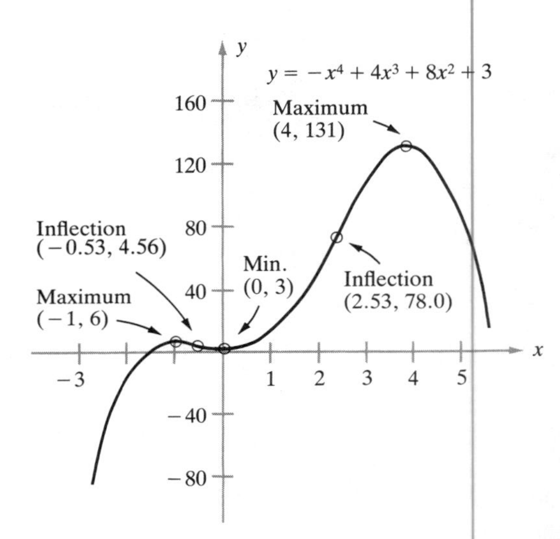

FIGURE E.39

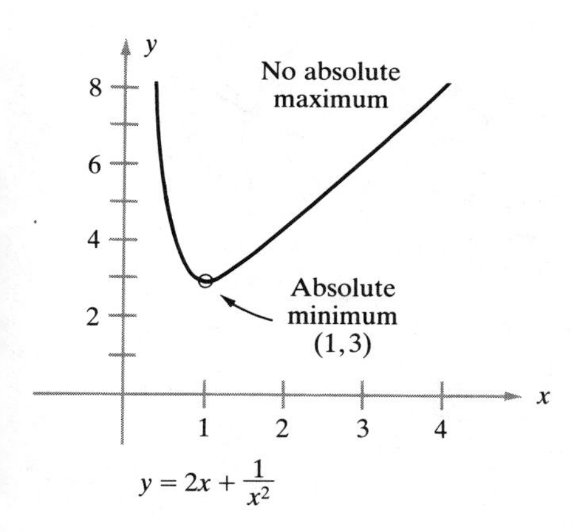

FIGURE E.40

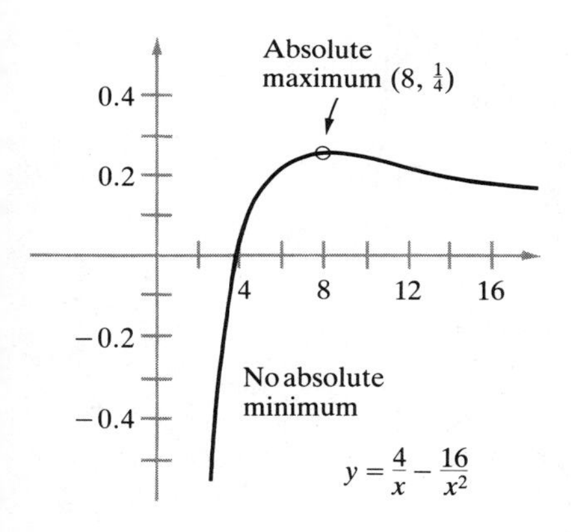

FIGURE E.41

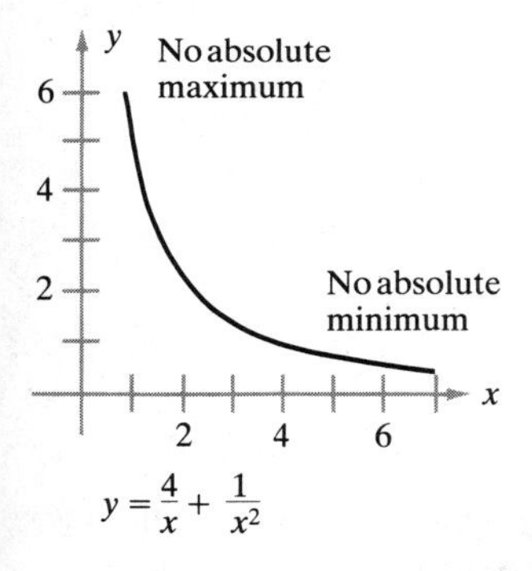

FIGURE E.42

22. See Figure E.43 **24.** See Figure E.44 **26.** See Figure E.45

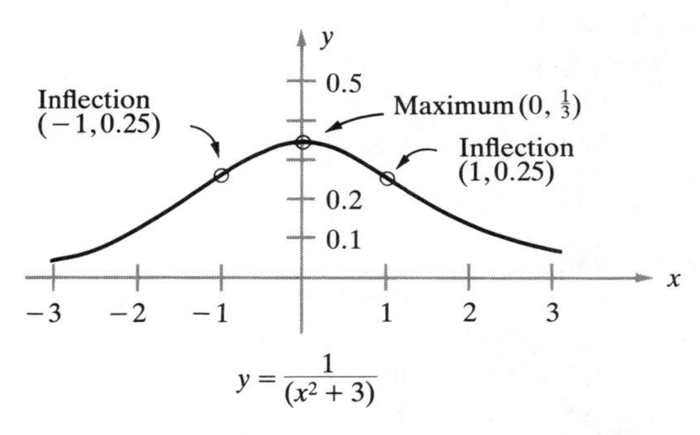

FIGURE E.43

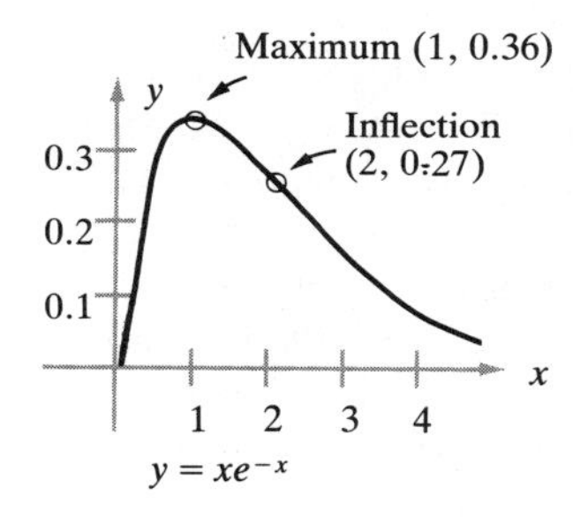

FIGURE E.44

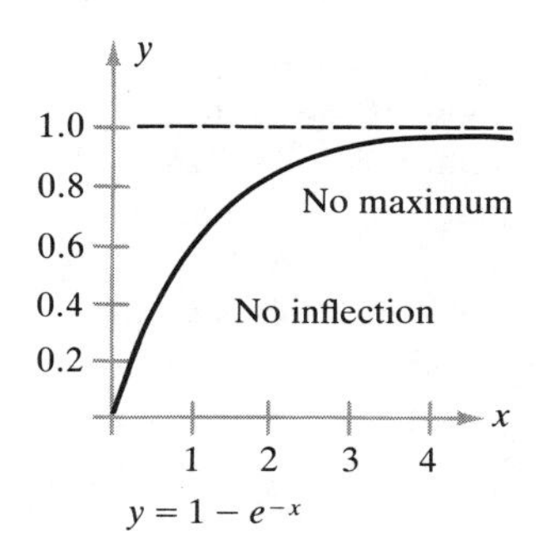

FIGURE E.45

Section 3.9

2. $y' = \frac{5}{7}$ **4.** $y' = x/4y$ **6.** $y' = (1 + 4x^3)/(1 + 3y^2)$ **8.** $y' = (2 - 3x^2y^3)/(3x^3y^2 - 3)$ **10.** $y' = (4 - 3x^2)/(1 - 3y^2)$; slope is -8; $y = -8x + 16$ **12.** $y' = (2y - 3x^2)/(3y^2 - 2x)$; slope is -1; $y = -x + 2$ **14.** $dr/dA = 1/8\pi r$ **16.** $dV/dt = 4\pi r^2 (dr/dt)$; if $r = 5$, $dV/dt = 10$ we get $dr/dt = 1/10\pi = 0.03183$ ft./min. **18.** Here $s^2 = x^2 + 36$, so $s' = xx'/s$; when $s = 10$ we have $x = \sqrt{64} = 8$, so that $x' = 1000$ mph **20.** $dP/dt = 720$ per year

22. Since $40^2 = x^2 + y^2$ we get $y' = -xx'/y$; when $x = 5$, $dx/dt = -3$, we have $y = \sqrt{1575} = 39.6863$ and $y' = 0.3780$ ft./sec. (increasing) **24.** $dy/dt = \frac{2}{5}$ ft./sec.

Section 3.10

2. $(\frac{1}{2} - x)(x^2 - x)^{-3/2}$
4. $\frac{1}{2}(x + 4)^{-1/2}(3 + 2x + x^2)^4 + 8(1 + x)(x + 4)^{1/2}(3 + 2x + x^2)^3$
6. $\frac{1}{2}(-5t^2 + 10t + 19)/(4 - t)^{1/2}$
8. $-48x^5(1 - x^6)^7 \ln x + (1 - x^6)^8/x$
10. $-\frac{2}{5}v(1 + v^2)^{-1}/(\ln(1 + v^2))^2$
12. $\frac{3}{2}(x - \sqrt{1 - x^2})^{1/2}(1 - x(1 - x^2)^{-1/2})$
14. $(-8x^3 - 8x - 2x)(1 + x^2)^{-2}e^{-4x^2+1}$
16. tangent line $y = x - 1$ **18.** $y = -\frac{1}{5}x + \frac{3}{5}$
20. $y = 1$ (horizontal) **22.** $dA/dt = 5$ **24.** Max. $P = 1410$ when $x = 240$; increases for $0 \le x \le 240$, decreases for $240 \le x \le 300$ **26.** Abs. max. 25 at $x = -1$ and 2; min. $\frac{275}{27}$ at $x = \frac{2}{3}$ **28.** Abs. max. $\frac{13}{3}$ at $x = \frac{2}{3}$, abs. min. 3 at $x = 1$ **30.** Abs. max. $1/e = 0.3679$ at $x = e$; abs. min. 0 at $x = 1$
32. Critical: $x = -4$ (loc. max.), $x = \frac{2}{3}$ (loc. min.)
34. Critical: $x = 1$ (loc. min.), $x = -2$ (second derivative test fails); sign diagram for y' reveals that $y' < 0$ on both sides of $x = -2$, so there is no local extremum there **36.** Critical: $x = -1$ (loc. min.)
38. Front edge $x = 300$, other side $y = 525$ ft.
40. 90 **42.** If $y =$ (altitude), $s =$ (direct distance), then $s^2 = 900 + y^2$ at all times, so $ds/dt = (y/s)\,dy/dt$; when $y = 40$, $dy/dt = 2$ we get $ds/dt = \frac{8}{5} = 1.6$ mi./sec. **44.** Critical: $x = 0$ ($y = 10$) and $x = 3$ ($y = -17$); infl. pts. at $x = 0$ ($y = 10$) and $x = 2$ ($y = -6$); see Figure E.46 **46.** Critical: $x = 2$ ($y = 4$); no infl. pts.; see Figure E.47
48. Undefined at $x = -1$; critical: $x = -2$ ($y = -4$) and $x = 0$ ($y = 0$); no infl. pts.; see Figure E.48

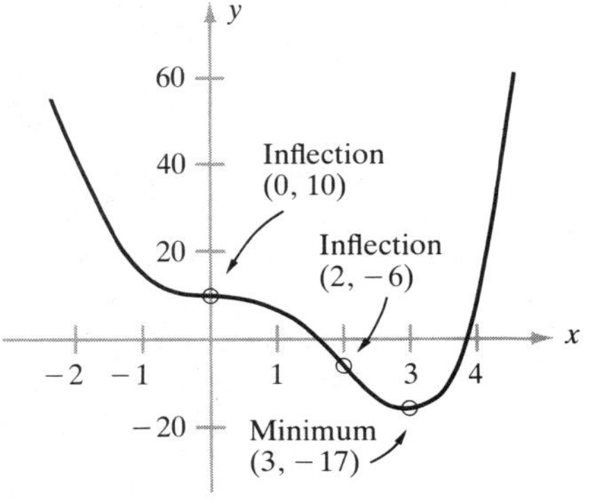

FIGURE E.46

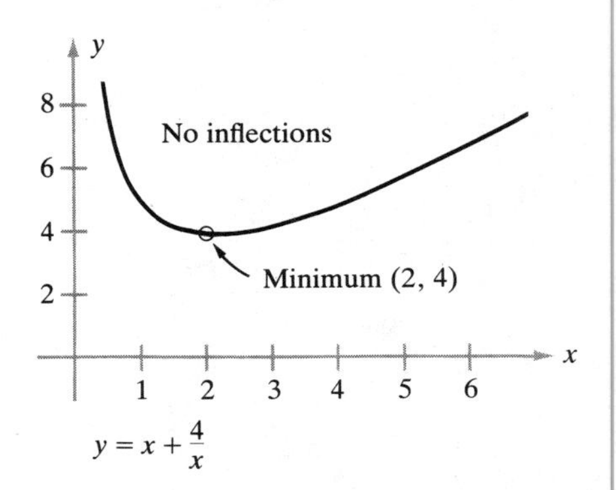

FIGURE E.47

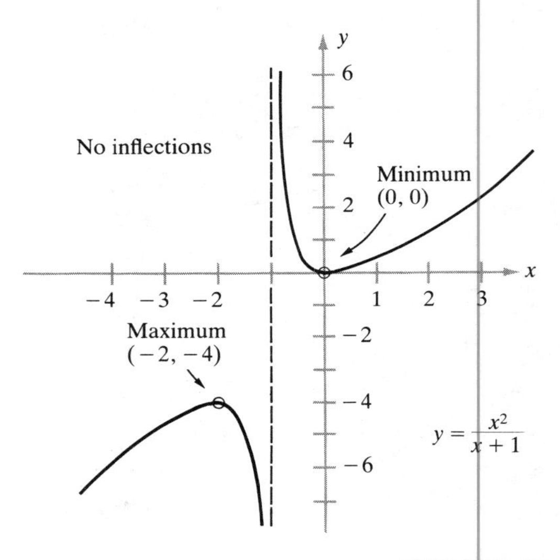

FIGURE E.48

Chapter 4

Section 4.1

2. (i) Int. 94.25, Total 1394.25; (ii) 188.50, 1488.50; (iii) 942.50, 2242.50 **4.** (i) 136.89; (ii) 138.59; (iii) 139.51; (iv) 140.16; (v) 140.49 **6.** (i) 10.96; (ii) 10.58 **8.** (i) 106; (ii) 106.18; for \$1000, multiply by 10; eff. rate 6.18% **10.** 2041.75 **12.** Don't buy;

worth buying if price is ≤ 4439.86 **14.** $P(1) = (1 - r)P$, $P(2) = (1 - r)P(1) = (1 - r)^2 P$, $P(3) = (1 - r)P(2) = (1 - r)^3 P$, etc. **16.** 2957.50, 812.20 **18.** 871,572.06 (total period: 350 years)

Section 4.2

2. 2.1170 **4.** 6.2152 **6.** 2.7180 **8.** 0.00004540 **10.** 0.04117 **12.** 1.6487 **14.** 0.13534 **16.** 0.4493 **18.** 13,099.64 **20.** 148.41 **22.** See Figure E.49

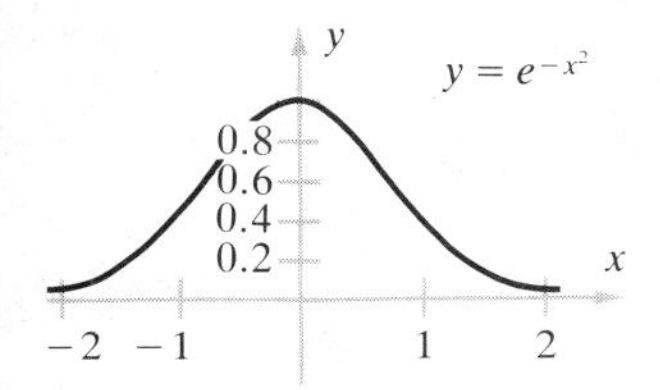

FIGURE E.49

24. 1.6993 **26.** 1.1447 **28.** 4.6502 **30.** 0.5025 **32.** 0.04621 **34.** 54.1832 **36.** −4.9628 **38.** 7316.16 **40.** Held for 15 years, $r = \ln(3)/13 = 0.08451$ or about 8.45% **42.** $t = 5.3\ln(2 \times 10^6)/\ln(2) = 110.94$ hours **44.** (i) $k = (\ln 2)/5 = 0.1386$; (ii) $k = (\ln 2)/T$ **46.** 30 years **48.** $t = 30\ln(\frac{100}{27})/\ln(2) = 56.7$ years

Section 4.3

2. (i) 6410.26; (ii) 5767.06; (iii) 5712.09 **4.** 20,004.19 **6.** $(3^{10} - 1)/2 = 29{,}524$ **8.** 510 **10.** 2.4661 **12.** 0.5702 **14.** 4.7481 **16.** 5.8585 **18.** $A = 2510.03$ **20.** Present value of 2 million at end of nth year $= 2[1 + \frac{1}{4}(0.06)]^{-4n} = 2[(1.015)^{-4}]^n = 2(0.94218)^n$; total present value $= 2(0.94218) + \cdots + 2(0.94218)^{20} = 22.6872$ million **22.** Present value of all payments (years 1 − 20) $= 100(e^{-0.07}) + \cdots + 100(e^{-0.07})^{20} = 1039.06$ **24.** Amount A should satisfy $A(e^{0.07}) + \cdots + A(e^{0.07})^{10} = 1380$, so that $A = 92.03$ **26.** 29.078 million; project seems more feasible because future dollars are not "discounted" as strongly **28.** $[1 - (\frac{1}{2})^{n+1}]/[1 - \frac{1}{2}]$ approaches 2 **30.** $(0.9)[1 - (0.9)^n]/[1 - 0.9] = 9[1 - (0.9)^n]$ approaches 9 **32.** $[1 - (\frac{1}{4})^{n+1}]/[1 - \frac{1}{4}] = \frac{4}{3}[1 - (0.25)^{n+1}]$ approaches $\frac{4}{3}$ **34.** $20[1 - (\frac{19}{20})^n]$ approaches 20

Section 4.4

2. See Figure E.50 **4.** $k = \ln(\frac{1}{2})/(4.5 \times 10^9) = -0.154 \times 10^{-9}$; amounts remaining are 0.8573,

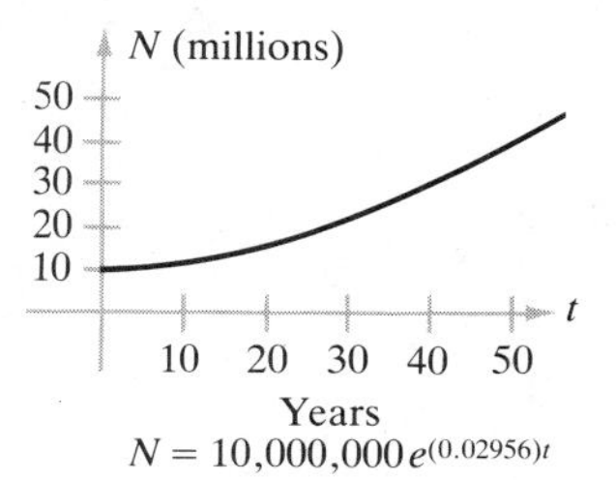

FIGURE E.50

0.2144, 2.0505×10^{-7} grams after 1, 10, 100 billion years **6.** If t = (elapsed time in years), then $0.99843 = Q(t)/Q(0) = e^{-(0.154 \times 10^{-9})t}$, so $\ln(0.99843) = -(0.154 \times 10^{-9})t$ or $t = 10.202$ million years **8.** $k = \frac{1}{5}\ln(0.90) = -0.021072$; if T is the half life then $\frac{1}{2} = e^{kT}$, $\ln(\frac{1}{2}) = -0.021072T$, or $T = 32.894$ years; if $0.05 = Q(t)/Q(0)$, we get $\ln(0.05) = -0.021072t$, or $t = 142.2$ years **10.** Doubling times: 69.66, 35.00, 23.45, 14.21 years **12.** $k = \ln(3)/3 = 0.3662$, $C = 10{,}000$; doubles in 1.8928 hours **14.** $t = 39.0$ years **16.** Here $C = Q(0)$. If time T is such that $Q(T) = 2Q(0)$, then $Q(t + T) = Q(0)e^{k(t+T)} = (Q(0)e^{kT})e^{kt} = Q(T)e^{kt} = 2Q(0)e^{kt} = 2Q(t)$ for any time t; $T = (\ln 2)/k$, independent of C

Section 4.5

2. $3e^{3x}$ **4.** $2e^{2x+1}$ **6.** $-1/x$ **8.** $2e^{2x}$ **10.** $1/x$ **12.** $2e^{-2x}$ **14.** $y' = (\ln 10)10^x$; $y' = (\ln 10)y$ **16.** $y' = -(\ln 3)3^{-x}$; $y' = -(\ln 3)y$ **18.** $y' = 1/x$; $y' = 1/e^y = e^{-y}$ **20.** $1000e^{-0.03t}$, decreases toward zero as t increases **22.** (i) 73,216.82; (ii) 46,685.10; (iii) 25,621.33 **24.** $P(t) = 5000[1 + (t^2/20)]e^{-0.07t}$; $P(0) = 5000$, $P(10) = 14{,}897.56$ **12.** 12 years; max. $P = 47861.61$ **28.** Max. $P = 88{,}142.28$ when $t = 24.470$ years

Section 4.6

2. 14,349.44 **4.** 12.75% **6.** 21.35 years **8.** $k = -\frac{1}{8}\ln 2 = -0.08664$ **10.** 2,785,999 **12.** 6876.09 **14.** $t = 5720\ln(0.82)/\ln(2) = 1637.7$ years **16.** $-\frac{1}{10}(x - 3)e^{-(x-3)^2/20}$ **18.** $(3/x)e^{3\ln x} = 3x^2$

20. See Figure E.51

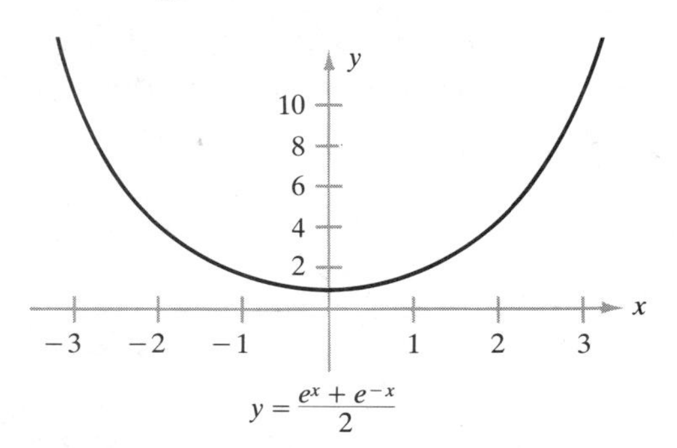

FIGURE E.51

Chapter 5

Section 5.1

2. $\frac{1}{5}x^5 + c$ **4.** $-x^{-1} + c$ **6.** $5x^{1/5} + c$ **8.** $-\frac{1}{4}x^{-4} + c$ **10.** $-e^{-x} + c$ **12.** $\frac{1}{3}e^{3x} + c$ **14.** $-2e^{-t/2} + c$ **16.** $\frac{1}{3}x^3 + c$ **18.** c **20.** $\frac{1}{2}\ln x + c$ **22.** $3x - (4/\sqrt{3})x^{1/2} + c$ **24.** $x - (1/x) + c$ **26.** $-0.01x^3 - 5x^2 + 4500x + c$ **28.** $4\ln x + 4e^{x/2} + c$ **30.** $5x - x^2 + c$ **32.** $\frac{1}{2}x^2 + (1/x) + c$ **34.** $\frac{2}{3}x^{3/2} - 5\sqrt{3}x + c$ **36.** $\frac{1}{2}x^2 + e^x + c$ **38.** $\frac{1}{3}x^9 - \frac{7}{4}x^4 - \frac{1}{4}x^2 + c$ **40.** $\frac{1}{3}x^3 - 2x + c$ **42.** $\frac{1}{4}x^4 - \frac{2}{5}x^{5/2} + c$
44. Direct differentiation of the right side
46. (i) x^4; (ii) e^x/x

Section 5.2

2. $f = \frac{1}{3}x^3 + 2x - \frac{1}{3}$ **4.** $V = 5t^2 + 37$ **6.** $f = \frac{1}{3}e^{3x} - \frac{1}{3}$ **8.** $R = 15x - 0.02x^2$ **10.** $P = 65x - 0.02x^2 - 0.005x^3 - 12{,}000$ **12.** $V(0) = 0$; $V(t) = 1000t + 20t^2$ **14.** $V = 15{,}000{,}000 - 3000t - 90t^2 - 30t^3$ **16.** 748.8 **18.** $N = 5x + 2x^2 + 30$
20. $A(t) = \frac{100}{3}(e^{0.06t} - 1)$; 17.399 from 1960 − 67, 27.404 from 1960 − 70, 16.286 from 1962 − 68
22. $s = 88t - 2t^2$; $s(5) - s(0) = 390$, $s(10) - s(0) = 680$, $s(10) - s(5) = 290$ **24.** $v = 15t$; $v(10) = 150$ ft./sec., $v(100) = 1500$ ft./sec.
26. $v(t) = -32t + v_0$, $h(t) = -16t^2 + v_0t + h_0$
28. $W(t) = \frac{1}{15}t + \frac{1}{60}t^2 + 0.4$; $w(16) = 5.733$ grams

Section 5.3

2. -8 **4.** $e^2 - 1 = 6.3891$ **6.** 4 **8.** $\frac{9}{2}$ **10.** $\frac{9}{10}$
12. $4\sqrt{3}$ **14.** 0 **16.** $\frac{32}{3}$ **18.** $-\frac{7}{2}$ **20.** $\frac{52}{5}$

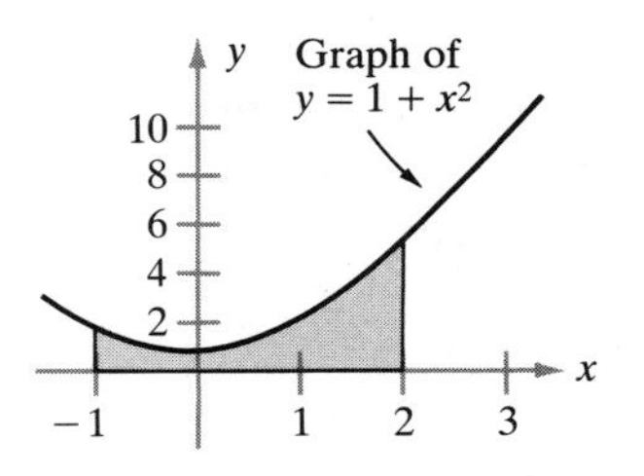

FIGURE E.52

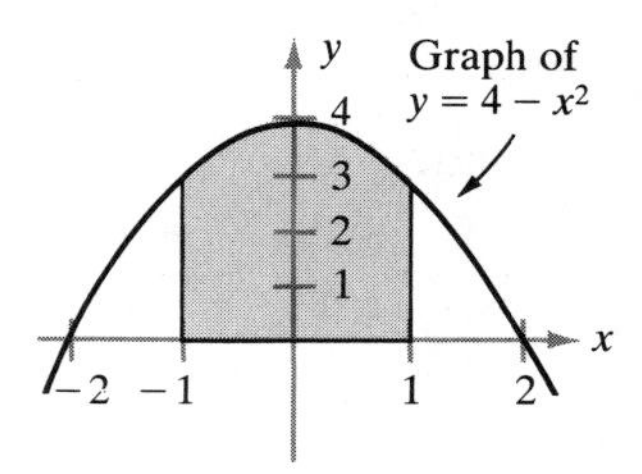

FIGURE E.53

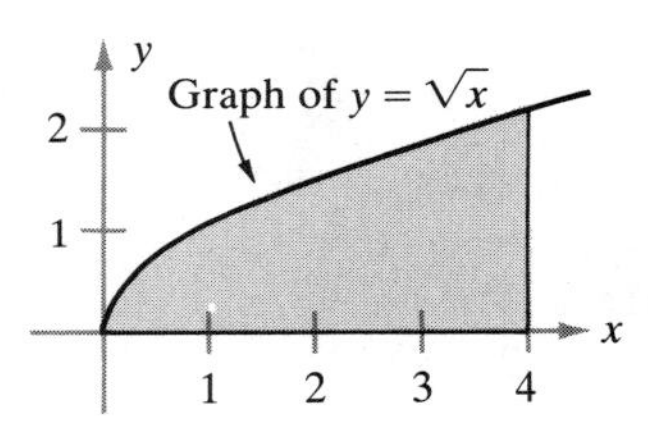

FIGURE E.54

22. 4.0597 **24.** Area = 6; see Figure E.52
26. Integral = $\frac{22}{3}$; see Figure E.53 **28.** Integral = $\frac{16}{3}$; see Figure E.54 **30.** 8 **32.** $\frac{1}{5}$ **34.** $\frac{14}{3}$ **36.** $\frac{40}{3}$
38. Fundamental theorem valid only if $f(x)$ is defined and continuous on the whole interval $-1 \leq x \leq 1$; $y = 1/x^2$ is undefined, and badly behaved, at $x = 0$

Section 5.4

2. See Figure E.55 **4.** See Figure E.56

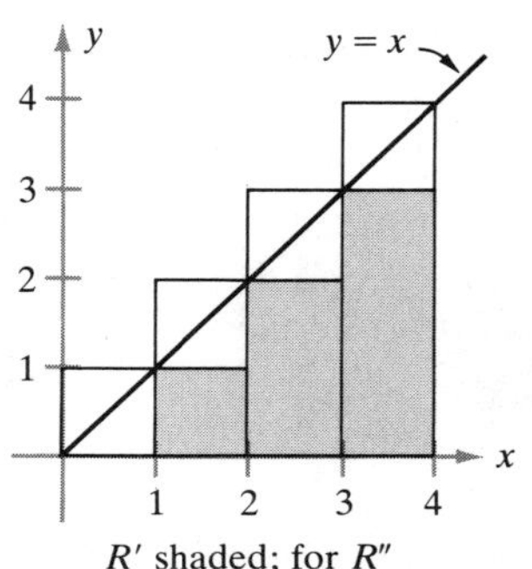

R′ shaded; for R″ add unshaded blocks

FIGURE E.55

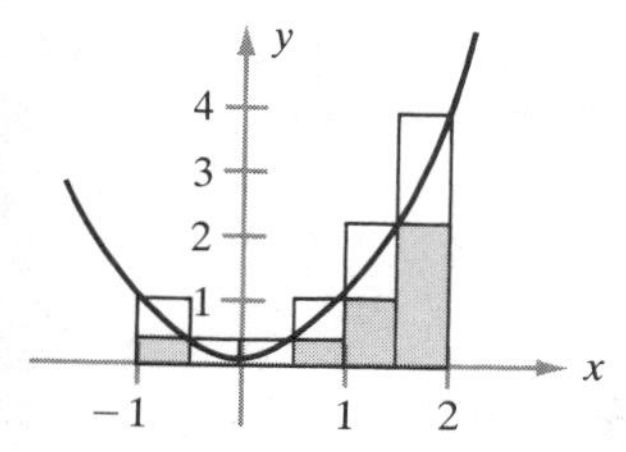

R′ shaded; for R″
add unshaded blocks

FIGURE E.56

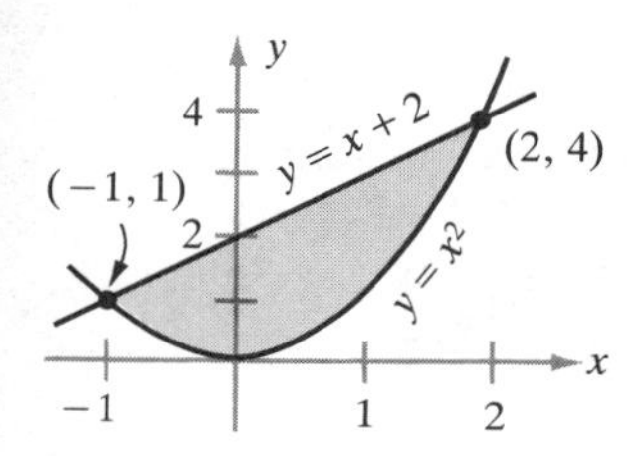

FIGURE E.57

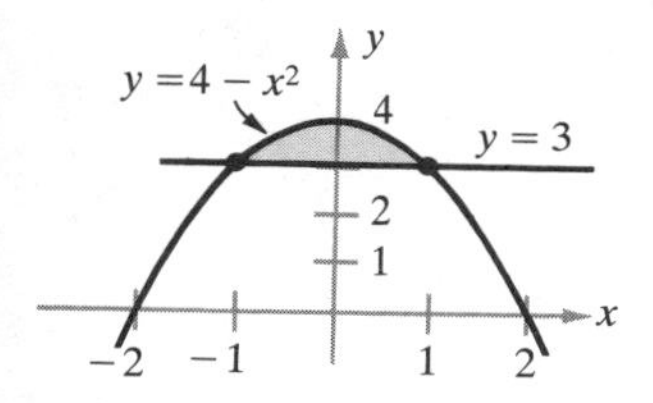

FIGURE E.58

6. Area(R') = Area(R'') = 15 **8.** Area(R') = 16.2, Area(R'') = 16.8, exact area = $\frac{33}{2}$ = 16.5 **10.** Area(R') = 8.5, Area(R'') = 12.5, exact area = $\frac{32}{3}$ = 10.6667 **12.** Area(R') = 2.1763, Area(R'') = 3.0373, exact area = 2.5829 **14.** For $n = 4$, areas are 15.5 and 12.5, for $n = 8$ they are 14.75 and 13.25 **16.** Area(R') = 0.72548, Area(R'') = 0.76762, average = 0.74655 **18.** Area(R') = 7.2985, Area(R'') = 7.5913, average = 7.4445 **20.** Area(R') = 0.4191 Area(R'') = 0.7727, average = 0.5959 **22.** Area(R') = 165, Area(R'') = 193, average = 179 miles is best estimate of mileage

Section 5.5

2. 8 **4.** $\frac{64}{3}$ **6.** f is upper; area = $\frac{8}{3}$ **8.** f is upper; area = $\frac{13}{12}$ **10.** Cross at $x = -1$, $x = 2$; f upper; area = $\frac{9}{2}$; see Figure E.57 **12.** Cross at $x = -1$, $x = 1$; f upper; area = $\frac{4}{3}$; see Figure E.58 **14.** Cross at $x = -1$, $x = 3$; g upper; area = $\frac{32}{3}$ **16.** $2 + 2\ln 3 = 4.1972$ **18.** $x^* = 10$, $p^* = 5$ **20.** $x^* = 16$, $p^* = 5$ **22.** p^* increases, x^* decreases **24.** $x_J^* = 250$, $p_J^* = 1.50$ **26.** Consumers = \$31,250; producers = \$187,500 **28.** $x^* = 60$, p^* = \$1,040,000; consumers = 14.4 million, producers = 9 million **30.** $x^* = 100$, $p^* = 50{,}000$; consumers = 2.5 million, producers = 1.0 million **32.** 7 **34.** $\frac{11}{2}$ **36.** $\frac{1}{2}\ln 3 = 0.5493$ **38.** 11 **40.** $\frac{1}{3}(1 - e^{-3}) = 0.3167$ **42.** 84°F **44.** 1.1973 **46.** 0.3 mile

Section 5.6

2. $\frac{1}{5}$ **4.** 0.09003 **6.** $\frac{1}{9}$ **8.** negative near $x = 4$ **10.** negative near $x = -1$ **12.** $k = \frac{1}{4}$ **14.** $k = 1/(\ln 3) = 0.9102$ **16.** $k = \frac{1}{20}$ **18.** $k = \frac{1}{3}$ **20.** 0.3125 **22.** 0.0837 **24.** 0.7901 **26.** $\bar{X}$ = 0.9660, fraction = 0.6364 **28.** $\bar{X} = 5.5$, $\sigma = 0.2887$ **30.** $\bar{X} = \frac{2}{3}$, $\sigma = 1/\sqrt{18} = 0.2357$ **32.** $\bar{X} = \frac{5}{7}$ = 0.7143, max. at $x = \frac{4}{5}$, $\sigma = 0.1597$ **34.** (i) 97.2%; (ii) $\bar{X} = 16.4$; (iii) $\sigma = 0.2236$

Section 5.7

2. $4x - \frac{1}{6}x^2 + c$ **4.** $\frac{1}{45}x^5 - \frac{4}{9}x^3 + 4x + c$ **6.** $\frac{4}{3}x^{3/2} - 2x^{1/2} + c$ **8.** $\frac{2}{5}x^{5/2} + \frac{4}{3}x^{3/2} + 6x^{1/2} + c$ **10.** $t + \frac{1}{7}e^{-7t} + c$ **12.** $-37.7884e^{-1.257t} + c$ **14.** -5 **16.** $-\frac{4}{3}$ **18.** $-\frac{1}{4}$ **20.** $\frac{1}{4}(e^{20} - 1) \approx 1.213 \times 10^8$ **22.** 108 ft./sec., 560 ft. **24.** $\frac{45}{4}$ **26.** 147.30 **28.** $\frac{4}{3}$ **30.** Area(R') = 3.923, Area(R'') = 4.236 average = 4.079; see Figure E.59 **32.** 7.2933 million **34.** (i) 0.2498; (ii) $e^{-1} = 0.3679$

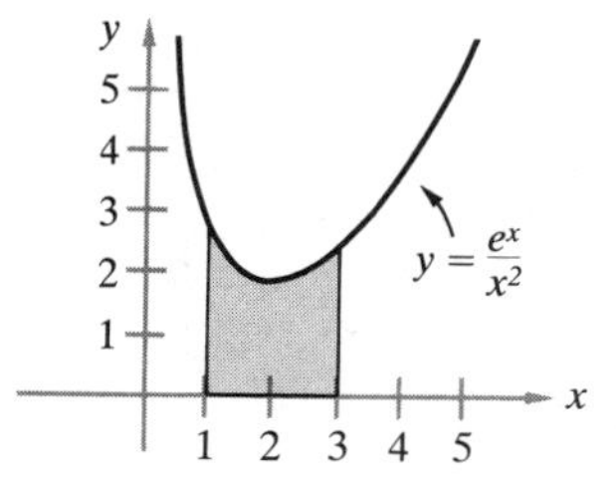

FIGURE E.59

Chapter 6

Section 6.1

2. $-\frac{2}{3}(1 - x^3)^{1/2} + c$ **4.** $\frac{2}{3}(x^2 + 4)^{3/2} + c$ **6.** $-\frac{1}{6}(3x + 2)^{-2} + c$ **8.** $e^{t+3} + c$ **10.** $\frac{1}{24}(x^3 - 3x + 1)^8 + c$ **12.** $\frac{1}{3}(x^2 + 4)^{3/2} + c$ **14.** $\frac{1}{2}(2 - x)^{-2} + c$ **16.** $\frac{1}{12}(3 - x^2)^{-6} + c$ **18.** $-\frac{1}{2}e^{-2x+1} + c$ **20.** $\sqrt{5} - 1$ **22.** $\frac{1}{6}(\ln 14 - \ln 5) = 0.1716$ **24.** $\frac{1}{2}(\ln 9 - \ln 5)$ = 0.2939 **26.** 0 **28.** $\ln(1 + \ln x) + c$

30. $x + \ln(x + 2) + c$ **32.** $\frac{1}{2}\ln(2x + 3) + c$
34. $-\frac{1}{2}e^{(1-x^2)} + c$
36. $-\frac{1}{7}(1 - x)^{-7} + \frac{1}{8}(1 - x)^{-8} + c$

Section 6.2

2. $-\frac{1}{3}xe^{-3x} - \frac{1}{9}e^{-3x} + c$ **4.** $(x - 1)e^x + c$
6. $(2 - x)e^x + c$ **8.** $(-\frac{3}{4} - \frac{1}{2}x)e^{-2x}$
10. $(3 - 7e^{-1}) = 0.4248$ **12.** $e + e^{-1} = 3.0862$
14. $x \ln x - x + c$ **16.** $-2x(1 - x)^{1/2} - \frac{4}{3}(1 - x)^{3/2} + c$ **18.** $\frac{1}{2}x^2 \ln(x^3) - \frac{3}{4}x^2 + c$
20. $(\frac{3}{2}x^2 + x)\ln x - \frac{3}{4}x^2 - x + c$ **22.** $-xe^{-x} - e^{-x} + c$ **24.** $(-x^3 - 3x^2 - 6x - 6)e^{-x} + c$
26. $(1/k)xe^{kx} - (1/k^2)e^{kx} + c$

Section 6.3

2. 0 **4.** $+\infty$ **6.** $+\infty$ **8.** $\frac{1}{2}$ **10.** $+\infty$ **12.** 1
14. $r - (1/r) \to +\infty$ as $r \to +\infty$; divergent **16.** 1
18. $\ln r \to +\infty$ as $r \to +\infty$; divergent **20.** $+\infty$
22. $+\infty$ **24.** $3(4)^{-1/3} = 1.8899$
26. (i) $(0.002)(\sqrt{3} - 1)$ (ii) 0.018; (iii) 1.998; (iv) $+\infty$
28. $\frac{3}{2}$ **30.** 1 **32.** $\frac{1}{14}$ **34.** $\frac{1}{4}$ **36.** $\frac{1}{2}\ln(4 + r^2) \to +\infty$ as r $\to +\infty$; divergent **40.** 0.5134, 0.03012
42. 0.2397

Section 6.4

2.–6. Routine calculations **8.** $\frac{1}{2}x^2 + c$ **10.** $\frac{1}{2}x^2 - x + c$ **12.** $y = Ce^{t^2/2}$ **14.** $y = (x^3 + c)^{1/2}$
16. $y = 2e^x$ **18.** $y = 5e^{1/2}e^{-x^2/2} = 5e^{(1-x^2)/2}$
20. $y = 10 - 10e^{0.3x}$ **22.** $N = (0.05t + 100)^2$, $N(100) = 11{,}025$ **24.** $U = 50 + 20e^{-0.2t}$, $U(t) =$ 60°F when $t = (\ln\frac{1}{2})/(-0.2) = 3.4657$ hrs.
26. $y = 10{,}000e^{-0.02t}$; number to replace $= y(0) - y(8) = 10{,}000(1 - e^{-0.16}) = 1478$
28. $y = 250Ce^{0.5t}/(1 + Ce^{0.5t})$ with $C = \frac{3}{47}$; $y(10) = 226.1$; $t = 2\ln(\frac{47}{3}) = 5.503$ days
30. $y = (1 + Ce^{2x})/(1 - Ce^{2x})$

Section 6.5

2. $\frac{1}{2}(4 - (x/3))^{-6} + c$ **4.** $-(2 - (x/3))^3 + c$
6. $(\ln x)(\ln(x^2)) - (\ln x)^2 + c = (\ln x)^2 + c$
8. $\frac{1}{3}x^3 \ln(x^3) - \frac{1}{3}x^3 + c$
10. $(-t^2 - 2t - 2)e^{-t} + c$
12. $-\frac{1}{5}(5x + 1)^{-1} + c$ **14.** $1 - 3e^{-2} = 0.5940$
16. 6.7041 **18.** $\frac{1}{12}$ **20.** $\frac{1}{64}$ **22.** $\frac{1}{4}$ **24.** $\frac{1}{5}e^{-35} = 1.261 \times 10^{-16}$ **26.** $\frac{7}{2}t^2 + c$ **28.** $y = (2x^2 - 6x + c)^{1/2}$ **30.** $y = (x + c)^{3/2}$

32. $y = (1 - 2x)^{-1/2}$ **34.** $y = -\ln(2 - x)$
36. $A = Ce^{0.06t}$; $A = 800e^{0.06t}$

Chapter 7

Section 7.1

2. 1 **4.** -3 **6.** 3.9 **8.** -932.5 **10.** -935.236
12. -909.45 **14.** All feasible except (iii), (vi)
16. 2, 3 **18.** $\frac{5}{6}$, 0 **20.** -86, 24 **22.** 0.001503. 0.3679 **24.** $C = 12.5x + 18.5y + 1000$; $R = 19.5x + 32.5y$; $P = 7x + 14y - 1000$ **26.** $V = x^2y$ **28.** $H = 8x^2 + 12xy$ (3 side walls only)
30. 3, 3, 3 **32.** 0, 15, -1 **34.** 0, 59.4843, 0
36. $V = xyz$ **38.** $C = 70xy + 35xz + 70yz$

Section 7.2

2.–10. See Figure E.60 **12.** (i) 80,000; (ii) $-175{,}000$; (iii) 80,000 **14.** Max. $P \approx 300{,}000$ at $x = 1750$, $y = 1750$ **16.** See Figure E.61

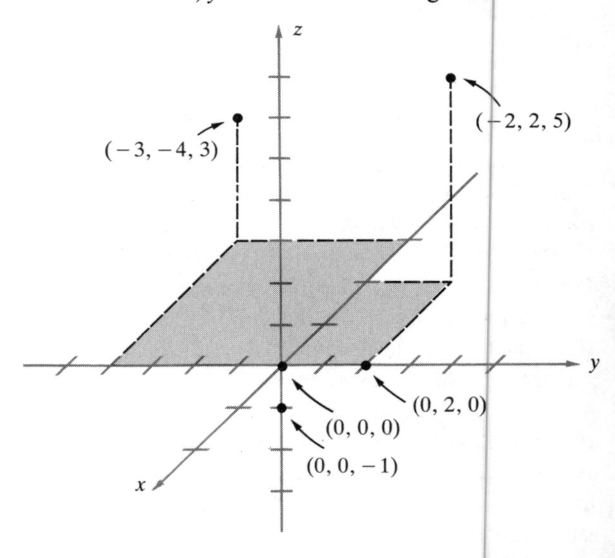

FIGURE E.60

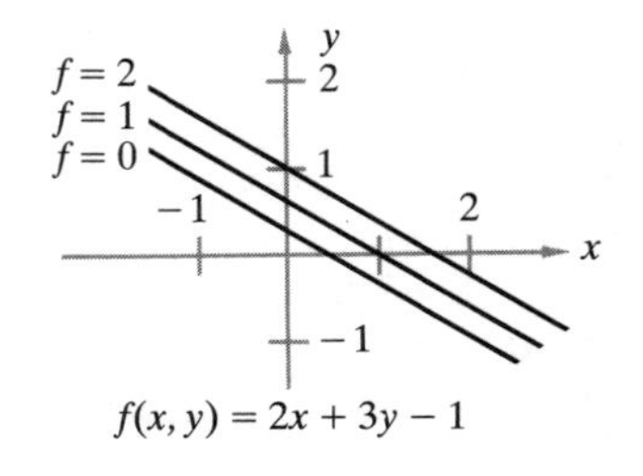

FIGURE E.61

18. See Figure E.62 **20.** See Figure E.63
22. Max. $P = 50$ at vertex (100, 100) of feasible set

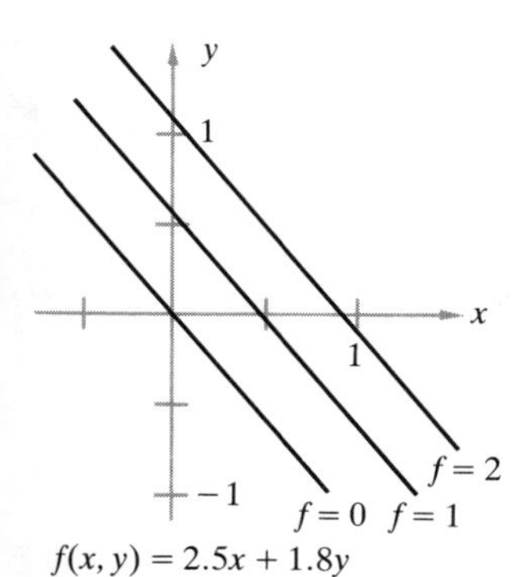

FIGURE E.62

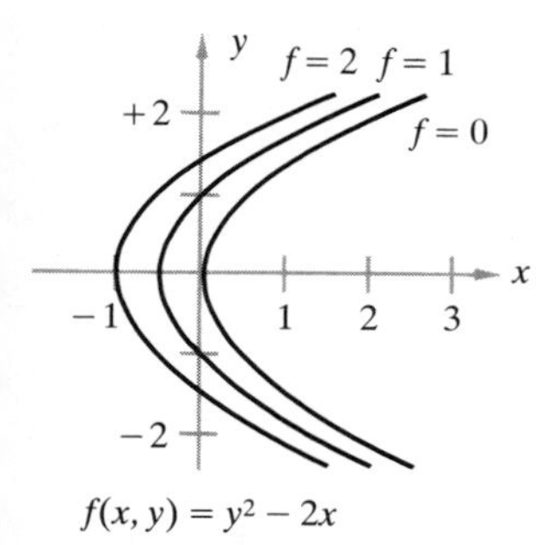

FIGURE E.63

Section 7.3

2. $\partial f/\partial x = 1 - 3x^2y + 3y^3$, $\partial f/\partial y = 3 - x^3 + 9xy^2$; (i) 1; (ii) 3; (iii) 1; (iv) -5 **4.** $\partial f/\partial x = \partial f/\partial y = 0$ **6.** $\partial f/\partial x = 0$, $\partial f/\partial y = -1$
8. $\partial f/\partial x = y + 2x$, $\partial f/\partial y = x - 2y$ **10.** $\partial f/\partial x = 10y$, $\partial f/\partial y = 10x + 6y^{-3}$ **12.** $\partial f/\partial x = (x + 1)(x^2 + 2x - y^2)^{-1/2}$, $\partial f/\partial y = -y(x^2 + 2x - y^2)^{-1/2}$ **14.** $\partial f/\partial x = (-x - 1)(x^2 + 2x + y^2 + 1)^{-3/2}$, $\partial f/\partial y = -y(x^2 + 2x + y^2 + 1)^{-3/2}$ **16.** $\partial f/\partial x = 1/(1 - y)$, $\partial f/\partial y = (1 + x)/(1 - y)^2$ **18.** $\partial f/\partial x = -2x/(4 + y^2)$, $\partial f/\partial y = -2y(1 - x^2)/(4 + y^2)^2$
20. $\partial f/\partial x = -32x - 4y$, $\partial f/\partial y = -4x - 50y$
22. $\partial f/\partial x = (2x + y)e^{x^2+xy}$, $\partial f/\partial y = xe^{x^2+xy}$
24. $\partial f/\partial x = \ln y$, $\partial f/\partial y = x/y$ **26.** $\partial f/\partial x = 5(x - 2y)^{-1}$, $\partial f/\partial y = -10(x - 2y)^{-1}$
28. $\partial f/\partial x = -x^{-2} - 3x^2$, $\partial f/\partial y = 1 - y^{-2}$
30. $\partial V/\partial r = \frac{2}{3}\pi rh$, $\partial V/\partial h = \frac{1}{3}\pi r^2$; at $(r, h) = (2, 10)$ values are $40\pi/3$ and $4\pi/3$
32. $\partial A/\partial x = \frac{1}{4}e^{-t/3}$, $\partial A/\partial t = -\frac{1}{12}xe^{-t/3}$
34. $A(100, 3) = 25e^{-1} = 9.1970$ units/liter
36. $\partial Y/\partial t = (0.6 + 0.386q - 0.093q^2)(2 - 0.1t)$, $\partial Y/\partial q = (0.386 - 0.186q)(2t - 0.05t^2)$
38. $P = kT/V$; since k is constant, $\partial P/\partial V = -kT/V^2$, $\partial P/\partial T = k/V$ **40.** $\partial P/\partial V < 0$
42. $\partial f/\partial x = 12y + 28z$, $\partial f/\partial y = 12x - 14yz^2$, $\partial f/\partial z = 28x - 14y^2z$; values are 88, 460, and -196 **44.** $\partial f/\partial x = yz$, $\partial f/\partial y = xz$, $\partial f/\partial z = xy$; values are $-8, 4, -2$ **46.** $\partial f/\partial x = y^2(1 + z^2)^{-1}$, $\partial f/\partial y = 2xy(1 + z^2)^{-1}$, $\partial f/\partial z = -2xy^2z(1 + z^2)^{-2}$; values are $\frac{4}{17}, -\frac{4}{17}, -\frac{32}{289}$
48. $\partial f/\partial x = 2xe^{x^2+y^2-z^2}$, $\partial f/\partial y = 2ye^{x^2+y^2-z^2}$, $\partial f/\partial z = -2ze^{x^2+y^2-z^2}$; values are $2e^{-11}$, $-4e^{-11}$, $-8e^{-11}$

Section 7.4

2. $\Delta f \approx -8(\Delta x) + 13(\Delta y)$, $f(1 + \Delta x, -2 + \Delta y) \approx -17 + \Delta f$; $f(0.9, -2.05) \approx -16.85$ **4.** $\Delta f \approx 8(\Delta x) + 16(\Delta y)$, $f(2 + \Delta x, 2 + \Delta y) \approx 16 + \Delta f$; $f(2.2, 1.9) \approx 16$ **6.** $\Delta f \approx -e^{-3}(\Delta x) - 2e^{-3}(\Delta y)$, $f(1 + \Delta x, 1 + \Delta y) \approx e^{-3} + \Delta f$; $f(1.2, 0.9) \approx 0.04979$ **8.** Area $= 170\pi$, $\Delta A \approx 10\pi(\Delta h) + 22\pi(\Delta d)$; (i) $\Delta A \approx 22\pi$; (ii) $\Delta A \approx 10\pi$; (iii) $\Delta A \approx -2\pi$ **10.** (i) $\partial P/\partial x = 0.1$, $\partial P/\partial y = 1.7$; (ii) yes; (iii) yes **12.** $\partial S/\partial p(1.5, 1000) = -465{,}702$, $\partial S/\partial a(1.5, 1000) = 6.8201$ **14.** $\Delta S \approx 6820$
16. $\partial P/\partial V = -2.034T/V^2$, $\partial P/\partial T = 2.034/V$
18. $\partial P/\partial V(10, 293) = -5.9596$, $\partial P/\partial T(10, 293) = 0.2034$ **20.** $\Delta P \approx -1.0882$

Section 7.5

2. none **4.** (0, 4) **6.** (0, 0) **8.** $(-\frac{1}{6}, \frac{1}{3})$ **10.** (0, 0)
12. (0, 6) **14.** $(-2, 1)$ **16.** (0, 1) and $(0, -3)$
18. (0, 0), (8, 2), and $(-8, -2)$ **20.** none **22.** (1, 0) and $(-1, 0)$ **24.** $(1, \frac{1}{2})$ and $(-2, \frac{1}{2})$ **26.** (0, 0)
28. Critical: (1, 3) and (1, 0); max. $f = 40$ at (1, 3)
30. Critical: (0, 0), $(0, -1)$, and (0, 2); min. $f = -32$ at (0, 2) **32.** $x = \$150$, $y = \$400$ **34.** $l = w = 6$, $h = 3$ **36.** $l = w = 100$, $h = 50$ ft. **38.** $(x, y) = (\frac{10}{3}, \frac{10}{3})$ **40.** $x = 2858.04$, $y = 1892.82$; max. $P = 1{,}434{,}540$

Section 7.6

2. Residuals $y_i - (-15.5x_i + 40)$ are $-12.5, 7, 4, -7.5$; sum of squares $= 277.5$ **4.** $y = -0.175x + 2.675$ **6.** $y = 1.225x + 3.025$
8. $y = -15.403x + 37.315$
10. $y = -0.2581x + 109.15$ mm
12. $y(30) = 101.4$ mm **14.** $y = -170x + 380$
16. $y = 0.4164x + 2.2806$ **18.** $y = 2.3360x + 467.68$ **20.** $y = 0.2392x - 2.3443$; 85.2°F

Section 7.7

2. All are zero **4.** $\partial^2 f/\partial x^2 = 24x$, $\partial^2 f/\partial x \partial y = 2$, $\partial^2 f/\partial y^2 = -12$ **6.** $\partial^2 f/\partial x^2 = 2e^{-y}$, $\partial^2 f/\partial x \partial y = -2xe^{-y}$, $\partial^2 f/\partial y^2 = (x^2 + 4)e^{-y}$ **8.** $\partial^2 f/\partial x^2 = -2y^2(x^2 - 2y^2)^{-3/2}$, $\partial^2 f/\partial x \partial y = 2xy(x^2 - 2y^2)^{-3/2}$, $\partial^2 f/\partial y^2 = -2x^2(x^2 - 2y^2)^{-3/2}$ **10.** $D = -8 - 8y - 4x^2$; $D(0, 0) = -8$ **12.** $D = -36xy - 1$; $D(-\frac{1}{3}, \frac{1}{3}) = 3$ **14.** $D = -5$ **16.** $D(1, 2) = -24$
18. $D(1, 2) = 0$ **20.** Critical: $(-\frac{5}{6}, \frac{1}{3})$, loc. min.
22. Critical: $(\frac{28}{15}, -\frac{11}{15})$, loc. min. **24.** Critical: $(-760, -1000)$, no extremum since $D < 0$
26. Critical: $(2/\sqrt{3}, 1/\sqrt{3})$ and $(-2/\sqrt{3}, -1/\sqrt{3})$, neither an extremum since $D < 0$ **28.** Critical: (0, 0), not an extremum; $(1, -1)$ loc. min.; $(-2, 2)$ loc. min. **30.** Critical: $(\frac{102}{17}, \frac{51}{17})$, loc. max.
32. To find critical points note that $\partial f/\partial y = 2xy - x^2 = x(2y - x) = 0$ when $x = 0$ or $x = 2y$. Separate examination of these cases yields (0, 1) and $(0, -1)$, neither an extremum **34.** $x = 3$, $y = 2$

Section 7.8

2. Max. $P = 200$ **4.** Max. $= \frac{1}{2}$ at $(1, \frac{1}{2})$ **6.** $x = \frac{3}{5}$, $y = \frac{4}{5}$, $\lambda = \frac{2}{5}$; min. is $f(\frac{3}{5}, \frac{4}{5}) = 11$ **8.** Critical values: $(x, y, \lambda) = (1/\sqrt{2}, 1/\sqrt{2}, \frac{1}{2})$, $(-1/\sqrt{2}, -1/\sqrt{2}, \frac{1}{2})$, $(1/\sqrt{2}, -1/\sqrt{2}, -\frac{1}{2})$, $(-1/\sqrt{2}, 1/\sqrt{2}, -\frac{1}{2})$; max. $= \frac{1}{2}$ at $x = 1/\sqrt{2}$, $y = 1/\sqrt{2}$ and at $x = -1/\sqrt{2}$, $y = -1/\sqrt{2}$
10. $x = 695$, $y = 305$ ($\lambda = -93.5$) **12.** $x = 100$, $y = 50$ ($x + y \neq 1000$ here) **14.** $x = 20$, $y = \frac{80}{3}$ ($\lambda = 3.30$) **16.** $\Delta P_{max} \approx \$6600$ **18.** $x = 30$, $y = 40$ ($\lambda = 0.4$) **20.** $\Delta H \approx 2000$ thermal units
22. There are three λ values: $\lambda = -2$, $\lambda = -2 \pm \sqrt{3}$; min. distance 0.1519 at $x = 1.3660$, $y = 1.8660$ (for $\lambda = -2 + \sqrt{3}$) **24.** $\lambda = \frac{1}{4}(6^{1/3})$, $x = 8\lambda$, $y = \frac{8}{3}\lambda$, $z = 4\lambda$; min. $f = 6(6^{1/3}) = 10.903$

Section 7.9

2. 4 **4.** $-\frac{1}{2}$ **6.** 24 **8.** Region is a rectangular block: length $= b$, width $= d$, height $= h$; integral $=$ volume $= bdh$ **10.** 72 **12.** 104
14. 140,000 cu. ft. **16.** 0.0245 cu. mi.

Section 7.10

2. (i) 0; (ii) $34^{-1/2} = 0.1715$; (iii) $-5(26^{-1/2}) = -0.9806$ **4.** (i) $C = 23.75x + 53.5y$; (ii) $R = 60x + 105y$; (iii) $P = 36.25x + 51.5y$ **6.** Increase
8. Negative **10.** (130, 60) **12.** No **14.** Yes
16. Max. at (4, 1), min. at (1, 5); see Figure E.64

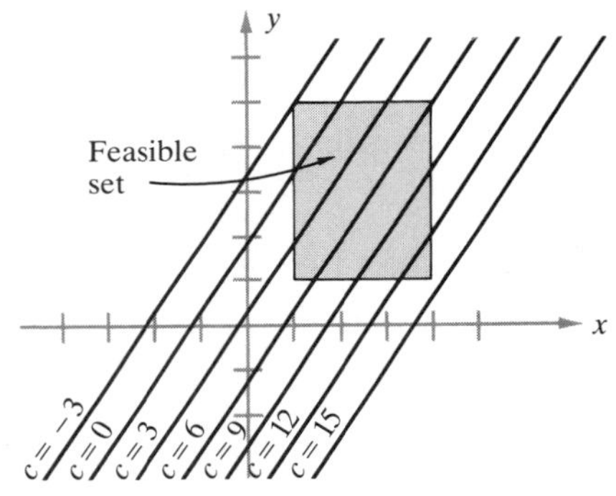

FIGURE E.64

18. $\partial f/\partial x = -(1 + y^2)^{-1}$, $\partial f/\partial y = -2y(1 - x)(1 + y^2)^{-2}$; values are $-\frac{1}{5}$, $-\frac{1}{5}$, and $\frac{4}{25}$
20. $\partial f/\partial x = 2x(x^2 + y^2)^{-1}$, $\partial f/\partial y = 2y(x^2 + y^2)^{-1}$; values are $\ln(8), \frac{1}{2}, \frac{1}{2}$ **22.** $\partial f/\partial x = (x + y^2)^{-1} - z + ze^{x+y}$, $\partial f/\partial y = 2y(x + y^2)^{-1} - z + ze^{x+y}$, $\partial f/\partial z = -x - y + e^{x+y}$ **24.** $\partial f/\partial x = 2x - 5y + 3$, $\partial f/\partial y = -5x + 4y$, $\partial^2 f/\partial x^2 = 2$, $\partial^2 f/\partial x \partial y = -5$, $\partial^2 f/\partial y^2 = 4$ **26.** $\partial f/\partial x = y(x + y)^{-2}$, $\partial f/\partial y = -x(x + y)^{-2}$, $\partial^2 f/\partial x^2 = -2y(x + y)^{-3}$, $\partial^2 f/\partial x \partial y = (x - y)(x + y)^{-3}$, $\partial^2 f/\partial y^2 = 2x(x + y)^{-3}$ **28.** $\Delta V \approx 11.2$ cu. in.
30. $(\frac{8}{3}, -\frac{7}{3})$ **32.** $(0, \frac{1}{2})$, $(5.2737, -2.1369)$, $(-0.02370, 0.5119)$ **34.** Base dimensions: $x = y = 20$, $z = 7.5$ **36.** $(-\frac{14}{5}, \frac{8}{5})$, not an extremum
38. $(\frac{1}{4}, -\frac{1}{4})$, loc. min. **40.** (0, 0), not an extremum
42. $y = -\frac{3}{2}x + \frac{19}{6}$ **44.** $y = 0.5x + 1.5$
46. $x = 1$, $y = 2$ ($\lambda = 4$); min. is $f(1, 2) = 19$

48. Max. $f = 2/\sqrt{3}$ at $(1/\sqrt{3}, -1/\sqrt{3})$, with $\lambda = 1/\sqrt{3}$; min. $f = -2/\sqrt{3}$ at $(-1/\sqrt{3}, 1/\sqrt{3})$, with $\lambda = -1/\sqrt{3}$ **50.** $x = 30{,}000$ and $y = 10{,}000$ **52.** $\frac{112}{3}$

Appendix 1

2. $\frac{17}{21}$ **4.** $\frac{5}{3}$ **6.** $-\frac{8}{35}$ **8.** $\frac{1}{6}$ **12.–22.** See Figure E.65

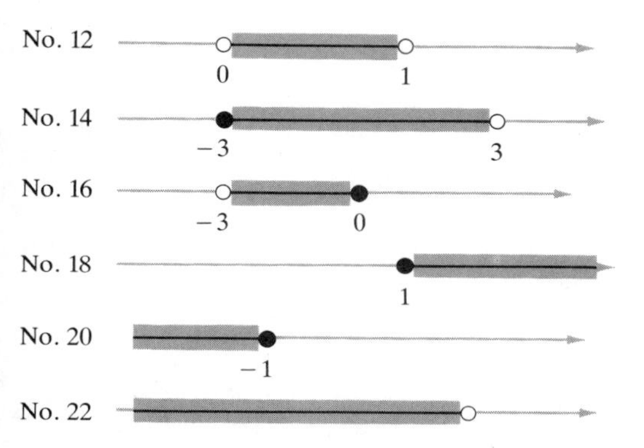

FIGURE E.65

24. Rent from A if $x < 150$ **26.** (i) $x > 931.0$; (ii) $x > 1103.4$ **28.–38.** See Figure E.66 **40.** Use alternative (ii); same for 12 pages **42.** 7 **44.** 3.142 **46.** $x = -2, 8$ **48.** $x = -9, -3$ **50.** $x = 2.18, 3.28$ **52.–56.** See Figure E.67 **58.** $2^{1/3}$ **60.** $7^{-1/2}$ **62.** $28^{-2/3}$ **64.** 1 **66.** 1 **68.** 0.04 **70.** 0.03125 **72.** 0.001 **74.** 10, 28, 80, 226 **76.** 10^{-6} **78.** 10^{6} **80.** 10^{3} **82.** 10^{-3} **84.** 0.5091×10^{-10} **86.** 5×10^{2} **88.** 9.61×10^{26} **90.** a^5 **92.** a^6 **94.** a^6c^6 **96.** (i) $a^2 + 2ab + b^2$; (ii) $a^3 + 3a^2b + 3ab^2 + b^3$; (iii) $a^4 + 4a^3b + 6a^2b^2 + 4ab^3 + b^4$; (iv) $a^2 - b^2$; (v) $a^2 - 2ab + b^2$

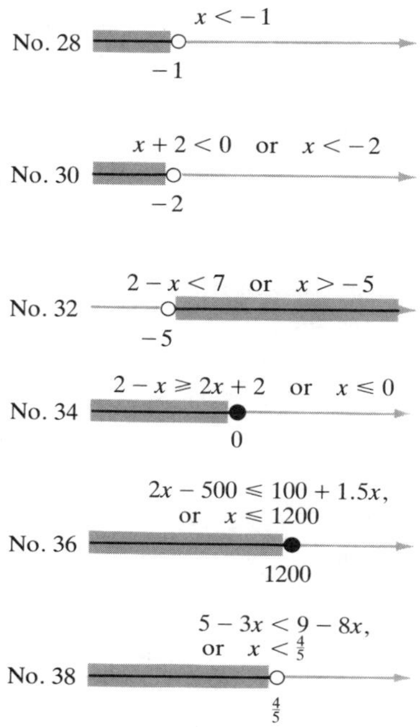

No. 28: $x < -1$ (−1)

No. 30: $x + 2 < 0$ or $x < -2$ (−2)

No. 32: $2 - x < 7$ or $x > -5$ (−5)

No. 34: $2 - x \geq 2x + 2$ or $x \leq 0$ (0)

No. 36: $2x - 500 \leq 100 + 1.5x$, or $x \leq 1200$ (1200)

No. 38: $5 - 3x < 9 - 8x$, or $x < \frac{4}{5}$ ($\frac{4}{5}$)

FIGURE E.66

No. 52: $|x - 1| \leq 2$, or $-1 \leq x \leq 3$ (−1, 1, 3; ←2→←2→)

No. 54: $|x - 3| \leq 5$, or $-2 \leq x \leq 8$ (−2, 3, 8; ←5→←5→)

No. 56: $|x + 6| > 1$ (−7, −6, −5; ←1→←1→)

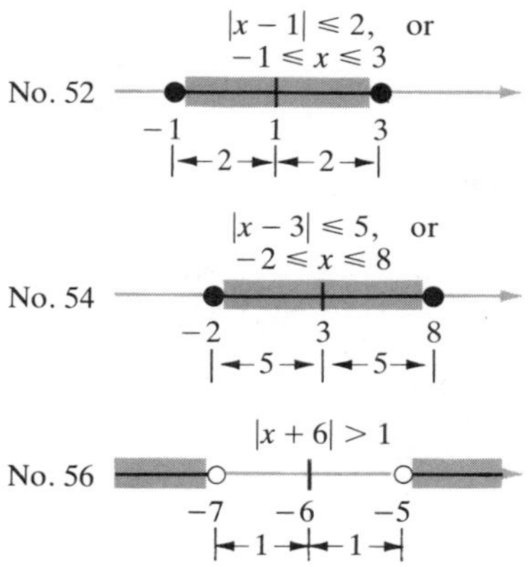

FIGURE E.67

Index

Q

R

S

T

U

V

W

B C D E F G H I J
5 6 7 8 9 0 1 2 3 4